Scottish Buil

Dundee
City Council
Leisure & Communities
CHANGING
FOR THE **FUTURE**

FOR REFERENCE ONLY

Scottish Building Standards in Brief

Ray Tricker and Rozz Algar

AMSTERDAM • BOSTON • HEIDELBERG • LONDON
NEW YORK • OXFORD • PARIS • SAN DIEGO
SAN FRANCISCO • SINGAPORE • SYDNEY • TOKYO
Butterworth-Heinemann is an imprint of Elsevier

Butterworth-Heinemann is an imprint of Elsevier
Linacre House, Jordan Hill, Oxford OX2 8DP
30 Corporate Drive, Suite 400, Burlington, MA 01803

British Library Cataloguing-in-Publication Data
A catalogue record for this book is available from the British Library

Library of Congress Cataloguing-in-Publication Data
A catalogue record for this book is available from the Library of Congress

ISBN: 978-0-7506-8558-0

For information on all Butterworth-Heinemann publications
visit our web site at http://books.elsevier.com

Typeset by Charon Tec Ltd., A Macmillan Company (www.macmillansolutions.com).

Printed and bound in the UK
08 09 10   10 9 8 7 6 5 4 3 2 1

Contents

About the authors   x

Foreword   xi

Preface   xiii

1   The Building (Scotland) Act 2003   1

1.1   What is the aim of the Building (Scotland) Act 2003   2
1.2   What does the Act contain?   2
1.3   Who polices the Act?   6
1.4   Are there any exemptions from Building Regulations?   6
1.5   When can the Building Regulations be relaxed?   6
1.6   What if the work does not comply with Building Regulations?   6
1.7   What are guidance documents?   7
1.8   What are Building Warrants?   7
1.9   What are Completion Certificates?   8
1.10   Where can I find out about Building Warrants and Completion
       Certificates in my areas?   9
1.11   What if a building is defective or dangerous?   9
1.12   What about scheduled monuments and listed buildings?   10
1.13   Do the local authority or Scottish Ministers have the
       right to enter my property?   10
1.14   Can occupants be evicted from buildings?   10
1.15   Can the authority compulsorily purchase the building?   10
1.16   Can the authority sell off materials from demolished
       buildings?   11
1.17   Can I appeal decisions made in relation to my building?   11
1.18   What about civil liability?   11
1.19   Who are the Building Standards Advisory Committee?   11
1.20   What is a Building Standards Assessment?   11
1.21   What are the Supplementary Regulations?   12
1.22   What about the rest of the United Kingdom?   15

2   The Building (Procedure) (Scotland) Regulations 2004   18

2.1   What is the purpose of the Building (Scotland) Regulations?   18
2.2   Why do we need the Building Regulations?   18
2.3   What building work is covered by the Building Regulations?   19
2.4   What is a conversion?   20
2.5   What are the requirements associated with the Building
       Regulations?   20

2.6	What are the Technical Handbooks?	20
2.7	Are there any exemptions?	25
2.8	What is the difference between 'exempt' and 'work not requiring a warrant'?	26
2.9	What happens if I do not comply with a Technical Handbook?	26
2.10	Verifiers	26
2.11	Building Warrants	27
2.12	Approved Certifiers	42
2.13	Energy Performance Certificates	44
2.14	Planning officers	45
2.15	Do I need to employ a professional builder?	45
2.16	Are there any occasions when the rules are relaxed?	45
2.17	Completion certificate	46
2.18	When do the Scottish Ministers get involved?	48
2.19	What matters can you appeal against?	49
2.20	What are Building Regulations compliance notices?	50
2.21	Can a Verifier impose 'continuing requirements'?	51
2.22	What regulations apply to Crown buildings?	51
2.23	What about historic buildings?	52
2.24	What about dangerous buildings?	52
2.25	What other authorities will be consulted?	55
2.26	What records does the local authority hold?	57
2.27	Other sources	58

3   The Requirements of the Building (Scotland) Regulations (2004) — **59**

3.1	Introduction	59
3.2	Section 0 – General Provisions	60
3.3	Section 1 – structure	67
3.4	Section 2 – fire	67
3.5	Section 3 – environment	69
3.6	Section 4 – safety	73
3.7	Section 5 – noise	76
3.8	Section 6 – energy	76
	Appendix 3A	81
	Appendix 3B	85
	Appendix 3C	86
	Category A	86
	Appendix 3D	90
	Appendix 3E	91

4   Planning Permission — **92**

4.1	Why do we have to have a planning system?	93
4.2	What is sustainable development?	93

4.3	Who controls planning permission?	93
4.4	So when do Scottish Ministers get involved?	94
4.5	What does the Scottish Planning Act cover?	94
4.6	What are Development Plans?	95
4.7	Who requires planning permission?	96
4.8	Do I really need planning permission?	96
4.9	Do I have to notify my neighbours of my intentions?	97
4.10	Will the council advertise my planning application?	97
4.11	Where can I find out about local planning applications?	98
4.12	Do the council always involve the public in all planning matters?	98
4.13	Why is planning permission necessary?	98
4.14	How should I set about gaining planning permission?	99
4.15	What types of planning permission are available?	99
4.16	Are there any concessionary fees and exemptions?	100
4.17	How do I apply for planning permission?	100
4.18	What information do I have to provide?	101
4.19	Applying online your council's website	102
4.20	What are Certificates of Site Ownership?	102
4.21	What are Certificates of Agricultural Holding?	102
4.22	What sort of plans will I have to submit?	103
4.23	What important areas should I take into consideration?	104
4.24	What is the planning permission process?	105
4.25	How do they reach their decision?	108
4.26	What are the main objections against granting planning permission?	108
4.27	What matters cannot be taken into account?	110
4.28	What are the most common stumbling blocks?	110
4.29	What if the council refuses the planning application?	111
4.30	What should I think about before I start work?	112
4.31	What could happen if you don't bother to obtain planning permission?	119
4.32	Who enforces planning control?	119
4.33	How much does a planning application cost?	122
4.34	Can I get any further assistance on planning issues?	122
Annex A:	What are the government's restrictions on planning applications?	126
5	**Requirements for Planning Permission and Building Regulations Approval**	**129**
5.1	Decoration and repairs inside and outside a building	131
5.2	Structural alterations inside	132
5.3	Replacing windows and doors	133
5.4	Electrical work	134
5.5	Plumbing	135

5.6	Central heating	136
5.7	Oil-storage tank	136
5.8	Planting a hedge	137
5.9	Building a garden wall or fence	137
5.10	Felling or lopping trees	138
5.11	Laying a path or a driveway	138
5.12	Building a hardstanding for a car, caravan or boat	140
5.13	Installing a swimming pool	141
5.14	Erecting aerials, satellite dishes, television and radio aerials, wind turbines and flagpoles	141
5.15	Advertising	142
5.16	Building a porch	143
5.17	Outbuildings	144
5.18	Garages	145
5.19	Building a conservatory	146
5.20	Loft conversions, roof extensions and dormer windows	148
5.21	Building an extension	149
5.22	Conversions and change of use	151
5.23	Building a new house	156
5.24	Infilling	157
5.25	Demolition	157
5.26	What is a defective building?	160
5.27	Where can I find out more?	161
6	**The Building (Scotland) Regulations**	**162**
6.0	Introduction	162
6.1	Design and construction	167
6.2	Foundations	197
6.3	Drainage	223
6.4	Ventilation	250
6.5	Basements and cellars	290
6.6	Floors	299
6.7	Walls	344
6.8	Ceilings	436
6.9	Roofs	450
6.10	Chimneys and fireplaces	481
6.11	Windows	509
6.12	Doors	524
6.13	Stairs and ramps	546
6.14	Lifts and platforms	581
6.15	Corridors	591
6.16	Lobbies	598
6.17	Entrances and access	606
6.18	Balconies	630
6.19	Water (and earth) closets, bathrooms and showers	635

6.20   Combustion appliances   658
6.21   Hot water storage systems   698
6.22   Heating systems   711
6.23   Electrical safety   730
6.24   Kitchens and utility rooms   759
6.25   Extensions   769
6.26   Conservatories   784
6.27   Garages and outbuildings   789
6.28   Residential care buildings and hospitals   801
6.29   Shopping centres   816
6.30   Building documentation and information to be provided   824

Bibliography   *833*
Index   *887*

About the authors

Ray Tricker (MSc, IEng, FIET, FCMI, FCQI, FIRSE) is the Principal Consultant of Herne European Consultancy – a company specializing in Integrated Management Systems – and an established Butterworth-Heinemann author (39 titles). He served with the Royal Corps of Signals (for a total of 37 years), during which time he held various managerial posts culminating in his being appointed as the Chief Engineer of NATO's Communication Security Agency (ACE COMSEC).

Most of Ray's work since joining Herne has centred on the European railways. He has held a number of posts with the Union International des Chemins de fer (UIC), for example, Quality Manager of the European Train Control System (ETCS), European Union (EU) T500 Review Team Leader, European Rail Traffic Management System (ERTMS) Users' Group Project Co-ordinator and HEROE (Harmonization of European Rail Rules) Project Co-ordinator, and currently (as well as writing books for Butterworth-Heinemann!) he is busy assisting small businesses from around the world (usually on a no-cost basis) to produce their own auditable Quality Management Systems to meet the requirements of ISO 9001:2000. He is also a Consultant to the Association of American Railroads (AAR) advising them on ISO 9001:2000 compliance, and was recently appointed as UKAS Technical Specialist for the assessment of Notified Bodies for the Harmonization of the trans-European high-speed rail system and as an independent consultant for the new Trinidad Rapid Rail Project.

Ray's co-author is Rozz Algar

Rozz has worked for many years in the advertising industry, originally in London as an Account Director for Saatchi & Saatchi and M&C Saatchi, managing multi-million-pound advertising budgets for key clients. On leaving London, Rozz established the marketing functions and activities for one of the country's foremost farming co-operatives, expanding their customer base from the traditional farming market to include other trade business and general consumers.

Throughout her career Rozz has managed large teams of staff, developing organizational strategy, and taken a lead role in managing complex projects for clients. Today Rozz works as Group Marketing Director for a top 20 UK Advertising Agency based in the West Country, responsible for the marcoms strategy of a multi-million- pound brand as well as overseeing the internal realization of this strategy amongst a diverse workforce of over 300 staff, internally focusing on quality management systems and resource development strategies.

Rozz and her husband Graham, a landscape gardener, are also very keen property developers and have spent the last 8 years renovating and extending their dream home in Devon.

Foreword

Subject to specified exemptions, all building work in Scotland is governed by the Building (Scotland) Regulations.

This is a statutory instrument, which sets out the minimum requirements and performance standards for the design and construction of buildings, and extensions to buildings. The current regulations are the Building (Scotland) Regulations 2004.

Although the regulations are comparatively short, they rely on their technical detail being available in two technical handbooks (one for domestic and one for non-domestic buildings) and a vast number of British, European and international standards, codes of practice, drafts for development, published documents and other non-statutory guidance documents.

The main problem, from the point of view of the average builder and/or DIY enthusiast, is that Regulations are too professional for their purposes. They cover every aspect of building, are far too detailed and contain too many options. All the builder or DIY person really requires is sufficient information to enable them to comply with the regulations in the simplest and most cost-effective manner possible.

Verifiers and Certifiers, acting on behalf of local authorities, are primarily concerned with whether a building complies with the requirements of the Building Regulations and to do this they need to 'see the calculations'. But how does the DIY enthusiast and/or builder obtain these calculations? Where can they find, for instance, the policy and requirements for loadbearing elements of a structure?

Builders, through experience, are normally aware of the overall requirements for foundations, drains, walls, central heating, air conditioning, safety, security, glazing, electricity, plumbing, roofing, floors, etc., but they still need a reminder when they come across a different situation for the first time (e.g. if they are going to construct a building on soft soil, how deep should the foundations have to be?).

On the other hand, the DIY enthusiast, keen on building his own extension, conservatory, garage or workshop, etc., usually has no past experience and needs the relevant information – but in a form that he can easily understand without having had the advantage of many years, experience. In fact, what he really needs is a rule of thumb guide to the basic requirements.

From a number of surveys it has emerged that the majority of builders and virtually all DIY enthusiasts are self-taught and most of their knowledge is gained through experience. When they hit a problem, it is usually discussed over a pint in the local pub with friends in the building trade as opposed to seeking professional help. What they really need is a reference book to enable them to understand (or remind themselves of) the official requirements.

The aim of our book, therefore, is to provide the reader with an in-brief guide that can act as an *aide-mémoire* to the current requirements of the Building (Scotland) Regulations. Intended readers are primarily builders and the DIY fraternity (who need to know the regulations but do not require the detail), but the book, with its ready reference and no-nonsense approach, will be equally useful to students, architects, designers, building surveyors and inspectors, etc.

 Note from the authors: If any reader has any thoughts about the contents of this book (such as areas where perhaps they feel we have not given sufficient coverage, omissions and/or mistakes, etc.) then **please** let us know by e-mailing us at rayrozz@herne.org.uk and we will make suitable amendments in the next edition of this book!

Preface

The Great Fire of London in 1666 was probably the single most significant event to shape today's legislation. The rapid growth of fire through co-joined timber buildings highlighted the need to consider the possible spread of fire between properties and this consideration resulted in the publication of the first building construction legislation in 1667 requiring all buildings to have some form of fire resistance.

Two hundred years later, the Industrial Revolution meant poor living and working conditions in ever expanding, densely populated urban areas. Outbreaks of cholera and other serious diseases, through poor sanitation, damp conditions and lack of ventilation, forced the government to take action and building control took on the greater role of health and safety through the first Public Health Act of 1875. This Act had a major revision in 1936 which led to the introduction of the Building (Scotland) Act 1959. Over the years this Act has been amended and updated and the current document for Scotland is the Building (Scotland) Act 2003. This Act is implemented by the Building (Procedure) (Scotland) Regulations 2004 (Statutory Instrument 2004 No. 406) which is made by the Scottish Ministers under powers delegated under the Building (Scotland) Act 2003.

 Note: Copies of the above documents are available from the Scottish Stationery Office (TSO), 71 Lothian Road, Edinburgh EH3 9AZ (Tel: 0870 606 5566, Fax: 0870 606 5588, www.tsoscotland.com) and through booksellers. Copies of the Act and the Regulations can also be viewed (and downloaded) on the OPSI website (www.opsi.gov.uk) whilst copies of the Technical Handbooks are available at www.sbsa.gov.uk

Aim of this book

The prime aim of this book is to provide builders and DIY people with an *aide-mémoire* and a quick reference to the requirements of the Scottish Building Regulations. This book provides a user-friendly background to the Building (Scotland) Act 2003 and its associated Building (Scotland) Regulations 2004. It explains the meaning of the Building Regulations, their current status, requirements, associated documentation and how local authorities and councils view their importance. It goes on to describe the content of the Technical Booklets published by the Scottish Building Standards Agency (an executive agency of the Scottish Executive, responsible for writing the Scottish Building Regulations) and, in a series of 'what ifs', provides answers to the most common questions that DIY enthusiasts and builders might ask concerning building projects.

The book is structured as follows:

Chapter 1 – The Building (Scotland) Act 2003
Chapter 2 – The Building (procedure) (Scotland) Regulations 2004
Chapter 3 – The Requirements of the Building (Scotland) Regulations (2004)
Chapter 4 – Planning permission
Chapter 5 – Requirements for planning permission and Building Regulations Approval
Chapter 6 – The Building (Scotland) Regulations.

This is then supported by a bibliography – a list of the most relevant British, European and International Standards, a full list of current statutory instruments that cover topics relevant to building works and details of other useful and relevant publications – useful names and addresses, and a full index.

The following symbols will help you get the most out of this book:

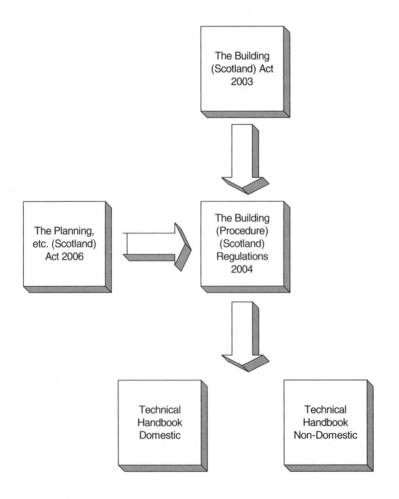

an important requirement or point

 a good idea or suggestion

 Note: Further amplification or information.

How do all the elements fit together?

At first glance the whole thing can be a bit of a minefield but essentially the elements of the Building Act, Building Regulations, Planning Permissions and Technical Handbooks flow from each other as Figure P1 attempts to show.

In a nutshell:

- The Building (Scotland) Act 2003 sets out the parameters under which all construction and conversion work of all buildings must be carried out.
- The Building (Scotland) Regulations then specify the minimum requirements in order to comply with the Act.
- The Planning, etc. (Scotland) Act 2006 sets out the statutory requirements for controlling all development of building works in Scotland.
- The Technical Handbooks give practical guidance on achieving the standards set in the Building Regulations.

1
The Building (Scotland) Act 2003

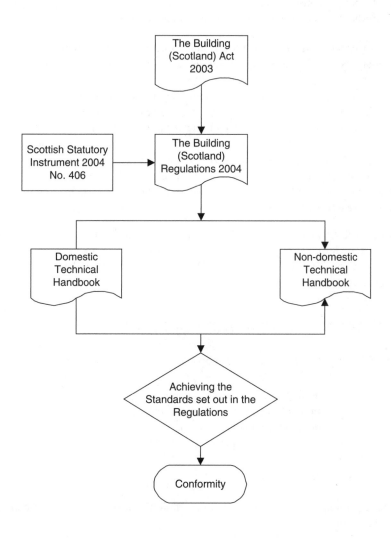

1.1 What is the aim of the Building (Scotland) Act 2003?

The Bill for this Act of Scottish Parliament was passed on 20 February 2003 and received Royal Assent on 26 March 2003. Although this Act retains the general framework of the 1959 Act, some of the building standard processes have been changed to ensure that they reflect existing and current practice.

The Building (Scotland) Act 2003 provides Scottish Ministers with the power to create Building Regulations to:

- secure the health, safety, welfare and convenience of persons in or about buildings;
- further the conservation of fuel and power;
- further the achievement of sustainable development.

It also provides them with the power to make procedural regulations, fees regulations and the other supporting legislation that is required to operate the system.

The Act is then supported by Building Regulations contained in The Building (Scotland) Regulations 2004 (Scottish Statutory Instrument 2004 No. 406) and the methods for implementing these requirements are contained in two Technical Handbooks (one for domestic work and one for non-domestic). These Handbooks provide practical guidance on achieving the standards set out in the Building Regulations, whilst the standards themselves describe the functions that a building should perform for conformity (such as 'providing resistance to the spread of fire').

 Note: The technical specifications of these standards are inspected by the European Commission (EC) to ensure that they meet the requirements of the Construction Products Directive (CPD) and to ensure that no barriers to trade in construction products are created.

1.2 What does the Act contain?

The Building (Scotland) Act 2003 consists of six parts:

      Part 1 Building Regulations
      Part 2 Approval of construction work, etc.
      Part 3 Compliance and enforcement
      Part 4 Defective and dangerous buildings
      Part 5 General
      Part 6 Supplementary

These parts are then broken down into a number of sections and subsections as shown in Table 1.1.

Table 1.1 The parts of the Building (Scotland) Act 2003

Part	Title	Section	Sub-section
1.	Building Regulations	Power to make Building Regulations	Building Regulations
			Continuing requirements
		Relaxation of Building Regulations	Relaxation of Building Regulations
		Guidance documents	Guidance documents for purposes of Building Regulations
			Compliance with guidance documents
		Building standards assessments	Building standard assessments
2.	Approval of construction work, etc.	Verifiers and Certifiers	Verifiers and Certifiers
		Building Warrants	Building Warrants
			Building Warrants: grant and amendment
			Building Warrants: extension, alteration and conversion
			Building Warrants: certification of design
			Building Warrants: reference to Ministers
			Building Warrants: further provision
			Building Warrants: limited life buildings
			Building Warrants: late applications
			Applications and grants: offences
		Completion Certificates	Completion Certificates
			Completion Certificates: acceptance and rejection
			Completion Certificates: certification of construction
			Completion Certificates: offences
			Occupation or use without Completion Certificates
		Imposition of continuing requirements by Verifiers	Imposition of continuing requirements by Verifiers
			Discharge and variation of continuing requirements imposed by Verifiers

(Continued)

Table 1.1 (Continued)

Part	Title	Section	Sub-section
3.	Compliance and enforcement	Buildings standards registers	Building standards registers
		Compliance and enforcement	Building Regulations compliance
			Continuing requirement enforcement notices
			Building Warrant enforcement notices
4.	Defective and dangerous buildings	Defective and dangerous buildings	Defective buildings
			Dangerous buildings
			Dangerous buildings notice
		Building Standards Advisory Committee	Building Standards Advisory Committee
		Functions of Ministers, local authorities, Verifiers and Certifiers	Exercise of local authority functions
			Procedure regulations
			Reports and information
			Scheduled monuments, listed buildings, etc.
5.	General	Scheduled monuments, listed buildings, etc.	
		Documents	Forms
			Service of notices, etc.
		Fees and charges	Fees and charges
		Entry, inspection and tests	Powers of entry, inspection and testing
			Work required by notice: right of entry
			Tests of materials
		Evacuation of buildings	Evacuation of buildings
			Unlawful occupation of evacuated buildings
		Execution of work	Expenses
			Compulsory purchase where owner cannot be found
			Sale of materials from demolished buildings

Appeals
Offences and liability
  Appeals
  Penalties for offences
  Offences by bodies corporate, etc.
  Criminal liability of trustees, etc.
  Civil liability
Inquiries
  Inquiries
Crown application
  Crown application
Orders and regulations
  Orders and regulations
Interpretation
  Meaning of 'Building'
  Interpretation

6. Supplementary

Supplementary
  Ancillary provision
  Modification of enactments
  Commencement and short title

Schedule 1 – Building Regulations
Schedule 2 – Verifiers and Certifiers
Schedule 3 – Procedure regulations: particular matters
Schedule 4 – Powers of entry, inspection and testing: further provision
Schedule 5 – Evacuation of buildings
Schedule 6 – Modification of enactments

1.3 Who polices the Act?

Under the terms of the Act, the Scottish Minister directs all local authorities to be responsible for ensuring that any building work being completed conforms to the requirements of the associated Building Regulations. They have the authority to:

- make you take down and remove or rebuild anything that contravenes a regulation;
- make you complete alterations so that your work complies with the Building Regulations;
- employ a third party (and then send you the bill!) to take down and rebuild non-conforming buildings or parts of buildings.

They can, in certain circumstances, even take you to court and have you fined – especially if you fail to complete the removal or rebuilding of the non-conforming work.

 The above authority to prosecute and order remedial work to be completed applies equally whether you are the actual owner or merely the occupier – so be warned!

1.4 Are there any exemptions from Building Regulations?

There are **NO** exemptions from Building Regulations allowable under this Act.

1.5 When can the Building Regulations be relaxed?

Building Regulations can be dispensed with or relaxed where the Scottish Ministers believe that a regulation or regulations in relation to a particular building or description of building is unreasonable (e.g. the requirement to provide access for a fire engine might be waived on a remote island where there is no fire engine). Individuals can make applications to the Scottish Ministers to relax the regulations.

 This is not restricted to just the owner of a building but others (such as prospective buyers) may also apply.

1.6 What if the work does not comply with Building Regulations?

If the local authority believes that the building does not comply with Building Regulations, they may issue a building notice (a 'Building Regulations compliance notice') requiring the owner, by a set date, to ensure that the building complies with relevant regulations. Should the owner not comply by the set date then the local authority can carry out the work specified in the notice and recover the costs from the owner!

1.7 What are guidance documents?

Guidance documents are issued for the purpose of providing practical guidance in relation to the requirements. The Scottish Ministers can issue, revise or withdraw guidance documents as and when required. Failure to comply with guidance documents does not in itself render a person liable to civil or criminal proceedings, **but** proof of compliance with them may be relied on as evidence that Building Regulations have not been contravened.

1.8 What are Building Warrants?

Building Warrants are required for **any** work on construction, demolition or conversion of any building as well as the provision of services, fittings or equipment in (or in conjunction with) any building to which Building Regulations apply. Carrying out work without a warrant, or not in accordance with the warrant, is an offence and punishable!

Verifiers are obliged to issue Building Warrants for construction, conversion, demolition or the provisions of services, fittings, etc., as long as the work to be carried out complies with the Building Regulations and as long as there is nothing to indicate that when constructed, the building or services will fail to comply with the regulations. Verifiers can ask for further information (including Completion Certificates) relating to the application and can delay issuing a warrant until they are completely satisfied that works will comply with Building Regulations.

Building Warrants are required for limited life buildings and must state the period of intended life of the building. An additional warrant is needed for the demolition of this limited life building.

 You can seek to extend the life span of the building by applying to the verifier.

1.8.1 Who are Verifiers and Certifiers?

The Scottish Ministers can appoint persons (these can be individuals, corporate bodies, public bodies or office holders) as:

- Verifiers;
- approved Certifiers of design;
- approved Certifiers of construction.

These Verifiers and Certifiers carry out the day-to-day work of inspecting and approving building works.

Verifiers are able to issue Building Warrants for:

- construction or demolition of buildings;
- the provision of services, fittings or equipment.

Certifiers are able (dependent on their appointment) to issue certificates for elements of design and construction which Verifiers will accept as valid when issuing a warrant. For example, a certificate might certify that an innovative design for the conversion of fuel and power fulfils the requirements of the Building Regulations.

The Scottish Ministers can set limitations on the functions that any of these Verifiers and Certifiers can carry out. For example, they may only be able to approve a scheme in a particular geographical area or type of building.

A full list of those approved to work as Verifiers and Certifiers along with the limitations and restrictions set against them is available for public inspection at all 'reasonable times' from the local authority concerned.

1.8.2 What are Certifications of Design?

Certifications of Design are used as additional substantiation to the Verifier as part of the Building Warrant application. Issued by Certifiers, these certificates certify that the design or construction complies with Building Regulations and the Verifiers must accept the certificate as conclusive.

 Any Certifier who issues a misleading or false certificate can be liable for prosecution.

1.8.3 What if I carry out the work without a Building Warrant?

The local authority can take out an enforcement action if it appears to them that the work has been carried out without a warrant or not in accordance with an applicable warrant. The enforcement action can require the person to submit a Completion Certificate, obtain a warrant, to secure work that complies with the warrant or obtain an amendment to the Building Warrant to cover the non-compliance and set a time limit for this to be carried out. The enforcement notice may specify that any work has to be suspended until the notice has been complied with.

If the work has not been carried out by the specified date the authority may carry out the work and recover any costs of doing so from the owner.

1.9 What are Completion Certificates?

Completion Certificates are required for **all** work relating to Building Warrants and must be submitted by the Certifier. This Completion Certificate certifies that the works have been carried out in accordance with the Building Warrant and comply with Building Regulations. If Building Regulations have changed since the Building Warrant was issued, the work must comply with Building Regulations at the time the warrant was issued, as opposed to any later version in force.

The Verifier must accept or reject the certificate only after reasonable inquiry and when they are satisfied that work has been carried out in accordance with the warrant.

1.9.1 What is a certification of construction?

Certification of construction can be used in conjunction with a Completion Certificate to certify that a specific element of construction such as plumbing or electrical work complies with Building Regulations. The Verifier is obliged to accept the facts certified by an approved Certifier of construction.

1.9.2 Can we use or occupy the building without a Completion Certificate?

No, not unless the Verifier has granted permission for temporary occupation of the building!

1.10 Where can I find out about Building Warrants and Completion Certificates in my areas?

The local authority must maintain a register for their area of information about applications for Building Warrants, Completion Certificates and other matters required by regulations. These must be available for public inspection at reasonable times.

1.11 What if a building is defective or dangerous?

If the local authority believes that the building is either defective or dangerous they can issue a notice insisting that work be carried out to rectify the problem and a specific date by which these works should be carried out will be set. In respect of dangerous buildings, the local authority can, when it considers the action to be urgent, carry this out without prior notice to the owner. This can include demolition.

Whilst a warrant will be required for remedial work to be carried out to defective buildings, the local authority does not need a warrant for work carried out in relation to dangerous buildings, but in **all** cases a Completion Certificate is required.

1.12 What about scheduled monuments and listed buildings?

Where work is required in relation to scheduled monuments or listed buildings the local authority **must** consult the Scottish Ministers, the planning authority and other appropriate persons/authorities they think fit, prior to any notices or warrants being issued.

1.12.1 What if a scheduled monument or listed building is dangerous?

Even if the scheduled monument or listed building is considered to be dangerous, the local authority must consult with the relevant bodies 'as long as this is reasonably practicable'.

1.13 Do the local authority or Scottish Ministers have the right to enter my property?

Yes – in order to carry out tests on materials in relation to certain functions under the Act. The Act states that owners and occupiers 'must provide the authority with assistance' and it is an offence not to do so.

1.14 Can occupants be evicted from buildings?

Yes – the authority is required to ensure the health and safety of residents so if the building is deemed dangerous, or is to be demolished, they have a duty to evacuate the building. It is an offence to occupy a building that has been served with the relevant notice.

1.15 Can the authority compulsorily purchase the building?

Yes – if the owner cannot be found, the local authority can compulsorily purchase the building and its site in relation to dangerous buildings in order to recover the costs of doing so.

1.16 Can the authority sell off materials from demolished buildings?

Yes – the local authority can sell off materials from a demolition if the owner on whom the notice has been served fails to comply with a Building Warrant in order to recoup any costs owed to the authority.

1.17 Can I appeal decisions made in relation to my building?

Yes – you have the right to appeal against decisions made by the Verifiers in relation to refusal to issues Building Warrants, rejection of Completion Certificates, enforcement notices, etc. This appeal should be made within 21 days of the date of the decision or notice and be made to the sheriff. The decision of the sheriff on an appeal is final.

1.18 What about civil liability?

It is an aim of the Building Act that all building work is completed safely and without risk to people employed on the site or whilst visiting the site, etc. Any contravention of the Building Regulations that causes injury (or death) to any person is liable to prosecution in the normal way.

1.19 What is the Building Standards Advisory Committee?

The role of the Building Standards Advisory Committee (which was established under the previous 1959 Building Act) is to report to the Scottish Ministers on all aspects of the building standards system that may need attention. It considers developments in the industry, changes in public expectations or particular problems that arise. The Committee must be consulted on all proposed amendments to legislation or guidance documents.

Details of the current membership of the committee are provided on the Scottish Public Appointments page at: www.scotland.gov.uk

1.20 What is a Building Standards Assessment?

A Building Standards Assessment is an assessment of the extent to which the local authority considers that the building complies with Building

Regulations. Individuals can request that the local authority carry out a Building Standards Assessment (e.g. the owner of a building might request such an assessment on behalf of someone intending to purchase the building).

1.21 What are the Supplementary Regulations?

Part 6 of the Building Act contains six schedules whose function is to list the principal areas requiring regulation and to show how the Building Regulations are to be controlled by the local authority. These schedules are:

- Schedule 1 – Building Regulations
- Schedule 2 – Verifiers and Certifiers
- Schedule 3 – Procedure regulations: particular matters
- Schedule 4 – Powers of entry, inspection and testing: further provision
- Schedule 5 – Evacuation of buildings
- Schedule 6 – Modification of enactments.

1.21.1 What is Schedule 1 of the Building (Scotland) Act 2003?

This schedule makes provision in relation to the matters about which Building Regulations may stipulate. It:

- enables the Building Regulations to reference a document published by or on behalf of the Scottish Ministers;
- enables specified persons to express their approval for the purposes of satisfying Building Regulations;
- allows for special provision for buildings that may have limited lifespan (e.g. temporary premises);
- allows for particular types of buildings to be either completely or partially exempt from the requirements of the regulations (e.g. car ports, garden sheds);
- allows for certain things to be provided or done in connection with the building and/or specifies the manner in which the work is to be carried out. The matters referred to are:
  - preparation of sites;
  - strength and stability (including the safeguarding of adjacent buildings and services);
  - fire precautions (including resistance of structure to the outbreak and spread of fire, the protection of occupants and means of escape in the event of fire and the provision of facilities to assist fire fighting);
  - resistance to moisture and decay;
  - resistance to the transmission of sound;
  - durability;
  - resistance to infestation;
  - drainage;

- ventilation (including the provision of open space for it);
- day lighting (including the provision of open space for it);
- service, fittings and equipment (including broadband communication technology and other electronic services, fittings and equipment and services, fitting and equipment for the supply or use of gas or electricity);
- measures to ensure that pipes used in connection with the provision of water for domestic purposes are not fitted in a way that may contribute to the concentration of lead in such water exceeding the limit specified in Table B of Schedule 1 to the Water Supply (Water Quality) (Scotland) Regulations 2001 (S.S.I. 2001/207);
- measures affecting the emission of smoke, gases, fumes, grit, dust or other noxious or offensive substances;
- accommodation and ancillary equipment;
- access, including in particular access for disabled persons;
- suitability for use by disabled persons;
- prevention of danger and obstruction;
- security;
- reuse of building materials.

1.21.2 What is Schedule 2 of the Building (Scotland) Act 2003?

This schedule:

- makes provisions in connection to the appointment and termination of Verifiers and Certifiers;
- provides that Certifiers approved by virtue of membership of an approved scheme are subject to various limitations;
- places a duty on the Scottish Ministers to appoint a successor Verifier if the Verifier's appointment has been terminated (and for the successor to take over unfinished matters in relation to Building Warrants and Completion Certificates);
- makes provisions to avoid conflict of interest if the Verifier is also a Certifier or where the Verifier has an interest in the building.

1.21.3 What is Schedule 3 of the Building (Scotland) Act 2003?

Schedule 3 deals with procedural regulations such as:

- consultation of Verifiers;
- maintenance (by local authorities and Verifiers) of records of application and decisions on applications;
- duration of the validity of Building Warrants, etc.

1.21.4 What is Schedule 4 of the Building (Scotland) Act 2003?

Schedule 4 makes further provision about the rights of entry, inspection and testing conferred on the Scottish Ministers and local authorities. It:

- provides that entry may be demanded only at a reasonable time; that 3 days' notice of intended entry must be given to the occupier and, if owner is known, the owner of the premises, unless in the case is one of urgency;
- provides that the sheriff or JP can, if satisfied that if there are reasonable grounds to do so, authorize the relevant person (e.g. local authority) to exercise power and if need be force. The grounds mentioned are that:
  - the exercise of the power in relation to the premises has been refused;
  - such a refusal is reasonably apprehended;
  - the premises are unoccupied;
  - the occupier is temporarily absent from the premises;
  - the case is one of urgency;
  - an application for admission to the premises would defeat the object of the proposed entry.

 A warrant under this schedule continues to be in force until the purpose for which the warrant has been issued has been fulfilled

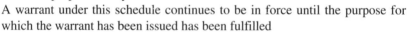

- provides that anyone exercising power of entry must provide written evidence of that entitlement;
- provides that anyone who enters a premises in the exercise of any power can take other people or equipment onto the premises as necessary subject to the conditions of the warrant;
- provides that any person exercising power of entry must on leaving the premises ensure it is effectually secured against unauthorized entry;
- provides that should the person exercising power of entry disclose or use information obtained on those premises with regard to any manufacturing process or trade secret they will be guilty of an offence.

1.21.5 What is Schedule 5 of the Building (Scotland) Act 2003?

Schedule 5 provides for the procedures to be followed when securing the removal from the building of any occupant who has failed to leave the building when required to do so. These procedures mean that:

- the local authority may apply for a warrant of ejection (this application must be accompanied by a certificate, certifying that the removal is required due to the building being in a dangerous state, is due for demolition, and that works need to be carried out that will endanger the occupant;
- the sheriff will be required to service a notice on the occupants;
- the sheriff's decision on the application is final.

The schedule makes provision for occupants who are evicted from the premises in these circumstances protecting their status and rights as tenants.

1.21.6 What is Schedule 6 of the Building (Scotland) Act 2003?

Schedule 6 amends and repeals provisions of various other Acts in consequence of the Act. These consist of:

- Ancient Monuments and Archaeological Areas Act 1979;
- Building (Scotland) Act 1959;
- Building (Scotland) Act 1970;
- Chronically Sick and Disabled Persons Act 1970;
- Civic Government (Scotland) Act 1982;
- Clean Air Act 1993;
- Control of Pollution Act 1974;
- Criminal Procedure (Scotland) Act 1995;
- Fire Precautions Act 1971;
- Fire Safety and Safety of Places of Sport Act 1987;
- Health and Safety at Work, etc. Act 1974;
- Housing (Scotland) Act 1986;
- Housing (Scotland) Act 1987;
- Local Government (Scotland) Act 1973;
- Local Government Act 1988;
- Local Government, etc. (Scotland) Act 1994;
- Local Government in Scotland Act 2003;
- Land Compensation (Scotland) Act 1973;
- Licensing (Scotland) Act 1976;
- Mines and Quarries (Tips) Act 1969;
- Planning (Consequential Provisions) (Scotland) Act 1997;
- Roads (Scotland) Act 1984;
- Safety of Sports Grounds Act 1975;
- Sewerage (Scotland) Act 1968;
- The Scotland Act 1998.

What happens if I contravene any of these requirements?

Where the offence is committed by a corporate body (including local authorities, Scottish partnerships and unincorporated associations other than a Scottish partnership) a director/officer/member of the authority/partner/person concerned with the management of the association as well as the association may be proceeded against and punished accordingly!

1.22 What about the rest of the United Kingdom?

The Building (Scotland) Act 2003 **only** applies to Scotland. England and Wales are controlled by the Building Act 1984. Northern Ireland is controlled by The Building Regulations (Northern Ireland) Order 1979. Their 'interoperability' is shown in Table 1.2.

Table 1.2 The interoperability of Building Regulations

Scotland		England and Wales		Northern Ireland	
Section 1	Structures	Part A	Structure	Technical Booklet D	Structure
Section 2	Fire	Part B	Fire safety	Technical Booklet E	Fire Safety
Section 3	Environment	Part C	Site preparation and resistance to contaminants and water	Technical Booklet C	Preparation of site and resistance to moisture
		Part D	Toxic substances	Technical Booklet B	Materials and workmanship
		Part F	Ventilation	Technical Booklet K	Ventilation
		Part G	Hygiene	Technical Booklet P	Sanitary appliances and unvented hot water storage systems
		Part H	Drainage and waste disposal	Technical Booklet J Technical Booklet N	Solid waste in buildings Drainage
Section 3	Environment	Part J	Combustion appliances and fuel storage systems	Technical Booklet L	Heat-producing appliances and liquefied petroleum gas installations
Section 4	Safety	Part K	Protection from falling, collision and impact	Technical Booklet H	Stairs, ramps and protection from impact
		Part M	Access and facilities for disabled people	Technical Booklet R	Access for facilities and disabled people
		Part P	Electrical safety		
Section 5	Noise	Part E	Resistance to the passage of sound	Technical Booklet G	Sound installation of dwellings
Section 6	Energy	Part L	Conservation of fuel and power	Technical Booklet F	Conservation of fuel and power

1.22.1 England and Wales

England and Wales are controlled by the Building Act 1984. The Building Regulations 2000 then set out the functional standards under this Act, which are further set out in a series of Approved Documents.

1.22.2 Northern Ireland

The Building Regulations (Northern Ireland) Order 1979 (as amended by the Planning and Building Regulations (Amendment) (NI) Order 1990) is the main legislation for Northern Ireland and the Building Regulations (Northern Ireland) 2000 then details the requirements for meeting this legislation.

Supporting publications (such as British Standards, BRE publications and/or Technical Booklets published by the Department) are then used to ensure that the requirements are implemented (*deemed to satisfy*).

The main procedural difference between the Scottish system and the others is that a Building Warrant is **still** required before work can start in Scotland.

1.22.3 Where do I find details of the specific elements of the relevant Building Regulations?

In all cases, the relevant legislation is further clarified in Approved Documents, Technical Handbooks, Booklets and Guidance Notes.

2

The Building (Procedure) (Scotland) Regulations 2004

Even when planning permission is not required, most building works, including alterations to existing structures, are subject to minimum standards of construction to safeguard public health and safety.

2.1 What is the purpose of the Building (Scotland) Regulations?

The Building (Procedure) (Scotland) Regulations are legal requirements laid down by Scottish parliament, based on the Building (Scotland) Act 2003. They are approved by parliament and deal with the **minimum** standards of design and building work for the construction of domestic, commercial and industrial buildings.

Building Regulations ensure that new developments or alterations and/or extensions to buildings are all carried out to an agreed standard that protects the health and safety of people in and around the building.

Building standards are enforced by the Verifier at your local council, but for matters concerning drainage or sanitary installations, you will need to consult their technical services department.

2.2 Why do we need the Building Regulations?

The Great Fire of London in 1666 was the single most significant event to have shaped today's legislation. The rapid growth of the fire through timber buildings built next to each other highlighted the need for builders to consider the possible spread of fire between properties when rebuilding work commenced. This resulted in the first building construction legislation that required all buildings to have some form of fire resistance.

During the Industrial Revolution (200 years later) poor living and working conditions in ever expanding, densely populated urban areas caused outbreaks of cholera and other serious diseases. Poor sanitation, damp conditions and lack of ventilation forced the government to take action and building control took on the greater role of health and safety through the first Public Health Act of 1875. This Act had two major revisions in 1936 and 1961 and led to the first set of national building standards – the Building Regulations 1965.

The current legislation (i.e. within Scotland) is the Building (Procedure) (Scotland) Regulations 2004 (Statutory Instrument 2004 No. 428) which is made by the Scottish Ministers under powers delegated by Scottish parliament under the Building (Scotland) Act 2003.

The Building Regulations are a set of minimum requirements designed to secure the health, safety and welfare of people in and around buildings and to conserve fuel and energy in Scotland (other legislation covers buildings in England, Wales and Northern Ireland). They are basic performance standards and the level of safety and acceptable standards are set out as guidance in the two Technical Handbooks (one for Domestic and the other for Non-domestic Buildings). Compliance with the detailed guidance of the Technical Handbooks is usually considered as evidence that the Regulations themselves have been met.

 Alternative ways of achieving the same level of safety, or accessibility, are also acceptable.

2.3 What building work is covered by the Building Regulations?

 The Building Regulations cover all NEW building work.

This means that if you want to put up a new building, extend or alter an existing one, convert a building or provide new and/or additional fittings in a building such as drains or heat-producing appliances, washing and sanitary facilities and hot water storage (particularly unvented hot water systems), the Building Regulations will probably apply.

It should be remembered, however, that although it may appear that the Regulations do not apply to some of the work you wish to undertake, the end result of doing that work could well lead to you contravening some of the Regulations. You should also recognize that some work – whether or not controlled – could have implications for an adjacent property. In such cases, it would be advisable to take professional advice and consult the local authority or a Verifier. Some examples are:

- removing a buttressed support to a party wall;
- underpinning part of a building;
- removing a tree close to a wall of an adjoining property;

- adding floor screed to a balcony which may reduce the height of a safety barrier;
- building parapets which may increase snow accumulation and lead to an excessive increase in loading on roofs.

2.4 What is a conversion?

A conversion is a specified change of occupation or use of a building, which will cause Building Regulations to apply – for example, changing a garage to a bedroom (a full list of specified types of work can be found in Appendix 3B in Chapter 3). If you are in any doubt it is advisable to speak to a Verifier.

 A Building Warrant is STILL required for a conversion even if building work is not carried out.

2.5 What are the requirements associated with the Building Regulations?

The Building (Scotland) Regulations contain a list of requirements (referred to as 'Part 1') that are designed to ensure the health and safety of people in and around buildings; to promote energy conservation; and to provide access and facilities for disabled people. In total there are six sections (1–6) to these requirements and these cover subjects such as structure, fire, environment, safety, noise and energy and are expressed in broad, functional terms in order to give designers and builders the maximum flexibility in preparing their plans. It is the responsibility of the Verifier to interpret the requirements.

2.6 What are the Technical Handbooks?

The Technical Handbooks contain practical and technical guidance on ways in which the requirements of each section of the Building (Scotland) Regulations can be met.

Each Technical Handbook reproduces the requirements contained in the Building Regulations relevant to the subject area. This is then followed by practical and technical guidance, with examples, on how the requirements can be met in some of the more common building situations.

 There may, however, be alternative ways of complying with the requirements to those shown in the Technical Handbooks and you are, therefore, under no obligation to adopt any particular solution in a Technical Handbook if you prefer to meet the relevant requirement(s) in some other way.

Just because a Technical Handbook has **not** been complied with, however, does not necessarily mean that the work is wrong. The circumstances of each particular case should be considered when an application is made to make sure that adequate levels of safety will be achieved.

 Note: The Building (Scotland) Regulations are constantly reviewed to meet the growing demand for better, safer and more accessible buildings as well as the need to reflect emerging harmonized European Standards. Building (Scotland) Regulations were last consolidated in SI 2004:428; however, every year the Technical Handbooks are reviewed and amended and the latest amendments to each section are clearly marked in a sub-section of the introduction to each section. This book has been written using the Technical Documents 2007.

The Technical Documents are in six sections (plus section 0 – General Considerations) and consist of:

Section 1 – Structure
Section 2 – Fire
Section 3 – Environment
Section 4 – Safety
Section 5 – Noise
Section 6 – Energy

You can buy a copy of the Technical Handbooks (and the Building Act 1984 if you wish) in most major bookshops or from the Scottish Building Standards Agency (SBSA), from the Scottish Stationery Office (TSO), 71 Lothian Road, Edinburgh EH3 9AZ (Tel.: 0870 606 5566, Fax: 0870 606 5588, www. tsoscotland.com). Occasionally, they are available from libraries.

Alternatively, you can download pdf copies of the Approved Documents from www.sbsa.gov.uk

2.6.1 Section 1 – structure

The intention of this section is to ensure that the structure of a building does not pose a threat to the safety of people in or around buildings and existing buildings as a result of:

- loadings;
- the nature of the ground;
- collapse or deformations;
- stability of the building and other buildings;
- climatic conditions;
- materials;
- structural analysis;
- details of construction;
- safety factors.

2.6.2 Section 2 – fire

Life safety is the paramount objective of fire safety. Domestic buildings should be designed and constructed in such a way that the risk of fire is reduced and, if a fire does occur, there are measures in place to restrict the growth of fire and smoke to enable the occupants to escape safely and for fire-fighters to deal with fire safely and effectively.

In the event of an outbreak of fire, it is important that the occupants are warned as soon as possible and guidance contained in standard 2.11 provides recommendations for the installation of alarm and detection systems in domestic buildings.

The Building Regulations are primarily concerned with the protection of people from the dangers inherent in buildings, rather than protecting the owners of buildings from any economic loss which might occur. However, although property protection is not covered by Building Regulations, the added benefit of some life safety measures will provide a degree of property protection. It is important, therefore, for designers and owners of buildings to understand that following Building Regulations guidance will not necessarily provide sufficient fire protection from the total destruction of the building and the subsequent economic loss. Although beyond the scope of this guidance, the adoption of good fire safety practices should be encouraged to reduce the risk of fire occurring in the first place. Fire prevention will not only save lives but also reduce environmental pollution.

The standards and guidance in this section of the Technical Handbooks are designed to work together to provide a balanced approach for fire safety. Where a building element, material, component or other part of a building is covered by more than one standard, the more demanding guidance should be followed.

In order to achieve these objectives, the building elements, materials, components or other parts of the building identified in the guidance should follow the appropriate performance levels that are recommended throughout the guidance.

Guidance in the Technical Handbook may not be appropriate for the following buildings or parts of a building as they are rarely designed and constructed:

a. those containing a basement storey with a storey area more than $200\,m^2$;
b. those containing a basement storey at a depth of more than 4.5 m;
c. those containing flats or maisonettes with a communal room more than $60\,m^2$;
d. those containing catwalks, openwork floors or escalators;
e. those containing places of special fire risk;
f. those with a storey at a height of more than 60 m.

In the case of sub-clauses above and in the case of a mixed use building containing non-domestic and domestic accommodation, reference should be made to the Technical Handbook for non-domestic buildings.

In the case of sub-clause f. above, the alternative approach described in clause 2.0.6 should be used.

2.6.3 Section 3 – environment

The intention of this section is to ensure that, as far as it is reasonably practicable, buildings do not pose a threat to the environment and dwellings and people in and around dwellings are not placed at risk as a result of:

- site conditions;
- hazardous and dangerous substances;
- the effects of moisture in various forms;
- an inadequate supply of air for human occupation of a dwelling;
- inadequate drainage from a building and from paved surfaces around a dwelling;
- inadequate and unsuitable sanitary facilities;
- inadequate accommodation and facilities in a dwelling;
- inadequately constructed and installed combustion appliances;
- inadequately constructed and installed oil storage tanks;
- inadequate facilities for the storage and removal of solid waste from a dwelling.

2.6.4 Section 4 – safety

Section 4 provides recommendations for the design of buildings that will ensure access and usability and reduce the risk of accident. The standards within this section:

- ensure accessibility to and within buildings and that areas presenting risk through access are correctly guarded;
- reduce the incidence of slips, trips and falls, particularly for those users most at risk;
- ensure that electrical installations are safe in terms of the hazards likely to arise from defective installations, namely fire and loss of life or injury from electric shock or burns;
- prevent the creation of dangerous obstructions, ensure that glazing can be cleaned and operated safely and reduce the risk of injury caused by collision with glazing;
- safely locate hot water and steam vent pipe outlets, minimize the risk of explosion through malfunction of unvented hot water storage systems and prevent scalding by hot water from sanitary facilities;
- ensure the appropriate location and construction of storage tanks for liquefied petroleum gas.

2.6.5 Section 5 – noise

The purpose of section 5 is to protect the residents of a dwelling from noise in other areas of the same building or an attached building. Recurrent noise can adversely affect the health of residents and inconvenience them by disrupting their everyday activities. In view of this, measures should be incorporated to reduce the transmission of the sounds of normal conversation, television, radio, music and domestic activities.

It is important to recognize that following the guidance in this section will not guarantee freedom from the transmission of all types of disturbing noise. Firstly, it does not address sound transmission between parts of the same dwelling. Secondly, it does not suggest that the construction should insulate against excessive noise from sources such as power drills, saws or sanders, noise from a hi-fi system inconsiderately played at full volume, or wall-mounted 'surround sound' flat panel loudspeakers. In addition, it does not address external sources of noise, such as aircraft, railways, road traffic or industry.

 Noise transmission to buildings or parts of buildings other than dwellings or external sources into a dwelling is **not** controlled by the Scottish Building Regulations.

Although noise transmission from external sources into a dwelling is not controlled by the Scottish Building Regulations, it may be managed through the land use planning system. Advice can be found in PAN 56 'Planning Advice Note: Planning and Noise', 1999.

Detailed guidance on noise issues relating to construction sites can be found in BS 5228 'Noise control on construction and open sites'.

Advice to consumers on dealing with noise problems is given in a leaflet – 'Sound advice on Noise: don't suffer in silence', 2001.

The Building Performance Centre at Napier University publishes guidance on good practice in improving sound insulation 'Housing and sound insulation: Improving existing attached dwellings and designing for conversions', 2006.

2.6.6 Section 6 – energy

The intention of section 6 is to ensure that effective measures for the conservation of fuel and power are incorporated in dwellings and buildings consisting of dwellings. In addition to energy conservation provisions for the building fabric and the building services it contains, a carbon dioxide emissions standard obliges a designer of new dwellings to consider buildings in a holistic manner. In view of this, localized or building-integrated Low and Zero Carbon Technologies (LZCT) (e.g. photovoltaics, active solar water heating, combined heat and power and heat pumps) can be used as a contribution towards meeting this standard. Although the focus is primarily on lowering carbon dioxide emissions from dwellings in use, the measures within this section will also address fuel poverty issues to a certain degree.

The standards and guidance given in this section of the Technical Handbooks are intended to achieve an improvement of around 18–25% fewer emissions than the previous standards; **however**, nothing here prevents a domestic building from being designed and constructed to be even more energy efficient and make greater use of LZCT! Where this occurs, both the monetary and environmental savings will be improved.

Section 6 should be read in conjunction with all the guidance to the Building (Scotland) Regulations 2004 but in particular section 3 of the Technical Handbooks (i.e. *Environment*) has a close affiliation with energy efficiency, regarding:

- heating of dwellings;
- ventilation of domestic buildings;
- condensation;
- natural lighting;
- combustion air and cooling air for combustion appliances;
- drying facilities;
- storage of woody biomass.

2.7 Are there any exemptions?

There are a number of broad categories of buildings that are exempt from Building Regulations. Some examples of small buildings associated with houses, flats or maisonettes that are exempt are listed below:

(1)   A detached single-storey building, with a floor area not more than $8\,m^2$, within the garden of a house, that:
   - is at least 1 m from the house or is closer than 1 m to the house but is at least 1 m from any boundary;
   - does not contain sleeping accommodation;
   - does not contain a flue; a fixed solid fuel, oil or gas appliance;
   - does not contain a sanitary fitting.
(2)   A detached single-storey building, with a floor area not more than $8\,m^2$, within the garden of a flat or maisonette that:
   - is at least 1 m from the flat or maisonette or 3 m from any other part of the building containing a flat or maisonette;
   - is at least 1 m from any boundary;
   - does not contain a flue, fixed solid fuel, oil or gas appliance;
   - does not contain a sanitary fitting.
(3)   A single-storey conservatory or porch with a floor area of not more than $8\,m^2$ that is attached to an existing house, and:
   - is at least 1 m from a boundary;
   - does not contain a fixed solid fuel, oil or gas appliance;
   - does not contain a sanitary fitting;
   - meets the regulations on safety glazing.
(4)   A single-storey greenhouse, carport or covered area, each with a floor area not more than $30\,m^2$, that is detached or attached to an existing house and:
   - does not contain a flue, fixed solid fuel, oil or gas appliance;
   - does not contain a sanitary fitting.
(5)   A paved area or hardstanding not more than $200\,m^2$ in area that:
   - is not part of any access route required by the regulations.

Note: A full list is given in Appendix 3C in Chapter 3 of this book.

2.8 What is the difference between 'exempt' and 'work not requiring a warrant'?

As opposed to 'exempt work', 'work not requiring a warrant' must still comply with the building standards set in the Building Regulations. Within the regulations schedule 3 sets out work not requiring a warrant; this list can be found in Chapter 3 Appendix 3C. This list is in two parts, work listed in part A must meet the standards in full but the work listed in part B need only be done in such a way that the completed result does not make the relevant part of the building worse in relation to the standards.

 Note: If you intend to carry out additional works that require warrant approval at the same time as works specified in schedule 3, the latter should also be included in the warrant application. The Building Warrant fee should be calculated taking all work into account.

 Certain buildings have moved from being completely exempt

Although schedule 3 (Appendix 3C) allows more minor works to be done without a warrant than previous legislation did, one effect of the 2004 regulations is to move certain building work from being completely exempt to being controlled by schedule 3. That is, this building work must now meet the regulation standards, but does not require a warrant. Transitional arrangements allowed small buildings and garages of $8–30\,m^2$ and replacement windows (i.e. work falling within types 3–5 and 20 of schedule 3) to be completed without meeting the standards **provided** construction was commenced on or before 1 June 2005, and was completed on or before 2 September 2005.

2.9 What happens if I do not comply with a Technical Handbook?

Not actually complying with a Technical Handbook (which is after all only meant as a guidance document) **doesn't** mean that you are liable to any civil or criminal prosecution. If, however, you have contravened a Building (Scotland) Regulation then not having complied with the recommendations contained in the Technical Handbooks may be held against you.

2.10 Verifiers

2.10.1 What do Verifiers do?

Verifiers are essentially the people appointed to oversee all building works in Scotland; they are responsible for issuing Building Warrants without which no building work should be undertaken, for inspecting building works to ensure that they meet Building Regulations, and for issuing completion certificates once they are completely satisfied that all work has been carried out in accordance with Regulations. Therefore, the importance of Verifiers should not be underestimated and you would be wise to ensure that you build a good relationship with your local authority Verifier!

 See Chapter 1 for the function of Verifiers.

2.10.2 Who sets out performance criteria for Verifiers?

The Scottish Ministers may give directions to Verifiers, of either a general or specific nature, as to how the Verifier should carry out their function. These directions may be to a particular Verifier, or to some, or to all.

At present, there is an agreement with Verifiers that performance is measured in five key business areas:

1. Public interest;
2. Private customer;
3. Internal business;
4. Continuous improvement;
5. Finance.

These key areas of business are collectively referred to as the 'balanced scorecard'. Verifier's, individual balanced scorecards contain a range of performance measures and areas for future service improvement.

2.10.3 Where can I find a list of Verifiers?

Currently, Verifiers are part of your local authority therefore contacting them will put you in touch with the right person for your area. However, the Act requires Scottish Ministers to keep lists of Verifiers available for public inspection at all reasonable times. Details of Verifiers and published balanced scorecards can be found on the agency website at: www.sbsa.gov.uk

2.11 Building Warrants

2.11.1 What is a Building Warrant?

A Building Warrant is the legal permission to start building work, or to convert or demolish a building. If you carry out work that requires a Building Warrant without first obtaining a warrant, you are committing an offence and could be liable to prosecution (or even a short holiday in one of HM Prisons!).

Verifiers are responsible for issuing Building Warrants and in assessing your application for a warrant, they must apply the standards set by the

Building Regulations **at the date of your application** (which in some cases could be an advantage to you!).

2.11.2 How do I obtain a Building Warrant?

If you are applying yourself, the application forms can be obtained from your Verifier, who at present is your local authority building standards department. The form (together with the submission of the appropriate plans, etc.) is the same whether you wish to alter, erect, extend, demolish or convert.

If you have appointed an architect, or other suitably qualified person to prepare your plans, the procedures should be known by them and you can ask them to act as your agent to apply on your behalf. This is recommended as the best course for people not experienced in building work.

Additional documents are available on the SBSA's website (www.sbsa.gov. uk) for use in support of a Building Warrant application, covering such matters as details showing methods of complying with Section 6 (Energy) of the Technical Handbook.

2.11.3 Do I need to inform my neighbours when I make application for warrant and do they have the right to object to the works shown in my application?

No. However, a warrant only shows compliance with the Building (Scotland) Regulations 2004. Where the proposed building work is likely to affect or involve a mutual part of a building or any part of another building, you may have other legal obligations and so it is advisable to inform any affected party. However, if you have already applied for planning permission you will have notified your neighbours of planned building works at this stage anyway.

2.11.4 What should be shown on the plans supporting my warrant application?

The plans should clearly show:

- the location and nature of your proposals and how they relate to any adjoining or existing building;
- the type of materials and products being used;
- the size of rooms;
- the position of appliances proposed;
- drainage details.

In addition, structural design certification and/or calculations plus an energy rating may be required as well as information on precautions being taken for the safety of the public during building or demolition works (e.g. keeping a building site secure).

The information required can be complex and it may be advisable to contact your Verifier who, on having your proposals explained to them, will be able to give you full advice on which requirements are applicable.

2.11.5 Do I have to get an architect to draw my plans?

No. The plans may be drawn by any competent person but they must be legible and clearly indicate the proposals. The means of complying with the Building Regulations should be clear and the materials and methods of construction specified. Full details of the drawings and additional information required to accompany a Building Warrant application is provided in the Procedural Handbook available from SBSA or via their website:www.sbsa. gov.uk/proceed_legislation/proc_handbook.htm

2.11.6 Is a charge made for the warrant service?

Yes. The level of fee is based on the estimated value of the work you propose as opposed to the actual cost of the work. The Verifier can advise you of the required fee.

 The fee is for the application and not just for the issue of warrant; it is not usually refundable.

F1:   The fee before discounts, for lodging a Building Warrant application, other than late Building Warrant applications

Value of works (£)	Warrant fee (£)	Value of works (£)	Warrant fee (£)
0–5,000	100	5,001–5,500	115
5,501–6,000	130	6,001–6,500	145
6,501–7,000	160	7,001–7,500	175
7,501–8,000	190	8,001–8,500	205
8,501–9,000	220	9,001–9,500	235
9,501–10,000	250	10,001–11,000	265
11,001–12,000	280	12,001–13,000	295
13,001–14,000	310	14,001–15,000	325
15,001–16,000	340	16,001–17,000	355
17,001–18,000	370	18,001–19,000	385
19,001–20,000	400	20,001–30,000	460
30,001–40,000	520	40,001–50,000	580
50,001–60,000	640	60,001–70,000	700
70,001–80,000	760	80,001–90,000	820
90,001–100,000	880	100,001–120,000	980
120,00 1–140,000	1,080	140,001–160,000	1,180
160,001–180,000	1,280	180,001–200,000	1,380
200,001–220,000	1,480	220,001–240,000	1,580
240,001–260,000	1,680	260,001–280,000	1,780
280,001–300,000	1,880	300,001–320,000	1,980
320,001–340,000	2,080	340,001–360,000	2,180
360,001–380,000	2,280	380,001–400,000	2,380
400,001–420,000	2,480	420,001–440,000	2,580
440,001–460,000	2,680	460,001–480,000	2,780
480,001–500,000	2,880	500,001–550,000	3,055
550,001–600,000	3,230	600,001–650,000	3,405
650,001–700,000	3,580	700,001–750,000	3,755
750,001–800,000	3,930	800,001–850,000	4,105
850,001–900,000	4,280	900,001–950,000	4,455
950,001–1,000,000	4,630	And for every 100,000 or part thereof	Add 250

F2	Application for Building Warrant for conversion only, that is, without any building work	£100

F3	Application for demolition only, that is, where there are no immediate plans for rebuilding	£100

F4   Application for amendment of Building Warrant:

(a)   where the new total estimated value is less than £50 the original, or is an increase of no more than £5,000

(b)   where the new total estimated value increases the amount for a building by more than £5,000 Warrant of the same value, that is, if the increase is £20,000, the fee will be £400.

F5	Application for an amendment to Building Warrant for demolition or conversion	£50

F6	Application to extend the period of validity of a Building Warrant	£50

F7   Where a late application for Building Warrant is made, or a Completion Certificate is submitted and there was no Building Warrant obtained when there should have been, the fee is increased by 25%
    The additional fee covers the increased difficulty the Council will have in establishing whether work that is already underway or completed complies with the plans, specifications and other information provided. The Council will require to inspect the work and disruptive surveys may be needed to establish what has been constructed. The resulting fees are detailed in F1 above and as follows:

7.1 Application for late Building Warrant, that is, where work is already started:

(a) application for a Building Warrant for the construction of a building or the provision of services, fittings and equipment in connection with a building (whether or not combined with an application for demolition)   125% of the fee in table of fees F1 above

7.2 Submission of a Completion Certificate where no Building Warrant was obtained for:

(a) the construction of a building or the provision of services, fittings or equipment (whether or not combined with an application for conversion or for Building Warrant for demolition)   Same as for a late application of the same value of works, that is 125% of the fee in table of fees F1 above

(b) application for demolition only, or for conversion only   £125

F8   A Building Warrant fee is discounted, when a Building Warrant application is registered, and where Certificates from Approved Certifiers of Design are presented.
    Where one or more such Certificates are presented with a Building Warrant application, by −10% for each Certificate that covers the whole of any section of the functional standards, and 1% for each Certificate covering a single item in any such section, up to a maximum of 5% for any one sectional subject to a maximum discount of 60% of the warrant fee.

Note that the discounts apply where a late application for Building Warrant is made. The discount is applied to the whole fee. Discounts also apply to the stages in a staged warrant where the discount is on the fee for the amendment of a Building Warrant (which in staged warrants should take account of the increased value of the work).

F9	A refund of the original Building Warrant fee (before any discounting) is made where one or more Certificates from an Approved Certifier of Construction are presented, as below:

Where one or more such Certificates are submitted with a Completion Certificate, by − 1% for each Certificate covering a defined trade or installation, up to a maximum of 20%, but subject to a minimum refund of £20.

Note that a refund is only made where a Council has been informed in writing of the intention to use the Approved Certifier of Construction in relation to any Certificate. This precludes refunds where a late submission of Completion Certificate is made. However, a Certificate from an Approved Certifier of Design may accompany a submission of a Completion Certificate without Building Warrant, and a discount of the fee is allowed for as described above.

F10	The Fee Regulations state that no fee shall be payable where the purpose of the work to which the application relates is to alter or extend a dwelling so that it is made suitable as a dwelling for a disabled person.

The fee is set at zero for works to alter or extend the existing dwelling occupied by a disabled person provided the works are solely for the benefit of the disabled person.

The relief therefore is not for disabled people in general, it relates specifically to works to provide facilities for disabled people as defined in the building standards. 'Disabled person' means a person with a physical, hearing or sight impairment, which affects that person's mobility or use of buildings.

The cost of works that do not require Building Warrant approval (e.g. decoration, floor coverings, etc.) do **not** need to be included in the estimated value of the works. However, temporary works and preliminaries relating to the permanent works required to comply with the Building Regulations should be included.

If the Verifier feels the estimate of value provided by the applicant is incorrect, they may check the amount by reference to established indices of building costs, for example, the RICS Building Cost Information Surveys of Tender Prices.

If the Verifier believes the value of the works should be higher than stated, the Verifier can refuse to consider a warrant application unless the value is increased and the appropriate fee paid.

There is no fee for works to alter or extend the dwelling of a disabled person provided the works are solely for the benefit of that person.

2.11.7 Are fee discounts available?

A warrant fee is discounted where one or more certificates from Approved Certifiers of design are presented with a warrant application, as below:

- 10% for each certificate that covers the whole of any section of the functional standards; and/or
- 1% for each certificate covering a single item in any such section, up to a maximum of 5% for any one section.

all subject to a maximum discount of 60% of the warrant fee.

When a local agreement is in place between the Verifier and the applicant for phased payment of the warrant fee, the discount is due on all the payments, provided a certificate was submitted with the warrant application.

A partial refund of the original warrant fee (before any discounting) is payable where one or more certificates from an Approved Certifier of construction are presented with a completion certificate, as below:

- 1% for each certificate covering an approved scheme; or
- 20% for a single certificate covering the construction of the entire building all subject to a maximum refund of 20% and a minimum refund of £20.

2.11.8 What happens after I have submitted my warrant application to the Verifier?

- Firstly, your application is recorded on the Building Standards Register. This allows the Verifier or any interested party to chart its course through the warrant process.
- Your application will then be assessed against the requirements of the Building (Scotland) Regulations 2004.
- The Verifier will then produce an assessment of where your application does not comply.

 If you are not able to make adjustments or answer all the matters raised, the Verifier may be able to help, or you may require to seek advice from building specialists, depending on the nature of your proposals. The procedures for the adjustment of plans may differ between Verifiers but, generally speaking, you will be able to discuss the assessment with your Verifier and arrange to have the plans amended:

- After your plans have been adjusted or amended to comply with the assessment, they will be re-assessed.
- If the plans comply with the regulations, the warrant will be granted.

 If you do not amend your plans, your application cannot be processed further and the Verifier will be required to consider formal refusal but, before a Verifier refuses an application, you will be given the opportunity to resolve any outstanding matters. If there is genuine doubt over an aspect of compliance, you may, with the agreement of the Verifier, seek a view from SBSA:

- If your Building Warrant is refused by the Verifier, you can appeal against this decision to the Sheriff Court.

2.11.9 How long does it take to get a Building Warrant?

This depends on the workload of the Verifier and the extent and quality of the drawings and details included with your application. Your Verifier should be able to give you an idea of the first response time; however, it should be remembered that, if the Verifier requires additional information, the time taken to grant a warrant will depend to some extent on the time taken to get the additional information to them.

It should also be remembered that spring and early summer tend to be busy periods; therefore, response times may not be as quick as at other times of the year.

2.11.10 Is there any type of building work that does not require a warrant?

Yes. The following building work does not require a warrant, **provided the work complies with the Building Regulations:**

Note: This is not an exhaustive list; a full list is given in Appendix 3C in Chapter 3.

(a) Any building work to or in a **house** that **does not** involve:
   - increasing the floor area;
   - the demolition or alteration of a roof, an external wall or an element of structure, for example, forming a doorway in a load-bearing wall;
   - underpinning;
   - any work adversely affecting a separating wall;
   - changing the method of wastewater discharge;
   - any work to a house having a storey, or creating a storey, at a height of more than 4.5 m (normally a three or more storey house).

For example, the alteration and refit of a kitchen or bathroom or forming an en-suite bathroom or shower room.

(b) A detached single-storey building having an area exceeding $8\,m^2$ but not exceeding $30\,m^2$, ancillary to and within the garden of a **house**, provided that the building:
   - is at least 1 m from the house, or is closer than 1 m to the house but is at least 1 m from any boundary;
   - does not contain a fixed solid fuel, oil or gas appliance;
   - does not contain a sanitary fitting.

For example, the construction of a detached shed, detached carport or detached garage.

(c)   A detached single-storey building having an area exceeding $8\,m^2$ but not exceeding $30\,m^2$, ancillary to and within the garden of a **flat or maisonette**, provided that the building:
   • is at least 1 m from the flat or maisonette or within 3 m of any other part of the building containing the flat or maisonette;
   • is at least 1 m from any boundary;
   • does not contain a fixed solid fuel, oil or gas appliance;
   • does not contain a sanitary facility.

For example, the construction or installation of a detached shed, detached carport or detached garage.

(d)   Any building work associated with a domestic scale fixed solid fuel, oil or gas appliance or other part of a heating installation that does not include work associated with a chimney, flue pipe or constructional hearth.
(e)   Any building work associated with a balanced flue serving a room-sealed appliance.
(f)   Any building work associated with the installation of a flue liner.
(g)   Any building work associated with refillable liquefied petroleum gas storage cylinders supplying, via a fixed pipework installation, combustion appliances use principally for providing space heating, water heating or cooking facilities.
(h)   Other minor work such as the provision of a single sanitary facility [other than a watercloset (WC)], installation of an extractor fan or, in a **dwelling**, the installation of a stairlift.
(i)   Additional insulation (other than insulation applied to the outer surface of an external wall), the construction of walls not exceeding 1.2 m in height, fences not exceeding 2 m in height, raised external decking at a height of no more than 1.2 m (other than where forming part of any access or escape route required by the regulations), and paved areas exceeding 200 m in area (other than where forming part of any access required by the regulations).
(j)   Replacement doors, windows and rooflights when the frame is also being replaced.

2.11.11 What about extensions?

Extensions to existing buildings must be constructed in such a way that the extension complies with all current standards that are relevant to the construction and use of the extension. In addition, the whole building must not, as a result of the extension, fail to comply with Building Regulations (if it complied originally) or fail to a greater degree if it failed to comply originally. If either of those conditions is not met the Verifier is required to refuse a warrant and certifiers of design must not certify such an extension. In addition:

• alterations to existing buildings must also be completed in compliance with the full current standards relevant to the alteration work;

- if an existing building is altered internally and access to the entrance is not affected, there is no requirement to provide access for disabled people;
- if a building is extended, there is no requirement to provide an accessible entrance for disabled people to the extension **but** if there is suitable access to the existing building it should continue into the extension;
- if a lift is added to a building, which might allow access for disabled people to an upper floor, there is no requirement to alter existing door or corridor widths on that floor, nor to add refuges on an escape stair;
- if an accessible WC is available in a dwelling, on the entrance level, there is no requirement for any new WC to meet the access standard **but** the accessible WC cannot be removed and replaced by one on say the first floor; if there is a WC on the entrance floor of a dwelling but it does not comply, it can be moved to another floor.

The Building Regulations define 'reasonably practicable' as 'having regard to all the circumstances, including the expense involved in carrying out the building work'. The best way to proceed is for an applicant to show how any failure of the building, as it is proposed to be after conversion, is being addressed. Except for historic buildings, it is only in exceptional cases that it is likely that taking no action is acceptable. The regulations do, however, require the building to be no worse than before, so the construction must either be left untouched or be altered in a way that is closer to compliance with the standards.

2.11.12 Do I require a Building Warrant for repair work?

A warrant is **not** required for repair or maintenance work where the fitting or equipment is being replaced, either totally or in part, by the same general type and the installation is to a standard no worse than at present. In other words, the replacement or repair work does not make the service, fitting or equipment worse than it was before. However, the existing facilities may be improved upon, for example, by installing double glazed units within existing window frames. Examples of such work may include the repair or maintenance of:

- a sanitary appliance or sink and branch soil or waste pipe;
- rainwater gutter or downpipe;
- solid fuel combustion appliance;
- electrical fixture, ventilation fan;
- chimney or flue outlet fitting or terminal;
- solid waste chute or container;
- kitchen fitments or other fitted furniture;
- ironmongery;
- flooring;
- wall and ceiling linings;
- cladding;
- covering or rendering either internally or externally.

The **repair** of a door, window or rooflight, including glazing would also be included in this grouping but not where the entire unit, including the frame, is being replaced.

2.11.13 What about conservatories?

As with many things this is potentially more complicated than you think and the SBSA has published a specialized guidance document specifically aimed at assisting in the design of conservatories to which the Building Regulations apply. This document ('Conservatory Guidance') may be purchased from the SBSA or is also available on the SBSA's website (www.sbsa.gov.uk). The advice contained within it may be applied to the entire structure or a single element or aspect.

2.11.14 What about warrants for demolition?

Warrants for demolition, while nominally valid for 3 years, are usually granted subject to a time limit from the commencement of the demolition to the completion of works. This period should be set by the Verifier in agreement with the applicant.

The information required on the existing building and on the method of demolition, is required by the Verifier so that he can judge what is the appropriate provision of protective works to ensure the safety of the general public.

2.11.15 Where can I obtain assistance in understanding the requirements?

A Verifier (at present the building standards department of your local council) will be able to provide assistance with:

- advice about how to incorporate the most efficient energy safety measures into your scheme;
- advice about the use of materials;
- advice on electrical safety;
- advice on fire safety measures (including safe evacuation of buildings in the event of an emergency);
- at what stages local councils need to inspect your work;
- deciding what type of application is most appropriate for your proposal;
- how to apply for Building Regulations approval;
- how to prepare your application (and what information is required);
- how to provide adequate access for disabled people;
- what your Building Regulation Completion Certificate means to you.

Note: A full list of local authority Building Standards Offices where assistance and further information may be obtained is available from SBSA (www.sbsa.gov.uk).

2.11.16 What sort of warrant can the Verifier issue?

A warrant may be subject to conditions, in the following ways:

- a staged warrant has conditions preventing work on later stages being carried out until an amendment of warrant has been issued;
- a relaxation related to an application may have conditions that must be met;
- a warrant may be subject to continuing requirements;
- a warrant may be granted subject to a limited-life condition;
- a warrant to demolish includes a condition setting a time period within which the demolition must be carried out.

 Note: To prevent applications that have run into difficulties lying unfinished, a Building Warrant application that has not been granted within 9 months of the date of the first report is automatically determined to be refused unless a longer period is agreed with the Verifier. A fresh application and fee are necessary if the application is to be reconsidered.

The first report should give clear warning of this procedure to the applicant and the Verifier may consider it prudent to issue a further letter closer to the 'deemed refusal' date, reminding the applicant of the impending action and advising what options are available to them. If the warrant has not been granted within the statutory time period or the period has not been extended, the Verifier may consider issuing formal notification of the 'deemed refusal'. The time period is extended by any time taken to obtain a view or a relaxation from Scottish Ministers or a response to a consultation request or if there is agreement between Verifier and applicant.

Where an application for amendment of Warrant affects work that has been certified, the certifier must re-assess and re-certify the work in question.

What happens after the warrant has been granted?

Immediately after the warrant has been granted you may start the work.

It should be noted, however, that a grant of warrant does not preclude the need to have any other legal requirements (such as planning permission) in place before starting work. There will be also other considerations (such as the position of existing services like gas pipes or power cables) which are outside the remit of the building standards system.

 It is the responsibility of those wishing to build to check with utility companies before work starts. You are also required to let the Verifier know when you are starting work and to inform the Verifier at certain stages of construction, so as to allow the Verifier the option of inspecting or attending tests, such as at drainage installation. You must inform the Verifier when all the work is completed.

2.11.17 How long is a warrant valid for?

A warrant is valid for 3 years from the date of issue and it is expected that **all** works covered by the warrant would be completed within this time.

However, the time period may be extended by applying to the Verifier before the expiry of the warrant. If the building works have not commenced before the warrant expires the Verifier would not normally extend the warrant.

2.11.18 What happens if I apply for the warrant 'late'?

Where work for which a Building Warrant is required has started without a warrant, an offence has been committed; however, in order to deal with these cases where work has started without a warrant (perhaps through ignorance of the law) the Act, allows a late application for warrant to be submitted at any time before the works on site are complete. If the works are complete a late completion certificate may be submitted.

2.11.19 What are staged warrants?

In some projects, particularly for commercial buildings, a building cannot be fully designed until the eventual occupant is identified. Specialist sub-contractors, who are often required to complete the detailed design of parts of the building, may also not be identifiable at the outset.

The Act allows for staged warrants to be granted in such cases **provided** that the applicant has agreed with the Verifier which later stages of work cannot commence until details of those stages are provided. The warrant for the whole project is then granted with a condition that work on the identified stages does not start until the necessary information has been submitted and an amendment of warrant for the next stage(s) granted. Thus, work on piling or foundations can start before the rest of the design is finalized.

 It is the responsibility of the applicant or his agent to apply for the amendment of warrant in good time to allow checking and approval so that site work can continue smoothly.

An application for a staged warrant as described above may include a certificate from an Approved Certifier of design if that is applicable to the stage in question. If the certificate covers the whole of a particular aspect of design (e.g. the structure) then subsequent application(s) for amendment will have to include updated certificate(s) which relate to the stage in question and must also take responsibility for the work up to that point.

 The fee for a staged warrant is payable in full at the time of the initial application, based on the estimated total value of the project.

Other than a design covering the same section of the Technical Handbooks, a discount is available for any certificate of design presented except where a certificate covers a new aspect of design (in which case it could attract a discount on the fee payable for the amendment). Any other arrangement for collecting the fee payable, for example, a phased payment rather than all with the original application is at the discretion of the Verifier.

The principle of staged warrants can be applied to work on existing buildings where, for instance, a specialist piece of work can only be fully designed after work has progressed to a stage where full dimensions are available. For example, a metal stair may only be finally detailed when alterations to floors are completed.

A completion certificate will not normally be accepted until all stages are complete; however, there are occasions when a completion certificate can be accepted before all works covered by a staged Building Warrant are complete. For example, individual certificates would be accepted in relation to a phased housing development where additional buildings are to be the subject of further stage(s).

It is not possible in a staged warrant application to have separate completion certificates for shell and for fit out. If an applicant wishes (perhaps for contractual arrangements) to split a shell from fit out work, separate warrant applications can be made for each and individual completion certificates accepted for each. However, as a completion certificate has to certify the building complies with the Building Regulations as well as the warrant drawings, a warrant granted for a shell should include a continuing requirement that effectively prevents occupation of the completed shell until the fit out is completed. Such a continuing requirement should quote the provisions of the regulations which the fit out has still to satisfy.

Although a Building Warrant is granted to a particular applicant, if a building is sold while under construction the effect of the warrant transfers to the new owner or indeed all the persons having an interest in the building. In other words, construction may continue and on completion the new owner must submit the completion certificate.

2.11.20 What are limited-life warrants?

Where a building is intended to be used for a limited period of time it may not be necessary to apply all the standards for a permanent building. The Act allows Building Regulations to specify which of the standards need not apply and lets the procedure regulations set a time limit. This limit is currently **5 years** for a 'limited-life warrant' and is measured from the date of acceptance of the completion certificate for the building or, if applicable, the date of any permission for temporary occupation or use.

It is important to stress that this type of warrant is **only** intended for buildings that will be removed by the end of the time limit and removal is a condition of the warrant.

It is possible for the owner to apply for an extension of the limit but this is at the discretion of the Verifier. It is an offence to fail to demolish (or otherwise remove) by the time limit set in the warrant unless an extension of time has been granted (not just applied for). Continued occupation of such a building after the time limit can be prevented by the Courts.

Note: An application to demolish is required at least 3 months before the limit is reached, unless otherwise agreed with the Verifier.

2.11.21 What happens if I decided I no longer want to do the work?

If you decide, once you have applied for a warrant, that you no longer want to do the work and therefore want to withdraw an application you can do so; however, there is **no** refund of fee, and the entry remains on the building standards register.

2.11.22 What if my warrant is refused?

Where a warrant application is refused, the Verifier ensures this information is sent to the local authority to be recorded in the building standards register. The principal plans and specifications, together with any amendments, including any certificates submitted in support of the application, are retained in the normal way but the remaining drawings and information are returned to the applicant in the format in which they were received, along with a formal refusal. An applicant or their duly authorized agent has the right to appeal to the Sheriff Court in respect of a Verifier's decision to refuse to grant a Building Warrant or if the warrant is deemed refused.

2.11.23 Is my building work subject to inspection while in progress?

Your Verifier may inspect while work is in progress to check that the warrant is being complied with. To help to facilitate this the Verifier must be notified, in writing, of the commencement of works within 7 days of the start date. The need for and frequency of inspections is assessed by the Verifier, taking into account matters such as the type of work, quality of information submitted, etc.

It must be stressed, however, that these inspections are to protect the public interest in terms of compliance with Building Regulations, not to ensure that all the work is constructed as the person paying for the work would want it. The Verifier is not responsible for checking the quality of work done or supervising the builders employed. Supervision of the building work should be the responsibility of whichever person is appointed by you for that purpose.

Should any Verifier be actively prevented from undertaking inspections of work, the 'relevant person' submitting a completion certificate for the work can expect to have it refused!

 Note: The Verifier is **not** responsible for inspecting work that is covered by an Approved Certifier of Construction.

2.11.24 What happens if the works on site differ from the approved plans?

You are committing an offence if you carry out work that differs from the approved plans issued as part of the Building Warrant. You should discuss in advance with the Verifier any changes to your warrant proposals before carrying these out.

A formal amendment to your warrant can be sought at any time during the period of the validity of the warrant and this amendment application will follow the same procedures as the initial application but the plans **must** show the changes you wish to make. Once approved, you can proceed on site with the change to your proposals. A fee is payable for an application for amendment of warrant and the Verifier will advise you of the amount.

Note: If any part of the original application is certified by an Approved Certifier of Design, you must ask the certifier to check the changes and, if necessary, re-certify that all works comply with the regulations.

2.11.25 What happens if I contravene the regulations?

You are committing **an offence if you start work, without a warrant, on work that requires a Building Warrant**. It is also an offence to contravene the requirements of Building (Scotland) Regulations 2004.

The guilt in these cases is quite far reaching. Those who will be seen to have committed of an offence if a warrant is not obtained before work is done are:

- any person carrying out the work (such as a self-build owner or tenant, a developer who is a builder, or a builder);
- any person for whom the work is to be done (such as an owner, tenant or developer who is not doing the work but has engaged a builder to do it);
- the owner (if the owner is different from the persons above).

For any one project, therefore, a variety of people may have committed an offence. These can include the owners, tenants or developers (who order the work to be done), and any person doing the work (including the builder), and in these cases either the offence is reported to the Procurator Fiscal or, more normally, a Building Warrant enforcement notice is issued.

Although a builder can be reported for committing an offence, if the builder is carrying out the work on behalf of a client then the builder cannot be served with such a notice.

You are committing an offence if you occupy or use a new building, or extension to an existing building, without first having submitted a completion certificate and it has been accepted by the Verifier.

The exception to this rule is if you receive permission from the Verifier for the temporary occupation or use of the building for a specified time. The local authority can take enforcement action in each instance and contravention of the regulations can incur a maximum fine of £5,000.

2.11.26 What happens when my building work is finished?

It is your responsibility to submit a completion certificate on the appropriate form. A completion certificate is required to confirm that a building has been constructed, altered or converted in accordance with the Warrant and the Building (Scotland) Regulations 2004.

 It is an offence to submit a false completion certificate or to occupy a building without a completion certificate being accepted by the Verifier.

The Verifier must make reasonable enquiry to establish that the work complies with the warrant. If he is satisfied that the work complies, a Verifier must accept the completion certificate.

 The Verifier must accept or reject (with reasons) the submission within 14 days.

2.12 Approved Certifiers

2.12.1 What is an Approved Certifier?

Approved Certifiers have government approval to certify that part of a design or construction complies with the Building (Scotland) Regulations 2004.

2.12.2 Why use an Approved Certifier?

The application should take less time to process because the Verifier does not need to check any work that is certified, although they do need to confirm the Approved Certifier's registration.

An Approved Certifier of design can provide a certificate saying that particular aspects of the work will comply with the Building Regulations and so less detail needs to be provided for that aspect as a Verifier is required to accept this certificate.

 You will also get a discount on the warrant fee if you submit a certificate of design from an Approved Certifier of Design with your application (but don't forget that the certifier will charge you for the certificate so it may even out in the end!). You will also get a small refund if you use an Approved Certifier of Construction to certify work for the completion certificate, **provided** you notify the Verifier before the start of work on site.

2.12.3 What will the Approved Certifier provide me with?

The information provided must allow the Verifier to clearly identify the scope of the certified work and allow consultation with other authorities when necessary. The information on the certificate of design should match that provided on the Building Warrant application form in respect of the:

- location of the project;
- description of works;
- description of the stage of work (if applicable).

The applicant must also provide enough information on the certified work to assist any site inspections the Verifier wishes to make. Using structural certification as an example, the Approved Certifier of design may have to provide the applicant with details of beam sizes, bearings, etc., for the Verifier, even where the design has been certified by an Approved Certifier of design for structure.

On some projects a specialist contractor may be providing a structural element. This could delay the ability to certify the whole building and obtain a warrant until the contractor designed details have been completed. To accommodate this, the structural certification scheme permits the certification of specific details to be undertaken some time after the certification of the general structural arrangement has been completed and a warrant (or staged warrant) has been granted.

The design certificate must be accompanied by a schedule (known as schedule 1) listing any structural items or details which the certificate does not cover in detail and, once the details have been finalized, the Approved Certifier issues the finalization notice (known as Model Form Q). If an Approved Certifier is used they must retain, in their own records, not just the information submitted but all the details relevant to the project. It should be noted that certificates of design or construction should only be issued in respect of works for which Building Warrant approval is required. That is to say, they should not be issued on works that are exempt from Building Regulations or works not requiring a warrant.

2.12.4 Where do I go to find an Approved Certifier?

The Certification Register (which is kept by the SBSA and can be accessed via the SBSA website) contains details of all approved certification schemes and the registration status of scheme members.

2.12.5 Who appoints certifiers?

The Scottish Ministers may appoint individuals or bodies, either public or private, as Approved Certifiers of design or construction. For full information on certification see the certification handbook, available on the agency website (www.sbsa.gov.uk).

Approved Certifiers are directly appointed on specified terms for a period set at appointment. This period may be varied or the appointment terminated on the grounds of a breach of any terms of the appointment, a failure to retrain when building standards legislation changes, or following an unsatisfactory audit.

Certification can only be undertaken by approved individuals and bodies that have demonstrated they meet the appropriate criteria. A body that is approved cannot undertake certification unless it employs at least one Approved Certifier holding any designation(s) appropriate to the scope of the scheme.

2.12.6 Is the work of certifiers audited?

The Scottish Ministers require that approval of individuals, bodies and scheme providers is renewed at intervals to be set on the basis of a risk assessment.

Renewals will be subject to auditing the continuing eligibility and performance of the individual, body or scheme provider.

Approval of certifiers and approval of certification schemes is subject to termination in the case of negligent certification, negligent approvals, auditing or other poor performance.

2.13 Energy Performance Certificates

2.13.1 What are Energy Performance Certificates (EPCs)?

An EPC is a document which records the amount of carbon dioxide a building produces and shows how energy efficient a building is. EPCs also include simple cost-effective home improvement measures that will help to save energy, reduce bills and cut carbon dioxide emissions. A sample EPC is shown below.

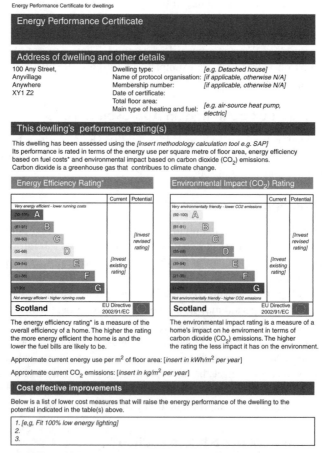

Energy Performance Certificate for dwellings

2.13.2 When do I need an EPC?

If you are building a new building you will need an EPC to be submitted along with your Completion Certificate. If you are using an architect, surveyor, etc., they will most likely be able to provide this for you.

Note: You will **not** need an EPC for a conversion of a property in order to comply with the Building Warrant, but this will be needed if you sell or rent out the property at a later date.

2.14 Planning officers

Before construction begins, planning officers determine whether the plans for the building or other structure comply with the Building Regulations and if they are suited to the engineering and environmental demands of the building site. Verifiers are then responsible for inspecting the structural quality and general safety of buildings.

2.15 Do I need to employ a professional builder?

Unless you have a reasonable working knowledge of building construction it would be advisable before you start the work to get some professional advice (e.g. from an architect, a structural engineer or a building surveyor) and/or choose a recognized builder to carry out the work. It is also advisable to consult the local authority building control officer or an approved inspector in advance.

2.16 Are there any occasions when the rules are relaxed?

There will always be cases when working with existing buildings that adopting solutions based on whatever is in the Technical Handbooks or other guidance documents will not be entirely suitable. Rather than deal with these cases by relaxations, as previously, it is now possible for the Verifier to use judgement as to whether a standard is adequately met.

In difficult cases, it will be possible to obtain a formal view from the SBSA to assist the decision or where it is the functional standard itself that is not applicable a relaxation may be requested from the SBSA. The approach expected from Verifiers is that where relaxation might have been accepted in the past it can be approved, if necessary with whatever compensating features that might previously have been requested under 'conditions of relaxation'. A similar approach applies to certifiers of design, although the option of a formal view is not available to them.

For historic buildings (i.e. *a building of special interest or significance*) the judgement of what is reasonably practicable must be made in a wider context. Section 35 of the Act gives specific protection to certain historic buildings in relation to statutory notices. However, a wider range of buildings may need sympathetic consideration. That interest or significance may be local or national and may be a consequence of, for example, the building's age, structure or location. It may result from its connection with a person or persons or with local or national events or industry; or from a combination of these or other factors. A historic building does not have to be listed by Scottish Ministers or lie within a conservation area to be deemed to have special interest or significance.

2.17 Completion certificate

2.17.1 Why do I need a completion certificate?

A completion certificate certifies that the local authority is satisfied that the work complies with the relevant requirements of the Building Regulations.

A completion certificate is a valuable document that should be kept in a safe place! Full rates for the property would also apply from that time onward. As Verifiers cannot regularly or sufficiently monitor all construction work to check that the regulation requirements are being met, the relevant person (i.e. the owner) should ensure that adequate procedures are in place to allow the completion certificate to be properly submitted.

2.17.2 How to submit a completion certificate

On completion of either work (which includes demolition) or conversion for which a warrant is required, the 'relevant person' must submit a completion certificate. The 'relevant person' (as defined in the Act) can be summarized as the:

- owner, tenant or developer doing the building work or conversion themselves; or
- owner, tenant or developer who has employed a builder to do work for them; or
- owner, where the tenant, developer or builder has not submitted the certificate when they should have done so.

 Note: The owner must always be named on the completion certificate, as the procedure regulations require an owner to be notified when a completion certificate is accepted or rejected.

Checks by Verifiers and Approved Certifiers help to protect the public interest but do not remove the relevant person's responsibility for ensuring compliance with the Building Regulations! A duly authorized agent may sign this completion certificate on behalf of the relevant person but this does not move the responsibility for compliance, which remains with the relevant person. The submission and accompanying information can be on paper, in an electronic format or in a combination as agreed with the Verifier and, where required must include a copy of the EPC(s) for the building(s).

The completion certificate form includes an annex that should be completed if any Approved Certifiers of construction have been used, and the original, signed certificate(s) should accompany the completion certificate.

Unlike the previous Building (Scotland) Act, the current Act does not require a certificate to be submitted with the completion certificate by the electrical installer but Verifiers will still, however, seek proof that the installation is properly done and unless the installation is covered by an Approved Certifier of construction, detailed evidence may be required.

Where there are several buildings covered by a warrant (excluding new dwellings or existing dwellings in different ownership), the regulations permit the submission of a completion certificate related to either each building or all buildings covered by the warrant.

On an estate of houses, a completion certificate must be submitted and accepted for **each** dwelling, provided the common services required by the Building Regulations for that dwelling have been completed. In other words, a completion certificate should not be accepted for a dwelling until it is connected to a suitable drainage system or until access to a suitable road is complete.

Matters not covered by a warrant, such as street lighting are, however, outside the scope of the certificate, and the final completion certificate for the last dwelling cannot be accepted until all the common items are complete.

In blocks of flats, the completion certificate for an individual flat cannot be accepted until the common areas affecting that flat are complete, including, for example, the access and facilities for fire-fighting. The completion certificate for the final flat cannot be accepted until works to all the common areas are complete. If an Approved Certifier of construction is used a certificate of construction should accompany each completion certificate submitted.

2.17.3 What if I make a late submission without a warrant?

Where work has been carried out, or a conversion made without a warrant when there should have been a warrant applied for, a completion certificate **must** still be submitted. However, in this case the certificate may only be accepted if it confirms that the work or conversion has been carried out in accordance with and now complies with, the Building Regulations applicable at **the time of the submission of the completion certificate**.

Note: There is no limit as to how long after the completion of the work a late completion certificate may be submitted but it is always the regulations at the time of submission that must be met.

2.17.4 Acceptance or rejection of completion certificate

A Verifier must accept a completion certificate if, after reasonable enquiry, they are satisfied the work/conversion certified complies with the relevant warrant and Building Regulations. Some product certificates, such as those accepted by local authorities in the previous system in relation to intumescent products, may continue to be accepted as evidence of compliance. Such documents should not be confused with certificates from Approved Certifiers where there is considerable protection given to both clients and Verifiers through the approval scheme.

A Verifier must provide a response as to whether a submitted completion certificate is acceptable within **14 days** or such longer period as has been agreed with the person making the submission.

2.17.5 Will the Verifier come and inspect on receipt of a completion certificate?

Following the submission of a completion certificate, the Verifier usually carries out a non-disruptive inspection of the work. However, as a Verifier need only make 'reasonable enquiry' before acceptance of the certificate there can be circumstances where a site inspection may not be needed. For example, if the Verifier has chosen to make previous inspections, these may have provided sufficient assurance to inform the Verifier's decision.

If they consider it necessary, a Verifier may require a 'materials test' to be completed or test results to be provided if tests have already been carried out. A 'materials test' can include a test of materials in combination, for example, a separating wall or even a whole building, as well as individual materials.

 Note: In the case of work covered by a certificate of construction, a Verifier may not request a materials test before accepting a completion certificate.

2.18 When do the Scottish Ministers get involved?

To assist Verifiers and applicants for warrant in cases where there is doubt on whether proposals satisfy the regulations or whether continuing requirements need to be imposed as conditions of the warrant, the Scottish Ministers may give a view on the matter. Verifiers must have regard to any view given when determining the application. It should be noted that a formal view will not be given on matters certified by an Approved Certifier of design.

2.18.1 Can you apply to the Scottish Ministers to relax the regulations?

For any particular building, a person may apply to the Scottish Ministers for a direction to either relax or dispense with a provision of the Building

Regulations. The Building Regulations designate certain provisions that may not be relaxed, although there is currently no designation in relation to the building standards themselves.

2.18.2 Type relaxations

The Scottish Ministers may also give a direction relaxing or dispensing with a provision of the Building Regulations for any description of building, that is, a type relaxation. This can be either given on application by any person or by the Scottish Ministers on their own behalf. Before giving a direction the Building Standards Advisory Committee must be consulted, along with any other interested party. The consultation will normally include the fire authorities the Scottish Association of Building Standards Managers (SABSM) will also normally be consulted; and Verifiers and local authorities are also consulted.

The direction may set conditions and may set a date for expiry. Any direction may also be revoked or varied by a further direction.

2.18.3 Type approvals

The 2003 Act does not make provision for type approvals. However, a Scottish Type Approval Scheme is provided by the SABSM [previously the Scottish Association of Chief Building Control Officers (SACBCO)] and anyone wishing to have a design approved for use on several sites, subject to checks on site specific matters such as foundations, can apply to SABSM. See their website at: www.sabsm.co.uk

This scheme allows more rapid and certain processing of applications by members of the scheme. At the time of publication, all local authorities are members and while this arrangement exists, all local authorities approved as Verifiers will recognize type approvals.

2.19 What matters can you appeal against?

The Act permits appeals to the Sheriff Court on the following matters:

(1) where the Scottish Ministers refuse an application to relax or dispense with a provision of the Building Regulations;
(2) where a Verifier refuses to grant or amend the terms of a warrant, including deemed refusals resulting when the Verifier has not made a decision within the specified time limits;
(3) where a Verifier refuses to extend the life of a limited-life building, including deemed refusals resulting when the Verifier has not made a decision within the specified time limits;
(4) where a Verifier rejects a completion certificate, including deemed rejections resulting when the Verifier has not made a decision within the specified time limits;

(5)   where a Verifier imposes continuing requirements;

(6)   where a Verifier refuses to discharge or vary a continuing requirement;

(7)   where a local authority serves a notice (regarding Building Regulations compliance, continuing requirement enforcement, Building Warrant enforcement, defective or dangerous buildings).

2.19.1 Are there any time limits for appeal?

An appeal must be made within **21 days** of a decision being issued or of a notice being served. Where a Verifier has not made a decision in relation to items 2, 3, 4 or 6 above, a decision is deemed to have been made by the period specified in the procedure regulations.

2.20 What are Building Regulations compliance notices?

Where the Scottish Ministers consider it essential that certain types of existing building should be required to comply with the current standards, they may direct local authorities to take action to secure that these buildings are made to comply. The local authorities must serve on the owners of identified buildings of the prescribed type a 'Building Regulations compliance notice'.

The local authorities may also serve a notice where it appears any other building of the prescribed type does not comply and this provision of the Act will apply if the initial identification was of specific buildings. This power is limited to the purposes specified in the Act for the Building Regulations and can be applied only to buildings of a type to which the Building Regulations apply.

In the following cases, where local authorities consider it necessary, a 'Building Warrant enforcement notice' may be served on the relevant person. The three cases are that:

(1)   work requiring a warrant has been or is being done without a warrant;

(2)   work has been or is being done that is not in accordance with a warrant that has been granted; or

(3)   a limited-life building has not been demolished by the expiry of the period for which a warrant has been granted.

Note: Work listed in schedule 3 of the Building Regulations as work not requiring a warrant may have a notice served if the work does not comply with any relevant requirement of the Building Regulations. This is because such work is only exempt if it does comply.

A Building Warrant enforcement notice may require:

- for ongoing work without a warrant, that a Building Warrant be obtained;
- for completed work without a warrant, that a completion certificate be submitted and the Verifier's acceptance obtained;
- for work not carried out in accordance with the warrant, that the work be altered to comply or that an amendment of warrant covering the changes be obtained;
- for a limited-life building that has not been demolished, that a warrant for the demolition be obtained and the building demolished.

Note: In the case where a completion certificate is required to be submitted, it will be a late submission treated like a warrant application and, unlike a normal Building Warrant application, there will be time to negotiate changes on site to comply if necessary before acceptance is given.

2.21 Can a Verifier impose 'continuing requirements'?

Under section 22 of the Act, when granting a warrant or accepting a completion certificate, a Verifier may impose a continuing requirement. Such a requirement imposes on a building owner a duty that the Verifier feels must be fulfilled after the completion of the building to ensure that the purposes of the Building Regulations are not frustrated.

This is not intended to cover matters that always rely on adequate maintenance, such as the operation of lifts or the recoating of woodwork, nor is it for matters under health and safety or fire precautions legislation, such as testing of alarms or (in general) matters that involve action on the part of the owner. It is intended for special cases where the arrangements agreed for complying with the functional standards might be frustrated by uncontrolled changes.

A typical example would be acceptance of a movable platform for cleaning windows subject to a continuing requirement that adequate access and hard surfaces are provided and then kept clear and properly surfaced thereafter.

2.22 What regulations apply to Crown buildings?

The Act binds the Crown but the secondary legislation can differentiate as to which procedures apply. It is intended to provide a separate handbook for Crown buildings and at that time either the procedure regulations will be amended or separate procedure regulations will be made. Determining the 'ownership' of a Crown building is done by reference to sub-section 53(5) of the Act and the final decision as to who is the owner lies with the Scottish Ministers.

2.23 What about historic buildings?

For the purposes of enforcement under the Act, the following are designated as historic buildings:

- scheduled ancient monuments;
- listed buildings;
- buildings subject to preservation orders;
- those buildings in conservation areas subject to control of demolition.

Ancient monuments are those included in the schedule of monuments compiled and maintained under section 1 of the Ancient Monuments and Archaeological Areas Act 1979.

Listed buildings are those included in a list of buildings of special architectural or historic interest compiled or approved under section 1 of the Planning (Listed Buildings and Conservation Areas) (Scotland) Act 1997, preservation orders are under section 3 of that Act and control of demolition is under section 66 of that Act.

Before serving notices on historic buildings there must be consultation with the Scottish Ministers (through Historic Scotland), the planning authority (where this is not the local authority) and anyone else the local authority considers necessary. Where the planning authority is within the same authority there should still be a properly recorded consultation.

 Note: The only qualification of this requirement is that, in relation to a dangerous building notice or urgent work needed to remove a danger, consultation is required 'if reasonably practicable.' This is intended to cover problems requiring immediate action; in these instances the relevant authorities should be notified subsequently of any such action.

There is also an important limitation on the notices. Except in an urgent case as described above, the effect of the notices on a person required to do work in relation to any of the enforcement notices is restricted to work that is not inconsistent with any provisions of the legislation designating the historic buildings.

2.24 What about dangerous buildings?

Building owners are responsible for preventing their buildings falling into a dangerous condition. The powers given to local authorities by the Act do not diminish this responsibility but are merely a 'safety net' that must be used to protect the public when it appears to a local authority that, for whatever reason, a building owner has failed in the duty to fulfil this responsibility.

The powers available to the local authority can be applied to any structure that meets the definition of building within the Act. Thus, for example, these powers can be used on a building that has not been subject to the Building Regulations or the Building Warrant process [such buildings are detailed in

schedules 1 and 3 of the Building (Scotland) Regulations 2004] if that building has fallen into a dangerous state.

 There is a duty to act

Should a local authority become aware of a building that constitutes a danger to persons in or about the building, to the public generally or to adjacent buildings or places, then it has a duty to act. Each dangerous building
case should be dealt with on its own merits, as the local authority considers
necessary.

There are a considerable number of factors which can influence the
approach to be taken:

- the nature and/or severity of the danger (structural decay or damage, fire
  damage, impact damage, missing building safety features, loose parts of
  building elements or building fixtures, sudden subsidence, etc.);
- the physical proportions and nature of the building (low-rise, medium-rise,
  high-rise, spire or tower, viaduct, etc.);
- the geographical location (city centre, rural, remote, etc.);
- the location of the danger (internal only, internal but could affect whole
  building, external affecting curtilage of building, external affecting a public area, ease of accessibility to do the emergency work, etc.);
- the extent of the danger (several affected buildings, only one building or
  part of building affected, several parts of one building element, or just one
  part, etc.);
- the person at risk from the danger (general public passing by, building
  occupiers, unauthorized persons frequenting the building, etc.);
- the time of year and/or day (school holidays, public holidays, night time,
  rush hour, etc.);
- the building owner(s) (finding the owner(s), the number of owners, the
  owner's willingness to recognize danger and resolve matters, the finan-
  cial/construction resources available to the owner, etc.);
- the building occupier(s) (the occupier's willingness to recognize danger,
  need to temporarily re-house occupiers, disruption to occupier's business,
  etc.);
- the attendance of other statutory bodies at or permanent surveillance of,
  the dangerous building (police, fire service, security guards, etc.);
- the availability of emergency contractors (demand for other emergency the
  repairs, public holidays, etc.);
- the local weather forecast (high winds, flooding, period of calm, etc.).

In some instances, the degree of risk cannot be established except by insti-
gating further exploratory work on the building, either using the resources of
the local authority or subcontracting the work to a specialist. Enabling equipment or work will sometimes be necessary and again reasonable expenses
incurred for this may be recovered from the building owner(s).

2.24.1 What are dangerous building notices?

Dangerous building notices will be issued to the owner of the building. In cases where there are multiple owners of a building, each owner or set of joint owners must receive a notice.

The notice contains the following information:

- the name and address of the owner;
- the address of the dangerous building;
- a list of any co-owners and their addresses;
- the commencement date for remedial work (appropriate timescale set by authority);
- the completion date for remedial work (appropriate timescale set by authority);
- the dangerous aspects of the building;
- the work necessary to comply with the notice, including any protective works and specialist supervision required and offering options where possible, for example, repair or demolish;
- notes on right of appeal;
- a warning on the consequences of failing to carry out the stipulated works.

Optional information might be:

- guidance notes for owners receiving a notice, including procedures for obtaining a Building Warrant if other work is to be done at the same time as the remedial work;
- guidance on the follow-up procedures that will be adopted by the local authority;
- location plan of the building.

An owner may appeal to the sheriff within **21 days** of the date of the notice but there is no further right of appeal against the decision of the sheriff.

2.24.2 What happens if I fail to comply with a dangerous building notice?

Where an owner has not fulfilled the requirements of a notice, the owner is guilty of an offence and in these circumstances the local authority can carry out or complete the necessary work and recover from the owner any expenses reasonably incurred. Most cases should, however, be dealt with to avoid this outcome.

There may be a variety of reasons why failure to comply is likely, particularly where a number of owners are involved. Although a prior notification before serving a notice is no longer required, in many cases it could be useful to indicate the intention to issue a notice.

A local authority may recover from the owner of a dangerous building any expenses reasonably incurred in carrying out work in relation to the building in the normal debt recovery way.

Where a demand for payment is made by a local authority in relation to work on a dangerous building, section 44 of the Act may limit the amount payable. It depends on both the interest that the person has in the building and the resources available but is limited to trustees, liquidators, etc.

Where a dangerous building is demolished by a local authority, section 46 of the Act permits the authority to sell the materials arising from the demolition. The local authority may offset the proceeds of the sale against the sum to be recovered. Any surplus must be accounted for to the owner or any other person having an interest in the building.

2.25 What other authorities will be consulted?

2.25.1 Sharing of information

At intervals of no more than 7 days, each Verifier submits (usually electronically) a list of warrant applications to the fire authority, planning authority, environmental health authority, highways authority, Scottish Environment Protection Agency (SEPA), Scottish Water and the council tax valuation assessor. The list contains the description of the work and is a way of sharing information and advising such bodies that building work in which they may have an interest may take place in the future. The bodies may then inspect any warrant applications they wish to see, at a time convenient to the Verifier. Any observations or objections that these bodies might have are then given in writing to the applicant, copied to the Verifier.

Where amendments to warrant are applied for, Verifiers should consult in cases where they consulted on the initial warrant or where the solution does not now follow the Technical Handbooks.

If you have chosen to elect to propose 'novel solutions' to achieving compliance with the building standards you will need to bear in mind that the Verifier may need more time to assess compliance, particularly where consultations have to take place.

2.25.2 Fire authorities

In some circumstances, consultation with the fire authorities is a requirement of the regulations; however, the consultation should not be confused with the fire authority's acceptance or rejection of the fire risk assessment required by the Fire (Scotland) Act 2005.

At application for warrant, Verifiers will consult on:

- non-domestic residential buildings;
- non-domestic, non-residential buildings (where the design does not follow section 2: Fire of the non-domestic Technical Handbook approved by the Scottish Ministers);
- domestic buildings with a storey at a height over 18 m;

- domestic buildings with a storey height over 7.5 m (but not over 18 m where the design does not follow section 2: Fire of the domestic Technical Handbook approved by the Scottish Ministers).

2.25.3 Highways department

If your warrant includes an access to a public road, it would be wise to consult the relevant department of the local authority. You will need to gain permission for the following: road openings, pavements, crossings, and any temporary occupation of roads during construction.

 Consultation requests to the highways department are less common than some other consultations in this chapter. Highways are usually asked for comments when planning permission is sought.

2.25.4 Scottish Environment Protection Agency

SEPA may require an authorization under the terms of the Water Environment (Controlled Activities) (Scotland) Regulations 2005 for the following:

- discharges of sewage effluent whether to ground via an infiltration system or to a watercourse;
- discharges of surface water run-off;

There are various levels of risk-based authorizations required and reference to the SEPA document 'A Practical Guide to The Water Environment (Controlled Activities) (Scotland) Regulations' 2005 provides advice.

Oil storage tanks with a capacity of 2,500 litres or more and/or serving a building, other than a dwelling, are required to comply with the requirements of the Water Environment (Oil Storage) (Scotland) Regulations 2006. Although authorization is not required from SEPA, they may take enforcement action in the event of non-compliance. Unlike the Building Regulations, the Water Environment (Oil Storage) (Scotland) Regulations 2006 compliance is not limited to the storage of heating fuel oil (consult the SEPA site for further information on the types of oil affected by these regulations).

 Domestic oil storage tanks with a capacity of less than 2,500 litres and serving dwelling houses are exempt from the requirements of the Water Environment (Oil Storage) (Scotland) Regulations 2006; however, they must comply with the relevant Technical Standards.

2.25.5 Scottish Water

Prior to applying for Building Warrant, you are advised to consult with Scottish Water:

- where a private drain discharges into a public sewer;
- where it is intended that a drain will be vested in Scottish Water;

- on the design and construction of disconnecting manholes and disconnecting chambers;
- where it is intended to build over sewers.

2.25.6 Police

Police liaison officers are available in some areas to give advice on the security aspects of a development. A Verifier will specifically ask for consultation where the security requirements of the 'Secure by Design' initiative may adversely impact on compliance with the Building Regulations, for example, locking of external doors.

2.25.7 Historic Scotland

Where a warrant application is for work on a building that is designated a 'historic building', the applicant 'is advised of the need to obtain any necessary permissions and if necessary consult 'Historic Scotland'.

Note: The Verifier may wish to check on the historic status of the building.

2.25.8 Care commission

Where a warrant application is for a building which will need to be registered by the Scottish Commission for the Regulation of Care, the applicant should consult the commission. The National Care Standards include some physical standards for premises that must be met.

2.26 What records do the local authority hold?

Your local authority is required to maintain a register for the geographical area of the authority including the following:

- applications for warrant and amendments to warrants;
- decisions on the applications;
- submissions of completion certificates;
- decisions on the submissions (acceptance or rejection);
- particulars of EPCs;
- particulars of notices.

Note: When local authorities are acting as Verifiers they do not need to keep other records.

2.27 Other sources

Most of the documents mentioned in this chapter (and other sections of this book) can be purchased from Scottish Stationery Office (TSO), 71 Lothian Road, Edinburgh EH3 9AZ (Tel.: 0870 606 5566, Fax: 0870 606 5588, www.tsoscotland.com) or from any main bookshop. Copies should also be available in public reference libraries.

3

The Requirements of the Building (Scotland) Regulations (2004)

3.1 Introduction

This chapter focuses on the two Technical Handbooks (one to cover domestic buildings and the other for non-domestic buildings) which provide guidance on how to comply with the Building (Scotland) Regulations 2004.

Each Handbook has seven sections. Although Section 0 (General Provisions) of each handbook is identical and the guidance for the remaining sections very similar for both domestic and non-domestic buildings, some differences do occur. For the purposes of this chapter, therefore, all information has been listed together and where differences between the two Technical Handbooks occur, this has been highlighted.

 The Building (Scotland) Regulations 2004 apply to all buildings and works where an application for warrant is made on or after 1 May 2005 and are restricted in terms of construction type, size and subsoil conditions to those commonly occurring in Scotland.

3.2 Section 0 – General Provisions

Regulation number	Title	Regulation	In a nutshell
1	Introduction		
2	Citation, commencement and interpretation	Regulation 2 sets out the defined terms within the regulations.	On completion of the works (and before the period specified in the building warrant expires) a completion certificate must be submitted.
3	Exempted buildings and services, fittings and equipment	(1) Regulations 8–12 shall not apply to any building or any services, fittings and equipment types described in schedule 1. (2) The provision of services, fittings and equipment to, or the demolition or removal of, exempted buildings is exempt; services, fittings and equipment to, or the demolition of, exempted services, fittings and equipment is exempt. (3) For the purposes of this regulation, for the avoidance of doubt, each such exempted type does not include any of the exceptions expressed in relation to that type.	• Schedule 1 is a full list of exempted buildings and services, fittings and equipment – see Appendix 3A There are exceptions to the exempted buildings listed in schedule 1. • These exemptions are not type related.
4	Changes in occupation or use of a building that cause regulations to apply	For the purposed of section 56 (1) (Interpretation) of the Act and these regulations the changes in occupation or use of buildings set out in schedule 2 shall be conversions to the extent specified by regulation 12.	• Schedule 2 is a list of conversions to which regulations will apply (e.g. the conversion or a building to become a dwelling – see Appendix 3B) • For conversions, it is the intention that the standards achieved in the converted building should be broadly similar to those achieved by entirely new buildings. • Where a change of use applies, the requirements of other legislation (for example regulations made under Health and Safety at Work or Licensing Legislation) must be taken into consideration.

5	Building, works, services and equipment not requiring a warrant	For the purposes of section 8 (8) (Building Warrant) of the Act, any work which consists solely of a building or conversion, including the provision of services or equipment, of a kind specified in schedule 3 shall meet the standards required by regulations 8–12 but shall not, subject to the exceptions and conditions require a warrant.	Building work can be done without the need to obtain a Building Warrant as long as it falls within the definitions in schedule 3 (see Appendix 3C). There are exceptions to each element listed in schedule 3 so it would always be worth checking first!
6	Limited-life buildings	For the purposes of paragraph 3 of schedule 1 of the Act (which enables special provision to be made for buildings intended to have a limited life) a period of 5 years is hereby specified.	This concession only apply to buildings which are not dwellings.
7	Measurements	For the purposes of these regulations, measurements shall be made calculated in accordance with schedule 4.	Schedule 4 provides a set of measurements that must be used for calculation purposes (see Appendix 3D).
8	Durability, workmanship and fitness of materials	• Work to every building designed, constructed and provided with services, fittings and equipment to meet a requirement of regulation 9–12 must be carried out in a technically proper and workmanlike manner, and the materials used must be durable, and fit for their intended purpose. • All materials, services, fittings and equipment used to comply with the requirements of regulations 9–12 must, so far as is reasonably practicable, be sufficiently accessible to enable necessary maintenance and repair work to be carried out.	Materials, fittings and components used in the construction of the building should be suitable for their purpose, correctly applied and sufficiently durable taking into account normal maintenance practices. The intention is the guidance is adequately flexible to accommodate new techniques as well as proven traditional practices. For example where a product bears a CE this will be accepted as meeting the requirement as long as used in the correct way.
9	Building Standards applicable to construction	Construction shall be carried out so that the work complies with the applicable requirements of schedule 5.	Regulation 9 and schedule 5 are the heart of the Building Standards system as they set out what must be achieved in building work. The standards are given in full along with the associated guidance in sections 1–6 of the Technical Handbooks.

(Continued)

Regulation number	Title	Regulation	In a nutshell
10	Building Standards applicable to demolition	Every building to be demolished must be demolished in such a way that all service connections to the building are properly closed off and any neighbouring building is left stable and watertight. When demolition work has been completed and, where no further work is to commence immediately, the person who carried out that work shall ensure that the site is (a) immediately graded and cleared; or (b) provided with such fences, protective barriers or hoardings as will prevent access thereto.	• Regulation 10 sets out the mandatory requirements when undertaking demolition work. • The Building Regulations do not control the method or process of demolition. This is the responsibility of the Health and Safety Executive and is covered under other legislation.
11	Building Standards applicable to the provision of services, fittings and equipment	Every service, fitting or piece of equipment provided so as to serve a purpose of these regulations shall be so provided in such a way as to further those purposes.	Every service, fitting or piece of equipment provided should be designed, installed and commissioned in such as way as to fulfil its intended purpose.
12	Building Standards applicable to conversions	Conversion shall be carried out so that the building as converted complies with the applicable requirements of schedule 6.	• Certain changes of use or occupation were defined as conversions in schedule 2 (see Appendix 3B) and are therefore subject to Building Regulations. • Regulation 12 states that in these cases they must meet the requirements set out in schedule 6 (see Appendix 3E). • It is recognized that this is not reasonably practicable in many existing buildings therefore Schedule 6 (Appendix 3E) also lists those standards where a lower level of provision may be sufficient.

It is important to discuss with the Verifier all elements where you believe it is not reasonably practicable to meet the standard. The Verifier's decision is final!

Regulation 13 requires that building sites be fenced off in such a way as to protect the public. It also provides powers to deal with building sites where work has for any reason ceased and the Health and Safety at Work etc. Act provisions are no longer applicable.

13 Provision of protective work

(1) No person shall carry out work unless the following provisions of this regulation are complied with.

(2) Subject to paragraph (3), where work is to be carried out on any building site or building which is within 3.6 m of any part of a road or other place to which members of the public have access (whether or not on payment of a fee or charge) there shall, prior to commencement of the work, be erected protective works so as to separate the building site or building or that part of the building site or building on which work is to be carried out from that road or other place.

(3) Nothing in paragraph (2) shall require the provision of protective works in any case where the local authority is satisfied that no danger to the public is caused, or is likely to be caused, by the work.

(4) The protective works referred to in the preceding paragraphs are all or any of:

(a) providing hoardings, barriers or fences;
(b) subject to paragraph (5), where necessary to prevent danger, providing footpaths outside such hoardings, barriers or fences with safe and convenient platforms, handrails, steps or ramps, and substantial overhead coverings;
(c) any other protective works which in the opinion of the local authority are necessary to ensure the safety of the public, all of such description, material and dimensions and in such position as the local authority may direct.

(Continued)

Regulation number	Title	Regulation	In a nutshell
		(5) Nothing in paragraph(4)(b) shall require the provision of a platform, handrail, step or ramp: (a) where no part of the existing footpath is occupied by the protective works or in connection with the work; or (b) where that part of an existing footpath remaining unoccupied affords a safe means of passage for people, and is of a width of not less than 1.2m or such greater width as the local authority may direct. (6) Any protective works shall be so erected as to cause no danger to the public and shall be maintained to the satisfaction of the local authority. (7) Subject to paragraph (8), any protective works shall be removed: (a) in the case of a building which has been constructed by virtue of a warrant, not more than 14 days or such longer period as the local authority may direct from the date of acceptance of the certificate of completion; and (b) in any other case, on completion of the work. (8) Nothing in paragraphs (1)–(7) of this regulation shall prohibit the removal of the protective works or any part thereof prior to the completion of the work where the local authority is satisfied that no danger to the public is caused or is likely to be caused as a result of their removal. (9) Any protective works shall be illuminated, and any such works which project on to or over that part of a road which is not a pavement or footpath shall be provided with such markings, as in the opinion of the local authority are necessary to secure the safety of the public.	

		(10) Where work has been carried out without the provision of protective works, or where work on a building site has stopped or a building site has been abandoned, a local authority may require the site owner to carry out protective works.	
14	Clearing of footpaths	Where any work is being carried out on a building site or building, any neighbouring footpath (including any footpath provided so as to form part of the protective works) shall be regularly cleaned and kept free of building debris and related materials by the person carry out the work, to the satisfaction of the local authority.	Footpaths adjacent to building sites shall be regularly cleaned and must be kept free of building debris and related materials.
15	Securing of unoccupied and partially completed buildings	• Subject to paragraph (2) a person carrying on work shall ensure that any building which is partly constructed or partly demolished or which has been completed but not yet occupied is, so far as reasonably practicable, properly secured or closed against an authorized entry at all times when work thereon is not in progress. • Nothing in paragraph (1) shall apply to any work where the local authority is satisfied that adequate supervision of the building is being or will be maintained for the purpose of securing the building.	All building sites where there are unfinished or partially complete works shall be kept safe and secure.
16	Relaxations	No direction may be given under section 3 (2) of the act in relation to regulations 1 to 3, 5 and 7.	• Regulation 16 sets out the regulations that cannot at any time be relaxed. The elements of the act are: • Building Regulations • Continuing requirements • Relaxation of the Building Regulations • Compliance with guidance documents • Building Standards assessments

(Continued)

Regulation number	Title	Regulation	In a nutshell
17	Continuing requirements	Subject to paragraph (2) the owners of buildings shall ensure that: (1) Every air-conditioning system within a building is inspected at regular intervals; and (2) Appropriate advice is given to the users of the buildings on reducing the energy consumption of such an air-conditioning system. This regulation shall apply to; • air-conditioning systems with a total effective output rating of less than 12 kW; or • air-conditioning systems solely for the processes within the building. (3) In terms of section 2 of the Building (Scotland) Act 2003 the provisions of paragraph (1) are designated provision in respect of which there is a continuing requirement imposed on the owners of the building.	All air-conditioning systems shall be regularly inspected, and appropriate advice given to the users of the buildings on reducing the energy consumption of such an air-conditioning system.

3.3 Section 1 – structure

Number	Title	Regulation
1.1	Functional Standard	Every building must be designed and constructed in such a way that the loadings that are liable to act on it, taking into account the nature of the ground, will not lead to: (a) the collapse of the whole or part of the building; (b) deformations which would make the building unfit for its intended use, unsafe, or cause damage to other parts of the building or to fittings or to installed equipment; or (c) impairment of the stability of any part of another building.
1.2	Disproportionate collapse	Every building must be designed and constructed in such a way that in the event of damage occurring to any part of the structure of the building the extent of any resultant collapse will not be disproportionate to the original cause.

3.4 Section 2 – fire

Number	Title	Regulation
2.1	Compartmentation	Every building must be designed and constructed in such a way that in the event of an outbreak of fire within the building, fire and smoke are inhibited from spreading beyond the compartment of origin until any occupants have had the time to leave that compartment and any fire containment measures have been initiated. This standard does not apply to domestic buildings.
2.2	Separation	Every building which is divided into more than one area of different occupation must be designed and constructed in such a way that in the event of an outbreak of fire within the building, fire and smoke are inhibited from spreading beyond the area of occupation where the fire originated.
2.3	Structural Protection	Every building must be designed and constructed in such a way that in the event of an outbreak of fire within the building, the load-bearing capacity of the building will continue to function until all occupants have escaped, or been assisted to escape, from the building and any mandatory fire containment measures have been initiated.

(*Continued*)

Number	Title	Regulation
2.4	Cavities	Every building must be designed and constructed in such a way that in the event of an outbreak of fire within the building, the unseen spread of fire and smoke within concealed spaces in its structure and fabric is inhibited.
2.5	Internal Linings	Every building must be designed and constructed in such a way that in the event of an outbreak of fire within the building, the development of fire and smoke from the surfaces of walls and ceilings within the area of origin is inhibited.
2.6	Spread to neighbouring buildings	Every building must be designed and constructed in such a way that in the event of an outbreak of fire within the building, the spread of fire to neighbouring buildings is inhibited.
2.7	Spread on external walls	Every building must be designed and constructed in such a way that in the event of an outbreak of fire within the building, or from an external source, the spread of fire on the external walls of the building is inhibited.
2.8	Spread from neighbouring buildings	Every building must be designed and constructed in such a way that in the event of an outbreak of fire in a neighbouring building, the spread of fire to the building is inhibited.
2.9	Escape	Every building must be designed and constructed in such a way that in the event of an outbreak of fire within the building, the occupants, once alerted to the outbreak of the fire, are provided with the opportunity to escape from the building, before being affected by fire or smoke.
2.10	Escape Lighting	Every building must be designed and constructed in such a way that in the event of an outbreak of fire within the building, illumination is provided to assist in escape.
2.11	Communication	Every building must be designed and constructed in such a way that in the event of an outbreak of fire within the building, the occupants are alerted to the outbreak of fire.

This standard applies only to a building which:

- is a dwelling;
- is a residential building; or
- is an enclosed shopping centre.

2.12	Fire Service Access	Every building must be accessible to fire appliances and fire service personnel.
2.13	Fire Service Water Supply	Every building must be provided with a water supply for use by the fire service.

This standard does not apply to domestic buildings.

Number	Title	Regulation
2.14	Fire Service Facilities	Every building must be designed and constructed in such a way that facilities are provided to assist fire-fighting or rescue operations.
2.15	Automatic life safety fire suppression systems	Every building must be designed and constructed in such a way that, in the event of an outbreak of fire within the building, fire and smoke will be inhibited from spreading through the building by the operation of an automatic life safety fire suppression system.

This standard applies only to a building which:

- is an enclosed shopping centre;
- is a residential care building;
- is a high rise domestic building; or
- forms the whole or part of a sheltered housing complex.

3.5 Section 3 – environment

Number	Title	Regulation
3.1	Site preparation – harmful and dangerous substances	Every building must be designed and constructed in such a way that there will not be a threat to the building or the health of people in or around the building due to the presence of harmful or dangerous substances.
3.2	Site preparation – protection from radon gas	Every building must be designed and constructed in such a way that there will not be a threat to the health of people in or around the building due to the emission and containment of radon gas.
3.3	Flooding and ground water	Every building must be designed and constructed in such a way that there will not be a threat to the building or the health of the occupants as a result of flooding and the accumulation of ground water.
3.4	Moisture from the ground	Every building must be designed and constructed in such a way that there will not be a threat to the building or the health of the occupants as a result of moisture penetration from the ground.
3.5	Existing drains	Every building must not be constructed over an existing drain (including a field drain) that is to remain active.

This standard does not apply where it is not reasonably practicable to reroute an existing drain.

(*Continued*)

Number	Title	Regulation
3.6	Surface water drainage	Every building, and hard surface within the curtilage of a building, must be designed and constructed with a surface water drainage system that will: (a) ensure the disposal of surface water without threatening the building and the health and safety of the people in and around the building; and (b) have facilities for the separation and removal of silt, grit and pollutants.
3.7	Wastewater drainage	(a) Every wastewater drainage system serving a building must be designed and constructed in such a way as to ensure the removal of wastewater from the building without threatening the health and safety of the people in and around the building; (b) that facilities for the separation and removal of oil, fat, grease and volatile substances from the system are provided; (c) that discharge is to a public sewer or public wastewater treatment plant, where it is reasonably practicable to do so; and (d) where discharge to a public sewer or public wastewater treatment plant is not reasonably practicable that discharge is to a private wastewater treatment plant or septic tank. Standard 3 does not apply to a dwelling.
3.8	Private wastewater treatment systems – treatment plants	Every private wastewater treatment plant or septic tank serving a building must be designed and constructed in such a way that it will ensure the safe temporary storage and treatment of wastewater prior to discharge.
3.9	Private wastewater treatment systems – infiltration systems	Every private wastewater treatment system serving a building must be designed and constructed in such a way that the disposal of the wastewater to ground is safe and is not a threat to the health of the people in or around the building.
3.10	Precipitation	Every building must be designed and constructed in such a way that there will not be a threat to the building or the health of the occupants as a result of moisture from precipitation penetrating to the inner face of the building. This standard does not apply to a building where penetration of moisture from the outside will result in effects no more harmful than those likely to arise from use of the building.

(Continued)

Number	Title	Regulation
3.11	Facilities in dwellings	Every building must be designed and constructed in such a way that: (a) the size of any apartment or kitchen will ensure the welfare and convenience of all occupants and visitors; and (b) an accessible space is provided to allow for the safe, convenient and sustainable drying of washing. 💡 This standard applies only to a dwelling.
3.12	Sanitary facilities	Every building must be designed and constructed in such a way that sanitary facilities are provided for all occupants of, and visitors to, the building in a form that allows convenience of use and that there is no threat to the health and safety of occupants or visitors.
3.13	Heating	Every building must be designed and constructed in such a way that it can be heated. 💡 This standard applies only to a dwelling.
3.14	Ventilation	Every building must be designed and constructed in such a way that the air quality inside the building is not a threat to the health of the occupants or the capability of the building to resist moisture, decay or infestation.
3.15	Condensation	Every building must be designed and constructed in such a way that there will not be a threat to the building or the health of the occupants as a result of moisture caused by surface or interstitial condensation. 💡 This standard applies only to a dwelling.
3.16	Natural lighting	Every building must be designed and constructed in such a way that natural lighting is provided to ensure that the health of the occupants is not threatened. 💡 This standard applies only to a dwelling.
3.17	Combustion appliances – safe operation	Every building must be designed and constructed in such a way that each fixed combustion appliance installation operates safely.
3.18	Combustion appliances – protection from combustion products	Every building must be designed and constructed in such a way that any component part of each fixed combustion appliance installation used for the removal of combustion gases will withstand heat generated as a result of its operation without any structural change that would impair the stability or performance of the installation.

(Continued)

Number	Title	Regulation
3.19	Combustion appliances – relationship to combustible materials	Every building must be designed and constructed in such a way that any component part of each fixed combustion appliance installation will not cause damage to the building in which it is installed by radiated, connected or conducted heat or from hot embers expelled from the appliance.
3.20	Combustion appliances – removal of products of combustion	Every building must be designed and constructed in such a way that the products of combustion are carried safely to the external air without harm to the health of any person through leakage, spillage or exhaust, nor permit the re-entry of dangerous gases from the combustion process of fuels into the building.
3.21	Combustion appliances – air for combustion	Every building must be designed and constructed in such a way that each fixed combustion appliance installation receives air for combustion and operation of the chimney so that the health of persons within the building is not threatened by the build-up of dangerous gases as a result of incomplete combustion.
3.22	Combustion appliances – air for cooling	Every building must be designed and constructed in such a way that each fixed combustion appliance installation receives air for cooling so that the fixed combustion appliance installation will operate safely without threatening the health and safety of persons within the building.
3.23	Fuel storage – protection from fire	Every building must be designed and constructed in such a way that: (a) An oil storage installation, incorporating oil storage tanks used solely to serve a fixed combustion appliance installation providing space heating or cooking facilities in a building, will inhibit fire from spreading to the tank and its contents from within, or beyond, the boundary. (b) A container for the storage of woody biomass fuel will inhibit fire from spreading to its contents from within, or beyond the boundary. 💡 This standard does not apply to portable containers.
3.24	Fuel storage – containment	Every building must be designed and constructed in such a way that: (a) An oil storage installation, incorporating oil storage tanks used solely to serve a fixed combustion appliance installation providing space heating or cooking facilities in a building will: reduce the risk of oil escaping from the installation; contain any oil spillage likely to contaminate any water supply, ground water, watercourse, drain or sewer; and permit any spill to be disposed of safely.

(Continued)

Number	Title	Regulation
		(a) The volume of woody biomass fuel storage allows the number of journeys by delivery vehicles to be minimized.
		💡 This standard does not apply to portable containers.
3.25	Solid waste storage	Every building must be designed and constructed in such a way that accommodation for solid waste storage is provided which:
		(a) permits access for storage and for the removal of its contents;
		(b) does not threaten the health of people in and around the building; and
		(c) does not contaminate any water supply, ground water or surface water.
		💡 This standard applies only to a dwelling.
3.26	Dungsteads and farm effluent tanks	Every building must be designed and constructed in such a way that there will not be a threat to the health and safety of people from a dungstead and farm effluent tank.

3.6 Section 4 – safety

Number	Title	Regulation
4.1	Access to buildings	Every building must be designed and constructed in such a way that all occupants and visitors are provided with safe, convenient and unassisted means of access to the building.
		💡 There is no requirement to provide access for a wheelchair user to:
		• a house, between either the point of access to or from any car parking within the curtilage of a building and an entrance to the house where it is not reasonably practicable to do so; or
		• a common entrance of a domestic building not served by a lift, where there are no dwellings.
4.2	Access within buildings	Every building must be designed and constructed in such a way that:
		(a) in non-domestic buildings, safe, unassisted and convenient means of access is provided throughout the building;
		(b) in residential buildings, a proportion of the rooms intended to be used as bedrooms must be accessible to a wheelchair user;

(Continued)

Number	Title	Regulation
		(c) in domestic buildings, safe and convenient means of access is provided within common areas and to each dwelling; (d) in dwellings, safe and convenient means of access is provided throughout the dwelling; and (e) in dwellings, unassisted means of access is provided to, and throughout, at least one level.
		There is no requirement to provide access for a wheelchair user: • in a non-domestic building not served by a lift, to a room, intended to be used as a bedroom, that is not on an entrance storey; or • in a domestic building not served by a lift, within common areas and to each dwelling, other than on an entrance storey.
4.3	Stairs and ramps	Every building must be designed and constructed in such a way that every level can be reached safely by stairs or ramps.
4.4	Pedestrian protective barriers	Every building must be designed and constructed in such a way that every sudden change of level that is accessible in, or around, the building is guarded by the provision of pedestrian protective barriers. This standard does not apply where the provision of pedestrian protective barriers would obstruct the use of areas so guarded.
4.5	Electrical safety	Every building must be designed and constructed in such a way the electrical installation does not: (a) threaten the health and safety of the people in, and around, the building; and (b) become a source of fire. This standard does not apply to an electrical installation: • serving a building or any part of a building to which the Mines and Quarries Act 1954 or the Factories Act 1961 applies; or • forming part of the works of an undertaker to which regulations for the supply and distribution of electricity made under the Electricity Act 1989.
4.6	Electrical Fixtures	Every building must be designed and constructed in such a way that electric lighting points and socket outlets are provided to ensure the health, safety and convenience of occupants and visitors. This standard applies only to domestic buildings where a supply of electricity is available.
4.7	Aids to communication	Every building must be designed and constructed in such a way that it is provided with aids to assist those with a hearing impairment.

(Continued)

Number	Title	Regulation
		This standard does not apply to domestic buildings.
4.8	Danger from accidents	Every building must be designed and constructed in such a way that:
		(a) people in and around the building are protected from injury that could result from fixed glazing, projections or moving elements on the building;
		(b) fixed glazing in the building is not vulnerable to breakage where there is the possibility of impact by people in and around the building;
		(c) both faces of a window and rooflight in a building are capable of being cleaned such that there will not be a threat to the cleaner from a fall resulting in severe injury;
		(d) a safe and secure means of access is provided to a roof; and manual controls for ventilation and for electrical fixtures can be operated safely.
		Standards 4.8(d) does not apply to domestic buildings.
4.9	Danger from heat	Every building must be designed and constructed in such a way that protection is provided for people in, and around, the building from the danger of severe burns or scalds from the discharge of steam or hot water.
4.10	Fixed seating	Every building which contains fixed seating accommodation for an audience or spectators, must be designed and constructed in such a way that a number of level spaces for wheelchairs are provided proportionate to the potential audience or spectators.
		This standard does not apply to domestic buildings.
4.11	Liquefied petroleum gas storage	Every building must be designed and constructed in such a way that each liquefied petroleum gas storage installation, used solely to serve a combustion appliance providing space heating, water heating, or cooking facilities, will:
		(a) be protected from fire spreading to any liquefied petroleum gas container; and
		(b) not permit the contents of any such container to form explosive gas pockets in the vicinity of any container.
		This standard does not apply to a liquefied petroleum gas storage container, or containers, for use with portable appliances.
4.12	Vehicle protective barriers	Every building accessible to vehicular traffic must be designed and constructed in such a way that every change in level is guarded.

3.7 Section 5 – noise

Number	Title	Regulation
5.1	Resisting sound transmission	Every building must be designed and constructed in such a way that each wall and floor separating one dwelling from another, or one dwelling from another part of the building, or one dwelling from a building other than a dwelling, will limit the transmission of noise to the dwelling to a level that will not threaten the health of the occupants of the dwelling or inconvenience them in the course of normal domestic activities provided the source noise is not in excess of that from normal domestic activities.

This standard does not apply to:

- fully detached houses; or
- roofs or walkways with access solely for maintenance, or solely for the use, of the residents of the dwelling below.

3.8 Section 6 – energy

Number	Title	Regulation
6.1	Carbon dioxide emissions	Every building must be designed and constructed in such a way that:

(a) the energy performance is calculated in accordance with a methodology which is asset-based, conforms with the European Directive on the Energy Performance of Buildings 2002/91/EC and uses UK climate data; and

(b) the energy performance of the building is capable of reducing carbon dioxide emissions.

This standard does not apply to:

- alterations and extensions to buildings;
- conversions of buildings;
- non-domestic buildings and buildings that are ancillary to a dwelling that are stand-alone having an area less than $50\,m^2$;
- buildings which will not be heated or cooled other than by heating provided solely for the purpose of frost protection; or
- limited life buildings which have an intended life of less than 2 years.

(*Continued*)

Number	Title	Regulation
6.2	Building insulation envelope	Every building must be designed and constructed in such a way that an insulation envelope is provided which reduces heat loss. This standard does not apply to: • non-domestic buildings which will not be heated, other than heating provided solely for the purpose of frost protection; • communal parts of domestic buildings which will not be heated, other than heating provided solely for the purposes of frost protection; or • buildings which are ancillary to dwellings, other than conservatories, which are either unheated or provided with heating which is solely for the purpose of frost protection.
6.3	Heating systems	Every building must be designed and constructed in such a way that the heating and hot water service systems installed are energy efficient and are capable of being controlled to achieve optimum energy efficiency. This standard does not apply to: • buildings which do not use fuel or power for controlling the temperature of the internal environment; • heating provided solely for the purpose of frost protection; or • individual solid-fuel or oil-firing stoves or open-fires, gas or electric fires or room heaters (excluding electric storage and panel heaters) provided as secondary heating in domestic buildings.
6.4	Insulation of pipes, ducts and vessels	Every building must be designed and constructed in such a way that temperature loss from heated pipes, ducts and vessels, and temperature gain to cooled pipes and ducts, is resisted. This standard does not apply to: • buildings which do not use fuel or power for heating or cooling either the internal environment or water services; • buildings, or parts of a building, which will not be heated, other than heating provided solely for the purpose of frost protection; • pipes, ducts or vessels that form part of an isolated industrial or commercial process; or • cooled pipes or ducts in domestic buildings.

(*Continued*)

Number	Title	Regulation
6.5	Artificial and display lighting	Every building must be designed and constructed in such a way that the artificial or display lighting installed is energy efficient and is capable of being controlled to achieve optimum energy efficiency.

This standard does not apply to:

- process and emergency lighting components in a building;
- communal areas of domestic buildings; or
- alterations in dwellings.

6.6	Mechanical ventilation and air conditioning	Every building must be designed and constructed in such a way that:

(a) the form and fabric of the building minimizes the use of mechanical ventilating or cooling systems for cooling purposes; and
(b) in non-domestic buildings, the ventilating and cooling systems installed are energy efficient and are capable of being controlled to achieve optimum energy efficiency.

This standard does not apply to buildings which do not use fuel or power for ventilating or cooling the internal environment.

6.7	Commissioning building services	Every building must be designed and constructed in such a way that energy supply systems and building services which use fuel or power for heating, lighting, ventilating and cooling the internal environment and heating the water, are commissioned to achieve optimum energy efficiency.

This standard does not apply to:

- major power plants serving the National Grid;
- the process and emergency lighting components of a building;
- heating provided solely for the purpose of frost protection; or
- energy supply systems used solely for industrial and commercial processes, leisure use and emergency use within a building.

6.8	Writing information	The occupiers of a building must be provided with written information by the owner:

(a) on the operation and maintenance of the building services and energy supply systems; and
(b) where any air-conditioning system in the building is subject to a time-based interval for inspection of the system.

(Continued)

Number	Title	Regulation
		This standard does not apply to:
		• major power plants serving the National Grid;
		• buildings which do not use fuel or power for heating, lighting, ventilating and cooling the internal environment and heating the water supply services;
		• the process and emergency lighting components of a building;
		• heating provided solely for the purpose of frost protection;
		• lighting, ventilation and cooling systems in a domestic building; or
		• energy supply systems used solely for industrial and commercial processes, leisure use and emergency use within a building.
6.9	Energy Performance Certificates	Every building must be designed and constructed in such a way that:
		(a) an energy performance certificate for the building is affixed to the building, indicating the approximate annual carbon dioxide emissions and energy usage of the building based on a standardized use of the building;
		(b) the energy performance for the certificate is calculated in accordance with a methodology which is asset-based, conforms with the European Directive 2002/91/EC and uses UK climate data; and
		(c) the energy performance certificate is displayed in a prominent place within the building.
		This standard does not apply:
		• to buildings which do not use fuel or power for controlling the temperature of the internal environment;
		• to non-domestic buildings and buildings that are ancillary to a dwelling that are stand-alone having an area less than $50\,m^2$;
		• to conversions, alterations and extensions to buildings other than alterations and extensions to stand-alone buildings less than $50\,m^2$ that would increase the area to $50\,m^2$ or more; and
		• to buildings involving the fit-out of the building shell which is the subject of a continuing requirement; or
		• to limited life buildings which have an intended life of less than 2 years.

(*Continued*)

Number	Title	Regulation
		Standard 6.9(c) only applies to buildings with a floor area of more than $1000\,m^2$, which are occupied by public authorities and institutions providing public services, which can be visited by the public.
		Note: We are reliably informed that it is intended that Scottish Ministers will direct local authorities to apply standard 6.9 to **all** existing buildings (i.e. those being sold or rented out) using section 25 (2) of the Building (Scotland) Act 2003.
6.10	Metering	Every building must be designed and constructed in such a way that each part of a building designed for different occupation is fitted with fuel consumption meters.
		This standard does not apply to:

- domestic buildings;
- communal areas of buildings in different occupation;
- district or block heating systems where each part of the building designed for different occupation is fitted with heat meters; or
- heating fired by solid fuel or biomass.

Appendix 3A

Exempted buildings and services, fittings and equipment

	Type	Description	Exception
Buildings controlled by other legislation	1	Any building in which explosives are manufactured or stored under a licence granted under the Manufacture and Storage of Explosives Regulations 2005.	
	2	A building erected on a site which is subject to licensing under the Nuclear Installations Act 1965.	A dwelling, residential building, office, canteen or visitor centre.
	3	A building included in the schedule of monuments maintained under section 1 of the Ancient Monuments and Archaeological Areas Act 1979.	A dwelling or residential building.
Protective Works	4	Protective works subject to control by regulation 13.	
Buildings or work not frequented by people	5	A building into which people cannot or do not normally go.	• A building within 6 m or the equivalent of its height (whichever is the less) of the boundary. • A wall or fence. • A tank, cable, sewer, drain or other pipe above or below ground for which there is a requirement in these regulations.
	6	Detached fixed plant or machinery or a detached building housing only fixed plant or machinery, the only normal visits to which are intermittent visits to inspect or maintain the fixed plant or machinery.	A building within 1 m of a boundary.
Agricultural and related buildings	7	An agricultural greenhouse or other of mainly translucent material used mainly for commercial growing of plants.	A building used to any extent for building retailing (including storage of goods for retailing) or exhibiting.

(Continued)

Type	Description	Exception
8	A single-storey detached building used for any other form of agriculture, fish farming or forestry.	A building used to any extent for retailing (including storage for retailing) or exhibiting. • A building exceeding 280 m² in area. • A building within 6 m or the equivalent of its height (whichever is the less) of a boundary. • A dwelling, residential building, office, canteen or visitor centre. • A dungstead or farm effluent tank.
9 Works of civil engineering construction	A work of civil engineering construction including a dock, wharf, harbour, pier, quay, sea defence work, lighthouse, embankment, river work, dam, bridge, tunnel, filter station or bed, inland navigation, reservoir, waterworks, pipe line, sewage treatment works, gas holder or main, electricity supply line and supports, any bridge embankment or other support to railway lines and any signalling or power lines and supports, and a fire practice tower.	A bridge or tunnel forming part of an escape route or an access route provided to meet a requirement of these. • A private sewage treatment works provided to meet a requirement of these regulations.
10 Buildings of a specialist nature	A building essential for the operation of a railway including a locomotive or carriage shed, or for the operation of any other work of civil engineering contained in type 9 of this schedule and erected within the curtilage of such a railway or work.	A signalling and control centre for a railway or dock. • A building to which the public is admitted, not being a building exempted by type 11 of this schedule. • A dwelling, residential building, office, canteen, or warehouse.

11	A single-storey detached road or rail passenger shelter or a telephone kiosk which in so far as it is glazed complies with the requirements of regulation 9 and paragraph 4.8 of schedule 5.	• A building having a floor area exceeding 30 m². • A building containing a fixed combustion appliance installation.
12	A caravan or mobile home within the meaning of the Caravan Sites and Control of Development Act 1960, or a tent, van or shed within the meaning of section 73 of the Public Health (Scotland) Act 1897.	Any wastewater disposal system serving a building of this type.
Small buildings	13	A detached single-storey building having an area not exceeding 8 m².	• A dwelling or residential building. • A building ancillary to and within the curtilage of a dwelling. • A building within 1 m of a boundary. • A building containing a fixed combustion appliance installation or sanitary facility. • A wall or fence.
Construction and development buildings	14	A building used only by people engaged in the construction, demolition or repair of any building or structure during the course of that work.	A building containing sleeping accommodation.
Temporary buildings	15	A building used in connection with the letting or sale of any building under construction until such time as the letting or sale of all related buildings is completed.	A building containing sleeping accommodation.
	16	A building which, during any period of 12 months, is either erected or used on a site – for a period not exceeding 28 consecutive days; or for a number of days not exceeding 60, and any alterations to such buildings.	

(Continued)

	Type	Description	Exception
Buildings ancillary to houses	17	A detached single-storey building ancillary to and within the curtilage of a house.	• A building exceeding 8 m² in area. • A building within 1 m of the house unless it is at least 1 m from any boundary. • A building containing sleeping accommodation. • A building containing a flue, a fixed combustion appliance installation or sanitary facility. • A wall or fence.
	18	A single-storey building attached to an existing house, which is ancillary to the house and consists of a conservatory or porch which insofar as it is glazed complies with the requirements of regulation 9 and paragraph 4.8 of schedule 5.	• A building exceeding 8 m² in area. • A building containing a flue, a fixed combustion appliance or sanitary facility. • A building within 1 m of the house.
	19	A single-storey building which is detached, or is attached to an existing house and which is ancillary to the house and consists of a greenhouse, carport or covered area.	• A building exceeding 30 m² in area. • A building containing a flue, a fixed combustion appliance installation or sanitary facility.
Buildings ancillary to flats or maisonettes	20	A detached single-storey building ancillary to and within the curtilage of a flat or maisonette.	• A building exceeding 8 m² in area. • A building within 1 m of the flat or maisonette or within 3 m of any other part of the building containing the flat or maisonette. • A building within 1 m of a boundary. • A building containing a flue, a fixed combustion appliance installation or sanitary facility. • A wall or fence. • A swimming pool deeper than 1.2 m.
Paved areas	21	A paved area or hardstanding.	• A paved area or hardstanding exceeding 200 m² in area. • A paved area forming part of an access to meet a requirement of these regulations.

Appendix 3B

Conversions to which the regulations apply

Type	Conversion
1	Changes in the occupation or use of a building to create a dwelling or dwellings or a part thereof.
2	Changes in the occupation or use of a building ancillary to a dwelling to increase the area of human occupation.
3	Changes in the occupation or use of a building which alter the number of dwellings in the building.
4	Changes in the occupation or use of a domestic building to any other type of building.
5	Changes in the occupation or use of a residential building to any other type of building.
6	Changes in the occupation or use of a residential building which involve a significant alteration to the characteristics of the persons who occupy, or who will occupy, the building, or which significantly increase the number of people occupying, or expected to occupy, the building.
7	Changes in the occupation or use of a building so that it becomes a residential building.
8	Changes in the occupation or use of an exempt building (in terms of schedule 1) to a building which is not so exempt.
9	Changes in the occupation or use of a building to allow access by the public where previously there was none.
10	Changes in the occupation or use of a building to accommodate parts in different occupation where previously it was not so occupied.

Appendix 3C

Description of building and work, including the provision of services, fitting and equipment, not requiring a warrant.

Type	Description	Exception

Category A
On condition that types 1–23 in all respects and/or in the manner of their fitting meet any standards required by the regulations

Type	Description	Exception
1	Any *work* to or in a *house*.	• Any work which increases the floor area of the house. • Any demolition or alteration of the roof, external walls or elements of structure. • Any work involving underpinning. • Any work adversely affecting a separating wall. • Any work involving a change in the method of wastewater discharge. • Work, not being work of types 3–26 below, to a house having a storey, or creating a storey, at a height of more than 4.5 m.
2	Any work to or in a non-residential building to which the public does not have access.	• A non-residential building within which there is a domestic or residential building. • Any work which increases the floor area of the building. • Any demolition or alteration of the roof, external walls or elements of structure. • Any work involving underpinning. • Any work adversely affecting a separating wall. • Any work involving a change in the method of wastewater discharge. • Work, not being work of types 3–26 below, to a building having a storey, or creating a storey, at a height of more than 7.5 m.

and, without prejudice to the generality of types 1 and 2 above:

Type	Description	Exception
3	A detached single-storey building, having an area exceeding 8 m² but not exceeding 30 m².	• A dwelling or residential building. • A building ancillary to, or within the curtilage of, a dwelling. • A building within 1 m of a boundary. • A building containing a fixed combustion appliance installation or sanitary facility. • A swimming pool deeper than 1.2 m.

(Continued)

Type	Description	Exception
4	A detached single-storey building, having an area exceeding 8 m² but not exceeding 30 m², ancillary to and within the curtilage of a house.	• A building within 1 m of the house unless it is at least 1 m from any boundary. • A building containing a fixed combustion appliance installation or sanitary facility. • A swimming pool deeper than 1.2 m.
5	A detached single-storey building, having an area exceeding 8 m² but not exceeding 30 m², ancillary to and within the curtilage of a flat or maisonette.	• A building within 1 m of the flat or maisonette or within 3 m of any other part of the building containing the flat or maisonette. • A building within 1 m of a boundary. • A building containing a fixed combustion appliance installation or sanitary facility. • A swimming pool deeper than 1.2 m.
6.	Any work associated with a combustion appliance installation or other part of a heating installation, not being work of types 7 or 8 below.	• Any work associated with a solid fuel appliance having an output rating more than 50 kW, an oil-firing appliance with an output rating more than 45 kW or a gas-fired appliance having a net input rating more than 70 kW. • Any work associated with a chimney, flue-pipe or constructional hearth. • Any work associated with an oil storage tank with a capacity of more than 90 litres, including any pipework connecting the tank to a combustion appliance providing space or water heating or cooking facilities. • Any work adversely affecting a separating wall or separating floor.
7	Any work associated with a balanced flue serving a room-sealed appliance.	
8	Any work associated with pipework, radiators, convector heaters and thermostatic controls for, or associated with, type 6 above.	
9	Any work associated with installing a flue liner.	
10	Any work associated with refillable liquefied petroleum gas storage cylinders supplying, via a fixed pipework installation, combustion appliances used principally for providing space heating, water heating or cooking facilities.	

(Continued)

Type	Description	Exception
11	Any work associated with the provision of a single sanitary facility, together with any relevant branch soil or waste pipe.	Any work associated with a water closet, waterless closet or urinal.
12	Any work associated with the relocation within the same room or space of any sanitary facility, together with any relevant branch soil or waste pipe.	
13	Any work associated with the provision of an extractor fan.	
14	Any work associated with a stairlift within a dwelling.	
15	Any work associated with the provision of a notice or other fixture for which there is no requirement provided in these regulations.	
16	Any work associated with an outdoor sign that is the subject to the Town and Country Planning (Control of Advertisements) (Scotland) Regulations 1984	
17	Any work associated with thermal insulating material to or within a wall, ceiling, roof or floor.	Any *work* associated with the *application* of thermal insulating material to the outer surface of an *external wall.*
18	A wall not exceeding 1.2 m in height, or a fence not exceeding 2 m in height.	
19	Any *work* associated with open raised external decking.	• Any decking at a height of more than 1.2 m. • Decking that forms part of any access provided to comply with the requirements in regulation 9 and paragraph 4.1 of schedule 5. • Decking that forms any *escape route* other than from a flat or *maisonette,* provided to comply with the requirements in regulation 9 and paragraph 2.9 of schedule 5.
20	A door, window or rooflight when the work includes replacing the frame.	
21	A paved area or hardstanding exceeding 200 m² in area. A paved area forming part of an access to meet a requirement of these regulations.	

(*Continued*)

Type	Description	Exception
22	An electrical installation, including a circuit for telecommunication, alarm purposes or for the transmission of sound, vision or data, which operates at extra-low voltage (not exceeding 50V alternating current or 120V direct current, measured between conductors or to earth) and which is not connected directly or indirectly to an electricity supply which operates at a voltage higher than either of those specified above.	
23	The construction of a ramp not exceeding 5m in length.	

Category B
On condition that this work, service, fitting or equipment is to a standard no worse than at present

24	Any *work* associated with the replacement of a fitting or equipment, in whole or in part, by another of the same general type, including a *sanitary facility* (together with any relevant branch soil or waste pipe), rainwater gutter or downpipe, solid fuel combustion appliance, electrical fixture, ventilation fan, *chimney* or *flue* outlet fitting or terminal, fire hydrant or main, lift or escalator, solid waste chute or container, *kitchen* fitments or other fitted furniture and ironmongery.	Any door, window or rooflight. Any oil firing or gas boiler.
25	Any work associated with the replacement in whole or in part, by material of the same general type, of flooring, lining, cladding, covering or rendering either internally or externally.	
26	Any work to a door, window or rooflight, including glazing which is not a complete replacement falling within type 20 above	

Appendix 3D

Measurements

Area

(1)   Measurement of area shall be taken to the innermost surfaces of enclosing walls or, on any side where there is no enclosing wall, to the outermost edge of the floor on that side.

(2)   A room excludes any built-in fixture extending from the floor to the ceiling.

(3)   In the case of a dwelling, a room excludes any part where the height is less than 1.5 m.

Height and depth

(4)   The height of:
- a building shall be taken to be the height from the surface of the ground to the underside of the ceiling of the topmost storey or, if the topmost storey has no ceiling, one-half of the height of the roof above its lowest part; and
- a storey above the ground or the depth of a storey below the ground shall be taken to be the vertical height or depth as the case may be from the ground to the upper surface of the floor of the storey, and the expressions 'a storey at a height' and 'a storey at a depth' shall be construed accordingly.

(5)   In the measurement of height or depth from ground which is not level the height or depth shall be taken to be the mean height or depth, except that:
- for the purpose of types 1, 2, 3, 4, 5, 18 or 19 of schedule 3; and
- for any other purpose where the difference in level is more than 2.5 m, the height or depth shall be taken to be the greatest height or depth.

General

(6)   Except where the context otherwise requires, measurements shall be horizontal and vertical.

Appendix 3E

Conversions

Every *conversion*, to which these regulations apply, shall meet the requirements of the following standards in schedule 5:

Section	Standards
Fire	2.1, 2.3, 2.5, 2.9, 2.10, 2.11, 2.13, 2.14, 2.15
Environment	3.5, 3.6, 3.7, 3.8, 3.9, 3.11, 3.12, 3.13, 3.14, 3.17, 3.18, 3.20, 3.21, 3.22, 3.23, 3.24, 3.25, 3.26, in section 3
Safety	4.5, 4.6, 4.7, 4.9, 4.11, 4.12
Noise	All
Energy	6.7, 6.8, 6.10

Every *conversion*, to which these regulations apply, shall meet the requirements of the following standards in schedule 5 in so far as is *reasonably practicable*, and in no case be worse than before the *conversion*:

Section	Standards
Structure	All
Fire	2.2, 2.4, 2.6, 2.7, 2.8, 2.12
Environment	3.1, 3.2, 3.3, 3.4, 3.10, 3.15, 3.16, 3.19
Safety	4.1, 4.2, 4.3, 4.4, 4.8, 4.10
Energy	6.2, 6.3, 6.4, 6.5, 6.6

4

Planning Permission

Before undertaking (most) building projects, you must first obtain the approval of local government authorities. The rules controlling planning permission, however, are complicated. Indeed, in some instances minor development may be classed as 'permitted development' (i.e. does not require permission) and so it is always advisable to seek the advice of your local council on all planning matters before you begin.

Many people (particularly householders) are initially reluctant to approach local authorities because, according to local gossip, they are 'likely to be obstructive'. In fact the reality of it is quite the reverse as their purpose is to protect all of us from irresponsible builders and developers and they are normally most sympathetic and helpful to any builder and/or DIY person who wants to comply with the statutory requirements and has asked for their advice.

There are two main controls that districts rely on to ensure that adherence to the local plan is ensured, namely planning permission and Building Regulation approval. Quite a lot of people are confused as to their exact use and whilst both of these controls are associated with gaining 'planning permission', actually receiving planning permission does not automatically confer Building Regulation approval and vice versa. You **may** require **both** before you can proceed. Indeed, there may be a variation in the planning requirements (and to some extent the Building Regulations) from one area to another. **Consequently, the information given on the following pages should be considered as a guide only and not as an authoritative statement of the law.**

 You do not require planning permission to carry out any internal alterations to your home, house, flat or maisonette, provided that it does not affect the external appearance of the building.

4.1 Why do we have to have a planning system?

It's a good question and there is a very good answer. Whilst many of us would instinctively think that the planning system is bureaucratic and time consuming, the planning system is there essentially to safeguard the interests of all of us and ensure that any development is sustainable in the future. The nature, quality and location of new developments of any sort are hugely significant as poorly designed or inappropriately located developments can damage the quality of life in a community, the aspect and general appeal of an area. What's more, they can be enormously costly to put right.

The planning system is there to encourage positive changes to your local area whilst protecting the locale, ensuring what makes a local area attractive (such as the countryside or the historic buildings) are protected. The system tries to balance the demands for development against the long-term interests of us all – but in a sustainable way.

4.2 What is sustainable development?

Whilst there is an ever growing demand for homes, shops and leisure facilities, these developments need to be balanced against the long-term needs of the people, the economy and the environment. This is called sustainable development.

The planning system looks to encourage sustainable development by influencing new buildings and changes in land use.

In order to do this, planning policies will support:

- reusing previously developed or derelict land;
- reusing buildings that make the local area attractive;
- ensuring that the new developments can be approached by foot, cycle and public transport as well as by car;
- preserving areas or buildings of important historic significance.

4.3 Who controls planning permission?

Planning decisions are usually made at a local level and are normally a matter for councils. Your local council has three main duties:

- Preparing Development Plans.
- Deciding on planning permission applications.
- Taking action against unapproved development.

In Scotland, local councils decide in excess of 40,000 applications every year and on average they grant permission to more than 90% of these, so you can see that they are certainly trying to work with, as opposed to against, development, be it large or small.

The Scottish Ministers do, of course, have an over-arching view and involvement with planning matters and maintain and develop law relating to planning. They are also available to local councils for guidance and advice; they approve all local council Structure Plans and make decisions on some major planning applications and appeals (see below).

4.4 So when do Scottish Ministers get involved?

Scottish Ministers do, of course, have the power to make decisions on planning issues, but only do so in a small proportion of instances, such as if a council wants to deviate from the approved Structure Plan (e.g. if they want to develop designated green belt land or have a financial or other interest).

Appeals against refusals, conditions attached to planning permissions or enforcement notices are logged with the Scottish Ministers and are considered by the Inquiry Reporter's Unit, which makes the majority of decisions, although Scottish Ministers can be involved directly in some major cases, such as issues of significant national importance or any they consider worthy of their intervention!

4.5 What does the Scottish Planning Act cover?

The Planning etc. (Scotland) Act 2006 is a fundamental reform of the Scottish planning system. It aims to be 'more inclusive and efficient and improve community involvement, to support the economy and help it grow in a sustainable way'.

The planning act is made up of 10 parts; some parts are already in force and some parts are still in consultation with new regulations coming into effect in the near future.

Part	Title	In a nutshell	Coming into force
1	National Planning Framework	Sets out the arrangements of the planning framework for spatial plan for Scotland and sets out the procedure for Parliamentary considerations	In force as from April 2007
2	Development Plans	Sets out the policy framework for development of land in the local authority areas. Local authorities will prepare and publish strategic and local Development Plans	Coming into force 2008 following consultation

(Continued)

Part	Title	In a nutshell	Coming into force
3	Development Management	How planning applications will be handled This selectively amends part 3 of the 1997 Act to bring in a range of improvements	Coming into force during 2008 following consultation on the general development procedure order and the general permitted development order
4	Enforcement	Introduce provisions for temporary stop notices to stop unauthorized development, fixed penalties for a range of breaches of planning control and enforcement charters	Coming into force in 2008 following consultation on draft regulations
5	Trees	Updates and amends the provision in the 1997 Act relating to the protection of trees through Tree Preservation Orders	Coming into force in 2008 after consultation on draft regulations
6	Correction of Errors	Allows the correction of errors in official decision letters	Coming into force in 2008
7	Assessment	Introduces new powers for Scottish Ministers to assess the performance of planning authorities carrying out their planning function	Likely to be in force in 2008
8	Financial Provisions	Amends existing provisions in the 1997 Act relating to fees and charges – enables grants to be made for advice and assistance in the planning system	Likely to be in force in 2008
9	Business Improvement Districts	Introduces provisions to allow local businesses to invest collectively in improvements to an area in which they operate	In force from April 2007
10	Miscellaneous and General	Contains provisions for National Scenic Areas which designate areas of outstanding landscape and cultural heritage, and explicit equal opportunities duty on Scottish Ministers and Planning Authorities. Repeals existing legislation and commencements of this act	Coming into force in 2008

4.6 What are Development Plans?

Development Plans are the ground rules on which all planning application decisions can be made. They contain policies on development and the use of land; they cover a wide range of issues relating to employment, recreation, housing, transport and the conservation of the environment and countryside.

Development Plans are made up of two parts – The Structure Plan and the Local Plan.

- **The Structure Plan** – is a long-term view about developing looking in general terms at the broad scale and location of future development. These Structure Plans are frequently put together in conjunction with neighbouring councils and always in consultation with the public. The Structure Plan is submitted to the Scottish Ministers for approval
- **The Local Plan** – sets out more detailed policies and proposals for a smaller area in order to guide development. These Local Plans are again done in consultation with the public and any other neighbouring councils or with interested parties. Local Plans must by their very nature be in line with the Structure Plan.

4.7 Who requires planning permission?

Planning permission is required for any 'development'. This covers a wide range of building and engineering work as well as changes to the use of land or a building. It is always advisable to discuss any planned development with your local Planning Officer before to commencing work to establish whether your development work will require permission or is deemed permitted development that does not need permission.

Note: As the rules covering permitted development are complicated we have not attempted to try and define what is or is not permitted development as we would invariably be inaccurate for some region or area!

4.8 Do I really need planning permission?

Most alterations and extensions to property and changes of use of land need to have some form of planning permission, which is achieved by submitting a planning application to the local authority. The purpose of this control is to protect and enhance our surroundings, to preserve important buildings and natural areas and strengthen the local economy.

However, not all extensions and alterations to dwelling houses require planning permission. Certain types of development are permitted without the need to make an official request and so it is always wise to contact the local authority before commencing any work.

Whether or not planning permission is required, good design is always important. Extensions and alterations should be in scale and in harmony with the remainder of the house. The builder should ensure that details such as window openings and matching materials are taken into account.

Householders are encouraged (by councils) to employ a skilled designer when preparing plans for extensions and alterations. Alternatively, the authority's

Planning Officers are able to offer general design guidance prior to the submission of your scheme.

Chapter 5 covers most aspects of building development for a property and indicates where planning permission is required and whether Building Regulations have to be adhered to. In **all** circumstances it is recommended that you first talk to your local Planning Officer before contemplating any work. The cost of a local phone call could save you a lot of money (and stress) in the long term!

4.9 Do I have to notify my neighbours of my intentions?

Yes – as part of the planning permission procedure, you are required to notify your neighbours. You are required to complete, sign and date a 'neighbour notification certificate' at the time of making your planning application. This is not something to be treated lightly or as an afterthought; in any case, getting your neighbours on board early will save potential hassles later!

For the purposes of planning applications there is a definition of which neighbours you should contact. For example, it may not be everyone in your street, but those closest to you, sharing boundaries with you, etc., and those who will be directly affected by the proposed extension/development. Your council will be able to offer you advice as to whom you must include.

When notifying your neighbours you must provide them with a Notice to Neighbour Form (available from your council), a copy of the location plan (showing the site of the proposed development) and a copy of the guidance notes in order that they know how they can contribute to the planning process. All the forms and guidance notes will be available either from your local council or via their website.

Your neighbours have a right to comment on the application within 14 days of the council receiving it.

 Note: It is important that the information you provide your neighbours is exactly the same as that given on the application form.

4.10 Will the council advertise my planning application?

Some (but not all) planning applications will be advertised in the local newspapers; this is up to the local council to decide on. However, all applications are available for viewing via your local council offices and a weekly list of applications is published, normally via the council's website.

4.11 Where can I find out about local planning applications?

Planning applications can be viewed at your council offices. Everyone has a right to view planning matters that affect them in order that their views are taken into consideration when the council is reaching its decision. Community councils receive a list of planning applications every week and have the right to ask to be formally consulted on some of them – usually those which have a 'genuine' potential impact on community interests.

4.12 Do the council always involve the public in all planning matters?

Councils must involve the public in the planning system; this is set out by law, and must involve them in two ways: in the preparation of Development Plans (see section 4.6 *What are Development Plans?*) available to be viewed at your local council offices, and in commenting on planning applications.

Councils must provide opportunities for the public to be involved in the development of structure and local plans. It is up to the specific council to decide the ways in which to involve the public, but they will, in all instances, advertise how and when you can be involved. Any objections are usually settled by negotiation with all involved parties, but if this is not possible opinions and objections may be considered in a 'local plan inquiry' before the council makes it final decisions.

Although you are only obliged to notify your neighbours prior to seeking planning permission, **everyone** has the right to comment on the proposed development plan that is likely to affect them; this does not only mean neighbours but the wider community as well. Your local council will notify the public how to make comment.

4.13 Why is planning permission necessary?

Planning permission is handled as part of the wider planning control process. This is administered by your local authority and the system 'exists to control the development and use of land and buildings for the best interests of the community'.

The process is intended to make the environment better for everyone and acts as a service to manage the types of constructions, modifications of premises and uses of land, and ensures the right mix of premises in any one vicinity (that individuals may plan to make) is maintained. The key feature of the process is to allow a party to propose a plan and for other parties to object if they wish to, or are qualified to.

4.14 How should I set about gaining planning permission?

If you are in the planning stages for your work and you know planning permission will be required, it is wise to get the plans passed before you go to any expense or make any decisions that you may find hard to reverse – such as signing a contract for work. If your plans are rejected, you will still have to pay your architect or whoever prepared your plans for submission but you won't have to pay any penalty clauses to the building contractor.

It is always best to submit an application in the early stages – if you try to be clever by submitting plans at the last minute (in the hope that neighbours will not have time to react) then you could be in for an expensive mistake! It's much better to do things properly and up-front.

An architect (surveyor or general contractor) can be asked to prepare and submit your plans on your behalf if you like, but as the owner and person requiring the development, it will be your name that goes on the application, even if all the correspondence goes between your architect and the planning department.

You don't have to own the land to make a planning application, but you will need to disclose your interest in the property. This might happen if you plan to buy land, with the intention of developing it, subject to planning approval. It would, therefore, be in your best interest to obtain the consent before the purchase proceeds.

To submit your application you will need to use the official forms, available from the local authority planning department. It's a good idea to collect these personally, as you may get the opportunity to talk through your ideas with a Planning Officer and in doing so probably get some useful feedback. You will also need to include detailed plans of the present and proposed layout as well as the property's position in relation to other properties and roads or other features.

New work requires details of materials used, dimensions and all related installations, similar to those required for Building Regulations.

4.15 What types of planning permission are available?

There are three types of planning permission available: outline, reserved and full, plus some concessionary fees and exemptions.

4.15.1 Outline

This is an application for a development 'in principle' without giving too much detail on the actual building or construction. It basically lets you know, in advance, whether the development is likely to be approved. Assuming permission is granted under these circumstances, you will then have to submit a further application in greater detail. In the main, this 'application in principle' only really applies to large-scale developments and you will probably be better off making a full application in the first place.

4.15.2 Reserved matters

This is the follow-up stage to an outline application to give more substance and more detail.

4.15.3 Full planning permission

This is the most widely used type of application and is for erection or alteration of buildings or changes of use. There are no preliminary or outline stages and when consent is granted it is for a specific period of time. If this is due to lapse, a renewal of limited permission can be applied for.

4.16 Are there any concessionary fees and exemptions?

There are some instances when no fee is payable, for example works to improve a disabled person's access to a public building or dwelling house, or a fresh application for development of the same character or description within 12 months of receiving permission. Your local council will able to provide you with a full list of exemptions and concessions.

4.17 How do I apply for planning permission?

In order to apply for planning permission you will need to complete the relevant planning application forms for your local area. The easiest way to find exactly what forms are required for your local area will be to go onto your local council website or contact them directly.

Most (if not all) councils have a website and whilst every council website is designed differently, from our research, we have found that most of these are very user-friendly sites and many have 'planning questionnaires' which clearly take you through the process and help you to decide whether you need planning permission and what forms to complete.

A planning application must include:

- correctly and fully completed forms dated and signed;
- copies of all plans (probably four, but check with your local council to confirm their requirement);
- correctly and fully completed certificates dated and signed;

- the correct planning fee;
- confirmation that you have notified all your neighbours.

Many people employ a professional to help them with planning permission as they have the knowledge and skills to send in the correct applications and drawings, and can negotiate any issues raised by the planning authority.

There are fees to pay for each application for planning permission and your local planning office can provide you with the relevant details. Table 4.1 gives an indication of the fees payable; however, it is advisable to double check as local variances may occur. You must make sure you have paid the correct fee – as permission can be refused if there is a discrepancy on fees paid.

4.18 What information do I have to provide?

Your local council will be able to provide you with all the relevant forms you need to complete – some have an official form to complete, others merely provide a list of the details you are required to provide but do not issue a specific form.

At the very least you will be required to provide the following information:

- Name and address.
- Agents name and address (if relevant).
- Address and location of the proposed development.
- Description of the proposed development.
- Proposed external materials.
- Proposed water and drainage supply.
- Certificate of site ownership.
- Certificate of Agricultural Holding (if relevant).
- Certificate of Neighbour Notification.
- Plans.
- Relevant fees.

Note: Every council will be able to provide you with guidance notes on how to fill in the relevant forms and what exactly you need to provide in your particular case.

It is in your own interest to provide plans of good quality and clarity and so it is probably advisable to get help from an architect, surveyor or similarly qualified person to prepare the plans and carry out the necessary technical work for you. You can obtain the necessary application forms from the planning department of your local council and you will find that this is laid out simply, with guidance notes to help you fill it in. Alternatively, you can ask a builder or architect to make the application on your behalf. This is sensible if the development you are planning is in any way complicated, because you will have to include measured drawings with the application form.

4.19 Applying online on your council's website

Most councils now have very user-friendly websites and many will accept planning applications electronically; **however**, the same amount of information and detail will still need to be provided, so this isn't a shortcut. It is advisable to check with your local council website to see if this is something that will be possible in your area.

Note: Don't forget to sign and date all the relevant forms as any incomplete or incorrect information will slow the whole process up and mean the council will not be able to proceed with your application.

4.20 What are Certificates of Site Ownership?

In order to complete your planning application you will need to complete Certificates of Site Ownership. You do not need to have a legal interest in the land to apply for planning permission, nor do you need to get consent from the owner; however, you are obliged to give notice to the owner of the land or agricultural tenant prior to making the planning application.

There are two types of certificate:

Certificate A   this is used if you own all the land to which the application applies

Certificate B   this is used if you are not the owner of the land, in which case you are legally required to notify the owners by issuing them with a completed copy of the Notice to Owners form (available from your council). Once you have served this notice you should list the names and addresses of the owners notified, along with the date on which notice was served

Note: If you do not know who owns the land you are required to place an advertisement in the local newspaper.

4.21 What are Certificates of Agricultural Holding?

As with Certificates of Site Ownership above, if there is an agricultural tenant on the land you are required to notify them of intended planning application. Once you have served this notice you must list the names and address of tenants together with the date on which notice was served along with your application. Again, if you don't know who the tenants are you will need to place an advertisement in the local newspaper.

4.22 What sort of plans will I have to submit?

Dependent on your local council the exact requirements for plans may vary; however, they are likely to require of some or all the following:

4.22.1 Location plan

The location plan indicates the development location and its relationship to neighbouring property and roads, etc. Minimum scale is 1:2500 (or 1:1250 in a built-up area). The land to which the application refers must be outlined in red ink. Adjacent land, if owned by the applicant, shall be outlined in blue ink.

4.22.2 Site layout plan

Using a scale of 1:200 the boundaries of the site should be outlined in red, any other land you own outlined in blue and the plan should also indicate access, buildings, trees and walls.

4.22.3 Neighbouring properties plan

At a suitable scale (probably 1:500), a plan will need to be provided to indicate all neighbouring properties that have been notified of the proposed development.

4.22.4 Block plan

For a new building a block plan is likely to be needed. This will give show all existing buildings, trees, walls and access(es) on the site, where appropriate, along with the full extent of the proposed development (including buildings to be erected, altered or demolished, new or altered access, etc.).

4.22.5 Building plans

The exact content of these building plans will depend on proposed building works but the plans should clearly show existing and proposed elevations and floor plans drawn to a scale of 1:50 or 1:100. These plans are normally very thorough and include types of material, colour and texture, the layers of foundations, floor constructions and roof constructions, etc.

In relation to applications for replacement windows, photographs of the existing windows to be replaced **will** be required.

4.22.6 Ordnance Survey map

If you live in a rural area you will also need to provide an Ordnance Survey map scale 1:10,000 of the local area for identification purposes.

Note: All dimensions on plans must be metric.

4.23 What important areas should I take into consideration?

The following are some of the most important areas that should be considered before you submit a planning application.

4.23.1 Advertisement applications

If your proposal is to display an advertisement, you will need to make a separate application on a special set of forms. Three copies of the forms and the relevant drawings must be supplied. These must include a location plan and sufficient detail to show the dimensions, materials and colour of the sign and its position. No certificate of ownership is required, **but** it is illegal to display signs on the property without the consent of the owner.

4.23.2 Conservation Area consent

Development within Conservation Areas is dealt with under the normal planning application process, except where the proposal involves some form of demolition.

If you live in a Conservation Area, you will need Conservation Area consent to do the following:

- substantial demolition of a building (there are a few exceptions and further information will be available from your council, for example if demolition would be considered partial rather than substantial then a planning application may be required instead);
- demolish a gate, fence, wall or railing or tree felling – again if demotion is only partial planning permission may be necessary rather than Conservation Area consent. Seek advice from your local council.

4.23.3 Listed building consent

You will need to apply for listed building consent if either of the following cases apply:

- you want to demolish a listed building;
- you want to alter or extend a listed building in a manner which would affect its character as a building of special architectural or historic interest.

You may also need listed building consent for any works to separate buildings within the grounds of a listed building. Check the position carefully with the council – it is a criminal offence to carry out work which needs listed building consent without obtaining it beforehand.

4.23.4 Trees

Many trees are protected by Tree Preservation Orders (TPOs), which mean that, in general, you need the council's consent to prune or fell them. In addition, there are controls over many other trees in Conservation Areas. In order to cut, lop, top or uproot a tree you will need to submit a Tree Work Notice application and await consent from your council prior to commencing work. Six weeks, notice should be given.

4.24 What is the planning permission process?

If you think you might need to apply for planning permission, then this is the process to follow (Figure 4.1):

Step 1
Contact the planning department of your council. Tell the planning staff what you want to do and ask for their advice.

Step 2
If they think you need to apply for planning permission, ask them for an application form. They will tell you how many copies of the form you will need to send back and how much the application fee will be. Ask if they foresee any difficulties which could be overcome by amending your proposal. It can save time or trouble later if the proposals you want to carry out also reflect what the council would like to see. The planning department will also be able to tell you if Building Regulations approval will also be required.

Step 3
Decide what type of application you need to make. In most cases this will be a full application but there are a few circumstances when you may want to make an outline application – for example, if you want to see what the council thinks of the building work you intend to carry out before you go to the trouble of making detailed drawings (but you will still need to submit details at a later stage).

Step 4
Send the completed application forms and supporting documents to your council, together with the correct fee. Each form must be accompanied by a plan of the site and a copy of the drawings showing the work you propose to carry out. (The council will advise you on what drawings are needed.)

Step 5
Within 3 days of receipt of your planning application your local planning department will acknowledge receipt of your application by letter (telling you who your Planning Officer will be), and publicly announce it – via letters to the neighbourhood parish council and anyone directly affected by the proposal;

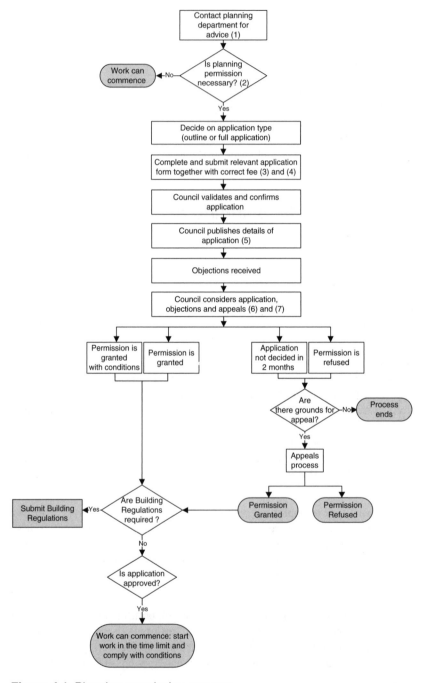

Figure 4.1 Planning permission process

publishing details of the application in the local press (if appropriate); notifying your neighbours and/or putting up a notice on or near the site. The council may also consult other organizations, such as the highway authority or the parish council, Heritage Scotland, etc.

A copy of the application will also be placed on the planning register at the council offices so that it can be inspected by any interested member of the public. Anyone can object to the proposal, but there is a limited period of time (usually 2 weeks) in which to do this and they must specify the grounds for objection.

Step 6

The Planning Office then has 2 months in which to process your application. Some applications are more difficult to process than others and need more time to consider. When this happens the Planning Office will contact you to agree with you an extended time period. Should you not agree to the extended time period you can appeal to the Scottish Executive on the grounds that the application has not been decided in the 2 months.

In order to progress your application it may be necessary for the Planning Officer to consult other services of the council or other agencies. Should they need to do this, this will commence within 7 working days of registering your application, and you will be notified in writing as to the progress as soon as possible. It is normal for these other departments or agencies to respond within 14 days; however, dependent on workload and complexity of you application it may take longer, but you will be kept informed at all times.

During this 2 month period the Planning Officer will conduct a site visit, consult law and guidance, look at local Development Plans, take into account letters of objection or support, etc.

Step 7

Should the Planning Officer believe that amendments to the plan will result in approval of your application they will contact you within 2 months of the date of registration.

Step 8

Your planning application will either be approved directly with a Planning Officer (in fact 75% of all planning applications are decided in this way) or may need to be heard at an Area Committee meeting. Should your application need to be heard at an Area Committee Meeting you will be advised of the date your application will be heard.

When your application is decided it will be sent to you within 3 working days of the decision being taken.

How long will the council take to decide?

It is the policy of Scottish Ministers to ensure that 80% of decisions are reached within 2 months. The council may choose to:

- grant planning permission without conditions;
- grant planning permission with conditions;
- refuse permission.

If the council cannot make a decision within 8 weeks then it must obtain your written consent to extend the period. If it has not done so, you can appeal to the Scottish Ministers. But appeals can take several months to decide and it may be quicker to reach agreement with the council.

Note: Should you have to reapply a second application is normally exempt from a fee.

4.25 How do they reach their decision?

The Planning Officer dealing with your application is responsible for gathering all the relevant information and compiling a case. He will conduct a site visit (it isn't normally necessary for you to be in attendance), consult law and guidance, check whether this application is in line with local Development Plans, get the views of any other council departments or agencies as necessary, take into account the views of locals in support or objecting to the application and consider all these things against precedent and the character of the local area.

Most cases will be decided by the Planning Officer directly but should this not be possible then your Planning Officer will write a report and make recommendations which will then be considered by the Area Committee.

4.26 What are the main objections against granting planning permission?

The following are some of the main objection areas that your application may meet.

4.26.1 The property is a listed building

Listed buildings are protected for their special architectural or historical value and a Listed Building Consent may be needed for alterations, but grants could be available towards repair and restoration!

If it is a listed building, it probably has some historic interest or special architectural interest and will have been listed by Historic Scotland on behalf of the Scottish Ministers. This could apply to houses, factories, warehouses and even walls or gateways. Most alterations which affect the external appearance or design will require listed building consent in addition to other planning consents.

If you think there is a chance that your property is listed you would be wise to contact your local council or Historic Scotland direct to check out the status of your property.

Note: Your property does not have to be in a conservation area to be listed so it is always worth double checking to be sure.

4.26.2 The property is in a conservation area

This is an area defined by the local authority, which is subject to special restrictions in order to maintain the character and appearance of that area. Again, other planning consents may be needed for areas designated as green belt, areas of outstanding natural beauty, national parks or sites of specific scientific interest. To find out if your property is in a conservation area you can either contact your local council or Historic Scotland.

4.26.3 The application does not comply with the local Development Plan

Local authorities are obliged to publish a development plan (which you should be able to view at your local council offices), which sets out policies and aims for future development in certain areas. These are to maintain specific environmental standards and can include very detailed requirements such as minimum or maximum dimensions of plot sizes, number of dwellings per acre, height and style of dwellings, etc. It is important to check what plans exist for your area, as proposals can meet with some fierce objections from residents protecting their environment.

4.26.4 The property is subject to a covenant

This is an agreement between the original owners of the land and the persons who acquired it for development. They were implemented to safeguard residential standards and can include things like the size of outbuildings, banning use of front gardens for parking cars, or even just the colours of exterior paintwork.

4.26.5 There is existing planning permission

A previous resident or owner may have applied for planning permission, which may not have expired yet. This could save time and expense if a new application can be avoided.

4.26.6 The proposal infringes a right of way

If your proposed development would obstruct a public right of way that crosses your property, you should discuss the proposals with the council at an early stage. The granting of planning permission will **not** give you the right to interfere with, obstruct or move the path. A path cannot be legally diverted or closed unless the council has made an order to divert or close it to allow the development to go ahead. The order must be advertised and anyone may object. You must not obstruct the path until any objections have been considered and the order has been confirmed. You should bear in mind that confirmation is not automatic; for example, an alternative line for the path may be proposed, but not accepted.

 If you are considering a planning application, you should consider all of the above questions. Normally your retained expert – architect, surveyor or builder – can advise and help you to get an application passed. Most information can be collected from your local planning department, or if you need to find out about covenants, look for the appropriate Land Registry entry.

4.27 What matters cannot be taken into account?

- Competition.
- Disturbance from construction work.
- Loss of property value.
- Loss of view.
- Matters controlled under other legislation such as Building Regulations (e.g. structural stability, drainage, fire precautions).
- Moral issues.
- Need for development.
- Private issues between neighbours (e.g. land and boundary disputes, damage to property, private rights of way, deeds, covenants).
- Sunday trading.
- The identity or personal characteristics of the applicant.

4.28 What are the most common stumbling blocks?

In no particular order of priority, these are:

- Adequacy of parking.
- Archaeology.
- Design, appearance and materials.
- Effect on listed building or Conservation Area.
- Government advice.
- Ground contamination.
- Hazardous materials.
- Landscaping.
- Light pollution.
- Local planning policies.
- Nature conservation.

- Noise and disturbance from the use (but not from construction work).
- Overlooking and loss of privacy.
- Previous planning decisions.
- Previous appeal decisions.
- Road access.
- Size, layout and density of buildings.
- The effect on the street or area (but not loss of private view).
- Traffic generation and overall highway safety.

4.29 What if the council refuses the planning application?

If the council refuses your planning application they **must** give clear reasons for doing so. You can appeal to the Inquiry Reporter's Unit of the Scottish Executive.

 Appeals must be made within 6 months of the local council's decision.

4.29.1 Who makes the final decision on the appeal?

The vast majority of appeals (some 98%) are given to a 'Reporter' delegated from the Inquiry Reports Unit and most appeals are dealt with under the written submissions procedure. In most instances decisions are issued by the Reporter without reference to the Minister as this is felt to be an efficient way of handing any appeal that does not raise issues of national importance. Should the issue raise concerns of national importance then the Reporter may 'recall' the issue to be determined by the Scottish Ministers. An appeal (where there is a substantial volume of objections) would not in itself give sufficient grounds for a recall, but each case is assessed individually. The designated Reporter will make a recommendation but the final decision lies with the Scottish Ministers or in some cases with officials of the Planning Divisions.

4.29.2 What if I am not happy with this final decision?

Any party aggrieved by a decision relating to an application, an appeal or recall, made by the Scottish Ministers may challenge the decision in the Court of Session. If this happens the application or appeal will be referred back to the Scottish Ministers to be determined afresh!

4.29.3 How long have I got to start work?

Once granted, planning permission is valid for 5 years.

 If the work is not begun within that time, you will have to apply for planning permission again.

4.30 What should I think about before I start work?

There are many kinds of alterations and additions to houses and other buildings which do not require planning permission. Whether or not you need to apply, you should think about the following before you start work.

4.30.1 What about design?

Everybody's taste varies and different styles will suit different types of property. Nevertheless, a well-designed building or extension is likely to be much more attractive to you and to your neighbours. It is also likely to add more value to your house when you sell it. It is, therefore, worth thinking carefully about how your property will look after the work is finished.

Extensions often look better if they use the same materials and are in a similar style to the buildings that are there already – but good design is impossible to define and there may be many ways of producing a good result. In some areas, the council's planning department issues design guides or other advisory leaflets that may help you.

4.30.2 What about crime prevention?

You may feel that your home is secure against burglary and you may already have taken some precautions such as installing security locks to windows. However, alterations and additions to your house may make you more vulnerable to crime than you realize. For example, an extension with a flat roof, or a new porch, could give access to upstairs windows which previously did not require a lock. Similarly, a new window next to a drainpipe could give access. Ensure that all windows are secure. Also, your alarm may need to be extended to cover any extra rooms or a new garage. The crime prevention officer at your local police station can provide helpful advice on ways of reducing the risk.

4.30.3 What about lighting?

If you are planning to install external lighting for security and/or other purposes, you should ensure that the intensity and direction of light does not disturb others. Many people suffer extreme disturbance due to excessive or poorly designed lighting. Always ensure that beams are **not** pointed directly at windows of other houses. Security lights fitted with passive infra-red detectors (PIRs) and/or timing devices should be adjusted so that they minimize nuisance to neighbours and are set so that they are not triggered by traffic or pedestrians passing outside your property.

4.30.4 What about covenants?

Covenants or other restrictions in the title to your property or conditions in the lease may require you to get someone else's agreement before carrying out some kinds of work to your property. This may be the case even if you do not need to apply for planning permission. You can check this yourself or consult a lawyer.

You will probably need to use the professional services of an architect or surveyor when planning a loft conversion. Their service should include considerations of planning control rules.

4.30.5 What about listed buildings?

Buildings are listed because they are considered to be of special architectural or historic interest and as a result require special protection. Historic Scotland, the body appointed by the Scottish Ministers, is responsible for overseeing the listing of buildings for the Scottish Ministers.

The term 'building' is defined very broadly and can include many structures both large and small, and can for example comprise walls, railings, sundials, statues, telephone boxes, bridges, dams, churches, castles, factories, farms and houses.

In principle buildings will be listed if they were erected before 1840 and are substantially intact. Later buildings are selected on the basis of individual character and quality. In assessing which buildings to list, Historic Scotland considers a range of values, including individual merit of the building, its role in the street or landscape, work of an important architect, connection to major historical events or figures, etc.

Listings can apply to whole buildings, or structures at that address. For example, a country house might be listed, but also other walls, outhouses, gates, etc., may also be listed. This is known as the 'curtilage' of a listing.

Whilst all listed buildings receive equal protection there are three categories of listing to reflect the degree of interest.

Category A Buildings of national or international importance, either architecturally or historically, or which are very fine, little altered examples of a period/style of building.

Category B Buildings of regional or more than local importance or which are major examples of a particular period/style of building.

Category C Buildings of lesser importance, lesser examples of a period or style, simple traditional buildings which group well with others in categories A and B or are part of a planned group, e.g. an estate or industrial complex.

At present there are over 47,000 listed buildings in Scotland, although this is constantly being updated.

You can find out if your building (or a building your are interested in) is listed by going on to the Historic Scotland website at www.historic-scotland.gov.uk which is a great source of information; All enquiries relating to listed buildings can be put to Historic Scotland Inspectorate, Listings Team, Room 2.20, Longmore House, Salisbury Place, Edinburgh EH9 1SH.

What are the owner's responsibilities for a listed building?

If you are the owner of a listed building or come into possession of one, you are tasked with ensuring that the property is maintained in a reasonable state of repair. The council may take legal action against you if they have cause to believe that you are deliberately neglecting the property, or have carried out works without consent. Enforcement action may be instigated.

There is no statutory duty to effect improvements, but you must not cause the building to fall into any worse state than it was in when you became its owner. This may necessitate some works, even if they are just to keep the building wind and watertight. However, you may need listed building consent in order to carry these works out!

A photographic record of the property when it came into your possession may be a useful asset, although you may also have inherited incomplete or unimplemented works from your predecessor, which you will become liable for!

If you are selling a listed building you may wish to indemnify yourself against future claims: speak to your solicitor.

4.30.6 What about conservation areas?

Tighter regulations apply to developments in conservation areas and to developments affecting listed buildings. Separate conservation area consent and/or listed building consent may be needed in addition to planning consent and Building Regulation consent.

Conservation areas are 'areas of special architectural or historic interest the character and appearance of which it is desirable to preserve or enhance'. As the title indicates these designations cover more than just a building or property curtilage and most local authorities have designated conservation areas within their boundary. Although councils are not required to keep any statutory lists, you can usually identify conservation areas from a Local Plan's 'proposals maps' and appendices. Some councils may keep separate records or even produce leaflets for individual areas.

The purpose of designating a conservation area is to provide the council with an additional measure of control over an area that they consider being of special historic or architectural value. This does not mean that development proposals cannot take place, or that works to your property will be automatically refused. It means, however, that the council will have regard to the effect of your proposals on the designation, in addition to their normal assessment. The council may also apply this additional tier of assessment to proposals that are outside the designated conservation area boundary, but which may potentially affect the character and appearance of the area. It should also be noted that just because you are in a conversation area it doesn't necessarily mean that you cannot build modern designed properties. However, the council will need to consider whether these fit in with the overall look of the area in terms of scale, form, alignment, materials and landscaping.

Notes

(1) Within a conservation area minor works such as erection of fences, gates and porches will probably require planning permission – these may be exempt from a fee, and it will be worth checking with your local council.
(2) If you live or work in a conservation area, grants may be available towards repairing and restoring your home or business premises.
(3) For major works you may need to involve an architect with experience of works affecting conservation areas.

What about trees in conservation areas?

Trees in conservation areas are generally treated in the same way as if they were protected by a Tree Preservation Order (TPO), i.e. it is necessary to obtain the council's approval for works to trees before they are carried out. There are certain exceptions (where a tree is dead or in a dangerous condition) but it is always advisable to seek the opinion of your council's tree officer to ensure your proposed works are acceptable. Even if you are certain that you do not need permission, notifying the council may save the embarrassment of an official visit if a neighbour contacts them to tell them what you are doing!

If you wish to lop, top or fell a tree within a conservation area you must give 6 weeks' notice, in writing, to the local authority. This is required in order that they can check to see if the tree is already covered by a TPO, or consider whether it is necessary to issue a TPO to control future works on that tree.

Contact your council's landscape or tree officer for further information.

4.30.7 What are Tree Preservation Orders?

Trees are possibly the biggest cause of upset in town and country planning and many neighbours fall out over tree-related issues. They may be too tall, may block out natural light, have overhanging branches, shed leaves on other property or the roots may cause damage to property. When purchasing a property the official searches carried out by your solicitor should reveal the presence of a TPO on the property or whether your property is within a conservation area within which trees are automatically protected.

However, not all trees are protected by the planning regulations system – but trees that have protection orders on them must not be touched unless specific approval is granted. Don't overlook the fact that a preservation order could have been put on a tree on your land before you bought it and is still enforceable.

Planning authorities have powers to protect trees by issuing a TPO and this makes it an offence to cut down, top, lop, uproot, wilfully damage or destroy any protected tree(s) without first having obtained permission from the local authority. All types of tree can be protected in this way, whether as single trees or as part of a woodland, copse or other grouping of trees.

 Protection does not, however extend to hedges, bushes or shrubs.

TPOs are recorded in the local land charges register which can be inspected at your council offices. The local authority regularly checks to see if trees on their list still exist and are in good condition. Civic societies and conservation groups also keep a close eye on trees. Before carrying out work affecting trees, therefore, you should check if the tree is subject to a TPO. If it **is**, you will need permission to carry out the work. A TPO will not necessarily prevent planning permission being granted for development. However, the council will take the presence of TPO trees into account when reaching their decision.

However, even with a preservation order it is possible to have a tree removed, if it is too decayed or dangerous, or if it stands in the way of a development, the local authority may consider its removal, but will normally want a similar tree put in or near its place.

 All trees in a conservation area are protected, even if they are not individually registered. If you intend to prune or alter a tree in any way you **must** give the local authority plenty of notice so they can make any necessary checks.

 If you have a tree on your property that is particularly desirable – either an uncommon species or a mature specimen – then you can request a preservation order for it. However, this will mean that in years to come you, and others, will be unable to lop it, remove branches or fell it unless you apply for permission.

What are my responsibilities with regard to TPOs?

Trees covered by TPOs remain the responsibility of the landowner, both in terms of any maintenance that may be required from time to time and for any damage they may cause. The council must formally approve any works to a TPO tree. If you cut down, uproot or wilfully damage a protected tree or carry out works such as lopping or topping which could be likely to seriously damage or destroy the tree then there are fines on summary conviction of up to £20,000, or, on indictment, the fines are unlimited. Other offences concerning protected trees could incur fines of up to £2,500.

What should I do if a protected tree needs lopping or topping?

Although there are certain circumstances in which permission to carry out works to a protected tree are not required, it is generally safe to say that you should always write to your council seeking their permission before undertaking any works. You should provide details of the trees on which you intend to do work, the nature of that work – such as lopping or topping – and the reasons why you think this is necessary. The advice of a qualified tree surgeon may also be helpful.

 You may, however, be required to plant a replacement tree if the protected tree is to be removed.

4.30.8 What about nature conservation issues?

Many traditional buildings, particularly farm buildings, provide valuable wildlife habitats for protected species such as barn owls and bats.

Planning permission will not normally be granted for conversion and reuse of buildings if protected species would be harmed. However, in many cases, careful attention to the timing and detail of building work can safeguard or re-create the habitat value of a particular building.

What about bats and their roosts?

Bats make up nearly one-quarter of mammalian species throughout the world. Some houses may hold roosts of bats or provide a refuge for other protected species. The Wildlife and Countryside Act 1981 and subsequent amending legislation gives special protection to **all** British bats because of their roosting requirements. Some parts of the Wildlife and Countryside Act 1981 have now been amended and in Scotland these are represented by Nature Conservation (Scotland) Act 2004. You would be advised to speak to your local advisor (e.g. remedial timber treatment, renovation, demolition and extensions) which is likely to disturb bats or their roosts. Scottish Natural Heritage (SNH) must then be allowed time to advise on how best to prevent inconvenience to both bats and householders. You can read more in 'Bats and People', which is available (free) by writing to Scottish Natural Heritage, Design & Publications, Battleby, Redgorton, Perth PH1 3EW or e-mailing pubs@snh.gov.uk

The stone barns and traditional buildings found in the UK have lots of potential bat roosting sites, the most likely places being gaps in stone rubble walls, under slates or within beam joints. These sites can be used throughout the year by varying numbers of bats, but could be particularly important for winter hibernation. As a result, the following points should be followed when considering or undertaking any work on a stone barn or similar building, particularly where bats are known to be in the area.

- A survey for the presence of bats should be carried out by a member of the local bat group (contact via SNH) before any work is done to a suitable barn during the summer bat breeding period.

- The pipistrelle (i.e. the smallest of the European bats) is often lurking in many strange places including in vases, under floorboards and between the panes of double glass.
- Any pointing of walls should not be carried out between mid-November and mid-April to avoid potentially entombing any bats. When walls are to be pointed, areas of the walls high up on all sides of the building should be left unpointed to preserve some potential roosting sites. If any bats are found whilst work is in progress, work should be stopped and your local SNH office contacted for advice on how to proceed.
- If any timber treatment is carried out, only chemicals safe for use in bat roosts should be used. A list of suitable chemicals is available from SNH on request. Any pre-treated timber used should have been treated using the CCA method (copper chrome arsenic) which is safe for bats.
- Work should not be commenced during the winter hibernation period (mid-November to mid-April). Any bats present during the winter are likely to be torpid (i.e. unable to wake up and fly away) and are therefore particularly vulnerable.

If these guidelines are followed, then the accidental loss of bat roosts and death or injury to bats will be reduced.

Although vampire bats feed primarily on domestic animals, they have been known to feed on sleeping humans on rare occasions! Vampire bats have chemicals in their saliva that prevent the blood they are drinking from clotting. They consume five teaspoons of blood each day. The vampire bat has been known to transmit rabies to livestock and to people.

Note: European Protected Species of Animals (of which bats are included) are highly legislated for and it is an offence to deliberately disturb such an animal:

- while it is occupying a structure or place which it uses for shelter and protection;
- while it is rearing or otherwise caring for its young;
- in a manner that is likely to significantly affect the local distribution or abundance of the species;
- in a manner that is likely to impair its ability to survive, breed or reproduce; to damage or destroy a breeding site or resting place of such an animal.

All this having been said it is still possible to remove bats which have become trapped in living spaces without the need to obtain a licence – but this is only if it is in the bats, own interest (for example to prevent the bat starving, dehydrating or becoming injured).

What about barn owls?

Barn owls also use barns and similar buildings as roosting sites in some areas. These are more obvious than bats and, therefore, perhaps easier to take into account. Barn owls are also fully protected by law and should not be disturbed during their breeding season. Special owl boxes can be incorporated into walls during building work, details of which can be obtained from Scottish Natural Heritage.

4.31 What could happen if you don't bother to obtain planning permission?

If you build something which needs planning permission without obtaining permission first, you may be forced to put things right later, which could prove troublesome and extremely costly. You might even have to remove an unauthorized building.

4.32 Who enforces planning control?

Your council is responsible for enforcement of all planning issues in your area. Your council has the power to issue notices showing what action is needed to correct a problem. If you are concerned that development work has taken place without permission you should and are actively encouraged to contact the planning department at your local council. Should a council be found not to have taken suitable action it can be reported to the Scottish Public Services Ombudsman.

Your council has statutory powers to investigate breaches; however, enforcement is discretionary, this is to say that the council has to consider if it is in the public interest to take enforcement action. It should be remembered that the purpose of planning enforcement is to resolve problems rather than punish mistakes, and any action has to be appropriate to the scale of the breach.

4.32.1 What are breaches of planning control?

Possible breaches of planning control can include:

- work being carried out without planning permission or consent;
- an unauthorized change of use;
- failure to comply with the conditions attached to a permission or consent;
- departure from approved plans or consent.

 A breach of planning control in itself is not a criminal offence; however, failure to comply with an Enforcement Notice is.

 It is a criminal offence to fail to act on a notice issued!

Priority will generally be given to breaches of a significant nature such as breaches of conditions for major development, irreversible damage to a listed building, authorized tree felling and matters relating to trees protected by Tree Preservation Orders and detrimental impact on amenity.

4.32.2 Why is enforcement needed?

Enforcement of planning rules helps to protect our countryside, our towns and us. It ensures that all development has been given the necessary permissions and that any development keeps in line with any conditions that may have been laid down at the time permissions were granted.

4.32.3 How do the public get involved?

The public are actively encouraged to report breaches in planning control. Should you suspect that a breach has been made, you are obliged to contact your council and provide the following information:

- the address of the property concerned;
- details of the suspected breach of planning control, with times and dates if relevant;
- your name, address and telephone number;
- an e-mail address if available;
- information on how the breach affects you;
- whether you wish for this information to be treated confidentially.

The council will always try, where possible, to treat all such information in strictest confidence unless it is not in the interest of the public to do so.

Note: Some complaints (such as neighbour disputes over boundaries) are not matters for the Planning Services Department; however, other departments within your council may be able to assist you.

4.32.4 Enforcement Notices

In a relatively small number of cases formal enforcement will be required. This starts with either an Enforcement or Breach of Condition Notice. Either of these notices will include the following information:

- a description of the breach of control;
- the steps that should be taken to remedy the breach;
- the timescales for taking these steps;
- the consequences of failure to comply;
- any rights of appeal, and how to lodge an appeal.

Appeals against enforcement are considered by Scottish Ministers.

Note: Anyone who has submitted information about a planning breach will be advised of any appeal.

Failure to comply with a notice may result in the planning authority taking further action; this can include: referring the case to the Procurator Fiscal for prosecution; carrying out work and charging the person for the costs; seeking a court interdict to stop or prevent a breach of the planning controls.

In order to do this your council has the powers to enter land to establish if a breach has taken place, check if you have complied with formal notices and check if a breach has been formally resolved.

 There is **no** right to appeal against a Breach of Condition notice.

4.32.5 Can an Enforcement Notice be issued against something that happened at some time in the past?

This will depend when the breach of planning occurred, but primarily:

There is a 4 year limit for any 'unauthorized operational development'.
Should there be unauthorized operational development (e.g. carrying out building, engineering, minor other operations and change of use to **a single dwelling house**) which occurred over 4 years ago, then the development becomes lawful and no enforcement action can be taken.

There is a 10 year limit for all other development.
This includes change of use (other than for a single dwelling house) and breaches of conditions.

After 10 years, the development becomes lawful unless no enforcement action has begun.

 Note: There is no time limit for breaches of Listed Building Control.

4.32.6 What types of notice are there?

Breach of Condition Notice

A Breach of Condition Notice is effective from the date it is served. This is used to enforce the conditions applied to any planning permissions. There is no right to appeal. Contravening the Breach of Condition Notice can result in prosecution and a fine of up to £1,000.

Enforcement Notice

This is generally used to deal with unauthorized development. An Enforcement Notice will specify the time period to take effect, the steps that must be taken to remedy the breach and the time for completion. There is a right to appeal, and the terms of the notice will be suspended subject to a decision being reached. Failure to comply with enforcement is an offence and may lead to a fine of up to £20,000. The Council may take action to correct the breach and you will be liable for the costs.

Listed Building Enforcement Notice

This must be served on the current owner, occupier or anyone else with an interest in the property. The notice will specify steps that must be taken to remedy the breach and the time limit for compliance. Failure to meet the requirements is an offence; what's more, it is a criminal offence to undertake unauthorized works to demolish, alter or extend a listed building and this can lead to **an unlimited fine or imprisonment**.

 There is a right to appeal against the notice.

Stop Notice

This is used in urgent and serious cases where activity must cease, particularly on grounds of public safety. When a Stop Notice is served the planning authority must also issue an Enforcement Notice which, as explained above, sets out the steps that must be taken to comply with planning regulations. There is no right to appeal against a Stop Notice, although of course you can appeal against the Enforcement Notice as above.

4.33 How much does a planning application cost?

A fee is required for the majority of planning applications and the council cannot deal with your application until the correct fee is paid. The fee is not refundable if your application is withdrawn or refused.

In most cases you will also be required to pay a fee when the work is commenced. These fees are dependent on the type of work that you intend to carry out. The fees outlined in Table 4.1 are typical of the charges made by local authorities during 2007.

 Note: Work to provide access and/or facilities for disabled people to existing dwellings is exempt from these fees.

4.34 Can I get any further assistance on planning issues?

If you are struggling with your planning applications or feel you need further assistance, there is an advice service called Planning Aid for Scotland which provides free professional advise to help you understand and engage more effectively with the planning system.

Table 4.1   Planning application fees

OUTLINE APPLICATIONS

(a) One dwelling house — £290

(b) Other types (except as specified below) — £290 for each 0.1 ha (or part thereof) of the site area **MAXIMUM** of £7,250

FULL APPLICATIONS

(a) Erection of dwellings — £290 for each dwellinghouse created subject to a **MAXIMUM** of £14,500

(b) Erection of buildings (other than dwellinghouses, extensions to dwellinghouses and buildings for agricultural purposes) —
i. £145 (No more than 40 m^2 of additional floorspace created)
ii. £290 (40 to 75 m^2 of additional floorspace created)
iii. £290 for each additional 75 m^2 (or part thereof) **MAXIMUM £14,500**
iv. £145 where no floorspace is to be created

(c) Enlargement, improvement or alteration to two or more dwellinghouses — £290

Householder Developments

(d) Alterations to an existing dwelling (enlargement, improvement, alteration) — £145

(e) The carrying out of operations including the erection of a building within the curtilage of an existing dwellinghouse for purposes ancillary to the enjoyment of the dwellinghouse — £145

(f) The erection or construction of gates, fences or walls or other means of enclosure along a boundary of the curtilage of an existing dwellinghouse — £145

(g) The construction of car parks, service roads and other means of access on land used for the purposes of a single undertaking, where the development is required for the purpose incidental to the existing use of the land — £145

(h) The erection on land used for the purposes of agriculture of those works, structures or buildings (excluded by virtue of Para 2(d) of Class 18 in Schedule 1 of the Town and Country Planning (General Permitted Development) (Scotland) Order 1992 from that class (other than buildings coming within category 2(j) below) —
(i)   NIL (less than 465 m^2 gross floor space)
(ii) £290 (465–540 m^2 gross floorspace)
(iii) where the area of gross floorspace to be created by the development exceeds 540 m^2, £290 for the first 540 m^2 and £290 for each 75 m^2 in excess of that figure subject to a **MAXIMUM** of £14,500

(Continued)

Table 4.1   (Continued)

(i) Plant and machinery (erection, alteration, replacement)	£290 per 0.1 ha (or part thereof) **MAXIMUM** £14,500
(j) Erection on land used for the purposes of agriculture, of glasshouses extended by virtue of paragraph 2(d) of Class 18 of the Town and Country Planning (General Permitted Development) (Scotland) Order 1992	(i) NIL (less than 465 m² gross floorspace) (ii) £1,675 (over 465 m² gross floorspace)
(k) The carrying out of any operations connected with exploratory drilling for oil or natural gas	£290 for each 0.1 ha of the site area subject to a **MAXIMUM** of £21,750
(l) Operations not within categories 1(a) and 1(b) and 2(a) to 2(k) above	In the case of operations for (i) the mining and working of minerals £145 for each 0.1 ha of the site area, subject to a **MAXIMUM** of £21,750 (ii) the mining and working of peat £145 for each 0.1 ha of the site area subject to a **MAXIMUM** of £2,175 (iii) any other purpose £145 for each 0.1 ha of the site area subject to a **MAXIMUM** of £1,450

RESERVED MATTERS APPLICATIONS

All reserved matters applications except as provided for below

	At the same rate as for full applications for planning permission
(a) Where earlier reserved matters application(s) have incurred fees not less than the fee which would have been payable in respect of the approval of all the matters reserved by the outline permission, in a single application	£290 per application
(b) Where an earlier reserved matters application has incurred fees at a rate lower than that prevailing at the date of the current application and subject to as above	£290 for the current application

NB. For a definitive explanation of these provisions please refer to the Statutory Instrument and Circular

ADVERTISEMENTS

All types	£145

CHANGES OF USE

(a) The change of use of a building to use as one or more separate dwellinghouses.	£290 for each additional dwellinghouse created **MAXIMUM** £14,500

(b) (i) The use of land for the disposal of refuse or waste material or for the deposit of material remaining after minerals have been extracted from the land

£145 for each 0.1 ha subject to a **MAXIMUM** £21,750

(ii) The use of land for the storage of minerals in the open

(c) The making of a material change in the use of a building or land, other than a material change of use within 5(a) and (b)

£290

CERTIFICATES OF LAWFUL USE OR DEVELOPMENT: (REGULATION 12)

Applications under Section 150 (1)(a) or (b) of the Act*

The amount payable is that in respect of an application for planning permission to institute the use or carry out the operations specified in the application

Applications under Section 150 (1)(c) of the Act*

£145

Applications under Section 151 (1) of the Act*

Half the amount that would normally have been payable under the main scale of fees

Applications under Section 151 (1)(a) where the use is one or more dwellinghouses

£290 per dwellinghouse to a **MAXIMUM** of £14,500

PRIOR APPROVAL (REGULATION 13)

Applications for determinations as to whether the prior approval of the Council will be required for development under Schedule 1 of the General Permitted Development (Scotland) Order 1992 (as amended)

£55

OTHERS

Fee payable to the City Council in respect of 'Bad Neighbour' developments under Section 34 of the Act*

£125

Application for planning permission for the development of land without complying with conditions

£145

Applications in respect of development already carried out

Fee as per the appropriate scale for the whole development

(a) Where the application had been carried out without permission

£145

(b) Other cases

This service is a volunteer-led service provided by fully qualified and experienced town planners based throughout Scotland who have a wide range of experience, knowledge and skills, which may be useful to you. They can give advice on a wide range of planning matters, ranging from renewable energy developments to the neighbour notification procedure, local and national planning policies. In addition they can help you prepare your statements and precognitions for Public Inquiries on both planning applications and Local Plans.

If you would like advice or information on any aspect of the planning system in Scotland go to the Planning Aid website http://www.planning-aid-scotland.org.uk and complete their user friendly enquiry form or call their helpline on 0845 603 7602.

Annex A: What are the government's restrictions on planning applications?

The main pieces of primary legislation in the field of town and country planning and their general content are:

The Town and Country Planning (Scotland) Act 1997
Defines the scope of town and country planning and sets out the general legislative framework for the preparation of structure and local plans and the administration of development control.

The Planning (Listed Buildings and Conservation Areas) (Scotland) Act 1997
Sets out the special controls in respect of buildings and areas of special architectural or historic interest.

The Planning (Hazardous Substances) (Scotland) Act 1997
Provides the statutory framework for regulating the storage of hazardous substances.

The Planning (Consequential Provisions) (Scotland) Act 1997
Deals with the changes to other legislation as a result of the consolidation of the main planning Act, e.g. the replacement of references to sections of the 1972 Act with the appropriate sections of the 1997 Act.

The main instruments of secondary planning legislation and their general content are:

The Town and Country Planning (Structure and Local Plans) (Scotland) Regulations 1983
Outline the procedures for the preparation, submission and approval or adoption of structure and local plans, and their alteration, repeal and replacement. SI 1983/1590.

The Town and Country Planning (Use Classes) (Scotland) Order 1997
Sets out 11 classes within which it is possible to change the use of land and buildings without the need to obtain planning permission. SI 1997/3061 amended by SI 1998/1196 and SSI 1999/1.

The Town and Country Planning (General Permitted Development) (Scotland) Order 1992
Grants a general planning permission for a wide range of minor developments and sets out the procedures for withdrawing these permitted development rights. SI 1992/223 amended by SI 1992/1078 and 2084, 1993/1036, 1994/1442, 2586, 2716 and 3294, 1996/252, 1266 and 3023, 1997/1871 and 3060 and 1998/1226.

The Town and Country Planning (General Development Procedure) (Scotland) Order 1992
Sets out the procedure to be followed in making and deciding planning applications, appeals and Certificates of Lawful Use or Development. There are separate regulations on procedures to be followed in appeal cases and for the conduct of public inquiries. SI 1992/224 amended by SI 1992/2083, 1993/1039, 1994/2585 and 3293, 1996/467 and 1997/749.

The Town and Country Planning (Simplified Planning Zones) (Scotland) Regulations 1987
Set out the procedures for making and altering simplified planning zones. SI 1987/1532.

The Environmental Impact Assessment (Scotland) Regulations 1999
Set out two scheduled categories of development:

- Planning applications for Schedule 1 projects (e.g. nuclear power stations) must be accompanied by an environmental statement.
- For Schedule 2 projects the planning authority will require an environmental statement if it believes the development is likely to have significant effects on the environment. SSI 1999/1.

The Town and Country Planning (Development by Planning Authorities) (Scotland) Regulations 1981
Set out the procedures which planning authorities must follow in respect of development which they propose to carry out in their area. SI 1981/829 amended by SI 1984/238.

The Planning (Hazardous Substances) (Scotland) Regulations 1993
Set out the substances and quantities to which hazardous substances controls apply as well as the procedures relating to obtaining hazardous substances consent, enforcement procedures and forms, etc. SI 1993/323 amended by SI 1996/252.

The Planning (Control of Major Accident Hazards) (Scotland) Regulations 2000
Amend the Planning system, particularly the hazardous substances consent regime, to implement the European Directive on the Control of Major Accident Hazards Involving Dangerous Substances. SSI 2000/179.

The Town and Country Planning (Enforcement of Control) (No. 2) (Scotland) Regulations 1992
Set out the procedures for the enforcement of planning control. SI 1992/2086.

The Town and Country Planning (Development Contrary to Development Plans) (Scotland) (No. 2) Direction 1994 Sets out the procedures which planning authorities must follow in dealing with applications for development contrary to Development Plans.

The Town and Country Planning (Notification of Applications) (Scotland) Direction 1997 Outlines the categories of development that should be notified to the Scottish Ministers in particular circumstances, the formal information to be submitted and the procedure to be followed.

The Town and Country Planning (Notification of Applications) (Scotland) Amendment Direction 1997 – Notification of Planning Applications Amends notification requirements regarding proposals for wind generators and introduces notification requirements for proposals affecting playing fields.

The Town and Country Planning (Notification of Applications) (Scotland) Amendment Direction 1998 – Notification of Planning Applications Changes the floor space requirements which trigger notification of retail proposals to the Scottish Ministers from 20,000 square meters to 10,000 square meters.

The Town and Country Planning (Notification of Applications) (Scotland) Amendment (No. 2) Direction 1998 Outlines the circumstances in which opencast coal and related mineral development proposals should be notified to the Scottish Ministers where the planning authority are minded to grant permission.

Other areas
As well as the legal requirement to make the planning register available for public inspection, councils will also allow the public to have access to all other relevant information such as letters of objection/support for an application or correspondence about considerations. Three clear days before any committee meeting, the file will normally be made available for public inspection and this file will remain available (i.e. for further public inspection) after the committee meeting. Although commercial confidentiality could well be a valid consideration, the council will not use it so as to prevent important information about materials and facilities also being available.

5

Requirements for Planning Permission and Building Regulations Approval

Before undertaking any building project, you must first obtain the approval of local government authorities. There are two main controls that districts rely on to ensure that adherence to the local plan is ensured, namely planning permission and Building Regulation approval.

Whilst both of these controls are associated with gaining planning permission, actually receiving planning permission does not automatically confer Building Regulation approval and vice versa.

 And don't forget that before you start any work you need to have a Building Warrant in place.

You **may** require **both** before you can proceed. Indeed, there may be variations in the planning requirements, and to some extent the Building Regulations, from one area of the country to another.

Provided, however, that the work you are completing does not affect the external appearance of the building, you are allowed to make certain changes to your home without having to apply to the local council for permission. These are called permitted development rights, but the majority of building works that you are likely to complete will still require you to have planning permission – so be warned!

The actual details of planning requirements are complex, but for most domestic developments, the planning authority is only really concerned with construction work such as an extension to the house or the provision of a new garage or new outbuildings that is being carried out. Structures like walls and fences also need to be considered because their height or siting might well infringe the rights of your neighbours and other members of the community. The planning authority will also want to approve any change of use, such as converting a house into flats or running a business from premises previously occupied as a dwelling only.

Planning consent **may** be needed for minor works such as television satellite dishes, dormer windows, construction of a new access, fences, walls and garden extensions. You are, therefore, advised to consult with development control staff before going ahead with such minor works (specific rules apply to listed, historic buildings etc.).

All new building work in Scotland, to which the Building Regulations apply, **must** have a building warrant prior to work commencing, except where the regulations permit. This current chapter attempts to highlight what permissions are needed for the most common alterations and extensions you may be planning to a building.

Note: There are a number of broad categories of buildings that are exempt from Building Regulations. Some examples of small buildings associated with houses, flats or maisonettes that are exempt are listed below:

(a) A detached single-storey building, with a floor area not more than $8\,m^2$, within the garden of a **house**, that:
  - is at least 1m from the house or is closer than 1m to the house, but is at least 1m from any boundary;
  - does not contain sleeping accommodation;
  - does not contain a flue: a fixed solid fuel, oil or gas appliance;
  - does not contain a sanitary fitting.

(b) A detached single-storey building, with a floor area not more than $8\,m^2$, within the garden of a **flat or maisonette** that:
  - is at least 1m from the flat or maisonette or 3 m from any other part of the building containing a flat or maisonette;
  - is at least 1m from any boundary;
  - does not contain a flue: fixed solid fuel, oil or gas appliance;
  - does not contain a sanitary fitting.

(c) A single-storey conservatory or porch with a floor area of not more than $8\,m^2$ that is attached to an existing **house**, and:
  - is at least 1m from a boundary;
  - does not contain a fixed solid fuel, oil or gas appliance;
  - does not contain a sanitary fitting;
  - meets the regulations on safety glazing.

(d) A single-storey greenhouse, carport or covered area, each with a floor area not more than $30\,m^2$, that is detached or attached to an existing **house** and:
  - does not contain a flue: fixed solid fuel, oil or gas appliance;
  - does not contain a sanitary fitting.

(e) A paved area or hardstanding not more than $200\,m^2$ in area that:
  - is not part of any access route required by the regulations.

A full list of exemptions is given in Appendix 3c in Chapter 3 of this book.

5.1 Decoration and repairs inside and outside a building

	Planning permission required	Building Warrant required
Do we need permission to carry out repairs inside and outside a building if the house is not altered?	No	Possibly
But what if it is a listed building?	Yes – consult your local authority	Yes – consult your local authority
What if the building is in a conservation area?	Yes – consult your local authority	Yes – consult your local authority
What if the alterations are major such as removing or part removing a load-bearing wall or altering the drainage system?	Yes	Yes

Generally speaking, you do **not** need to apply for planning permission:

- for repairs or maintenance;
- for minor improvements, such as painting your house or replacing windows;
- for internal alterations;
- for the insertion of windows, skylights or roof lights (but, if you want to create a new bay window, this will be treated as an extension of the house);
- for the installation of solar panels (which do not project significantly beyond the roof slope);
- to re-roof your house (but additions to the roof are treated as extensions to the house).

 Rules for listed buildings and houses in conservation areas may be different in some cases, however.

Occasionally, you may need to apply for planning permission for some of these works because your council has made direction withdrawing permitted development rights – if you are in any doubt it is always advisable to seek advice from your local authority.

Do I need a warrant to carry out repairs to my house, shop or office?
No – if the repairs are of a minor nature – e.g. replacing the felt to a flat roof, re-pointing brickwork or replacing floorboards.

Yes – if the repair work is major in nature – e.g. removing a substantial part of a wall and rebuilding it, or underpinning a building.

Do I need to apply for planning permission for internal decoration, repair and maintenance?
No.

Do I need to apply for planning permission for external decoration, repair and maintenance?
No – External work in most cases doesn't need permission, provided it does not make the building any larger.

Do I need approval to alter the position of a WC, bath, etc., within my house, shop or flat?
No – unless the work involves new or an extension of drainage or plumbing.

Do I need approval to alter in any way the construction of fireplaces, hearths or flues within my house, shop or flat?
Yes.

Do I need to apply for planning permission if my property is a listed building?
Yes.
If your property is a listed building, consent will probably be needed for **any** external work, especially if it will alter the visual appearance, or use alternative materials. It may be necessary to get consent for internal works, so in the case of listed buildings it is always best to ask first before you do anything significant. You also may need planning permission to alter, repair or maintain a gate, fence, wall or other means of enclosure.

 It may be possible to get a grant to help with the costs of high-quality repairs through the Historic Buildings Repair Grant Scheme – you would be wise to discuss this with your local authority or Historic Scotland.

Do I need to apply for planning permission if my property is in a conservation area?
Yes.
If the building undergoing repair or decoration is in a conservation area, or comes under any type of covenant restricting changes you will probably need planning permission. You may also be restricted to replacing items such as roof tiles with the approved material, colour and texture, and have to use cast iron guttering rather than plastic, etc.

5.2 Structural alterations inside

	Planning permission required	Building Warrant required
Do I need permission to make structural alterations inside the house, if the use of the house is not be altered?	No	Possibly – there is no need to obtain a warrant for simple works; however, you must meet Building Regulations
What if the alterations are major such as removing or part removing a load-bearing wall or altering a drainage system?	Yes	Yes
What if it is an office or a shop?	Yes	Yes

Do I need approval to make internal alterations within my house?
Yes.
If the alterations are to the structure such as the removal or part removal of a load-bearing wall, joist, beam or chimney breast, or would affect fire precautions of a structural nature either inside or outside your house. You also need approval if, in altering a house, work is necessary to the drainage system or to maintain the means of escape in case of fire.

Do I need approval to make internal alterations within my shop or office?
Yes.

Do I need approval to insert cavity wall insulation?
Yes.

Do I need approval to apply cladding?
Yes.
If you live in a conservation area, a national park or an area of outstanding natural beauty. You will also need to apply for planning permission before cladding the outside of your house with stone, tiles, artificial stone, plastic or timber.

If you are in any doubt about whether you need to apply for permission, you should contact your local authority planning department before commencing any work to your property.

5.3 Replacing windows and doors

	Planning permission required	Building Warrant required
Do I need permission to replace windows and doors?	No	No – however, work must meet the requirements of Building Regulations
What if they project beyond the foremost wall of the house facing the highway?	Yes	Probably – seek advice from your local council
My property is a listed building?	Yes	Probably – seek advice from your local council
My property is in a conservation area?	Yes	Probably – seek advice from your local council
Do I need permission if it is an office or shop?	Yes	Probably – seek advice from your local council

Do I need approval to install replacement windows in my house, shop or office?
No – provided:

- the window opening is not enlarged (if a larger opening is required, or if the existing frames are load-bearing, then a structural alteration will take place and approval is required);

- you do not remove those opening windows which are necessary as a means of escape in case of fire;
- the replacement of a window in an existing building is carried out by a competent person.

Although permission may not be needed, you will of course need to ensure that these items meet the standards in full.

 Replacement windows and doors in Scottish dwellings are expected to achieve U-values 10% lower than the rest of the UK. Soft coat Low E glass with a 16 mm cavity containing an inert gas will be necessary in most cases.

Do I need approval to replace my shop front?
Yes.
Anyone who installs replacement windows or doors has to comply with strict thermal performance standards and when a property is sold, the purchaser's surveyors will normally ask for evidence of this.

 Note: Further information is available from your local building control or from the Glass and Glazing Federation (GGF) website www.ggf.org.uk

5.4 Electrical work

	Planning permission required	Building Warrant required
Do I need permission to carry out electrical works?	No	Possibly – it will depend on the specific works. Work will need to comply with regulations set out in the Technical Handbooks

Do I need approval to replace electric wiring?
No – but:

- you must comply with Building Regulations, guidance on which is set out in the Technical Handbooks;
- you should meet the recommendations of the IET Wiring Regulations [i.e. BS 7671(latest edition)].

 Note: Your contract with the electricity supply company will have conditions about electrical safety which must not be broken. In particular, you should **not** interfere with the company's equipment which includes the cables to your consumer unit up to and including the separate isolator switch if provided.

Do I need approval to replace an existing electrical fitting?
No.
Non-notifiable work (such as replacing an electrical fitting) can be completed by a DIY enthusiast (family member or friend), but **still** needs to be installed

in accordance with the manufacturer's instructions and done in such a way that they do not present a safety hazard.

This work does *not* need to be notified to a local authority building control body (unless it is installed in an area of high risk such as a kitchen or a bathroom, etc.), *but* all DIY electrical work (unless completed by a qualified professional) will still need to be checked, certified and tested by a competent electrician.

Do I need approval to install a new electrical circuit?
Probably.

Any work that involves adding a new circuit to, in or around a dwelling will need to be either notified to the building control body (who will then inspect the work) or needs to be carried out by a competent person who is registered under a government approved self-certification scheme.

Work involving any of the following will also have to be notified:

- consumer unit replacements;
- electric floor or ceiling heating systems;
- extra-low voltage lighting installations (other than pre-assembled, CE-marked lighting sets);
- garden lighting and/or power installations;
- installation of a socket outlet on an external wall;
- installation of outdoor lighting and/or power installations in the garden or that involves crossing the garden;
- installation of new central heating control wiring;
- solar photovoltaic (PV) power supply systems;
- small-scale generators (such as micro-CHP units).

Note: Where a person who is **not** registered to self-certify intends to carry out the electrical installation, then a Building Warrant application will need to be submitted together with the appropriate fee, based on the estimated cost of the electrical installation. The building control body will then arrange to have the electrical installation inspected at first fix stage and tested upon completion.

5.5 Plumbing

	Planning permission required	Building Warrant required
Do I need permission to carry out replacement of existing plumbing?	No	No
What if I need to alter any internal or external drainage?	No	Probably – consult your local verifier
What about for unvented hot water systems?	No	Yes

Do I need approval to install hot water storage within my house, shop or flat?

Yes.

If the water heater is unvented (i.e. supplied directly from the mains without an open expansion tank and with no vent pipe to atmosphere) and has storage capacity greater than 15 litres.

5.6 Central heating

	Planning permission required	Building Warrant required
Do I need permission to alter central heating in my house?	No	No if electric
		Yes if gas, solid fuel or oil

Do I need approval to alter the position of a heating appliance within my house, shop or flat?

- *Gas*: Yes, unless the work is supervised by an approved installer under the Gas Safety (Installation and Use) Regulations 1984.
- *Solid fuel*: Yes.
- *Oil*: Yes.
- *Electric*: Yes, unless the work is carried out by a competent person.

Note: Hot water storage systems should meet the recommendations of BS7206:1990 or be the subject of an approval by a notified body.

The Home Energy Efficiency Scheme (Scotland) Regulations 2006 came into force on 1st January 2007. These regulations sets out the rules for the making of grants available for insulation and energy efficiency works and to install, repair or replace central heating systems. These Regulations extend the central heating programme to allow for the upgrading of partial or inefficient systems to persons entitled to claim the guarantee element of pension credit and they extend eligibility for insulation and energy efficiency measures to families with disabled children. As with all these things, the regulations are a minefield and if you think that you may be entitled to apply for a grant, you should contact you local council who should be able to offer more advice.

5.7 Oil-storage tank

	Planning permission required	Building Warrant required
Do I need permission for an oil storage tank?	No – provided that the tank is less than 2,500 litres in capacity for a single domestic site	No – provided that the tank is less than 2,500 litres in capacity for a single domestic site

Oil storage tanks, and the pipes connecting them to combustion appliances, should be constructed and protected so as to reduce the risk of the oil escaping and causing pollution in accordance with Building Regulations on Environment.

 Note: The Water Environment (Oil Storage) (Scotland) Regulations 2006 came into force on 1st April 2006. These regulations are different from those in England and will apply to any domestic oil tank larger than 2,500 litres (as opposed to 3,500 in England) as well as other oil storage facilities including storage of waste oil and storage of oil in buildings. If you think this applies to you, it would be advisable to seek advice. For more information on oil storage regulations go to www.sepa.org.uk or www.netregs.gov.uk

 SEPA (Scottish Environment Protection Agency) can offer help and advice to ensure that you comply. Failure to comply with regulations is a criminal offence and could result in a fine of up to £40,000.

5.8 Planting a hedge

	Planning permission required	Building Warrant required
Do I need permission to plant a hedge?	No – unless it obscures view of traffic at a junction or access to a main road	No

You do *not* need planning permission for hedges or trees.

 Note: However, if there is a condition attached to the planning permission for your property which restricts the planting of hedges or trees (for example, on an 'open plan' estate or where a sight line might be blocked), you will need to obtain the council's consent to relax or remove the condition before planting a hedge or tree screen. If you are unsure about this, you can check with the planning department of your council.

Hedges should not be allowed to block out natural light, and the positioning of fast growing hedges should be checked with your local authority. Recent incidents regarding hedging of the fast growing Leylandii trees have led to changes in the planning rules, where hedges previously had no restrictive laws.

5.9 Building a garden wall or fence

	Planning permission required	Building Warrant required
Do I need permission to erect a garden wall or fence?	Yes – if it is more than 1 m (3 ft 3″) high and is a boundary enclosure adjoining a highway or if it is more than 2 m (6 ft 6″) high elsewhere	No

Do I need approval to build or alter a garden wall or boundary wall?
No – subject to size.

You will, however, need to apply for planning permission if:

- your house is a listed building or in the curtilage of a listed building;
- the fence, wall or gate would be over 1 m high and next to a highway used for vehicles; or over 2 m high elsewhere.

In normal circumstances, the only restriction on walls and fences is the height allowed – which is 2 m or no more than 1 m if the walls or fence is near a highway or road junction, where its height might obscure a driver's view of other traffic, pedestrians or road users.

If there is a valid reason (e.g. for security purpose) for a wall or fence higher than the prescribed dimensions, then it is possible to get planning consent. If it has no effect on other people's valid interests and does not impair any amenity qualities in an area, there is no reason why a request should be refused.

Some walls, however, have historic value and they, as well as arches and gateways, can be listed. Modifications, extensions and removal of these must have planning consent.

5.10 Felling or lopping trees

	Planning permission required	Building Warrant required
Do I need permission to fell or lop a tree?	No – unless the trees are protected by a Tree Preservation Order or you live in a conservation area	No

Many trees are protected by Tree Preservation Orders (TPOs), which mean that, in general, you need the council's consent to prune or fell them. Nearly all trees in conservation areas are automatically protected.

5.11 Laying a path or a driveway

	Planning permission required	Building Warrant required
Do I need permission to lay a path or driveway?	No – unless it provides access to a main road	No

Do I need to apply for planning permission to install a pathway?
Generally *no* – but you may need approval from the highways department if the pathway crosses a pavement.

Do I need to apply for planning permission to lay a driveway?
No – unless it adjoins the main road.

Driveways

Provided a pathway or drive does not meet a public thoroughfare, you will not need planning consent. Currently there are no restrictions on the area of land around your house that you can cover with hard surfaces, but as flooding continues to be a problem across the UK, there is much talk about the need for permeable surfaces to allow excess water to drain away, so watch this space!

You will need to apply for planning permission only if the hard surface is not to be used for domestic purposes and is to be used instead, for example, for parking a commercial vehicle or for storing goods in connection with a business.

In the case of hardstanding you do not need permission to gain access to it within the confines of your land, but you would need permission for a hardstanding leading on to a public highway.

 Note: You must obtain the separate approval from the highways department of your council if you want to make access to a roadway or if a new driveway would cross a pavement or verge. The exception is if the roadway is unclassified and the drive or footway is related to a development that does not require planning permission. Your local authority highways department will be able to tell you if a road is classified or unclassified. If the road is classified then, depending on the volumes of traffic, it is harder to get permission. The busier the road the less likely a new driveway or footway will be allowed to meet it.

If a driveway crosses a pedestrian access, pavement or roadside verge, then the planning department will gain approval from the highways department. If this is the case, highways approval is required in addition to planning consent. The basic principle is to maintain safety and eliminate hazards.

You will also need to apply for planning permission if you want to make a new or wider access for your driveway onto a trunk or other classified road. The highways department of your council can tell you if the road falls into this category.

Pathways

Pathways do not normally need planning permission and you can lay paths however you like in the confines of your own property. The exception is for any path making access to a highway or public thoroughfare, in which case certain safety aspects arise. You may also need permission if your building is listed or is in a conservation area, so the style and size is suitable for the area.

If a pathway crosses a pedestrian access, pavement or roadside verge, then the planning department will gain approval from the highways department. If this is the case, highways approval is required in addition to planning consent. The basic principle is to maintain safety and eliminate hazards.

 If you are trying to construct a path or driveway across peat, you will need to ensure that you excavate down until you reach a solid base or your structure will be unstable. However, if the peat exceeds 500 mm in depth this may be impractical. In these cases the use of a synthetic geotextile would provide the foundation you need to allow the path to 'float' over the peat.

5.12 Building a hardstanding for a car, caravan or boat

	Planning permission required	Building Warrant required
Do I need permission to build a hardstanding for a vehicle or boat?	No – provided that it is within your boundary and is not used for a commercial vehicle	No

Do I need to apply for planning permission to build a hardstanding for a car?

No – provided that it is within your boundary and is not used for a commercial vehicle.

 Check local council rules.

Access from a new hardstanding to a highway requires planning consent. The exception is if the roadway is unclassified and the access to the hardstanding is related to a development that does not require planning permission. Your local authority highways department will be able to tell you if a road is classified or unclassified. If the road is classified then, depending on the volumes of traffic, it is harder to get permission. The busier the road the less likely a new driveway or footway will be allowed to meet it.

If the access crosses a pedestrian thoroughfare, pavement or roadside verge, then the planning department will gain approval from the highways department. If this is the case, highways approval is required in addition to planning consent. The basic principle is to maintain safety and eliminate hazards.

For a hardstanding on your own land, you do not need permission to gain access to it within the confines of your land, but you would need permission for a hardstanding leading on to a public highway.

There are different rules depending on what you use a hardstanding for. Planning permission is generally not needed, provided there are no covenants limiting the installation of hardstanding for parking of cars, caravans or boats. There are still rules for commercial parking, however (e.g. taxis or commercial delivery vans), and a 'change of use' as a trade premises would probably need to be granted for this to be allowed.

You should also check if there are any local covenants limiting changes in access to your premises or for hardstanding and parking of vehicles on it. If in doubt, contact the relevant local authority planning department for specific advice.

Do I need to apply for planning permission to build a hardstanding for a caravan and/or boat?

Some local authorities do not allow the parking of caravans or boats on driveways or hardstandings in front of houses. Check what the local rules are with your planning department, and if there's no restriction then you don't need to apply for permission.

There are no laws to prevent you, or your family from making use of a parked caravan while it's on your land or drive, but you cannot actually live in it as this would be classed as an additional dwelling. In addition, you cannot use a parked caravan for business use as this would constitute a change of use of the property.

If you want to put a caravan on your land to lease out as holiday accommodation or for friends or family to stay in while they visit you, then this would require planning permission. Rules on siting of static caravans or mobile homes are quite stringent.

5.13 Installing a swimming pool

	Planning permission required	Building Warrant required
Do I need permission to install a swimming pool?	Possibly – it would be wise to consult with your local authority	Yes – if it is an indoor pool

Swimming pools and saunas are subject to special requirements specified in Part 6 of BS 7671:2001.

 In the world of ever increasing concern about water usage, you could consider using recycled rain water to fill your pool. Recycled rain water can be diverted to domestic pools by installing a simple valve system and/or existing rainwater downpipes can be modified to include a diverter valve. A hose leading to the pool is then connected to the valve and when it rains, gravity then carries rain water from the roof directly into the pool.

5.14 Erecting aerials, satellite dishes, television and radio aerials, wind turbines and flagpoles

	Planning permission required	Building Warrant required
Do I need permission to erect a satellite dish?	No – unless you live in a flat or wish to put up more than two dishes	No
What if I live in a conservation area or my building is listed?	Possibly – consult your local authority	No
What if I want to erect a stand-alone antenna or mast?	Yes – if it is more than 3 m high	No
What if I want to erect a wind turbine either on my building or standalone?	Possibly – consult your local authority	No

Do I need to apply for planning permission to erect satellite dishes, television and radio aerials, wind turbines and flagpoles?
No – unless it is a stand-alone antenna, flagpole or mast greater than 3 m in height.

Note: Flagpoles, etc., erected in your garden are treated under the same rules as outbuildings, and cannot exceed 3 m in height.

There is no need for planning permission for attaching a maximum of two small satellite dishes (i.e. not exceeding 50 cm in diameter) on a single family dwelling. However, if it rises significantly higher than the roof's highest point then it may contravene local regulations or covenants.

You should get specific advice if you plan to install a large satellite dish or aerial, such as a short-wave mast, as the rules differ between authorities, and if you are a leaseholder, remember that you may need to obtain permission from the landlord.

Planning permission may be required to install a wind turbine (either free standing or attached to a dwelling) and it is recommended that you seek the advice of your local planning authority in relation to wind turbines as permissions vary depending on your region.

Conservation areas, national parks, etc., have specific local rules on aerials and satellite dishes, so you will need to approach your local planning department to find out the particular rules for your area (e.g. in a conservation area, is it not appropriate to place a dish on the front elevation or in a prominent position on the building).

If your house is a listed building, then you will more than likely need listed building consent to install a satellite dish on your house.

Note: Commercial satellite dishes and telecommunication equipment satellite dishes are considered telecommunications equipment and most dishes used for business purposes exceed 95 cm and will require planning permission. Telecommunications equipment installed by a registered operator, however, is exempt from telecommunications legislation, but these operators are required to notify the local council of their intention to commence work, and the council has 28 days in which to respond if they believe the proposal will require a full planning application.

5.15 Advertising

	Planning permission required	Building Warrant required
Do I need permission to erect an advertisement?	Not in most circumstances	Possibly – consult your local verifier

Do I need to apply for planning permission to erect an advertising sign?
Advertisement signs on buildings and on land do sometimes need planning consent and whilst some smaller signs and non-illuminated signs may not

require this consent, it is always advisable to check with development control staff first.

You are allowed to display certain small signs at the front of residential premises such as election posters, notices of meetings, jumble sales, car for sale, etc., but business types of display and permanent signs may require planning permission – they could even come under the category of 'advertising control' for which planning consent is always required.

There are 14 classes of signs that *do not* need permission; however, if your sign does not fall within one of these 14 classes you will need advertisement consent. For full details of the 14 classes of permitted signs, consult local authorities. Advertisements that you will need consent for include:

- most hoardings;
- free-standing advertisements and hoardings on walls;
- most illuminated signs;
- fascias and projecting signs which are more than 4.6 m above ground level;
- advertisements on gable ends;
- adverts in areas of historical, architectural or cultural importance;

Note: Any outdoor advertisement must be kept clean and tidy and in a safe condition, and must not obscure any official or safety signage.

You must make sure that you have the permission of the owner and other users of the land to display the advertisement even if you do have planning permission in place.

Temporary notices relating to local events, such as fêtes and concerts, may be displayed for a short period – although if you think that your advertisement will fall into one of the categories above, it would be worthwhile for you to seek the view of the local authority.

It is illegal to post notices on empty shops windows, doors and buildings, and also on trees. This is commonly known as 'fly posting' and can carry heavy fines.

5.16 Building a porch

	Planning permission required	Building Warrant required
Do I need permission to build a porch?	No – unless:	Possibly – if area exceeds 30 m² (35.9 yd²)
	The porch will extend beyond the line of the building within 20 m of a public road	
	or	
	The permitted development allowances have already been exceeded	

Do I need planning permission for a porch?
Possibly – depending on its size and position.

Most porches would come under *permitted development* and in these cases you will not need to apply for permission. However, you will need to apply for planning permission if:

- the porch would be less than 20 m away from a highway (which includes all public roads, footpaths, bridleways and byways);
- the permitted developments allowances have already been used up (for example if previous extensions have already taken place);
- permitted development rights are restricted because your house is listed or is in a conservation area, national park or area of outstanding natural beauty;
- permitted development rights have been removed by a condition of the original planning permission (e.g. some new developments).

The regulations are quite complicated and depend on previous works on the site, if any, so you should always check with development control staff.

Note: Even if a building warrant is not required due to the size of the porch, all building work should be carried out in accordance with Building Regulations.

Even if planning permission is not required, it would be advisable to discuss the proposed work with your neighbours before going ahead with the work!

5.17 Outbuildings

	Planning permission required	Building Warrant required
Do I need permission to build an outbuilding?	Possibly – erecting outbuildings can be a minefield and it is best to consult the local planning officer before commencing work	Yes

Many kinds of buildings and structures can be built in your garden or on the land around your house without the need to apply for planning permission. These can include sheds, garages, greenhouses, accommodation for pets and domestic animals (e.g. chicken houses), summer houses, swimming pools, ponds, sauna cabins, enclosures (including tennis courts) and many other kinds of structure.

Outbuildings intended to go in the garden of a house do not normally require any planning permission, so long as they are associated with the residential amenities of the house and a few requirements are adhered to such as position and size. However:

- they must **not** exceed 4 m in height for a ridged-roof or 3 m for a flat-roofed structure;

- the structure may not be in front of the building line nor within 20 m of nearest road;
- the structure must not cover more than 30% of the area around the house;
- the structure must not exceed 4 m^2 in area if within 5 m of the house.

Always try and keep the structure appropriate to the house so that it does not completely dominate the general visual amenities.

If your house is listed or is in a conservation area, national park or area of outstanding natural beauty, then you will more than likely need to obtain planning consent. If in doubt, contact the relevant local authority planning department for specific advice.

External water storage tanks

Many years ago, the demand for external tanks for capturing rainwater made their installation quite commonplace, but today the need for extra storage tanks is quite rare – unless you are in a rural position.

If you are considering installing an external water tank you should seek guidance from your local authority, especially if the tank is to be mounted on a roof.

Fuel storage tanks

Storage of oil, or any other liquids, especially petrol, diesel and chemicals, is strictly controlled and would not be allowed on residential premises. If you are considering installing an external oil storage tank for central heating use, then no planning permission is required, provided its capacity is no more than 2,500 litres.

You will need to apply for planning permission in the following circumstances:

- You want to install a storage tank for domestic heating oil with a capacity of more than 2,500 litres.
- You want to install a storage tank, which would be nearer to any *highway* than the nearest part of the *'original house'*, unless there would be at least 20 m between the new storage tank and any highway. (The term **'highway'** includes public roads, footpaths, bridleways and byways.
- You want to install a tank to store liquefied petroleum gas (LPG) or any liquid fuel other than oil.

Erecting any type of outbuilding can be a potential minefield and it is always best to consult with the local planning officer before commencing the work.

5.18 Garages

	Planning permission required	Building Warrant required
Do I need permission to build a garage?	Probably	Yes

Do I need approval to build a garage extension to my house, shop or office?

Probably – if the garage is to be within 5 m of the house whether attached or detached from the property, this is considered an extension and both planning permission and a Building Warrant will be required.

Should your garage be more than 5 m from the house, then this is considered to be an 'outbuilding' (see Section 5.17 for details).

5.19 Building a conservatory

	Planning permission required	Building Warrant required
Do I need permission to build a conservatory?	Possibly – you can extend your house by building a conservatory, provided that the total of both previous and new extensions does not exceed the permitted volume	Probably

Do I need permission to erect a conservatory?

Possibly – conservatories and sun lounges attached to a house are classed as extensions and thus the rules on permissible development apply. If you want a conservatory or sun lounge separated from the house, this needs planning consent under similar rules for outbuildings.

Another thing to keep in mind is your neighbour's reaction – always keep them informed of what's happening and be prepared to alter the plans you had for locating the building if they object – it's better in the long run, believe me.

Will you therefore need planning permission? Generally no, as the building is classed as a 'portable building', *nevertheless* it is your responsibility to check with your local planning office.

What is defined as a conservatory?

A conservatory is a building attached to a dwelling and having a door separating it from that dwelling and having not less than three-quarters of the area of its roof and not less than one half of the area of its external walls made of translucent material.

To meet the definition of conservatory, it must also:

- be attached to the ground storey of houses (but not flats or maisonettes) and only when those houses have foundations of traditional concrete strips, or when the ground conditions allow for the use of traditional strip foundations;
- have an internal floor area of more than 8 m², but not more than 20 m²;
- be connected to the mains drainage system, and the invert level at the point of connection to the house drainage system is not more than 1,000 mm;
- have an effective roof area of not more than 35 m² (including any house roof area to be drained via the conservatory gutters and down pipe);

- be fixed directly to masonry walling at least 100 mm thick;
- be at least 1 m from any boundary;
- be at a distance from the boundary (in metres) of at least one-sixth of the area of glazing and frame (in square metres) facing that boundary (e.g. if a side of the conservatory has an area of glass and frame of $9\,m^2$, then the conservatory must be at least $9 \div 6 = 1.5\,m$ from the boundary which that glazing faces);
- have glazing to all sides (excluding the back/house wall), onto which the roof is fixed;
- meet the thermal insulation requirements (see Chapter 6) whether the conservatory is to be heated or not;
- not be built on or above ground that is contaminated;
- not contain a chimney, flue pipe, fixed combustion appliance installation, washbasin, sink, bath, shower, urinal, watercloset or waterless closet, and are not be built over any form of underground drain;
- only serve the dwelling on to which the conservatory is to be built;
- have a floor level the same as, or not more than 600 mm below the floor level of the house;
- not be built over an existing rain water down pipe;
- not be built over any existing escape windows.

Do I need a Building Warrant?

A Building Warrant must be obtained before you start any work where the conservatory is more than $8\,m^2$ in floor area. A warrant is also required for conservatories below this size if they contain a chimney, fixed combustion appliance, washbasin, sink, bath, shower, urinal, watercloset or waterless closet, or are closer than 1 m to a boundary.

Whilst a Building Warrant may not be required for your specific conservatory **all** building works should nevertheless meet the Building Standards requirements!

 A document providing guidance on how to meet the Building Regulations for simple conservatories which are built on to existing houses has been developed and is available from the SBSA-www.sbsa.gov.uk/pdfs/conservatory.pdf

The regulations are quite complicated and depend on previous works on the site. A conservatory may need no permission at all, some permission, or indeed many permissions of various types. If you are in doubt you should contact your local authority building standards department for advice.

Although you will need to apply for a Building Warrant, the conservatory guidance document mentioned above can be used as part of your warrant application.

 Note: Conservatories will also need to have:

- heating systems equipped with independent temperature and on/off controls;
- thermal elements that have the appropriate U-values;
- glazed elements that comply with the standards.

5.20 Loft conversions, roof extensions and dormer windows

	Planning permission required	Building Warrant required
Do I need permission to convert a loft, roof extension?	No – provided the volume of the house is unchanged and the highest part of the roof is not raised	Yes
Do I need permission to put in dormer window?	Yes – *ALL* dormer windows require planning permission	Yes
What if I wish to put in another type of window in the loft space?	No – skylights are permitted as long as they do not protrude more than 10 cm above the plane of the roof	Yes

Do I need approval for a loft conversion?
Possibly – see below.

Do I need to apply for permission put in a dormer window?
Yes – ALL dormer windows do need planning permission as they alter the roof line.

Do I need to apply for permission to insert roof lights or skylights?
No – as long as they do not protrude more than 10 cm above the plane of the roof.

Do I need to apply for planning permission to extend or add to my house?
Yes – in the following circumstances:

- if the new works exceed the permitted development rights;
- if your property is a special or designated area, e.g. conservation area;
- if the extension will require the raising of the ridge height;
- if the loft conversion/extension is to flats.

Do I need to apply for planning permission to alter a roof or to re-roof my house?
You will need to apply for planning permission if you live in a conservation area, a national park or an area of outstanding natural beauty or you want to build an extension to the roof of your house or any kind of addition that would materially alter the shape of the roof.

Roofs are expected to match those of the surrounding area, so consider this if you live in a protected area. Some areas require that the colour and style of the roof covering matches the original, and the pitch and construction should be the same. If you plan to save the expense of matching the roof, by opting for a flat roof, be sure that your local authority will accept this. Often high flat roofs are not desirable, due to the appearance of the house elevation. Provided the alterations to your roof do not make a noticeable change or don't increase its height you would normally not need to obtain planning permission.

 You do not normally need to apply for planning permission to re-roof your house or for the insertion of roof lights or skylights.

In the case of re-roofing, if the tiles are the same type then no approval is needed. If the new tiling or roofing material is substantially heavier or lighter than the existing material, or if the roof is thatched or is to be thatched where previously it was not, then an approval under Building Regulations is probably required.

 Always check your deeds to ensure that there are no restrictive covenants

 Note: Loft conversions create an extra storey so escape provisions have to be made in case of fire. This could include escape windows, smoke alarms and fire-resisting self-closing doors.

 It is always a good idea to let your mortgage and insurance company know when you do the work.

5.21 Building an extension

	Planning permission required	Building Warrant required
Do I need permission to build an extension?	Possibly not – if you extend your house by building an extension, provided that the total of both previous and new extensions does not exceed the permitted volume	Yes

Do I need approval to build an extension to my house?
Yes – if it would 'materially alter the appearance of the building'.

Major alterations and extensions nearly always need approval. However, some small extensions such as porches, garages and conservatories may be 'permitted development' and, therefore, do not need planning consent.

 Building extensions is always a potential minefield and it is best to consult the local planning officer before contemplating any work!

What is a permitted development?
Permitted developments are pre-determined amounts of development work that can be carried out on a building without the need for planning permission. Permitted developments not requiring permission are as follows:

Terraced houses and houses in conservation areas:

- extensions not exceeding $16\,m^2$ or 10% of the area of the original house whichever is greater up to a maximum of $30\,m^2$;

- extensions not exceeding the highest point of the existing roof;
- extensions not extending beyond the building line facing the road if within 20 m of the road.

Other houses:
- extensions not exceeding 24 m² or 20% of the floor area of the original house whichever is the greater up to a maximum of 30 m²;
- extensions not exceeding the highest point of the existing roof;
- extensions not extending beyond the building line facing the road if within 20 m of the road.

Garages within 5 m (16 ft) of the house are considered extensions.

Porches are treated like any other extension and so volume rules above apply. Planning permission **will** be required if the porch extends beyond the building line within 20 m of a public road.

 Note: Flats do not benefit from *permitted development* and planning permission is required for **ANY** extension.

Thus:
- you **will** need to apply for planning permission to extend or add to your house;
- you may also require planning permission if your house has previously been added to or extended;
- you may also require planning permission if the original planning permission for your house imposed restrictions on future development (i.e. permitted development rights may have been removed. This is often the case with more recently constructed houses).

 An important point to remember is that the volume of **other buildings** which belong to your house (such as a garage or shed) will count against the volume allowances! In some cases, this can include buildings that were built at the same time as the house or that existed on 1 July 1948.

 Note: If a building is extended, or undergoes a material alteration, the completed building *must* comply with the relevant requirements of the Building Regulations or, where this is not feasible, be 'no more unsatisfactory than before'.

5.21.1 Extensions to non-domestic buildings

 In general terms *all* commercial/non-domestic extensions require planning permission.

However, some extensions to factories and warehouses do benefit from permitted development rights and therefore in these cases planning permission is not required. These are:

- an extension with a floor area less than 1,000 m² (measured externally), below 25% volume of the original building;
- an extension not higher than the existing building at any point;
- an extension that is in conjunction with the existing use;

- an extension that does not materially affect the appearance of the building;
- an extension that is not within 5 m of any boundary;
- an extension that does not reduce parking or other arrangement for vehicles, e.g. turning, refuse collection arrangements or other servicing requirements.

 Note: Different rules apply for agricultural buildings.

New non-domestic buildings which have been constructed as a shell under one building warrant for later fit out under a separate warrant should meet the maximum U-values for building elements of the insulation envelope (see Chapter 6).

If a building has a total useful floor area greater than $1,000\,m^2$ and the proposed building work includes:

- an extension;
- the initial provision of any fixed building services;
- an increase to the installed capacity of any fixed building services;

then 'consequential improvements' should be made to improve the energy efficiency of the whole building. These could include:

- upgrading all thermal units which have a high U-value;
- replacing all existing windows (less display windows), roof windows, rooflights or doors (excluding high-usage entrance doors) within the area served by the fixed building service with an increased capacity;
- replacing any heating system that is more than 15 years old;
- replacing any cooling system that is more than 15 years old;
- replacing any air handling system that is more than 15 years old;
- upgrading any general lighting system that serves an area greater than $100\,m^2$ which has an average lamp efficacy of less than 40 lamp-lumens per circuit watt;
- installing energy metering;
- upgrading existing LZC energy systems if they provide less than 10% of the building's energy demand.

5.22 Conversions and change of use

	Planning permission required	Building Warrant required
Do I need permission to convert or change the use of a building?	Possibly – it is advisable to check with your local authority	Yes – even if no building work is carried out
	Yes – for flats – even where construction works may not be intended	
	Yes – for shops and offices unless no building work is envisaged	

Note: A Building Warrant is required for conversion of a building even if no building work is carried out.

What is the difference between conversion and change of use?
Very little (!) and for the purposes of meeting Building Regulations, *all* are considered as conversions.

Conversions to which the regulations apply are:
- changes in the occupation or use of a building to create a dwelling or dwellings or a part thereof;
- changes in the occupation or use of a building ancillary to a dwelling to increase the area of human occupation;
- changes in the occupation or use of a building which alter the number of dwellings in the building;
- changes in the occupation or use of a domestic building to any other type of building;
- changes in the occupation or use of a residential building to any other type of building;
- changes in the occupation or use of a residential building which involve a significant alteration to the characteristics of the persons who occupy, or who will occupy, the building, or which significantly increase the number of people occupying, or expected to occupy, the building;
- changes in the occupation or use of a building so that it becomes a residential building;
- changes in the occupation or use of an exempt building (in terms of schedule 1) to a building which is not so exempt;
- changes in the occupation or use of a building to allow access by the public where previously there was none;
- changes in the occupation or use of a building to accommodate parts in different occupation where previously it was not so occupied.

Do I need approval to convert my house into flats?
Yes – even where construction works may not be intended.

Do I need approval to convert part or all of my shop or office to a flat or house?
Probably – where building work is proposed you will probably need approval if it affects the structure or means of escape in case of fire. But you should check with the local fire authority and the county council, to see whether a fire certificate is actually required.

You will probably also need planning permission whether or not building work is proposed.

What standards apply to conversions?

What standards will I have to meet?

Every conversion, to which these regulations apply, *shall meet the requirements* of the following standards:

Section	Standards
Fire	2.1, 2.3, 2.5, 2.9–2.11, 2.13–2.15
Environment	3.5–3.9, 3.11–3.14, 3.17, 3.18, 3.20–3.26
Safety	4.5–4.7, 4.9, 4.11, 4.12
Noise	All
Energy	6.7, 6.8, 6.10

Every conversion, to which these regulations apply, *shall meet the requirements* of the following standards in *so far as is reasonably practicable*, and in no case be worse than before the conversion:

Section	Standards
Fire	All
Environment	2.2, 2.4, 2.6, 2.7, 2.8, 2.12
Safety	3.1–3.4, 3.10, 3.15, 3.16, 3.19
Noise	4.1–4.4, 4.8, 4.10
Energy	6.2–6.6

Conversion of a building brings different requirements from those that apply to alterations. It is not just the planned work that must meet the standards but when converted the whole building is required to meet all the standards.

There are two important qualifications because few existing buildings can reasonably be altered to meet all aspects of current standards.

Firstly, schedule 6 to regulation 12 lists those standards which can be met by improving a building to as close to the full requirement as is reasonably practicable.

Secondly, it is important to remember that functional standards allow some flexibility in deciding what is acceptable.

Historic buildings also require sensitive application of the standards.

 Note: The Building Regulations define 'reasonably practicable' as '... having regard to all the circumstances, including the expense involved in carrying out the building work'. The way to proceed is that an applicant has to show how any failure of the building, as it is proposed to be after conversion, is being addressed. Except for historic buildings, it is only in exceptional cases that it is likely that taking no action is acceptable, but it will be normal for many of the identified standards that the full current requirements will not be met. The regulations do, however, require the building to be no worse than before, so the construction must either be left untouched or be altered in a way that is closer to compliance with the standards. The individual sections of the Technical Handbooks to the Building Regulations give guidance on what is likely to be an acceptable standard.

 It is an offence to occupy a converted building without first having received a completion certificate. See Chapter 2 for more details.

5.22.1 Material changes of use

Where there is a material change of use of a **whole** building to a hotel, boarding house, institution, public building or a shop (restaurant, bar or public house), the building must be upgraded, if necessary, so as to comply with current regulations.

If an existing building undergoes a change of use so that **part** of it can be used as a hotel, boarding house, institution, public building or a shop, the work being carried out must ensure that:

* people can gain access from the site boundary and any on-site car parking space;
* sanitary conveniences are provided in that part of the building or it is possible for people (no matter their disability) to use sanitary conveniences elsewhere in the building.

5.22.2 Buildings suitable for conversion

Most local plans stipulate that conversion proposals 'should relate to buildings of traditional design and construction which enhance the natural beauty of the landscape' as opposed to 'non-traditional buildings, buildings of inappropriate design, or buildings constructed of materials which are of a temporary nature'.

5.22.3 Isolated buildings

Planning permission will not normally be granted for the conversion or re-use of isolated buildings. Exceptionally, permission may be given for such buildings to be used for small-scale storage or workshop uses or for camping purposes.

An isolated building is normally:

* a building, or part of a building, standing alone in the open countryside;
* a building, or part of a building, comprised within a group which otherwise occupies a remote location having regard to the disposition of other buildings within the locality, to the character of the surroundings, and to the nature and availability of access and essential services.

Assessing whether or not a particular building should be regarded as isolated may not always be straightforward and, in such instances, early discussion with a planning officer at the National Park Authority is advised.

5.22.4 Structural condition

Buildings proposed for conversion should be large enough to accommodate the proposed use without the necessity for major alterations, extension or re-construction. In cases of doubt regarding the structural condition of any particular building, the authority may require the submission of a full structural survey to accompany a planning application. The authority can advise on this requirement and, if necessary, on persons who are suitably qualified to undertake such work and who practise locally.

Planning permission will not normally be granted for re-construction if substantial collapse occurs during work on the conversion of a building.

5.22.5 Workshop conversions

Redundant farm buildings and buildings of historic interest are often well suited to workshop use and such conversions normally require minimal alterations. Potential problems of traffic generation and un-neighbourliness can usually be addressed by the imposition of appropriate conditions.

The local authority will generally favourably consider proposals that make good use of traditional buildings by promoting local employment opportunities. In some instances grants may be available from other agencies to assist the conversion of buildings to workshop use.

5.22.6 Residential conversions

When reviewing proposals for converting a traditional building, the local authority will pay particular attention to the overall objectives of the housing policies of the local plan. If land that can be used for a new housing development is limited, residential conversions can make a valuable contribution to the local housing stock and support the social and economic well-being of rural communities.

The local plan will require that residential conversions should, in most instances, contribute to the housing needs of the locality. Permission for such conversions are, in some districts, only granted subject to a condition restricting occupancy to local persons – where 'local persons' are normally defined as persons working, about to work, or having last worked in the locality or who have resided for a period of 3 years within the locality.

5.22.7 An old building or historic building

Throughout Scotland, there are many under-used or redundant buildings, particularly farm buildings which may no longer be required, or suitable, for agricultural use. Such buildings of normally weathered stone and slate, contribute substantially to the character and appearance of the landscape and the built environment. Their interest and charm stem from an appreciation of the functional requirements of the buildings, their layout and proportions, the type of building materials used and their display of local building methods and skills.

In most cases traditional buildings are best safeguarded if their original use can be maintained. However, with changing patterns of land use and farming methods, changes of use or conversion may have to be considered.

The conversion or re-use of traditional buildings may, in the right location, assist in providing employment opportunities, housing for local people or holiday accommodation. Applicants and developers are encouraged to refer to the Local Plan for comprehensive guidance and to seek advice from a planning officer if further assistance is necessary.

All councils place the highest value on good design and proposals. Those that fail to respect the character and appearance of traditional buildings will not be permitted. Sensitive conversion proposals should ensure that existing ridge and eaves lines are preserved; new openings avoided as far as possible; traditional matching materials are used; and the impact of parking and garden areas is minimized. Buildings that are listed as being of 'special architectural or historic interest' require skilled treatment to conserve internal and external features.

In many instances, traditional buildings that are of simple, robust form with few openings may only be suitable for use as storage or workshops. Other uses, such as residential, may be inappropriate.

 Note: The SBSA has a comprehensive two-part guide on 'conversion of traditional buildings' and how the standards should be applied to these buildings. For more information go to www.sbsa.gov.uk/tech_handbooks/traditional_buildings.htm

5.23 Building a new house

	Planning permission required	Building Warrant required
Do I need permission to build a new house?	Yes	Yes

Do I need planning permission to erect a new house?
Yes.

 ALL new houses or premises of ANY kind require planning permission.

Private individuals will normally only encounter this if they intend to buy a plot of land to build on, or buy land with existing buildings that they want to demolish to make way for a new property to be built.

In all cases, unless you are an architect or a builder, you **must** seek professional advice. If you are using a solicitor to act on your behalf in purchasing a plot on which to build, he will include the planning questions within all the other legal work, as well as investigating the presence of covenants and existing planning consent, together with other constraints or conditions.

The architect, surveyor or contractor you hire will then need to take into account the planning requirements as part of their planning and design procedures. They will normally handle planning applications for any type of new development.

If you are hiring a professional (or more than one – say a building contractor to do the work and a surveyor or architect to plan and design) be sure to find out exactly who does what and that approval is obtained before going to too much expense, should a refusal arise!

5.24 Infilling

	Planning permission required	Building Warrant required
Do I need permission to infill on a piece of land?	Possibly	Yes

Can I use an unused, but adjoining, piece of land to build a house (e.g. build a new house on land that used to be a large garden)?
Often there may be no official grounds for denying consent, but residents and individuals can impose quite some delay. It is worth testing the likelihood of a successful application by talking to the neighbours and judging opinions.

Planning consent is often quite difficult to obtain in these cases as this sort of development normally causes a lot of opposition as it is in a settled residential area and people do not like change.

New developments will undoubtedly also need to follow Building Regulations. This, and all site visits from inspectors, is normally arranged by your building contractor.

 There are plenty of substantial building projects that don't require any planning permission; however, it is undoubtedly a good idea to consult a range of people before you consider any work.

5.25 Demolition

	Planning permission required	Building Warrant required
Do I need permission to demolish a house?	Yes – if it is a listed building or in a conservation area, if the whole house is demolished	Yes

You must have good reasons for knocking a building down, such as making way for rebuilding or improvement (which in most cases would be incorporated in the same planning application).

Penalties can be quite severe for demolishing something illegally!

You may not need to make a planning application to demolish a listed building or to demolish a building in a conservation area. However, you may need listed building or conservation area **consent** – but in these cases it is advisable to speak to your local authority.

Elsewhere, you will **not** need to apply for planning permission:

- to demolish a building such as a garage or shed of less than $50 \, m^3$;
- if the demolition is urgently necessary for health and safety reasons;
- if the demolition is required under other legislation;
- where the demolition is on land that has been given planning permission for redevelopment;
- to demolish a gate, fence, wall or other means of enclosure.

In all other cases (such as demolishing a house or block of flats) the council may wish to agree the details of how you intend to carry out the demolition and how you propose to restore the site afterwards. You will need to apply for a formal decision on whether the council wishes to approve these details. This is called a 'prior approval application' and your council will be able to explain what it involves.

 You are **not allowed** to begin any demolition work (even on a dangerous building) **unless** you have given the local authority notice of your intention and either this has been acknowledged by the local authority or the relevant notification period has expired. In this notice you will have to:

- specify the building to be demolished;
- state the reason(s) for wanting to demolish it;
- show how you intend to demolish it.

Copies of this notice will have to be sent to:
- the local authority;
- the occupier of any building adjacent to the building;
- the local gas supplier;
- the area electricity board in whose area the building is situated.

 This regulation does not apply to the demolition of an internal part of an occupied building, or a greenhouse, conservatory, shed or prefabricated garage (that forms part of that building) or an agricultural building.

5.25.1 What about dangerous buildings?

If a building, or part of a building or structure, is in such a dangerous condition (or is used to carry loads that would make it dangerous) then the local authority may apply to make an order requiring the owner:

- to carry out work to avert the danger;
- to demolish the building or structure, or any dangerous part of it, and remove any rubbish resulting from the demolition.

Note: The local authority can also make an order restricting its use until such time a court is satisfied that all necessary works have been completed.

If the building or structure poses a potential danger to the safety of people, the local authority will take the appropriate action to remove the danger. The local authority has powers to require the owners of buildings or structures to remedy the defects or they can direct their own contractors to carry out works to make the building or structure safe. In addition, the local authority may provide advice on the structural condition of buildings during fire-fighting to the fire brigade.

If you are concerned that a building or other structure may be in a dangerous condition, then you should report it to the local council.

Emergency measures

In emergencies the local authority can make the owner take immediate action to remove the danger, or they can complete the necessary action themselves. In these cases, the local authority is entitled to recover from the owner such expenses reasonably incurred by them. For example:

* fencing off the building or structure;
* arranging for the building/structure to be watched.

 Note: These works are controllable by the local authority under sections 29 and 30 of the Building (Scotland) Act 2003.

5.25.2 Can I be made to demolish a dangerous building?

If the local authority considers that a building is so dangerous that it should be demolished, they are entitled to issue a notice to the owner requiring the owner/occupier to demolish it.

The demolition notice must specify:
* the reason for the requirement;
* the standard to which the demolition is to be carried out (including any standard to which the site of the demolished house must be cleared);
* the period within which the demolition must be carried out.

The period specified must be the period beginning with the date from which the notice has effect within which the local authority reasonably considers that the demolition can be completed (but must not, in any case, be a period of less than 21 days).

 Before complying with this notice, the owner must give the local authority 48 h notice of commencement.

If the owner of a house fails to comply with a work notice or a demolition notice, the local authority may carry out:

* the work or the demolition required by the notice;
* any other work which, in the course of carrying out work or demolition authorized the local authority finds to be required to meet the local action

plan in relation to any house identified in it, or to bring any house which the local authority considers to be sub-standard into, and keeping it in, a reasonable state of repair.

Note: Before carrying out any work, the local authority must give 21 days notice of its intention to do so unless the local authority considers the situation is urgent.

However, the local authority may not carry out any work unless

- the period within which the work or demolition requires to be carried out has ended;
- the owner has given notice to the local authority – of being unable to comply with the work notice or demolition notice because of a lack of necessary rights (of access or otherwise) despite having taken reasonable steps for the purposes of acquiring those rights, or stating that the owner considers that carrying out the work or demolition required is likely to endanger any person.

Note: These works are controllable by the local authority under sections 33–36, 40 and 41 of the Housing (Scotland) Act 2006.

5.25.3 Replacing a demolished building

If you decide to demolish a building, even one that has suffered fire or storm damage, it does not automatically follow that you will get planning permission to build a replacement.

5.26 What is a defective building?

A defective building is one that is deemed by the local authority to unsafe or not fit for habitation. In these cases the local authority may issue a Defective Building Notice to compel the owner to bring the building back into a fit state of repair.

5.26.1 What is a Defective Building Notice?

A local authority may serve on the owner of a building a 'Defective Building Notice' requiring the owner to rectify such defects in the building as the notice may specify. The defects which may be specified in a defective building notice are defects which require rectification in order to bring the building into a reasonable state of repair having regard to its age, type and location.

The Defective Building Notice will specify: a date not less than 7 days after the date of service of the notice by which the owner must have begun the work required by the notice; a date not less than 21 days after that date by which the owner must have completed that work; and may specify different dates for the commencement and completion of different work.

A Defective Building Notice may specify particular steps which the local authority requires the owner to take in complying with the notice. (Nothing in this section, however, affects the requirement to obtain a building warrant – where one would normally be required for any work required to comply with a Defective Building Notice.)

 Should the owner fail to carry out the work specified, not only are they committing an offence, but the local authority may carry out the work necessary to complete the work required by the notice and may recover from the owner any expenses reasonably incurred by it in doing so!

It should be noted that the local authority may, at any time:

- withdraw a Defective Building Notice;
- waive or relax any requirement of such a notice, including substituting a later date for a date specified.

 Note: The withdrawal of a Defective Building Notice does not affect the power of the local authority to issue a further such notice.

5.27 Where can I find out more?

Many local authority websites have very use interactive questionnaires to help you determine if you need planning permission and/or a building warrant for your proposed work.

 As with all these things we advise you seek the views and assistance of your local authority.

6

The Building (Scotland) Regulations

6.0 Introduction

The Building (Scotland) Act 2003 provides Scottish Ministers with the power to create Building Regulations to:

- secure the health, safety, welfare and convenience of persons in or about buildings;
- further the conservation of fuel and power;
- further the achievement of sustainable development.

It also provides the power to make procedural regulations, fee regulations and the other supporting legislation that is required to operate the system.

The current Building Regulations are contained in The Building (Scotland) Regulations 2004 (Scottish Statutory Instrument 2004 No. 406). In support of these Building Regulations are two Technical Handbooks (one covering domestic buildings and the other non-domestic buildings) and these provide practical guidance on achieving the standards set out in the Building (Scotland) Regulations.

These standards describe the functions that a building should perform (such as 'providing resistance to the spread of fire') and for conformity. The technical specifications of these standards are inspected by the European Commission (EC) to ensure that they meet the requirements of the Construction Products Directive (CPD) and to ensure that no barriers to trade in construction products are created.

Each Handbook has six sections as follows:

Section	Title	Relevant EU CPD
Section 1	Structure	Mechanical resistance and stability
Section 2	Fire	Safety in case of fire
Section 3	Environment	Hygiene, health and the environment
Section 4	Safety	Safety in use
Section 5	Noise	Protection against noise
Section 6	Energy	Energy, economy and heat retention

Plus:

- Appendix A: Defined terms
- Appendix B: List of standards and other publications
- A full index.

The Technical Handbooks are then supported by The Building (Procedure) (Scotland) Regulations 2004 (Scottish Statutory Instrument 2004 No. 428) which clarifies the intent of the Building (Scotland) Regulations.

Note: It is intended that the Technical Handbooks will be updated annually and replacement pages published for any guidance which has been altered.

These Technical Handbooks replace the Technical Standards (6th Amendment) from 01/05/05 and they are available from the Stationery Office (TSO) or (as a free download) from the Scottish Building Standards Agency (SBSA) at www.sbsa.gov.uk

6.0.1 What is the scope of the two Technical Handbooks?

Domestic buildings

The guidance in the Domestic Technical Handbook covers the following types of buildings:

(a) Buildings with masonry walls:
  - domestic buildings, but restricted to houses not more than three storeys without basement storeys;
  - extensions with eaves heights not more than 3 m to low rise domestic buildings including garages and outbuildings;
  - single-storey, single-leaf buildings forming a garage or outbuilding within the curtilage of a dwelling.
(b) Buildings with timber-frame walls:
  - domestic buildings, but restricted to houses not more than two storeys without basement storeys;
  - extensions with eaves heights not more than 3 m to low rise domestic buildings.

Non-domestic buildings

The guidance in the Non-domestic Technical Handbook covers (but is not restricted to) the following:

- Large dwellings;
- Residential care homes;
- Hospitals;
- Enclosed shopping centres;
- Assembly buildings;
- Entertainment buildings (e.g. theatres and cinemas);
- Factories;
- Offices.

6.0.2 Who has to comply with the Building (Scotland) Regulations?

The duty to comply with the Building Regulations lies with the owner, or in some cases the client, for the work. Before work can begin, a Building Warrant must be obtained. For some building work and provision of services (see below), a warrant is not required.

But even so, the Building (Scotland) Regulations will still apply.

6.0.3 What about alternative solutions?

Although the Building (Scotland) Regulations are mandatory, the choice of actually how to comply with them lies with the building owner. There is no obligation to adopt any particular solution that is contained in any of these guidance documents (especially if you prefer to meet the relevant requirement in some other way), but if the guidance contained in the Technical Handbooks is followed, in full, then this would be accepted by the verifier as indicating that the Building Regulations have been complied with.

If you have **not** followed the guidance, then that will be seen as evidence tending to show that you have not complied with the requirements and it will then be up to you, the builder, architect and/or client to demonstrate that you have satisfied the requirements of the Building Regulations.

This compliance may be shown in a number of ways such as using:

- a product bearing CE marking (in accordance with the CPD (89/106/EEC) as amended by the CE Marking Directive (93/68/EEC) as implemented by the CPD 1994 (SI 1994/3051);
- an appropriate technical specification (as defined in the CPD – 89/106/EEC);
- a recognized British Standard;
- a British Board of Agrément Certificate;
- an alternative, equivalent national technical specification from any member state of the European Economic Area or Turkey;
- a product covered by a national or European certificate issued by a European Technical Approval issuing body.

6.0.4 What happens if the building is a mixture of domestic and non-domestic?

Where a building is used for both domestic **and** non-domestic purposes (such as a caretakers flat in an office building or a room used as an office in a sheltered housing complex), then the appropriate parts from each Technical Handbook will have to be used so as to ensure the standards are complied with in full.

As a rule of thumb, however, where a building or part of a building falls into more than one category, it should be designed to meet the most stringent recommendations.

6.0.5 What about materials and workmanship?

As stated in the Building Regulations, '*All building work should be carried out with proper materials and in a workmanlike manner*'.

6.0.6 What materials can I use?

Other than the two exceptions below, provided that the materials and components you have chosen to use are from an approved source and are of approved quality (CE marking in accordance with the CPD (89/106/EEC), the Low Voltage Directive (73/23/EEC and amendment 93/68/EEC) and the EMC Directive (89/336/EEC) as amended by the CE Marking Directive (93/68/EEC)), then the choice is fairly unlimited.

Short-lived materials

Even if a plan for building work complies with the Building Regulations, if this work has been completed using short-lived materials (i.e. materials that are, in the absence of special care, liable to rapid deterioration), the local authority can:

- reject the plans;
- pass the plans subject to a limited use clause (on expiration of which they will have to be removed);
- restrict the use of the building.

Unsuitable materials

If, once building work has begun, it is discovered that it has been made using materials or components that have been identified as being unsuitable materials, the local authority has the power to:

- reject the plans;
- fix a period in which the offending work must be removed;
- restrict the use of the building.

 If the person completing the building work fails to remove the unsuitable material or component(s), then that person is liable to be prosecuted and, on summary conviction, faces a heavy fine.

6.0.7 What about technical specifications?

Building Regulations may be made for specific purposes such as:

- health and safety;
- welfare and convenience of disabled people;
- conservation of fuel and power;
- prevention of waste or contamination of water.

These are aimed at furthering the protection of the environment, facilitating sustainable development or the prevention and detection of crime.

Although the main requirements for health and safety are now covered by the Building Regulations, there are still some requirements contained in the Workplace (Health, Safety and Welfare) Regulations 1992 that may need to be considered as they could contain requirements which affect building design. For further information see Workplace (Health, Safety and Welfare) Regulations 1992. Approved Code of Practice L24, published by HSE Books 1992 (ISBN 0 7176 0413 6).

Standards and technical approvals, as well as providing guidance, address other aspects of performance such as serviceability and/or other aspects related to health and safety not covered by the Regulations.

When a Technical Handbook makes reference to a named standard, the relevant version of the standard is the one listed at the end of that particular Technical Handbook. However, if this version of the standard has been revised or updated by the issuing standards body, the new version may be used as a source of guidance; provided it continues to address the relevant requirements of the Regulations.

6.0.8 What about independent certification schemes?

Within the United Kingdom, there are many product certification schemes that certify compliance with the requirements of a recognized standard or document that is suitable for the purpose and material being used. Certification Bodies which approve such schemes will normally be accredited by UCAS.

6.0.9 What about European Pre-standards (ENV)?

The British Standards Institution (BSI) will be issuing Pre-standard (ENV) Structural Eurocodes as they become available from the European Standards Organization, Comité Européen de Normalization Electrotechnique (CEN).

DD ENV 1992-1-1: 1992 Eurocode 2: Part 1 and DD ENV 1993-1-1: 1992 Eurocode 3: Part 1-1 General Rules and Rules for Buildings in concrete and steel have been thoroughly examined over a period of several years and are considered to provide appropriate guidance when used in conjunction with their national application documents for the design of concrete and steel buildings, respectively.

When other ENV Eurocodes have been subjected to a similar level of examination, they may also offer an alternative approach to Building Regulation compliance and, when they are eventually converted into fully approved EN standards, they will be included as referenced standards in the guidance documents.

If a national standard is going to be replaced by a European harmonized standard, then there will be a coexistence period during which either standard may be referred to. At the end of the coexistence period the national standard will be withdrawn.

Although covering similar generic requirements, the guidance contained in Approved Documents (ADs) supporting the English and Welsh Building Regulations, whilst seeming to be equivalent, **might not** be entirely suitable as an alternative solution to a system-specific Scottish requirement.

6.1 Design and construction

The Building (Scotland) Regulations are primarily concerned with two types of construction: domestic buildings and industrial (i.e. non-domestic) buildings. Each type of construction requires a unique team to design, build, operate and maintain (DBOM) the project.

Although the vast majority of building construction projects are small renovations, such as the addition of a conservatory or the renovation of an existing room (and frequently the owner of the property acts as labourer, paymaster and design team for the entire project), all building construction projects include some elements in common and the following are minimum requirements for the design and construction of buildings.

6.1.1 Requirements

Domestic and non-domestic

> *Every building must be designed and constructed in such a way that:*
>
> - *the loadings that are liable to act on it, taking into account the*   *1.1*
>   *nature of the ground, will not lead to:*
>   - *the collapse of the whole or part of the building;*
>   - *deformations which would make the building unfit for its*
>   *intended use, unsafe, or cause damage to other parts of the*
>   *building or to fittings or to installed equipment;*
>   - *impairment of the stability of any part of another building;*
> - *in the event of damage occurring to any part of the structure*   *1.2*
>   *of the building, the extent of any resultant collapse will not be*
>   *disproportionate to the original cause.*
>
> *Every building, which is divided into more than one area of different*
> *occupation, must be designed and constructed in such a way that in*   *2.3*
> *the event of an outbreak of fire within the building:*
>
> - *the load-bearing capacity of the building will continue to*
>   *function until all occupants have escaped, or been assisted to*
>   *escape, from the building and any fire containment measures*
>   *have been initiated;*
> - *the unseen spread of fire and smoke within concealed spaces in*   *2.4*
>   *its structure and fabric is inhibited;*

- *the development of fire and smoke from the surfaces of walls and ceilings within the area of origin is inhibited;*   *2.5*
- *the spread of fire to neighbouring buildings is inhibited;*   *2.6*
- *the spread of fire on the external walls of the building is inhibited;*   *2.7*
- *the occupants, once alerted to the outbreak of the fire, are provided with the opportunity to escape from the building, before being affected by fire or smoke;*   *2.9*
- *illumination is provided to assist in escape;*   *2.10*
- *the occupants are alerted to the outbreak of fire.*   *2.11*

Every building must be accessible to fire appliances and fire service   *2.12*
personnel, and every building must be designed and constructed in such a way that:

- *facilities are provided to assist fire-fighting or rescue operations;*   *2.14*
- *in the event of an outbreak of fire within the building, fire and*   *2.15*
  smoke will be inhibited from spreading through the building by the operation of an automatic life safety fire suppression system;
- *in the event of an outbreak of fire in a neighbouring building,*   *2.8*
  the spread of fire to the building is inhibited.

Additional mandatory requirements for non-domestic buildings

Every non-domestic building must be:

- *designed and constructed in such a way that in the event of an*   *2.1*
  outbreak of fire within the building, fire and smoke are inhibited from spreading beyond the compartment of origin until any occupants have had the time to leave that compartment and any mandatory fire containment measures have been initiated;
- *provided with a water supply for use by the fire service.*   *2.13*

6.1.2 Meeting the requirements

Structure

By its very nature, the structure of a building is fundamental to ensuring the safety of people in or around new and existing buildings, and can be affected by a number of environmental factors such as temperature, snow, wind, driving rain and flooding, and the impact of climate change – both inside and outside the building.

To prevent the collapse, excessive deformation or disproportionate collapse of buildings, the climatic conditions in Scotland should be carefully considered in the assessment of loadings and in the structural design of buildings.

The structure of a building shall not pose a threat to the safety of people in or around buildings and existing buildings as a result of:

D 1.02
ND 1.02

- loadings;
- the nature of the ground;
- collapse or deformations;
- stability of the building and other buildings;
- climatic conditions;
- materials;
- structural analysis;
- details of construction;
- safety factors.

All buildings shall be designed and constructed in such a way that the loadings that are liable to act on it, taking into account the nature of the ground, will not lead to:

D 1.1.0
ND 1.1.2

- the collapse of the whole or part of the building;
- deformations which would make the building unfit for its intended use, unsafe, or cause damage to other parts of the building or to fittings or to installed equipment;
- impairment of the stability of any part of another building.

The design and construction of a building should be carried out in accordance with the following British Standards (Table 6.1).

Table 6.1   Design and construction standards

Type of construction	Standard
For foundations	BS 8004: 1986
For structural work of reinforced, pre-stressed or plain concrete	BS 8110-1: 1997, BS 8110-2: 1985 and BS 8110-3: 1985
For structural work of steel	BS 5950-1: 2000, BS 5950-2: 2001, BS 5950-5: 1998, BS 5950-6: 1995, BS 5950-7: 1992 and BS 5950-8: 2003
For structural work of composite steel and concrete construction	BS 5950-3.1: 1990 and BS 5950-4: 1994
For structural work of aluminium	BS 8118-1: 1991 and BS 8118-2: 1991

Note: For the purpose of section 7.2 of BS 8118-1: 1991, the structure should be classified as a safe-life structure

Type of construction	Standard
For structural work of masonry	BS 5628-1: 2005, BS 5628-2: 2005 and BS 5628-3: 2005
For structural work of timber	BS 5268-2: 2002, BS 5268-3: 2006 and BS 5268-6.1: 1996
For earth retaining structures (e.g. basements)	BS 8002: 1994
For structural design of low rise buildings	BS 8103-1: 1995, BS 8103-2: 2005, BS 8103-3: 1996 and BS 8103-4: 1995

The design and construction of the building shall take into account:

- all aspects of the nature of the ground;
- ground movement caused by:
  - swelling, shrinkage or freezing of the subsoil;
  - landslip or subsidence (other than subsidence arising from shrinkage);
- recorded conditions of ground instability (e.g. caused by landslides, disused mines or unstable strata).

D 1.1.4
ND 1.1.4

The design and construction of buildings shall:

- ensure that it will not collapse during its design lifetime;
- ensure that the building does not fail or collapse in normal use owing to deformations of the building;
- take into account their potential impact on existing buildings.

D 1.1.0
ND 1.1.0

Conversions

The building when converted, shall:

- meet the requirements of the Building (Scotland) Regulations;
- be no worse than before the conversion.

D 1.1.0
ND 1.1.0

Disproportionate collapse

A building which is said to be susceptible to disproportionate collapse is one where the effects of accidents and, in particular, situations where damage to small areas of a structure or failure of single elements could lead to collapse of major parts of the structure.

Every building must be designed and constructed so that if the structure of the building is damaged, any resultant collapse shall be in proportion to the original cause.

D 1.2
ND 1.2

Note: This requirement was introduced in the United Kingdom following the Ronan Point disaster in 1968.

Buildings should be provided with a level of robustness by subjecting them to a risk analysis process (taking into account both the risk of the hazard and its consequences – see Table 6.2) and by ensuring that additional measures are then made available according to the level of risk and consequences of the building collapsing.

D 1.2.1
ND 1.2.1

Table 6.2 Building risk groups

Risk	Building type group	Additional measures
1	• Agricultural and agricultural-related buildings • Carports, conservatories and greenhouses • Domestic garages and other small single-leaf buildings not more than one storey • Houses not more than four storeys	Provided the building has been designed and constructed in accordance with the Technical Handbooks and associated standards and is in normal use, no additional measures are required.
2A	• Five-storey houses • Assembly buildings (other than educational buildings), entertainment buildings and buildings accessible to the general public less than two storeys and each storey area less than 2,000 m² • Factories (class 2) not more than three storeys • Flats and maisonettes not more than four storeys • Hotels not more than four storeys • Offices not more than four storeys • Shared residential accommodation, residential care buildings and other residential buildings all not more than four storeys • Shops/enclosed shopping centres less than three storeys and each storey less than 2,000 m² • Single-storey educational buildings	Provide effective horizontal ties, or effective anchorage of suspended floors to all framed and load-bearing wall constructions.

(Continued)

Table 6.2   (Continued)

Risk	Building type group	Additional measures
2B	• Assembly buildings (other than educational buildings), entertainment buildings and other buildings accessible to the general public all not more than two storeys and all with each storey area more than $2,000\,m^2$ but not more than $5,000\,m^2$ • Educational buildings; more than one storey but not more than 15 storeys • Flats and maisonettes more than two storeys but not more than 15 storeys • Hotels, shared residential accommodation, residential care buildings and other residential buildings all more than four storeys but not more than 15 storeys • Shops and enclosed shopping centres less than three storeys and with each storey area more than $2,000\,m^2$ • Shops and enclosed shopping centres more than three storeys but not more than 15 storeys • Hospitals not more than three storeys • Offices more than four storeys but not more than 15 storeys • Open-sided car parks and storage buildings (class 2) more than six storeys	Provide effective horizontal ties, or effective anchorage of suspended floors to all framed and load-bearing wall constructions, together with effective vertical ties, in all supporting columns and walls.
3	• Domestic buildings not covered in Risk Groups 1, 2A and 2B • Every non-domestic building not covered in risk groups 1, 2A and 2B • Grandstands accommodating more than 5,000 spectators • Storage buildings (class 1), factories (class 1) and places of special fire risk	Complete a systematic risk assessment of the building taking into account all foreseen normal hazards and any possible abnormal hazards. Then select appropriate protective measures.

Stability

Buildings should be stable whatever the combinations of dead load, imposed load and wind loading in terms of their individual structural elements, their interaction together and the overall stability as a structure.

The overall size and proportions of the building should be limited in accordance with the specific guidance for each form of construction.	D 1B1a
A layout of internal walls and external walls forming a robust three-dimensional box structure in plan should be constructed in accordance with the specific guidance for each form of construction.	D 1B1b

Foundations

Foundations should be designed so that the loadings (i.e. dead loads, imposed loads and wind loads) that are transmitted from the building to the subsoil will not cause undue settlement.	D 1C0

Size of domestic and timber-framed buildings

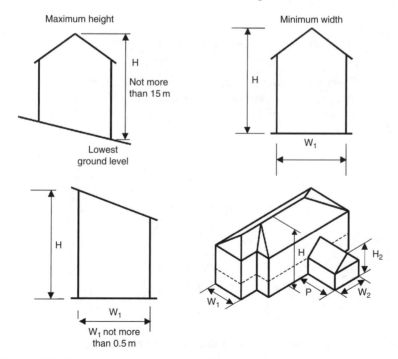

Figure 6.1 Dimensions of domestic and timber-framed buildings

For domestic buildings (no greater than three storeys high) and timber-framed buildings:

The maximum height of the building (H) measured from the lowest finished ground level adjoining the building to the highest point of any wall or roof should be not more than:	D 1D2a D 1E3a
• 15 m for domestic buildings; • 10 m for duo-pitched timber-framed buildings; • 5.5 m for mono-pitch or flat roofs on timber-framed buildings. The height of a domestic or timber-framed building (H) should be not more than twice the least width of the building W_1.	D 1D2a D 1E3b
The height of the wing H_2 of a domestic or timber-framed building should be not more than twice the least width of the wing (W_2) when the projection P is more than twice the width W_2.	D 1D2a D 1E3c

Size of extensions to domestic buildings

For single-storey, single-leaf extensions to domestic buildings (including garages and outbuildings):

The height H should be not more than the limits shown in Figure 6.2.	D 1D2b
Note: H is measured from the top of the foundation or from the underside of the floor slab where this provides effective lateral restraint.	

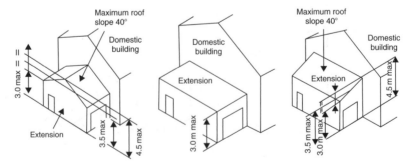

Figure 6.2 Dimensions of extensions to domestic buildings

Size of single-storey, single-leaf buildings

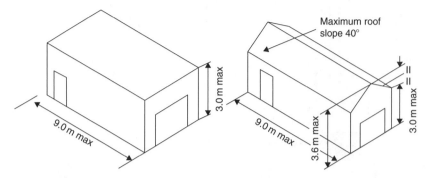

Figure 6.3 Dimensions of single-storey, single-leaf buildings

For single-storey, single-leaf buildings forming a garage or an outbuilding within the curtilage of a dwelling:

- the height H should be not more than 3 m;   D 1D2b
- the length should be not more than 9 m.

 Note: H is measured from the top of the foundation or from the underside of the floor slab where this provides effective lateral restraint.

Size of apartments

Every apartment should be of a size that will accommodate at   D 3.11.1
least a bed, a wardrobe and a chest of drawers.

Notes:
1. Activity spaces for furniture may overlap.
2. A built-in wardrobe space of equal size may be provided as an alternative to a wardrobe.

Enhanced apartment

Smaller apartments or those with an unusual shape may limit how space within can be used.

At least one apartment on the principal living level of a dwelling should be of a size and form that allows greater flexibility of use.

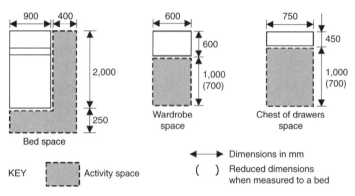

Bed space

KEY [Activity space]

Dimensions in mm

( )   Reduced dimensions
      when measured to a bed

Notes:
1. Activity spaces for furniture may overlap.
2. A built-in wardrobe space of equal size may be provided as an alternative to a wardrobe.

Figure 6.4 Apartment dimensions

Enhanced apartments should: D 3.11.2

• have a floor area of at least 12 m² and a length and width of
  at least 3.0 m;
  – this area should exclude any space less than 1.8 m in height
    and any portion of the room designated as a kitchen;
• contain an unobstructed manoeuvring space of at least 1.5
  by 1.5 m² or an ellipse of at least 1.4 by 1.8 m, which may
  overlap with activity spaces.

Maximum floor area

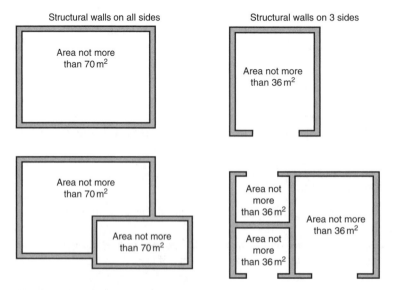

Figure 6.5 Maximum floor area

The maximum floor area that may be enclosed by:

- structural walls on all sides shall not be greater than $70\,m^2$;   D 1D3
- a structural wall on one side shall not be greater than $36\,m^2$.   D 1E4

Method used to measure storeys, walls, panels and building heights
Building heights are measured from the lowest finished ground level to the highest point of the roof as shown in Figure 6.6. Panel heights are measured from the underside of the bottom rail to the top of the top rail.

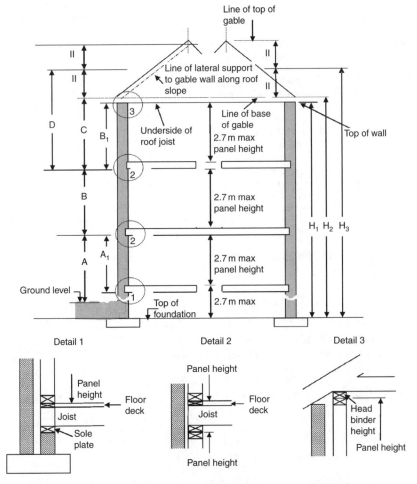

Figure 6.6 Rules for measuring the heights of storeys and walls

Measuring storey heights		Measuring wall heights	
A	is the ground storey height if the ground floor is a suspended timber floor or a structurally separate ground floor slab.	H^1	is the height of a wall that does not include a gable.
A1	is the ground storey height if ground floor is a suspended concrete floor bearing on the external wall.	H^2	is the height of a separating wall which may extend to the underside of the roof.
B	is the intermediate storey height.	H^3	is the height for a wall (except a separating wall) which includes a gable.
B1	is the top storey height for walls which do not include a gable.		
C	is the top storey height where lateral support is given to gable at both ceiling level and along the roof slope.		
D	is the top storey height for walls which include a gable where lateral support is given to the gable only along the roof slope.		

Size and proportions of openings

Other than windows and a single-leaf door that meet the requirements shown in Figure 6.7:

No more than two major openings (maximum height 2.1 m, maximum width 5.0 m) are allowed in any one wall of the building or extension, either a single opening or the combined width of two openings.	D 1D40
No other openings shall be within 2.0 m of a wall containing a major opening.	D 1D40
The total size of openings in a wall not containing a major opening should not exceed 2.4 m².	1D40
There should not be more than one opening between piers.	D 1D40
The distance from a window or a door to a corner should be at least 390 mm unless there is a corner pier.	D 1D40

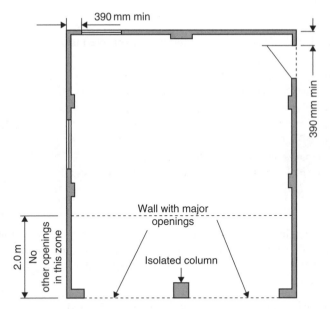

Figure 6.7 Size of openings in small single-storey, single-leaf buildings

Timber-frame walls

In Scotland, more and more use is being made of small, timber-framed buildings with external masonry cladding as they can more easily be designed to take into account loading conditions and/or restrictions on dimensions and openings.

These type of buildings typically consist of full-height timber wall panels for each storey built on to the floor below and with intermediate floors built on top of the wall panels. The roof is constructed on top of the top storey wall panels with the masonry cladding connected to the timber panels by wall ties.

When constructing a building that is not more than two storeys:

• the building should be rectangular or square shape in plan.	D 1E2c
• the roof construction should be:	D 1E2g
– flat, raised tie or collared roofs;	
– duo- or mono-pitch trussed rafters with 15–45° pitch and dead load not more than 1.04 kN/m² on the slope.	

Construction materials

Construction materials and methods for simple platform timber-framed buildings are restricted to those materials, timber strength classes, specifications and dimensions **which are most commonly used in Scotland.**	D 1E28

Fire

The overall aim of fire safety precautions is to ensure that:

• buildings should be designed and constructed in such a way that the risk of fire is reduced;	D 2.0.1
	ND 2.0.1
• if a fire does occur, measures shall be in place:	D 2.0.1
– to restrict the growth of fire and smoke;	ND 2.0.1
– to warn occupants as soon as possible;	
– to enable the occupants to escape safely;	
– to enable fire-fighters to deal with the fire safely and effectively.	

Fire safety engineering is suitable for solving problems concerning the design of the building which, although meeting the requirements of the Building (Scotland) Regulations, is still problematic. Factors that should be taken into account include:

- the ability of a structure to resist the spread of fire and smoke;
- the anticipated probability of a fire occurring;
- anticipated fire severity;
- consequential danger to people in or around the building.

Structural protection

Buildings must be designed and constructed in such a way that in the event of an outbreak of fire within the building, the load-bearing capacity of the building will continue to function until all occupants have escaped, or been assisted to escape, from the building and any mandatory fire containment measures have been initiated.	D 2.3
	ND 2.3
In order to prevent the premature collapse of the load-bearing structural elements of a building, appropriate levels of fire resistance duration should be provided to all elements of structure.	D 2.3.0
	ND 2.3.0
Structural fire protection shall minimize the risk to the occupants and fire-fighters (who may be engaged in fire-fighting or rescue operations).	D 2.3.0
	ND 2.3.0
During a fire, the elements of structure should continue to function and remain capable of supporting and retaining the fire protection to floors, escape routes and fire access routes, until all occupants have escaped (assisted to escape by staff or been rescued by the fire service).	D 2.3.0
	ND 2.3.0

Spread to neighbouring buildings

In order to reduce the danger to the occupants of other buildings, one building should be isolated from another by either construction or distance – the distance between a building and its relevant boundary being dictated by the amount of heat that is likely to be generated in the event of fire, influenced by the extent of openings, or other unprotected areas in the external wall of the building.

Detached non-domestic buildings (in the same occupation)

There may be a risk of fire spread between buildings, even when on land in the same occupation. A notional boundary therefore is used to determine the safe distance between buildings or compartments in this situation.

In order to establish whether a notional boundary calculation is necessary, the size of the opposing buildings or compartments should be established first (see Figure 6.8). Where the combined area of building A plus building B exceed the maximum allowable area of any compartment, then a notional boundary calculation is necessary.

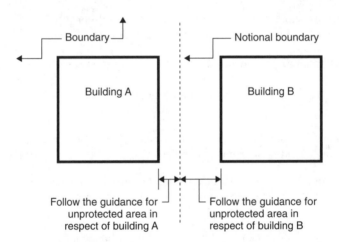

Figure 6.8 Notional boundary

The designer is free to set this notional boundary at any point between the two buildings under consideration, and this distance will determine the amount of unprotected area. The following applies:

The buildings should be separated by a distance not less than the sum of the distances calculated in respect of each building relative to a notional boundary.	ND 2.6.5
The roofs of the buildings should be separated by a distance not less than the sum of the distances provided in relation to the notional boundary in respect of each roof.	ND 2.6.5

Where a notional boundary is to be established between
two buildings, one of which is an existing building, for the
purposes of this guidance, the existing building should be
regarded as if it were a new building.

ND 2.6.5

Environment

These days, when you hear people talk about 'the environment', they are often
referring to the overall condition of our planet, or how healthy it is. From a
building perspective, the environment is everything that makes up our surroundings and affects our ability to live comfortably and in a sustainable manner.

Every building must be designed and constructed in such a
way that there will not be a threat to the building, the occupant
or the health of people in or around the building:

D 3.1

- due to the presence of harmful or dangerous substances;
- due to the emission and containment of radon gas;

ND 3.1

D 3.2
ND 3.2

- as a result of flooding and the accumulation of groundwater;

D 3.3
ND 3.3

- as a result of moisture penetration from the ground

D 3.4
ND 3.47

Buildings must **not** be constructed over an existing drain
including a field drain that is to remain active.

D 3.5
ND 3.5

Every building, and hard surface within the curtilage of a
building, must be designed and constructed with a surface
water drainage system that will:

D 3.6
ND 3.6

- ensure the disposal of surface water without threatening
  the building and the health and safety of the people in and
  around the building;
- have facilities for the separation and removal of silt, grit
  and pollutants.

Every private wastewater treatment plant or septic tank serving
a building must be designed and constructed in such a way that
it will ensure the safe temporary storage and treatment of wastewater prior to discharge.

D 3.8
ND 3.8

Every private wastewater treatment system serving a building
must be designed and constructed in such a way that the
disposal of the wastewater to ground is safe and is not a threat
to the health of the people in or around the building.

D 3.9
ND 3.9

Every building must be designed and constructed in such a way that:

D 3.10
ND 3.10

- there will not be a threat to the building or the health of the occupants as a result of moisture from precipitation penetrating to the inner face of the building;
- the size of any apartment or kitchen will ensure the welfare and convenience of all occupants and visitors;

  D 3.1
  ND 3.11

- an accessible space is provided to allow for the safe, convenient and sustainable drying of washing;
- sanitary facilities are provided for all occupants of, and visitors to, the building in a form that allows convenience of use and that there is no threat to the health and safety of occupants or visitors;

  D 3.12
  ND 3.12

- the air quality inside the building is not a threat to the health of the occupants or the capability of the building to resist moisture, decay or infestation;

  D 3.14
  ND 3.14

- each fixed combustion appliance installation operates safely;

  D 3.17
  ND 3.17

- any component part of a fixed combustion appliance installation used for the removal of combustion gases will withstand heat generated as a result of its operation without any structural change that would impair the stability or performance of the installation;

  D 3.18
  ND 3.18

- any component part of a fixed combustion appliance installation will not cause damage to the building in which it is installed by radiated, convected or conducted heat or from hot embers expelled from the appliance;

  D 3.19
  ND 3.19

- the products of combustion are carried safely to the external air without harm to the health of any person through leakage, spillage or exhaust, nor permit the re-entry of dangerous gases from the combustion process of fuels into the building;

  D 3.20
  ND 3.20

- each fixed combustion appliance installation receives air for combustion and operation of the chimney so that the health of persons within the building is not threatened by the build-up of dangerous gases as a result of incomplete combustion;

  D 3.21
  ND 3.21

- each fixed combustion appliance installation receives air for cooling so that the fixed combustion appliance installation will operate safely without threatening the health and safety of persons within the building;

  D 3.22
  ND 3.22

- an oil storage installation, incorporating oil storage tanks used solely to serve a fixed combustion appliance installation providing space heating or cooking facilities in a building, will inhibit fire from spreading to the tank and its contents from within, or beyond the boundary;

  D 3.23
  ND 3.23

- a container for the storage of woody biomass fuel will inhibit fire from spreading to its contents from within, or beyond the boundary;
- an oil storage installation, incorporating oil storage tanks used solely to serve a fixed combustion appliance installation providing space heating or cooking facilities in a building will:
  - reduce the risk of oil escaping from the installation;
  - contain any oil spillage likely to contaminate any water supply, groundwater, watercourse, drain or sewer;
  - permit any spill to be disposed of safely;
- a container for the storage of woody biomass fuel will inhibit fire from spreading to its contents from within, or beyond the boundary;
- the volume of woody biomass fuel storage allows the number of journeys by delivery vehicles to be minimized;
- there will not be a threat to the health and safety of people from a dungstead and farm effluent tank.

	D 3.24
	ND 3.24
	D 3.23
	ND 3.23
	D 3.24
	ND 3.24
	D 3.26
	ND 3.26

Additional requirements for domestic buildings

Every building must be designed and constructed in such a way that:

- it can be heated;
- there will not be a threat to the building or the health of the occupants as a result of moisture caused by surface or interstitial condensation;
- natural lighting is provided to ensure that the health of the occupants is not threatened;
- accommodation for solid waste storage is provided which:
  - permits access for storage and for the removal of its contents;
  - does not threaten the health of people in and around the building;
  - does not contaminate any water supply, groundwater or surface water.

	D 3.13
	D 3.15
	D 3.16
	D 3.25

Additional requirements for non-domestic dwellings

Every wastewater drainage system serving a building must be designed and constructed in such a way as to ensure the removal of wastewater from the building without threatening the health and safety of the people in and around the building.   ND 3.7

Hazard identification and assessment

A preliminary desk-top study should be carried out to provide information on the past and present uses of the proposed building site and surrounding area that may give rise to contamination.

Risk assessment should be specific to each building site and take into account the presence of source, pathways and receptors at a particular building site.	D 3.1.3 ND 3.1.3

Site preparation – harmful and dangerous substances

Every building must be designed and constructed in such a way that there will not be a threat to the building or the health of people in or around the building due to the presence of harmful or dangerous substances.	D 3.1 ND 3.1

Site preparation – protection from radon gas

Every building must be designed and constructed in such a way that there will not be a threat to the health of people in or around the building due to the emission and containment of radon gas.	D 3.2 ND 3.2

Flooding and groundwater

Every building must be designed and constructed in such a way that there will not be a threat to the building or the health of the occupants as a result of flooding and the accumulation of groundwater.	D 3.3 ND 3.3

Heating in dwellings

Heating, ventilation and thermal insulation in dwellings should be considered as part of a total design that takes into account all heat gains and losses. Failure to do so can lead to inadequate internal conditions, e.g. condensation and mould and the inefficient use of energy due to overheating.

Every building must be designed and constructed in such a way that it can be heated.	D 3.13
Every dwelling should have some form of fixed heating system, or alternative that is capable of maintaining a temperature of 21°C in at least one apartment and 18°C elsewhere, when the outside temperature is −1°C.	D 3.13.1

Ventilation

Ventilation of a building is required to prevent the accumulation of moisture that could lead to mould growth, and pollutants originating from within the building that could become a risk to the health of the occupants.

Every building must be designed and constructed in such a way that the air quality inside the building is not a threat to the health of the occupants or the capability of the building to resist moisture, decay or infestation.	D 3.14 ND 3.14
A building should have provision for ventilation by either: • natural means; • mechanical means; • a combination of natural and mechanical means.	D 3.14.1 ND 3.14.1
Ventilation should have the capability of: • providing outside air to maintain indoor air quality sufficient for human respiration; • removing excess water vapour from areas where it is produced in significant quantities, such as kitchens, utility rooms, bathrooms and shower rooms to reduce the likelihood of creating conditions that support the germination and growth of mould, harmful bacteria, pathogens and allergens; • removing pollutants that are a hazard to health from areas where they are produced in significant quantities, such as non-flued combustion appliances; • rapidly diluting pollutants and water vapour, where necessary, that are produced in apartments and sanitary accommodation.	D 3.14.1 ND 3.14.1
Ventilation should be to the outside (i.e. external) air.	D 3.14.1 ND 3.14.1

Safety

Safety has been defined by the International Standards Organization as '*a state of freedom from unacceptable risks of personal harm*'. This recognizes that:

• no activity is absolutely safe or free from risk;
• no building can be absolutely safe and some risk of harm to users may exist in every building.

Building standards seek to limit risk to an acceptable level by identifying hazards in and around buildings that can be addressed through the Building (Scotland) Regulations.

The Building (Scotland) Regulations provide recommendations for the design of buildings that will ensure access and usability, and reduce the risk of accident. Their aim is to:

- ensure accessibility to and within buildings and that areas presenting risk through access are correctly guarded;
- reduce the incidence of slips, trips and falls, particularly for those users most at risk;
- ensure that electrical installations are safe in terms of the hazards likely to arise from defective installations, namely fire and loss of life or injury from electric shock or burns;
- prevent the creation of dangerous obstructions, ensure that glazing can be cleaned and operated safely and to reduce the risk of injury caused by collision with glazing;
- safely locate hot water and steam vent pipe outlets, and minimize the risk of explosion through malfunction of unvented hot water storage systems and prevent scalding by hot water from sanitary facilities;
- ensure the appropriate location and construction of storage tanks for liquefied petroleum gas.

Note: It is unlawful to discriminate against a person on the grounds of that person's disability and steps should be taken to provide against this happening. The Disability Discrimination Act (DDA) requires service providers to consider any requirement where a physical feature places a disabled person at a disadvantage and to take reasonable steps to remedy the situation.

The DDA applies generally, with some limited exclusions, to all non-domestic buildings including:

- places of employment;
- any building used to provide goods or services to the public;
- places of education.

The Disability Discrimination (Providers of Services) (Adjustment of Premises) (Amendment) Regulations 2005 offers a limited exemption to physical elements of a building designed and constructed in compliance with the Building (Scotland) Regulations. Under this exemption, it is not considered reasonable for a service provider to be required to make further adjustment to compliant elements for a period of 10 years from the date that work was completed.

At the time of publication, it is not yet known whether this exemption will remain in place from May 2007.

Every building must be designed and constructed in such a way that:	
• all occupants and visitors are provided with safe, convenient and unassisted means of access to the building.	D 4.1 ND 4.1
• in non-domestic buildings, safe, unassisted and convenient means of access is provided throughout the building;	D 4.2 ND 4.2

- in residential buildings, a proportion of the rooms intended to be used as bedrooms must be accessible to a wheelchair user;
- in domestic buildings, safe and convenient means of access is provided within common areas and to each dwelling;
- in dwellings, safe and convenient means of access is provided throughout the dwelling;
- in dwellings, unassisted means of access is provided to, and throughout, at least one level;
- every level can be reached safely by stairs or ramps; D 4.3
  ND 4.3
- every sudden change of level that is accessible in or around D 4.4
  the building is guarded by the provision of pedestrian ND 4.4
  protective barriers;
- the electrical installation does not: D 4.5
  - threaten the health and safety of the people in and ND 4.5
    around the building;
  - become a source of fire;
- people in and around the building are protected from D 4.8
  injury that could result from fixed glazing, projections or ND 4.8
  moving elements on the building;
- fixed glazing in the building is not vulnerable to breakage where there is the possibility of impact by people in and around the building;
- both faces of a window and rooflight in a building are capable of being cleaned such that there will not be a threat to the cleaner from a fall resulting in severe injury;
- protection is provided for people in and around the D 4.9
  building from the danger of severe burns or scalds from ND 4.9
  the discharge of steam or hot water;
- each liquefied petroleum gas storage installation used D 4.11
  solely to serve a combustion appliance providing space ND 4.11
  heating, water heating or cooking facilities will:
  - be protected from fire spreading to any liquefied petroleum gas container;
  - not permit the contents of any such container to form explosive gas pockets in the vicinity of any container.

Every building accessible to vehicular traffic must be D 4.12
designed and constructed in such a way that every change in ND 4.12
level is guarded.

Buildings should be designed to consider safety and the D 4.0.1
welfare and convenience of building users. ND 4.0.1

Additional requirements for non-domestic buildings

Every building must be designed and constructed in such a way that: ND 4.6

- electric lighting points and socket outlets are provided to ensure the health, safety and convenience of occupants and visitors;
- it is provided with aids to assist those with a hearing impairment;   ND 4.7
- people in and around the building are protected from injury that could result from fixed glazing, projections or moving elements on the building;   ND 4.8
- fixed glazing in the building is not vulnerable to breakage where there is the possibility of impact by people in and around the building;
- both faces of a window and rooflight in a building are capable of being cleaned such that there will not be a threat to the cleaner from a fall resulting in severe injury;
- a safe and secure means of access is provided to a roof; and manual controls for ventilation and for electrical fixtures can be operated safely.

Every building which contains fixed seating accommodation for an audience or spectators must be designed and constructed in such a way that a number of level spaces for wheelchairs are provided proportionate to the potential audience or spectators.   ND 4.10

Noise

There are currently no Building Standards to protect the occupants or users of a non-domestic building from noise (within or without), but the need may well arise for such standards at a later date.

Every building must be designed and constructed in such a way that each wall and floor separating one dwelling from another, or one dwelling from another part of the building, or one dwelling from a building other than a dwelling, will limit the transmission of noise to the dwelling to a level that will not threaten the health of the occupants of the dwelling or inconvenience them in the course of normal domestic activities provided the source noise is not in excess of that from normal domestic activities.   D 5.1
ND 5.1

Performance testing

Use of the performance testing approach is particularly useful where the separating or flanking construction is of innovative design and for conversions where flanking transmission may be significant.

Performance values are given in terms of two acoustic parameters; one related to airborne sound, the other related to impact sound.

The airborne sound insulation characteristic of a wall or floor is the sound pressure level difference between the room containing the noise source and the receiving room. The larger the difference, the higher the level of airborne sound insulation.

Impact sound insulation is the sound pressure level in the receiving room.

Recommended performance values for separating walls and separating floors are given in Table 6.3.

Table 6.3 Recommended performance values for separating walls and separating floors

Airborne sound (minimum values)
Minimum values of weighted standardized level difference as defined in BS EN ISO 717-1: 1997:

	Mean value (dB)	Individual value (dB)
Walls	53	49
Floors	52	48

Impact sound (minimum values)
Maximum values of weighted standardized level difference as defined in BS EN ISO 717-2: 1997:

	Mean value	Individual value
Floors	62 dB	65 dB

Notes: Annex 5.A to the Building (Scotland) Regulations describes methods for calculating the mass of masonry wall leafs (mortar joints, *in situ* concrete, screeds, slabs and composite floor bases).

Annex 5.B describes methods for the selection of resilient materials used for soft coverings.

Annex 5.C describes methods of measurement and test procedures.

The scheme operated by Robust Details Ltd for the English and Welsh Building Regulations is also a useful measurement method, but at the time of writing this book, this scheme has not yet been fully reviewed in relation to construction practice in Scotland.

Energy

The current edition of the Building (Scotland) Regulations focuses on the reduction of carbon dioxide emissions arising from the use of heating, hot water and lighting in buildings, and its guidance sets an overall level for maximum carbon dioxide emissions in buildings incorporating a range of parameters which will influence energy use.

 This means that for **all** new buildings **designers are now obliged** to consider energy as a complete package, rather than only looking at individual elements such as insulation or boiler efficiency favouring, and localized or building-integrated Low and Zero Carbon Technologies (LZCT) (e.g. photovoltaics, active solar water heating, combined heat and power and heat pumps) can be used as a contribution towards meeting this standard.

Every building must be designed and constructed in such a way that:	D 6.1 ND 6.1
• the energy performance is calculated in accordance with a methodology which is asset based, conforms with the European Directive on the Energy Performance of Buildings 2002/91/EC and uses the UK climate data; • the energy performance of the building is capable of reducing carbon dioxide emissions.	
Every building must be designed and constructed in such a way that:	D 6.2 ND 6.2
• an insulation envelope is provided which reduces heat loss; • the heating and hot water service systems installed are energy efficient and are capable of being controlled to achieve optimum energy efficiency;	D 6.3 ND 6.3
• temperature loss from heated pipes, ducts and vessels, and temperature gain to cooled pipes and ducts, are resisted;	D 6.4 ND 6.4
• the artificial or display lighting installed is energy efficient and is capable of being controlled to achieve optimum energy efficiency;	D 6.5 ND 6.5
• the form and fabric of the building minimizes the use of mechanical ventilating or cooling systems for cooling purposes; • in non-domestic buildings, the ventilating and cooling systems installed are energy efficient and are capable of being controlled to achieve optimum energy efficiency;	D 6.6 ND 6.6
• energy supply systems and building services which use fuel or power for heating, lighting, ventilating and cooling the internal environment and heating are efficient;	D 6.7 ND 6.7
• each part of a building designed for different occupation is fitted with fuel consumption meters.	D 6.10 ND 6.10

Maximum *U*-values

The design of a dwelling should involve extensive use of building-integrated or localized LZCT levels of thermal insulation.	D 6.2.1 LD 6.2.1
The maximum *U*-values for building elements of the insulation envelope should not be greater than those shown in Table 6.4.	D 6.2.1 LD 6.2.1

Table 6.4   Maximum *U*-values for building elements of the insulation envelope

Type of element	(a) Area-weighted average *U*-value (W/m²K) for all elements of the same type	(b) Individual element *U*-value (W/m²K)
Wall*	0.30	0.70
Floor*	0.25	0.70
Roof	0.20 (0.25 for non-domestic buildings)	0.35
Windows, doors and rooflights	2.2	3.3

Note: *Excluding separating walls and separating floors where thermal transmittance should be ignored.

Resisting heat loss through thermal bridging

The insulation envelope of the dwelling (or building consisting of dwellings) should be constructed in such a way that there are no substantial thermal bridges or gaps where the layers of insulation occur.	D 6.2.3 ND 6.2.4

Limiting air infiltration

All building fabric will allow a certain degree of air leakage, and it is virtually impossible to make the insulation envelope 100% airtight. However: • if it is necessary to either vent or ventilate the building fabric to the outside air (e.g. to allow moisture due to either precipitation or condensation to escape), then this should be designed into the construction; • measures should be introduced, however, to reduce unwanted air leakage and thereby prevent an increase in energy use within the heated part of the building.	D 6.2.4 ND 6.2.5

Air-tightness testing

At the time of publication, air-tightness testing need only be carried out when better than routine air-tightness levels are declared at the building warrant application stage.

Conversion of historic buildings

With historic buildings, the energy efficiency improvement measures that should be invoked by conversion can be more complex as the majority of them have visual features that are not only worth preserving, but can be difficult to replace or replicate during conversions.

Although no specific guidance is provided on this subject in the Building (Scotland) Regulations, any improvements to the fabric insulation of the building will often depend on whether or not the installation work can be carried out using a non-disruptive method (e.g. insulating the ceiling of an accessible roof space).

Extensions to the insulation envelope

When constructing an extension, the majority of the work will be new and seldom will there be the need to construct to a lesser specification as is sometimes the case for alteration work (Table 6.5, Figure 6.9).

Table 6.5 Maximum *U*-values for building elements of the insulation envelope

Type of element	(a) Area-weighted average U-value (W/m^{2}K) for all elements of the same type	(b) Individual element U-value (W/m^{2}K)
Wall	0.27	0.70
Floor	0.22	0.70
Pitched roof (insulation between ceiling ties or collars)	0.16	0.35
Flat roof or pitched roof (insulation between rafters or roof with integral insulation)	0.20	0.35
Windows, doors, rooflights	1.8	3.3

Where the insulation envelope of a non-domestic building is extended, the new building fabric should be designed in accordance with Table 6.5.   ND 6.2.10

Where the insulation envelope of a domestic building is extended, the area of rooflights windows, doors, rooflights and roof windows should be limited to 25% of the floor area of the extension.   D 6.2.9

Where the insulation envelope of a dwelling or a building D 6.2.9
consisting of dwellings is extended, the new building fabric
should be designed in accordance with Table 6.5.

 For ease of understanding, the *U*-values (area-weighted
average *U*-values) in Table 6.5 are summarized in
Figure 6.9.

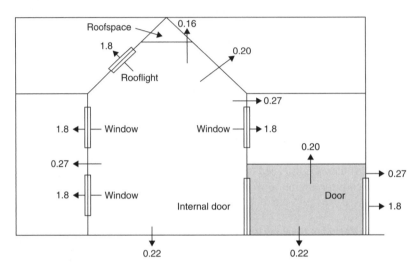

Figure 6.9 Maximum *U*-values for building elements of the insulation
envelope

 Note: The extension is the shaded portion; the existing dwelling is in elevation behind.

 The *U*-values for the elements involved in the work D 6.2.9
may be varied provided that the area-weighted overall
U-value of all the elements in the extension is no
greater than that of a 'notional' extension (see Annex 6.B of the
Domestic Technical Handbook for an example of this approach).

Alterations to the insulation envelope

For alterations, it is more than likely that the existing construction will be
from a different era in building regulation terms and in certain cases (e.g.
historic buildings) it may be necessary to adopt alternative energy efficiency
measures which relate to the amount of alteration work being undertaken.

Alterations that involve increasing the floor area and/or bringing parts of the
existing building that were previously outwith the insulation envelope into the
heated part of the dwelling are considered as extensions and/or conversions.

When alterations are carried out, attention should still be paid
to limiting infiltration thermal bridging at junctions and around
windows, doors and rooflights and limiting air infiltration.

D 6.2.11
ND 6.2.12

Stand-alone buildings

Where the area of a communal room or other heated
accommodation associated with a block of dwellings is less
than $50\,m^2$, these rooms or accommodation may be treated
as a stand-alone building.

D 6.2.13

CHPQA Quality Index (CHP(QI))

The Combined Heat and Power Assessment guide (CHP(QI)) is a registration and certification scheme which serves as an indicator of the energy efficiency and environmental performance of a CHP (Combined Heat and Power) scheme, relative to the generation of the same amounts of heat and power by separate, alternative means.

The required minimum (CHP(QI)) for all types of CHP
should be 105.

ND 6.3.3

Commissioning building services

Commissioning [i.e. in terms of achieving the levels of energy efficiency that the component manufacturers expect from their product(s)] should also be carried out with a view to ensuring the safe operation of the system.

Every building must be designed and constructed in such
a way that energy supply systems and building services
which use fuel or power for heating, lighting, ventilating and
cooling the internal environment and heating the water are
commissioned to achieve optimum energy efficiency.

D 6.7
ND 6.7

Domestic buildings

A heating, hot water service, ventilating or cooling system
in a dwelling should be inspected and commissioned in
accordance with manufacturers' instructions to ensure
optimum energy efficiency.

D 6.7.1

Non-domestic buildings

A building services installation in a building should be
inspected and commissioned in accordance with manufacturers' instructions to ensure optimum energy efficiency.

ND 6.7.1

The building and services should have facilities such as test points, inspection hatches and measuring devices to enable inspection, testing and commissioning to be carried out.   ND 6.7.1

Written information

Correct use and maintenance of building services equipment is essential if the benefits of enhanced energy efficiency are to be realized from such equipment. To achieve this, it is essential that user and maintenance instructions together with all other relevant documentation are available to the occupier of the building.

The occupiers of a building must be provided with written information by the owner: D 6.8
ND 6.8

- on the operation and maintenance of the building services and energy supply systems;
- where any air-conditioning system in the building is subject to a time-based interval for inspection of the system.

For a domestic building, written information concerning the operation and maintenance of the heating and hot water service system (together with any decentralized power generation equipment) should be made available for the use of the occupier.   D 6.8.1

For non-domestic buildings, a logbook containing information of energy system operation and maintenance (such as building services plant and controls) to ensure that the building user can optimize the use of fuel, shall be provided.   ND 6.8.1

 CIBSE Technical Memorandum 31 (TM31) provides guidance on the presentation of a logbook, and the logbook information should be presented in this or a similar manner.

Every building must be designed and constructed in such a way that: D 6.9
ND 6.9

- an energy performance certificate for the building is affixed to the building, indicating the approximate annual carbon dioxide emissions and energy usage of the building based on a standardized use of the building;
- the energy performance for the certificate is calculated in accordance with a methodology which is asset-based, conforms with the European Directive 2002/91/EC and uses UK climate data;
- the energy performance certificate is displayed in a prominent place within the building.

Note: We are reliably informed that it is intended that Scottish Ministers will direct local authorities to apply Standard 6.9 to **all** existing buildings (i.e. being sold or rented out) using Section 25(2) of the Building (Scotland) Act 2003.

6.2 Foundations

To support the weight of the structure, most brick-built buildings are built on a solid base called foundations (Figure 6.10).

Timber-framed houses, on the other hand, are usually built on a concrete foundation with a 'strip' or 'raft' construction to spread the weight (Figure 6.11).

Potential problems
There may be known and/or recorded conditions of ground instability, such as geological faults, landslides or disused mines, or unstable strata of similar nature which affect or may potentially affect a building site or its environs.

There may also be:

- unsuitable material including vegetable matter, topsoil and pre-existing foundations;
- contaminants on or in the ground covered, or to be covered, by the building and any land associated with the building;
- groundwater.

These conditions should be taken into account before proceeding with the design of a building or its foundations.

What about hazards?
Hazards associated with the ground may include:

- chemical and biological contaminants;
- gas generation from biodegradation of organic matter;
- naturally occurring radioactive radon gas and gases produced by some soils and minerals;
- physical, chemical or biological;
- underground storage tanks or foundations;
- unstable fill or unsuitable hardcore containing sulphate;
- the effects of vegetable matter including tree roots.

In the most hazardous conditions, only the total removal of contaminants from the ground to be covered by the building can provide a complete remedy. In other cases remedial measures can reduce the risks to acceptable levels. These measures should only be undertaken with the benefit of expert advice, and where the removal would involve handling large quantities of contaminated materials, then you are advised to seek expert advice.

Figure 6.10 Brick-built house – typical components

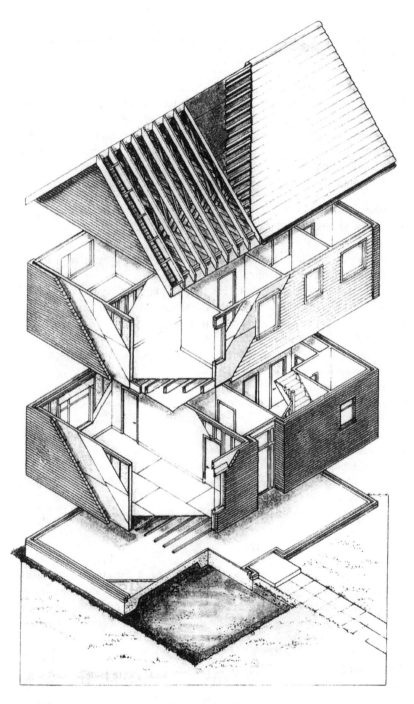

Figure 6.11 Timber-framed house – typical components

Even when these actions have been successfully completed, the ground to be covered by the building will still need to have at least 100 mm of concrete laid over it.

What about contaminated ground?

Potential building sites which are likely to contain contaminants can be identified at an early stage from planning records or from local knowledge (e.g. previous uses). In addition to solid and liquid contaminants, problems can arise from natural contamination such as methane and radioactivity. The following lists examples of sites that are most likely to contain contaminants:

- asbestos works;
- ceramics, cement and asphalt manufacturing works;
- chemical works;
- dockyards and dockland; engineering works (including aircraft manufacturing, railway engineering;
- works, shipyards, electrical and electronic equipment manufacturing works);
- gas works, coal carbonization plants and ancillary by-product works;
- industries making or using wood preservatives;
- landfill and other waste disposal sites;
- metal mines, smelters, foundries, steelworks and metal finishing works;
- munitions production and testing sites;
- nuclear installations;
- oil storage and distribution sites;
- paper and printing works;
- petrol filling stations;
- power stations;
- railway land, especially larger sidings and depots;
- road vehicle fuelling, service and repair – garages and filling stations;
- scrap yards;
- sewage works, sewage farms and sludge disposal sites;
- tanneries;
- textile works and dye works.

If any signs of possible contaminants are present, then the local authority's Environmental Health Officer should be told at once. If he confirms the presence of any of these contaminants (see Table 6.7) then he will require their removal or action to be completed before any planning permission for building work can be sought.

What about radon?

Radon is a naturally occurring radioactive colourless and odourless gas which is formed in small quantities by radioactive decay wherever uranium and radium are found. It can move through the subsoil and then into buildings, and exposure to high levels over long periods increases the risk of developing lung cancer. Some parts of the country (in particular the Highlands of Scotland) have higher natural levels than elsewhere and precautions against radon may be necessary.

 Note: Guidance on the construction of dwellings in areas susceptible to radon has been published by the Building Research Establishment as a Report ('Radon: guidance on protective measures for new dwellings').

What about contaminants?

Landfill gas is generated by the action of anaerobic micro-organisms on bio-degradable material in landfill sites and generally consists of methane and carbon dioxide together with small quantities of Volatile Organic Compounds (VOCs) which give the gas its characteristic odour. It can migrate under pressure through the subsoil and through cracks and fissures into buildings (Table 6.6).

Table 6.6   Examples of possible contaminants

Signs of possible contaminants	Possible contaminant
Vegetation (absence, poor or unnatural growth)	Metals Metal compounds Organic compounds Gases (landfill or natural source)
Surface materials (unusual colours and contours may indicate wastes and residues)	Metals Metal compounds Oily and tarry wastes Asbestos Other mineral fibres Organic compounds including phenols Combustible material including coal and coke dust Refuse and waste
Fumes and odours (may indicate organic chemicals)	Volatile organic and/or sulphurous compounds from landfill or petrol/solvent spillage Corrosive liquids Faecal animal and vegetable matter (biologically active)
Damage to exposed foundations of existing buildings	Sulphates
Drums and containers (empty or full)	Various

Methane and carbon dioxide can also be produced by organically rich soils and sediments such as peat and river silts, and a wide range of VOCs can be present as a result of petrol, oil and solvent spillages.

Site investigation

A site survey is now the recommended method for determining how much unsuitable material should be removed before commencing building work and this will normally consist of a number of well-defined stages, for example:

Planning stage	Scope and requirements.
Desktop study	Historical, geological and environmental information about the site.
Site reconnaissance design or walkover survey	Identification of actual and potential physical hazards of the main investigation.
Main investigation and reporting	Intrusive and non-intrusive sampling and testing to provide soil parameters.

Risk assessment

The site investigation may identify certain risks which will require a risk assessment, of which there are three types:

(1) Preliminary (once the need for a risk assessment has been identified, and depending on the situation and the outcome);
(2) Generic Quantitative Risk Assessment (GQRA);
(3) Detailed Quantitative Risk Assessment (DQRA).

Each risk assessment should include a:

Hazard identification	Developing the conceptual model by establishing contaminant sources, pathways and receptors (this is the preliminary site assessment which consists of a desk study and a site walkover in order to gather sufficient information to obtain an initial understanding of the potential risks).
	An initial conceptual model for the site can then be based on this information.
Hazard assessment	Identifying what pollutant linkages may be present and analysing the potential for unacceptable risks.
Risk estimation	Establishing the scale of the possible consequences by considering the degree of harm that may result and to which receptors.
Risk evaluation	Deciding whether the risks are acceptable or unacceptable – review all site data to decide whether estimated risks are unacceptable.

6.2.1 Requirements

Structure

Every building must be designed and constructed in such a   *1.1*
way that the loadings that are liable to act on it (taking into
account the nature of the ground) will not lead to:

- the collapse of the whole or part of the building;
- deformations which would make the building unfit for its
  intended use, unsafe, or cause damage to other parts of the
  building or to fittings or to installed equipment;
- impairment of the stability of any part of another building.

Every building must be designed and constructed in such a   *1.2*
way that in the event of damage occurring to any part of the
structure of the building the extent of any resultant collapse
will not be disproportionate to the original cause.

Fire

Every building must be designed and constructed in such   *2.3*
a way that in the event of an outbreak of fire within the
building, the load-bearing capacity of the building will continue
to function until all occupants have escaped, or been assisted
to escape, from the building and any fire containment measures
have been initiated.

Noise

Every building must be designed and constructed in such a   *5.1*
way that each wall and floor separating one dwelling from
another, or one dwelling from another part of the building, or
one dwelling from a building other than a dwelling, will limit
the transmission of noise to the dwelling to a level that will
not threaten the health of the occupants of the dwelling or
inconvenience them in the course of normal domestic activities
provided the source noise is not in excess of that from normal
domestic activities.

Environment

Every building must be designed and constructed in such a way   *3.1*
that there will not be a threat to the building, the occupants
and/or the health of people in or around the building:

- *due to the presence of harmful or dangerous substances;*
- *due to the emission and containment of radon gas;*   *3.2*
- *as a result of flooding and the accumulation of groundwater;*   *3.3*
- *as a result of moisture penetration from the ground;*   *3.4*
- *as a result of moisture from precipitation penetrating to the*   *3.10*
  inner face of the building;
- *as a result of moisture caused by surface or interstitial*   *3.15*
  condensation.

*Buildings must **not** be constructed over an existing drain*   *3.5*
including a field drain) that is to remain active.

Every building must be designed and constructed in such a   *3.14*
way that the air quality inside the building is not a threat to
the health of the occupants or the capability of the building to
resist moisture, decay or infestation.

Every building, and hard surface within the curtilage of a   *D 3.6*
building, must be designed and constructed with a surface
water drainage system that will:

- *ensure the disposal of surface water without threatening*
  the building and the health and safety of the people in and
  around the building; and
- *have facilities for the separation and removal of silt, grit*
  and pollutants.

6.2.2 Meeting the requirement

Hazard identification and assessment

By its very nature, the structure of a building is fundamental to ensuring the safety of people in or around new and existing buildings and can be affected by a number of environmental factors such as temperature, snow, wind, driving rain and flooding, and the impact of climate change – both inside and outside the building.

To prevent the collapse, excessive deformation or the disproportionate collapse of buildings, the climatic conditions in Scotland should be carefully considered in the assessment of loadings and in the structural design of buildings.

A preliminary desk top study should be carried out to provide information on the past and present uses of the proposed building site and surrounding area that may give rise to contamination.

The preliminary investigation can assist in the design of the exploratory and detailed ground investigation.

Generic assessment criteria may provide an indication of where further consideration of risk to receptors is required.	D 3.1.3 ND 3.1.3
Risk assessment should be specific to each building site and take into account the presence of source, pathways and receptors at a particular building site.	D 3.1.3 ND 3.1.3

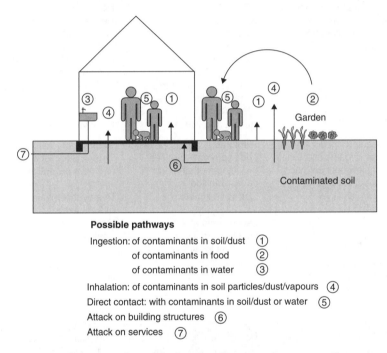

Possible pathways

Ingestion: of contaminants in soil/dust   ①
            of contaminants in food   ②
            of contaminants in water   ③
Inhalation: of contaminants in soil particles/dust/vapours   ④
Direct contact: with contaminants in soil/dust or water   ⑤
Attack on building structures   ⑥
Attack on services   ⑦

Figure 6.12 Conceptual model of a site showing a source–pathway–receptor

Risk estimation and evaluation

Where the desk study, records or local knowledge of previous use identifies land that may contain, or give rise to harmful or dangerous substances, planning permission will normally be subject to conditions.

These conditions may be imposed to ensure that the development proposed for the land will not expose future users or occupiers, or any building or services, to hazards associated with the contaminants.

There may be occasions when land containing harmful or dangerous substances has not been identified at the planning stage, and the presence of contaminants is only suspected later. Some signs of the possible presence of contaminants are given in Table 6.7 together with the possible contaminant and the probable remedial action recommended.

Table 6.7   Possible contaminants and actions

Signs of possible contamination	Possible containment	Probable remedial action recommended
Vegetation (absence, poor or unnatural growth)	Metals, metal compounds Organic compounds, gases	None Removal or treatment
Surface materials (unusual colours contours may indicate wastes and residues)	Oil and tarry wastes Asbestos (loose) Other fibres	None Removal, filling, sealing or treatment None
	Organic compounds including phenols	Removal, filling, sealing or treatment
	Potentially combustible material including coal and coke dust	Removal, inert filling or treatment
	Refuse and waste	Removal or treatment
Fumes and odours (may indicate organic chemicals at very low concentration)	Flammable, explosive, toxic and asphyxiating gases including methane and carbon dioxide	Removal or treatment. The construction is to be free from unventilated voids
	Corrosive liquids	Removal, filling, sealing or treatment
	Faecal, animal and vegetable matter (biologically active)	Removal, filling or treatment
Drums and containers (whether full or empty)	Various	Removal with all contaminated ground

If the presence of any of the contaminants listed in Table 6.7 is confirmed, some form of remedial action will be required (this is listed against each contaminant) and in all cases, these courses of action assume that the ground to be covered by the building will have at least 100 mm of *in situ* concrete cover.	D 3.1.5 ND 3.1.5

Note: The likely effects of flooding on materials and elements of the building and the various forms of construction and measures to reduce the risk of flood damage in dwellings should be taken into consideration by the developers.

There is a range of options for managing the risk of land that is contaminated. These can include removal or treatment of the contaminant source or breaking the pathway by which contaminants can present a risk to receptors (Table 6.8).

Table 6.8 Risk management techniques

Removal	The contaminant itself and any contaminated ground to be covered by the building should be taken out to a depth of 1 m (or less if the verifier agrees) below the level of the lowest floor. The contaminant should then be taken away to a place to be named by the local authority.
Filling	The ground to be covered by the building should be determined on a site-specific basis but it is normally to a depth of 1 m with a material which will not react adversely with any contaminant remaining and may be used for making up levels.
Sealing	An impermeable barrier is laid between the contaminant and the building and sealed at the joints, around the edges and at the service entries.
Ground treatment	Treatment processes can be biological, chemical or physical and be undertaken either *in situ* (contaminants are treated in the ground) or *ex situ* (contaminated material is excavated and then treated before being returned). The processes convert the contaminant into a neutral form or render it harmless.

Other considerations

All buildings shall be designed and constructed in such a way that the loadings that are liable to act on it, taking into account the nature of the ground, will not lead to:

D 1.1.0
ND 1.1.0

- the collapse of the whole or part of the building;
- deformations which would make the building unfit for its intended use, unsafe, or cause damage to other parts of the building or to fittings or to installed equipment;
- impairment of the stability of any part of another building.

The design and construction of the building shall take into account:

D 1.1.4
ND 1.1.4

- all aspects of the nature of the ground;
- ground movement caused by:
  - swelling, shrinkage or freezing of the subsoil;
  - landslip or subsidence (other than subsidence arising from shrinkage);
- recorded conditions of ground instability (e.g. caused by landslides, disused mines or unstable strata).

Loadings

The building shall be designed to transmit loads safely to the ground.

D 1.1.0
ND 1.1.0

A building should be capable of resisting all loads acting on it as a result of its intended use and geographical location.   D 1.1.1
ND 1.1 1

A building should be designed with margins of safety to ensure that the mandatory functional standard has been met.   D 1.1.1
ND 1.1 1

The foundations of buildings should be designed to sustain and transmit loadings to the ground so that there will be no ground movement which could impair the stability of the building.   D 1.1.1
ND 1.1 1

The loads to which a building will be subjected should be calculated in accordance with the following British Standards (Table 6.9).

Table 6.9   Loading standards

Loading	Relevant standard
For dead loads and imposed loads (excluding roof loads)	BS 6399-1: 1996
For imposed roof loads	BS 6399-3: 1988
For wind loads	BS 6399-2: 1997
For loading of any building for agriculture	BS 5502-22: 2003
For earth-retaining structures (e.g. basements)	BS 8002: 1994

Combined dead load and imposed load should be not more than 70 kN/m at base of the wall.   D 1E42

Note: Timber-frame walls should commence above the ground level and are, therefore, not subject to lateral loads other than from wind.

Stability of existing buildings

The design and construction of a new building shall take into account their potential impact on existing buildings.   D 1.1.0
ND 1.1 0

The design and construction of new buildings shall take into account the effect on existing buildings caused by: D 1.1.5
ND 1.1 5

- dead and imposed loads from the new building;
- wind loads including funnelling effects from the new building;
- pressure bulb extending below existing buildings;
- changes in groundwater level;
- loss of fines during pumping operations or climatic conditions.

The design and construction of buildings shall ensure that: • the building will not collapse during its design lifetime; • the building does not fail or collapse in normal use owing to deformations of the building.	D 1.1.0 ND 1.1 0
The ground shall safely support the building.	D 1.1.0 ND 1.1 0

Disproportionate collapse

A building which is said to be susceptible to disproportionate collapse is one where the effects of accidents and, in particular, situations where damage to small areas of a structure or failure of single elements could lead to collapse of major parts of the structure.

Every building must be designed and constructed so that if the structure of the building is damaged, any resultant collapse shall be in proportion to the original cause.	D 1.2 ND 1.2

Note: For further details about disproportionate collapse, see Section 6.1 of this book, 'Design and construction'.

Stability

Buildings should be stable whatever the combinations of dead load, imposed load and wind loading in terms of their individual structural elements, their interaction together and the overall stability as a structure.

The overall size and proportions of the building should be limited in accordance with the specific guidance for each form of construction.	1B1a
A layout of internal walls and external walls forming a robust three-dimensional box structure in plan should be constructed in accordance with the specific guidance for each form of construction.	1B1b
Internal walls and external walls should be connected by masonry bonding or via mechanical connections.	1B1c

Ground liable to flooding

Every building must be designed and constructed in such a way that there will not be a threat to the building or the health of the occupants as a result of flooding and the accumulation of groundwater.	D 3.3 ND 3.3

All proposed building sites should be appraised initially to ascertain the risk of flooding of the land and an assessment made as to what effects the development may have on adjoining ground.

D 3.3.1
ND 3.3.1

Ground below and immediately adjoining a building that is liable to accumulate floodwater or groundwater requires treatment to be provided against the harmful effects of such water.

D 3.3.1
ND 3.3.1

Treatment could include a field-drain system:

- to increase the stability of the ground;
- to avoid surface flooding;
- to alleviate subsoil water pressures likely to cause dampness to below-ground accommodation;
- to assist in preventing damage to foundations of buildings;
- to prevent frost heave of subsoil that could cause fractures to structures such as concrete slabs.

The selection of an appropriate drainage layout will depend on the nature of the subsoil and the topography of the ground.

Developers should be aware of the dangers from possible surface water run-off from their building site to other properties, and procedures should be in place to overcome this occurrence.

D 3.3.1
ND 3.3.1

Note: The installation of field drains or rubble drains may overcome the problem.

Site preparation

Surface soil and vegetable matter can be detrimental to a building's structure if left undisturbed within the building footprint.

Before work commences, all unsuitable material including turf, vegetable matter, wood, roots and topsoil should be removed from the ground to be covered by the building, and the ground immediately adjoining the building, to a depth of at least that which will prevent later growth that could damage the building.

D 3.1.1
ND 3.1.1

The solum (prepared area within the containing walls of a building) should be treated to prevent vegetable growth and reduce the evaporation of moisture from the ground to the inner surface of any part of a dwelling that it could damage.

D 3.1.1
ND 3.1.1

The solum should be brought to an even surface and any upfilling should be of hard, inert material.	D 3.1.1 ND 3.1.1
To prevent water collecting under the building, the solum should not be lower than the highest level of the adjoining ground.	D 3.1.1 ND 3.1.1
Any part of the under-building that is in contact with the ground, such as on sloping ground, should be tanked.	D 3.1.1 ND 3.1.1
Where the site contains fill or made ground, consideration should be given to its compressibility and its collapse potential.	D 3.1.1 ND 3.1.1
Thought should be given to foundation design to prevent the damaging effect of differential settlement.	D 3.1.1 ND 3.1.1
Because of their hazardous qualities, any ground below and immediately adjoining a building should have the hazards removed or made safe.	D 3.1.2 ND 3.1.2

Moisture from the ground

Water is the prime cause of deterioration in building materials and constructions, and the presence of moisture encourages growth of mould that is injurious to health. Groundwater can penetrate building fabric from below, rising vertically by capillary action.

Every building must be designed and constructed in such a way that:	D 3.4
• there will not be a threat to the building or the health of the occupants as a result of moisture penetration from the ground;	ND 3.4
• rising damp neither damages the building fabric nor penetrates to the interior where it may constitute a health risk to occupants;	D 3.4.0 ND 4.4.0
• there will not be a threat to the building or the health of the occupants as a result of moisture from precipitation penetrating to the inner face of the building;	D 3.10 ND 3.10

Structures below ground, including basements

Rain penetration occurs most often through walls exposed to the prevailing wet winds and, unless there are adequate damp proof courses and flashings, etc., materials in parapets and chimneys can collect rainwater and deliver it to other parts of the dwelling below roof level.

> Every building must be designed and constructed in such a   D 3.10
> way that there will not be a threat to the building or the health   ND 3.10
> of the occupants as a result of moisture from precipitation
> penetrating to the inner face of the building.
>
> This requirement does not apply to a building where
> penetration of moisture from the outside will result in
> effects not more harmful than those likely to arise from
> use of the building.

Radon

Radon is a naturally occurring, radioactive, colourless and odourless gas that is formed where uranium and radium are present. It can move through cracks and fissures in the subsoil, and so into buildings. Where this gas occurs under a dwelling, the external walls contain it and the containment of radon can build up inside the dwelling over the long term, posing a risk to health. To reduce the risk, all new dwellings, extensions and alterations built in areas where there might be radon concentration, may need to incorporate precautions against radon gas.

> Every building must be designed and constructed in such a   D 3.2
> way that there will not be a threat to the health of people in or   ND 3.2
> around the building due to the emission and containment of
> radon gas.
>
> If a dwelling is located on ground designated as a 'radon   D 3.2.2
> affected area' protective work should be undertaken to prevent   ND 3.2.2
> excessive radon gas from entering the dwelling.
>
> Radon protective measures should be provided in accordance   D 3.2.2
> with the guidance contained in BRE publication BR376 –   ND 3.2.2
> *'Radon: guidance on protective measures for new dwellings in*
> *Scotland'.*

Workplaces are less of a risk than dwellings because, generally speaking, people spend less time at work than at home, and workplaces generally have better ventilation provision. In addition, large buildings tend to be mechanically ventilated which will result in the dilution of radon gas.

Whilst the national reference level of $400\,Bq/m^3$ (Bq is a measuring unit for radioactivity) in workplaces makes it easier to stay within that level than the lower level of $200\,Bq/m^3$ for dwellings, research has shown that an impervious membrane with securely welted joints laid over the full area of the building will help to reduce radon ingress into the building.

Construction

Foundations should be designed so that the loadings (i.e. dead loads, imposed loads and wind loads) that are transmitted from the building to the subsoil will not cause undue settlement. (See Table 6.10 for examples of typical loads on foundations.)   1C0

Table 6.10   Loads on foundations

Number of stories	Wall type	Roof span (m)	Floor span (m)	Loading (kN/m)
3	Masonry cavity	12	6	80
3	Masonry cavity	7.5	6	70
2	Masonry cavity	12	6	60
2	Masonry cavity	7.5	6	50
2	Timber frame	7.5	6	40
1	Masonry cavity	7.5	6	30
1	Timber frame	7.5	6	30
1	Single-leaf masonry	5	5	20

The minimum depth of foundations should be either: 1C2

- to the selected rock or soil bearing stratum;
- 450 mm to the underside of foundations;
- 600 mm to the underside of foundations where clay soils are present (depending on which is the greater).

The design of foundations should comply with **all** of the following:

The foundations should be situated centrally under the wall.   1C3a

The foundations should be strip foundations or (i.e. as an alternative) trench-fill foundations.   1C3b
1C3f

The foundation width should not be less than the dimension W_F shown in Table 6.11 and at least as wide as the supported wall.   1C3c

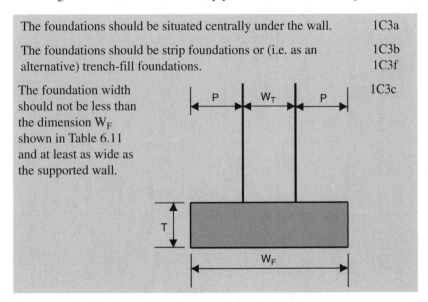

Table 6.11 Minimum width of strip footings

Type of ground (including engineering fill)		Condition of ground	Applicable field test	Total load of load-bearing walling not more than (kN/m)					
				Minimum width of strip foundation W_F (mm)					
				20	30	40	50	60	70
I	Rock	Not inferior to sandstone, limestone or firm chalk	Requires at least a pneumatic or other mechanically operated pick for excavation	At least equal to the width of the wall					
II	Gravel or sand	Medium dense	Requires a pick for excavation. Wooden peg 50mm^2 in cross-section, hard to drive beyond 150mm	250	300	400	500	600	650
III	Clay or sandy clay	Stiff	Can be indented slightly by thumb	250	300	400	500	600	650
IV	Clay or sandy clay	Firm	Thumb makes impression easily	300	350	450	600	750	850
V	Sand, silty sand or clayey sand	Loose	Can be excavated by a spade. Wooden peg 50mm^2 in cross-section can be easily driven in	400	600	x	x	x	x
VI	Silt, clay, sandy clay or silty clay	Soft	Finger pushed in up to 10mm	450	600	x	x	x	X
VII	Silt, clay, sandy clay or silty clay	Very soft	Finger easily pushed in up to 25mm	x	x	x	x	x	x

In chemically non-aggressive soils, concrete should be composed of Portland cement to BS EN 197-1 & 2: 2000 and fine and coarse aggregate conforming to BS EN 12620: 2002, and the mix should either be: 1C3d

- 50 kg of Portland cement to not more than 100 kg ($0.05\,m^3$) of fine aggregate and 200 kg ($0.1\,m^3$) of coarse aggregate;
- Grade ST2 or Grade GEN1 concrete to BS 8500-2: 2002.

 In chemically aggressive soils, follow the guidance provided in BS 8500-1: 2002 and BRE Special Digest 1.

The minimum thickness, T, of the concrete foundation should be is 150 mm or P (whichever is the greater).   1C3f

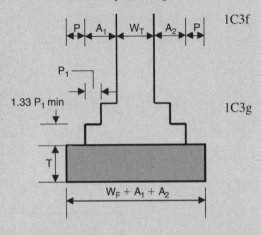

Footings with regular offsets should have a depth of at least 1.33 times the projection P_1, with the overall width not less than the sum of W_F plus the offset dimensions A_1 and A_2.   1C3g

Walls should be central on the foundation.   1C3g

For stepped elevation foundations: 1C3i

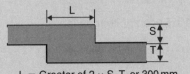

L = Greater of 2 × S, T, or 300 mm

- the height of steps, S, should be not more than the foundation thickness, T;
- overlap, L, should be twice the step height, S, the foundation thickness, T, or 300 mm (whichever is the greatest);

Foundations for piers, buttresses and chimneys should be as shown:

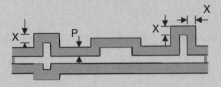

 where X must be not less than P.

Eccentric foundations

Note: The following guidance is limited to:

- single-storey buildings of 4.5 m maximum height where a wall is to be constructed either against a boundary or against an existing wall and where it is not possible to construct the wall centrally on the foundation;
- masonry cavity or timber-frame walls with masonry outer leaf with either a flat or pitched roof;
- similar ground conditions to types I–VI from Table 6.11 below both the existing and new foundations;
- foundations complying with **all** of the requirements of these Regulations.

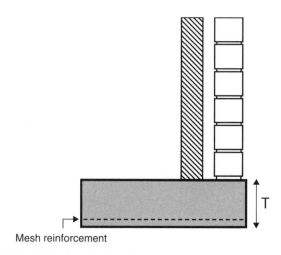

Mesh reinforcement

Figure 6.13 Eccentric foundations

The following requirements should be considered in the design   1C4
of eccentric foundations:

- the minimum foundation width, W_F, should be read from Table 6.11;
- the minimum foundation thickness, T, for the minimum foundation widths listed in Table 6.11 should be at least 200 mm;
- steel mesh reinforcement (e.g. A142) should be placed with 50 mm cover from the base of the foundation;
- where the wall and its foundation are to be constructed against an existing wall then the foundation should be treated as if it were an extension to an existing building (see Clause 1C5).

Extensions to existing buildings

Note: The following guidance is limited to:

- extensions not greater than of two storeys connected to existing buildings;
- masonry cavity or timber-frame walls with a masonry outer leaf and with either a flat or pitched roof;
- similar ground conditions to types I–VI from the Table 6.11 below both the existing and new foundations;
- extension foundations complying with **all** the requirements of these Regulations.

Where the depth of the existing foundation is less than:

- the selected rock or soil bearing stratum;
- 450 mm to the underside of foundations;
- 600 mm to the underside of foundations where clay soils are present (depending on which is the greater):

- the depth of the extension foundation should match the depth of the existing foundation at the interface and step down as shown in Figure 6.14.   1C5
- the initial stepdown in the underside of the new foundation should not commence until the horizontal distance from the vertical face of the existing foundation is at least the foundation thickness, T.

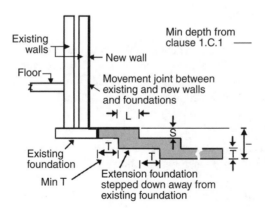

Figure 6.14 Foundations for extensions

To minimize the risk of differential settlement occurring between the extension and the existing structure, the following should be considered:

- movement joints should be placed between the existing   1C5
  and new foundations, and walls to accommodate any
  differential settlement;
- for soil types I–III the strip foundation widths should be as
  per Table 6.11;
- for soil types IV, V and VI the strip foundation widths
  shown in Table 6.11should be increased by 25%

 Additional information is provided in BRE GBG 53.
'Foundations for low-rise building extensions'.

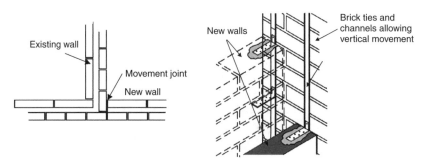

Figure 6.15 Movement joints and brick ties

Masonry cladding

Masonry cladding should be constructed on to the building   1E37
foundation and tied back to the timber-frame structure with a
cavity width of 50 mm between the inside face of the masonry
cladding and the outer face of the timber-frame wall.

Composite action:

- sheathings and linings shall be nailed to all perimeter and   1E38
  intermediate timber members as shown in Figure 6.16, as
  follows:
- sheathing edges should be backed by and nailed to timber   1E38a
  framing at all edges;
- where sheathing is nailed to studs, the nails should be at   1E38b
  least 7 mm from the edge of the board or the face of the
  stud;

- for plasterboard linings, nails should be at least 10 mm from formed board edges and at least 13 mm from ends of the board, at centres not more than 150 mm;   1E38c
- internal walls which are lined with plasterboard should be connected to the wall studs at the same perimeter of nail centres as for external sheathing material;   1E38d
- fixing of perimeter studs to sheathing (1) should be at the centres (see Figure 6.16);   1E38e
- fixing of intermediate studs to sheathing (2) should be at not more than twice the centres of the perimeter nailing (see Figure 6.16).   1E38f

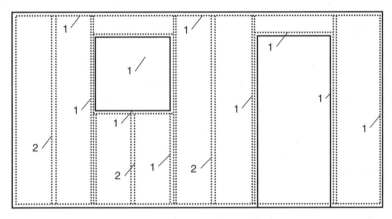

Figure 6.16 Perimeter and intermediate nailing diagram

Wall panel connections

To ensure that wall panels are able to resist overturning forces they should be combined to form lengths of wall as follows:

- tops and bottoms of individual wall panels should be linked by head binders and sole plates, respectively, that are continuous across panel joints including at junctions of the same dimensions as the top and bottom rails;   1E39a
- sole plates should be secured to either the concrete floor slab or the header joists in the case of a timber ground floor or the header joists of the intermediate floor.   1E39b

Nailing and fixing schedule

Table 6.12   Nailing and fixing schedule

Item	Recommended fixing
Foundations	
Sole plate to under building	Mechanical fixings at 600 mm centres rated at 4.7 kN shear resistance
Holding down straps providing at least 3.5 kN of resistance	Stainless steel strap 30 mm × 2.5 mm attached to stud by 6 no. 3.36 mm × 65 mm ring shank nails at 2.4 m centres, at every opening and at the end studs of a wall attaching the strap to the stud and placing the L-shaped end of the strap under the masonry cladding, creating the holding down resistance

Treatment of building elements adjacent to the ground

Floors, walls or other building elements adjacent to the ground should be constructed so as to prevent moisture from the ground reaching the inner surface of any part of a building that it could damage.	D 3.4.1 ND 3.4.1
Floors, walls or other building elements adjacent to the ground should be constructed as shown in Table 6.13.	D 3.4.1–D 3.4.4 ND 3.4.1–ND 3.4.4

Table 6.13   Construction of concrete and timber floors

Ground supported concrete floors

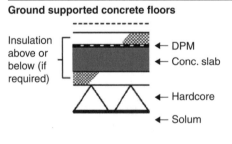

- the solum is brought to a level surface
- hardcore bed 100 mm thick of clean broken brick or similar inert material
- concrete slab 100 mm thick with insulation, if any, laid above or below the slab; with or without a screed or floor finish.
- damp-proof membrane above or below the slab or as a sandwich; jointed and sealed to the damp proof course or damp-proof structure in walls, columns and other adjacent elements

(continued)

Table 6.13   (Continued)

Suspended concrete floors 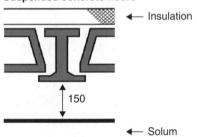	• the solum is brought to an even surface • any up filling to be of hard, inert material. • suspended concrete floor of *in situ* or precast concrete slabs or beams • permanent ventilation of the under floor space direct to the outside air by ventilators in two external walls on opposite sides of the building • the ventilated space to be 150 mm to the underside of the floor slab or beams
Suspended timber floors 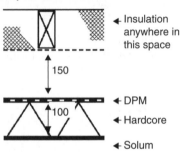	• the solum is brought to an even surface • any up filling to be of hard, inert material • hardcore bed 100 mm thick of clean broken brick; or concrete 50 mm thick laid on 0.25 mm (1,000 gauge) polyethylene sheet; or concrete 100 mm thick; so that in any case the top surface is not below that of the adjacent ground • suspended timber floor with or without insulation • floor joists carried on wall-plates supported as necessary by sleeper walls with a DPC under the wall-plates • permanent ventilation of the under floor space direct to the outside air by ventilators in two external walls on opposite sides of the building

Existing drains

Buildings must **not** be constructed over an existing drain (including a field drain) that is to remain active.	D 3.5 ND 3.5
Where it is proposed to construct a building over the line of an existing sewer, the sewer should be re-routed around the building.	D 3.5.0 ND 3.5.0

Permission will be required from the Water Authority for any work that is to be carried out to a public sewer or where it is necessary to build over a public sewer.	D 3.5.0 ND 3.5.0
A survey should be carried out to establish the geography and topography of the building site and ascertain whether there are any existing field drains.	D 3.5.1 ND 3.5.1
Where a building site requires that an existing drain (including a field drain) must remain active and be re-routed or retained, particular methods of construction and protection should be carefully considered.	D 3.5.1 ND 3.5.1
Where a building is erected over a private drain, including a field drain that is to remain active, the drain should be re-routed if reasonably practicable or re-constructed in a manner appropriate to the conditions of the site.	D 3.5.2 ND 3.5.2

In non-domestic buildings, it would be unreasonable for drains to be re-routed around a limited-life building, but care should be taken that no undue loading is transmitted to the drain that might cause damage (ND 3.5.2).

Every drain or sewer should be protected (e.g. by providing barriers) from damage by construction traffic and heavy machinery.	D 3.5.3 ND 3.5.3
Heavy materials should not be stored over drains or sewers.	D 3.5.3 ND 3.5.3
It is recommended that manholes are not located within a dwelling.	D 3.5.3 ND 3.5.3
Where a drain or sewer passes through, under or close to structures (including a manhole or inspection chamber), a detail should be devised to allow sufficient flexibility to avoid damage of the pipe due to movement.	D 3.5.4 ND 3.5.4
Where drains or sewers pass, similar precautions should be considered.	D 3.5.4 ND 3.5.4
Disused sewers or drains (which provide ideal nesting sites for rats) should be disconnected from the drainage system as near as possible to the point of connection.	D 3.5.5 ND 3.5.5
Sewers and drains less than 1.5 m from the surface and in open ground should be, as far as reasonably practicable, removed. Other pipes should be capped at both ends and at any point of connection, to ensure that rats cannot gain entry.	D 3.5.5 ND 3.5.5

Surface water drainage

Climate change is expected to result in more rain in the future and it is essential that this is taken into account in today's buildings.

Every building and hard surface within the curtilage of a building must be designed and constructed with a surface water drainage system that will:	D 3.6 ND 3.6
• ensure the disposal of surface water without threatening the building and the health and safety of the people in and around the building; • have facilities for the separation and removal of silt, grit and pollutants.	
It is essential that surface water from buildings is:	D 3.6.0 ND 3.6.0
• removed quickly and safely without damage to the building or danger to people around the building and does not pose a risk to the environment by flooding or pollution.	

6.3 Drainage

All plans for building work need to show that drainage of refuse water (e.g. from sinks) and rainwater (from roofs) has been adequately catered for. Failure to do so will mean that these plans will be rejected by the local authority.

All plans for buildings must include at least one water or earth closet **unless** the local authority is satisfied that one is not required (for example in a large garage separated from the house).

If you propose using an earth closet, the local authority cannot reject the plans unless they consider that there is insufficient water supply to that earth closet.

What are the rules about drainage?

Drains should either be connected with a sewer (unless the sewer is more than 120 ft away or the person carrying out the building work is not entitled to have access to the intervening land) or be able to discharge into a private wastewater treatment plant or septic tank, cesspool, settlement tank or other tank designed for the reception and/or disposal of foul matter from buildings.

The local authorities view this requirement very seriously and will need to be satisfied that:

- satisfactory provision has been made for drainage;
- all cesspools, private sewers, septic tanks, drains, soil pipes, rain water pipes, spouts, sinks or other appliances are adequate for the building in question;
- all private sewers that connect directly or indirectly to the public sewer are not capable of admitting subsoil water;

- the condition of a cesspool is not detrimental to health, or does not present a nuisance;
- cesspools, private sewers and drains previously used, but now no longer in service, do not prejudice health or become a nuisance.

This requirement can become quite a problem if it is not recognized in the early planning stages and so it is always best to seek the advice of the local authority. In certain circumstances, the local authority might even help to pay for the cost of connecting you up to the nearest sewer!

The local authority has the authority to make the owner renew, repair or clean existing cesspools, sewers and drains, etc.

Can two buildings share the same drainage?
Usually the local authority will require every building to be drained separately into an existing sewer but in some circumstances they may decide that it would be more cost-effective if the buildings were drained in combination. On occasions, they might even recommend that a private sewer is constructed.

What about ventilation of soil pipes?
All authorities will require that all soil pipes from water closets are properly ventilated and that no use is being made of:

- an existing or proposed pipe designed to carry rain water from a roof to convey soil and drainage from a sanitary convenience;
- an existing pipe designed to carry surface water from a premises to act as a ventilating shaft to a drain or a sewer conveying foul water.

What happens if I need to disconnect an existing drain?
If, in the course of your building work, you need to:

- reconstruct, renew or repair an existing drain that is joined up with a sewer or another drain;
- alter the position of an existing drain that is joined up with a sewer or another drain;
- seal off an existing drain that is joined up with a sewer or another drain;

then, provided that you give due notice (usually 48 h) to the local authority, the person undertaking the reconstruction may break open any street for this purpose.

You do not need to comply with this requirement if you are demolishing an existing building.

Can I repair an existing water closet or drain?
Repairs can be carried out to water closets, drains and soil pipes, but if that repair or construction work is prejudicial to health and/or a public nuisance, then the person who completed the installation or repair is liable, on conviction, to a heavy fine.

In some areas, a 'water closet' can **also** be taken to mean a urinal.

Can I repair an existing drain?

Only in extreme emergencies are you allowed to repair, reconstruct or alter the course of an underground drain that joins up with a sewer, cesspool or other drainage method (e.g. septic tank).

If you have to carry out repairs, etc., in an emergency, then make sure that you do **not** cover over the drain or sewer without notifying the local authority of your intentions!

 Under NO circumstances may buildings be constructed over an existing drain (including a field drain) that is to remain active.

6.3.1 Requirements

Fire

Every building must be provided with a water supply for use by the fire service.	*2.13*

 Note: This standard does not apply to domestic buildings.

Environment

Every building must be designed and constructed in such a way that there will not be a threat to the building or the health of occupants or people in or around the building:

• *due to the presence of harmful or dangerous substances;*	*3.1*
• *as a result of flooding and the accumulation of ground water;*	*3.3*
• *as a result of moisture penetration from the ground.*	*3.4*

Buildings must **not** be constructed over an existing drain (including a field drain) that is to remain active.   *3.5*

Every building, and hard surface within the curtilage of a building, must be designed and constructed with a surface water drainage system that will: *3.6*

• *ensure the disposal of surface water without threatening the building and the health and safety of the people in and around the building; and*
• *have facilities for the separation and removal of silt, grit and pollutants.*

*Every wastewater drainage system serving a building must   3.7
be designed and constructed in such a way as to ensure the
removal of wastewater from the building without threatening
the health and safety of the people in and around the building;*

- *that facilities for the separation and removal of oil,
  fat, grease and volatile substances from the system are
  provided;*
- *that discharge is to a public sewer or public wastewater
  treatment plant, where it is reasonably practicable to do so;*
- *where discharge to a public sewer or public wastewater
  treatment plant is not reasonably practicable that discharge
  is to a private wastewater treatment plant or septic tank.*

*Every private wastewater treatment plant or septic tank serving   3.8
a building must be designed and constructed in such a way that*

- *it will ensure the safe temporary storage and treatment of
  wastewater prior to discharge;*
- *the disposal of the wastewater to ground is safe and is not a   3.9
  threat to the health of the people in or around the building.*

*Every building must be designed and constructed in such a way   3.12
that:*

- *sanitary facilities are provided for all occupants of, and
  visitors to, the building in a form that allows convenience
  of use and that there is no threat to the health and safety of
  occupants or visitors;*
- *there will not be a threat to the health and safety of people   3.26
  from a dungstead and farm effluent tank.*
- *an oil storage installation, incorporating oil storage   3.24
  tanks used solely to serve a fixed combustion appliance
  installation providing space heating or cooking facilities
  in a building will: reduce the risk of oil escaping from the
  installation; contain any oil spillage likely to contaminate
  any water supply, ground water, watercourse, drain or
  sewer; and permit any spill to be disposed of safely;*
- *the volume of woody biomass fuel storage allows the
  number of journeys by delivery vehicles to be minimized;*
- *there will not be a threat to the building or the health of   3.10
  the occupants as a result of moisture from precipitation
  penetrating to the inner face of the building.*

For domestic buildings

Every building must be designed and constructed in such a way   D 3.25
that accommodation for solid waste storage is provided which:

- *permits access for storage and for the removal of its contents;*
- *does **not** threaten the health of people in and around the building; and*
- *does **not** contaminate any water supply, ground water or surface water.*

Every building must be designed and constructed in such a way   D 3.11
that:

- the size of any apartment or kitchen will ensure the welfare and convenience of all occupants and visitors; and
- an accessible space is provided to allow for the safe, convenient and sustainable drying of washing.

6.3.2 Safety

Every building must be designed and constructed in such a   4.1
way that:

- *all occupants and visitors are provided with safe, conven-ient and unassisted means of access to the building;*   4.2
- in non-domestic buildings, safe, unassisted and convenient means of access is provided throughout the building;
- in residential buildings, a proportion of the rooms intended to be used as bedrooms must be accessible to a wheelchair user;
- in domestic buildings, safe and convenient means of access is provided within common areas and to each dwelling;
- in dwellings, safe and convenient means of access is provided throughout the dwelling;
- in dwellings, unassisted means of access is provided to, and throughout, at least one level.

6.3.3 Meeting the requirements

Existing drains

Buildings must not be constructed over an existing drain   D 3.5
(including a field drain) that is to remain active.   ND 3.5

Where it is proposed to construct a building over the line of   D 3.5.0
an existing sewer, the sewer should be re-routed around the   ND 3.5.0
building.

Permission will be required from the Water Authority for any work that is to be carried out to a public sewer or where it is necessary to build over a public sewer.

A survey should be carried out to establish the geography and topography of the building site and to ascertain whether there are any existing field drains.

D 3.5.1
ND 3.5.1

Where a building site requires that an existing drain (including a field drain) must remain active and be re-routed or retained particular methods of construction and protection should be carefully considered.

D 3.5.1
ND 3.5.1

Re-routing of drains

Where a building is erected over a private drain, including a field drain that is to remain active, the drain should be re-routed if reasonably practicable or re-constructed in a manner appropriate to the conditions of the site.

D 3.5.2
ND 3.5.2

Note: In non-domestic buildings, it would be unreasonable for drains to be re-routed around a limited-life building, but care should be taken that no undue loading is transmitted to the drain that might cause damage.

Drainage system outside a building

A drainage system outside a dwelling should be constructed and installed in accordance with the recommendations in BS EN 12056-1: 2000, BS EN 752-3: 1997 (amendment 2), BS EN 752-4: 1998 and BS EN 1610: 1998.

D 3.7.3
ND 3.7.3

Note: Reducing the bore of a drain in the direction of flow may lead to blockages and is not recommended.

Health and safety legislation requires that manual entry to a drain or sewer system is **only** undertaken where **no** alternative exists, and remotely operated equipment will become the normal method of access.

D 3.7.3
ND 3.7.3

Re-construction of drains

During construction, it should be ensured that the assumptions made in the design are safeguarded or adapted to changed conditions.	D 3.5.3 ND 3.5.3
Every drain or sewer should be protected (e.g. by providing barriers) from damage by construction traffic and heavy machinery.	D 3.5.3 ND 3.5.3
Heavy materials should not be stored over drains or sewers.	D 3.5.3 ND 3.5.3
It is recommended that manholes are **not** located within a dwelling	D 3.5.3 ND 3.5.3

Drains passing through structures

Where a drain or sewer passes through, under or close to structures (including a manhole or inspection chamber) a detail should be devised to allow sufficient flexibility to avoid damage of the pipe due to movement.	D 3.5.4 ND 3.5.4
Where drains or sewers pass, similar precautions should be considered.	D 3.5.4 ND 3.5.4

Surfaces to accessible routes

Surface elements such as drainage gratings and manhole covers, uneven surfaces (such as cobbles) or loose-laid materials (such as gravel) should not create a trip or entrapment hazard.	D 4.1.4 ND 4.1.4

Sealing disused drains

Disused sewers or drains (which provide ideal nesting sites for rats) should be disconnected from the drainage system as near as possible to the point of connection.	D 3.5.5 ND 3.5.5
Sewers and drains less than 1.5 m from the surface and in open ground should, as far as reasonably practicable, be removed. Other pipes should be capped at both ends and at any point of connection, to ensure that rats cannot gain entry.	D 3.5.5 ND 3.5.5

Surface water drainage

Climate change is expected to result in more rain in the future and it is essential that this is taken into account in today's building.

Every building, and hard surface within the curtilage of a building, must be designed and constructed with a surface water drainage system that will:	D 3.6 ND 3.6

- ensure the disposal of surface water without threatening the building and the health and safety of the people in and around the building;
- have facilities for the separation and removal of silt, grit and pollutants.

It is essential that surface water from buildings is: D 3.6.0 ND 3.6.0

- removed quickly and safely without damage to the building or danger to people around the building and does not pose a risk to the environment by flooding or pollution;
- cleared quickly from all access routes to buildings, particularly with elderly and disabled people in mind.

Surface water drainage from dwellings

Every building should be provided with a drainage system to remove rainwater from the roof, or other areas where rainwater might accumulate, without causing damage to the structure or endangering the health and safety of people in and around the building. D 3.6.1 ND 3.6.1

Methods other than gutters and rainwater pipes may be utilized to remove rainwater from roofs. D 3.6.1 ND 3.6.1

If an eaves drop system to allow rainwater to drop freely to the ground is used, it should be designed taking into account: D 3.6.1 ND 3.6.1

- the protection of the fabric of the dwelling from ingress of water caused by water splashing on the wall;
- the need to prevent water from entering doorways and windows;
- the need to protect persons from falling water when around the dwelling;
- the need to protect persons and the building fabric from rainwater splashing on the ground or forming ice on access routes;
- the protection of the building foundations from concentrated discharges from gutters.

Gutters and rainwater pipes may be omitted from a roof at any height provided it has an area of not more than $8\,m^2$ and no other area drains onto it.	D 3.6.1 ND 3.6.1

Surface water run-off from small paved areas

Surface water should be free draining to a pervious area, such as grassland, provided the soakage capacity of the ground is not overloaded.	D 3.6.6 ND 3.6.6
Surface water discharge should not be adjacent to the building, where it could damage the foundations.	D 3.6.6 ND 3.6.6

Surface water drainage of paved surfaces

A wastewater drainage system should be tested to ensure the system is laid and is functioning correctly.	D 3.7.8 ND 3.7.10
To avoid ponding of water on paved surfaces, particularly in winter where ice can form, paved surfaces that are accessible to pedestrians should be drained quickly and efficiently.	D 3.6.2 ND 3.6.2
Every building should be provided with a drainage system to remove surface water from paved surfaces, such as an access route that is suitable for disabled people, without endangering the building or the health and safety of people in and around the dwelling.	D 3.6.2 ND 3.6.2
The paved surface should be so laid as to ensure rainwater run-off is not close to the building.	D 3.6.2 ND 3.6.2
A paved surface, such as a car park, of less than $200\,m^2$ is unlikely to contribute to flooding problems and may be designed to have free-draining run-off.	D 3.6.2 ND 3.6.2

Surface water discharge

Surface water discharged from a building and a hard surface within the curtilage of a building should be carried to a point of disposal that will not endanger the building, environment or the health and safety of people around the building.	D 3.6.3 ND 3.6.3

Surface water discharge should be to:

- a SUDS; or
- a soakaway; or
- a public sewer; or
- an outfall to a watercourse, such as a river, stream or loch or coastal waters; or
- a storage container with an overflow discharging to any of the four options above.

D 3.6.3
ND 3.6.3

Discharge from a soakaway should not endanger the stability of the building.

D 3.6.3
ND 3.6.3

Every part of a soakaway should be located at least 5 m from a building and from a boundary in order that an adjoining plot is not inhibited from its full development potential.

D 3.6.3
ND 3.6.3

Discharges into a drainage system

Where a discharge into a traditional drainage system contains silt or grit, for example from a hard standing with car wash facilities:

- there should be facilities for the separation of such substances;
- removable grit interceptors should be incorporated into the surface water gully pots to trap the silt or grit.

D 3.6.9
ND 3.6.9

Where a discharge into a drainage system contains oil, grease or volatile substances (e.g. from a vehicle repair garage) there should be facilities for the separation and removal of such substances.

ND 3.6.9
ND 3.7.8

The use of emulsifiers to break up any oil or grease in the drain is **not** recommended as they can cause problems further down the system.

Sustainable Urban Drainage Systems

Sustainable Urban Drainage Systems (SUDS) are built to manage surface water run-off and are used in conjunction with good management of the land to prevent pollution.

A SUDS technique for surface water drainage should be provided in accordance with the guidance contained in 'Sustainable Urban Drainage Systems: design manual for Scotland and Northern Ireland'.

D 3.6.4
ND 3.6.4

Note: The maintenance of a SUDS within the curtilage of a building is the responsibility of the building owner.

Soakaway serving single dwellings and small extensions

Soakaways have been the traditional method of disposal of surface water from buildings and paved areas where no mains drainage exists.

There should be individual soakaways for each building.	D 3.6.5 ND 3.6.5
A soakaway serving a single dwelling, a small building or an extension should be tested in accordance with the percolation test method.	D 3.6.5 ND 3.6.5

Wastewater drainage

This following guidance applies to wastewater systems that operate essentially under gravity but may also be used for pipework connecting to a private wastewater treatment plant or septic tank.

Every wastewater drainage system serving a building must be designed and constructed in such a way as to ensure the removal of wastewater from the building without threatening the health and safety of the people in and around the building, and: • that facilities for the separation and removal of oil, fat, grease and volatile substances from the system are provided; • that discharge is to a public sewer or public wastewater treatment plant, where it is reasonably practicable to do so; • where discharge to a public sewer or public wastewater treatment plant is not reasonably practicable that discharge is to a private wastewater treatment plant or septic tank.	D 3.7 ND 3.7

 Note: Some sewers carry wastewater and surface water in the same pipe. In such cases it may be appropriate to install a drainage system within the curtilage of a building as a separate system.

Connection to a public sewer

Where a private drain discharges into a public sewer (normally at the curtilage of a building) access should be provided for maintenance and this is usually via a disconnecting inspecting chamber immediately inside the curtilage of the house.	D 3.7.4 ND 3.7.4
It is preferable that a chamber is provided for individual houses but where this is not practicable, a shared disconnecting chamber should be provided.	D 3.7.4

A disconnecting chamber (or manhole where the depth is   D 3.7.4
more than 1.2 m) should be provided in accordance with the   ND 3.7.4
requirements of Scottish Water in whom it is likely to be
vested.

The disconnecting chamber, or manhole, for a block of   D 3.7.4
individually owned flats or maisonettes should be located
as close to the building as is reasonably practicable as the
drain will become a public sewer once it passes outwith the
footprint of the building.

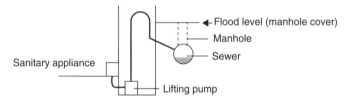

Figure 6.17 Disconnection chamber

Combined sewers

Some sewers (called combined sewers) carry wastewater and surface water in
the same pipe. These systems are **NOT** recommended today as they are more
likely to surcharge during heavy rains.

A separate drainage system carrying wastewater and surface   D 3.7.5
water should be constructed within the curtilage of a
building.

Private wastewater treatment systems – treatment plants

A wastewater treatment system is an effective, economical way of treating
wastewater from buildings. It consists of two main components: a watertight
underground tank into which raw sewage is fed and a system designed to discharge the wastewater safely to the environment without pollution.

This is normally an infiltration field through which wastewater is released
to the ground, but when ground conditions are not suitable, a discharge to a
watercourse or coastal waters may be permitted.

Every private wastewater treatment plant or septic tank   D 3.8
serving a building must be designed and constructed in such   ND 3.8
a way that it will ensure the safe temporary storage and
treatment of wastewater prior to discharge.

Treatment plants

Where it is not reasonably practicable to connect to a public sewer or a public wastewater treatment plant then discharge should be to a private wastewater treatment plant or septic tank.

 Domestic use of detergents and disinfectants is not considered to be detrimental but excessive use may have a harmful effect on the performance of the sewage treatment works.

A private wastewater treatment plant and septic tank should be designed, constructed and installed in accordance with:	D 3.8.1 ND 3.8.1

- the recommendations of BS EN 12566-1: 2000, for a prefabricated septic tank; or
- the recommendations of BS 6297: 1983; or
- the conditions of certification by a Notified Body.

The settlement tank of a private wastewater plant and a septic tank should have a securely sealed, solid cover that is capable of being opened by one person using standard operating keys.	D 3.8.2 ND 3.8.2

Inspection and sampling

A private wastewater plant and septic tank should be provided with a chamber for the inspection and sampling of the wastewater discharged from the tank.	D 3.8.3 ND 3.8.3
The owner should carry out inspection at regular intervals and a chamber should be provided in accordance with Figure 6.18.	D 3.8.3 ND 3.8.3

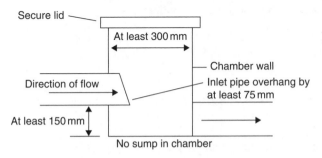

Figure 6.18 Section through inspection chamber

Location of a treatment plant

Research has shown that there are no health issues that dictate a safe location of a treatment plant or septic tank relative to a building. However, damage to the foundations of a building has been shown to occur where leakage from the tank has occurred. In the unlikely event of there being leakage, it is sensible to ensure that any water-bearing strata direct any effluent away from the dwelling. To prevent any such damage therefore:

Every part of a private wastewater plant and septic tank should be located at least 5 m from a building.	D 3.8.4 ND 3.8.4
Every part of a private wastewater plant and septic tank should be located at least 5 m from a boundary in order that an adjoining plot is not inhibited from its full development potential.	D 3.8.4 ND 3.8.4

Discharges from septic tanks and treatment plants

The Scottish Environment Protection Agency (SEPA) will require an authorization where mains drainage is not available and discharge of sewage effluent is to ground via an infiltration system or to a watercourse, loch or coastal waters.	D 3.8.5 ND 3.8.5

Access for desludging

Wastewater treatment plants should be **inspected monthly** to check they are working correctly.	D 3.8.6 ND 3.8.6
The effluent in the outlet from the tank should be free flowing.	D 3.8.6 ND 3.8.6
A private wastewater treatment plant and septic tank should be provided with an access for desludging.	D 3.8.6 ND 3.8.6
The frequency of desludging will depend upon the capacity of the tank and the amount of waste draining to it from the building.	D 3.8.6 ND 3.8.6

> If you require advice on desludging frequencies, speak to the tank manufacturer or the desludging contractor.

The desludging tanker should be provided with access to a working area that: • will provide a clear route for the suction hose from the tanker to the tank; • is not more than 25 m from the tank where it is not more than 4 m higher than the invert level of the tank; • is sufficient to support a vehicle axle load of 14 tonnes.	D 3.8.6 ND 3.8.6

Labelling

Every building with a drainage system discharging to a private wastewater treatment plant or septic tank should be provided with a label to alert the occupiers to such an arrangement.

D 3.8.7
ND 3.8.7

The label should describe the recommended maintenance necessary for the system and should include the wording shown in Figure 6.19:

D 3.8.7
ND 3.8.7

The drainage system from this property discharges to a wastewater treatment plant (or septic tank, as appropriate).

The owner is legally responsible for routine maintenance and to ensure that the system complies with any discharge consent issued by SEPA and that it does not present a health hazard or a nuisance.

Figure 6.19 Labelling of private wastewater treatment plant

This label should be located adjacent to the gas or electricity consumer unit or the water stopcock.

D 3.8.7
ND 3.8.7

Private wastewater treatment systems – infiltration systems

Subject to discharge authorization from SEPA, wastewater from treatment systems can discharge either to land via an infiltration system or to water-courses, lochs or coastal waters (Figure 6.20).

Every private wastewater treatment system serving a building must be designed and constructed in such a way that the disposal of the wastewater to ground is safe and is not a threat to the health of the people in or around the building.

D 3.9
ND 3.9

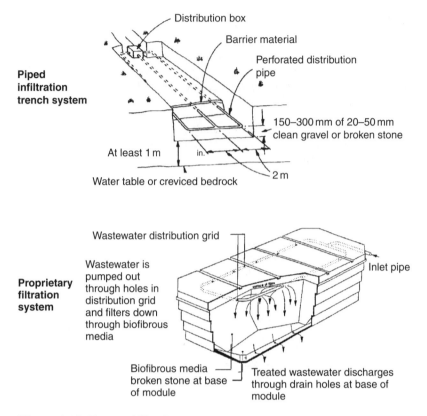

Figure 6.20 Types of filtration systems

Assessing the suitability of the ground

An infiltration system serving a private wastewater treatment plant or septic tank or for greywater should be constructed in ground suitable for the treatment and dispersion of the wastewater discharged.

A ground assessment and soil percolation test should be carried out to determine the suitability of the ground using the three-step procedure described in Table 6.14.	D 3.9.1 ND 3.9.1

Design of infiltration fields

An infiltration system serving a private wastewater treatment plant or septic tank should be designed and constructed to suit the conditions as determined by the ground into which the treated wastewater is discharged.	D 3.9.2 ND 3.9.2

Table 6.14   Ground assessment and soil percolation test

1 Carry out a preliminary ground assessment	• Consult SEPA, Verifier and the Environmental Health Officer as required • Consult SEPA's latest groundwater protection policy • Identify underlying geology and aquifers • Is ground liable to flooding? • Nature of the sub-soil and groundwater vulnerability • Implication of plot size • Proximity of underground services • Ground topography and local drainage patterns • Is water abstracted for drinking, used in food processing or farm dairies? • Implication for, and of, trees and other vegetation • Location of surface waters and terrestrial ecosystems
2 Dig a trial hole to determine the position of the water table and soil conditions	The trial hole should be: • a minimum of 2 m deep; or • a minimum of 1.5 m below the invert of the proposed distribution pipes: and • it should be left covered for a period of 48 h before measuring any water table level
3 Determine the type of infiltration system and the area of ground required	Carry out a percolation test

Location of infiltration fields – pollution

An infiltration system serving a private wastewater treatment plant or septic tank should be located to minimize the risk of pollution.	D 3.9.4 ND 3.9.4
An infiltration field should be located: • at least 50 m from any spring, well or borehole used as a drinking water supply; and • at least 10 m horizontally from any watercourse (including any inland or coastal waters), permeable drain, road or railway.	D 3.9.4 ND 3.9.4

Location of infiltration fields – damage to buildings

Every part of an infiltration system serving a private wastewater treatment plant or septic tank should be located at least 5 m from a building.	D 3.9.5 ND 3.9.5

An infiltration system should also be located at least 5 m
from a boundary in order that an adjoining plot is not
inhibited from its full development potential.

D 3.9.5
ND 3.9.5

Precipitation

Rain penetration occurs most often through walls exposed to the prevailing
wet winds and unless there are adequate damp proof courses and flashings,
etc., materials in roofs, parapets and chimneys can collect rainwater and
deliver it to other parts of the dwelling below roof level.

Every building must be designed and constructed with
adequate drainage and in such a way that there will not be
a threat to the building or the health of the occupants as
a result of moisture from precipitation penetrating to the
inner face of the building.

D 3.10
ND 3.10

This standard does not apply to a building where
penetration of moisture from the outside will result
in effects no more harmful than those likely to arise
from use of the building.

A floor, wall, roof or other building element exposed to
precipitation, or wind-driven moisture, should prevent
penetration of moisture to the inner surface of any part of
a building so as to protect the occupants and to ensure that
the building is not damaged.

D 3.10.1
ND 3.10.1

Moisture from the ground

Water is the prime cause of deterioration in building materials and constructions and the presence of moisture encourages growth of mould that is injurious to health. Drainage (i.e. groundwater) can penetrate building fabric from
below, rising vertically by capillary action.

Every building must be designed and constructed in such
a way that there will not be a threat to the building or the
health of the occupants as a result of moisture penetration
from the ground.

D 3.4
ND 3.4

Buildings need to be constructed in such a way that rising
damp neither damages the building fabric nor penetrates to
the interior where it may constitute a health risk to occupants.

D 3.4.0
ND 4.4.0

Flooding and ground water

The likely effects of flooding on materials and elements of the building and the various forms of construction and measures to reduce the risk of flood damage in dwellings should be taken into consideration by the developers.

Every building must be designed and constructed in such a way that there will not be a threat to the building or the health of the occupants as a result of flooding and the accumulation of groundwater.	D 3.3 ND 3.3
All proposed building sites should be appraised initially to ascertain the risk of flooding of the land and an assessment made as to what effects the development may have on adjoining ground.	D 3.3.1 ND 3.3.1
Ground below and immediately adjoining a building that is liable to accumulate floodwater or groundwater requires treatment to be provided against the harmful effects of such water.	D 3.3.1 ND 3.3.1

Treatment could include a field-drain system:

- to increase the stability of the ground;
- to avoid surface flooding;
- to alleviate subsoil water pressures likely to cause dampness to below-ground accommodation;
- to assist in preventing damage to foundations of buildings;
- to prevent frost heave of subsoil that could cause fractures to structures such as concrete slabs.

The selection of an appropriate drainage layout will depend on the nature of the subsoil and the topography of the ground.

Developers should be aware of the dangers from possible surface water run-off from their building site to other properties and procedures should be in place to overcome this occurrence.	D 3.3.1 ND 3.3.1
The installation of field drains or rubble drains may overcome the problem.	

Rainwater harvesting

Rainwater harvesting systems allow surface water run-off from dwellings or hardstanding areas to be collected, processed, stored and distributed, thereby reducing the demand for potable water, the load on drainage systems and surface water run-off that can lead to incidents of flooding.

Rain, as it falls on buildings, is soft, clear and largely free of contaminants. During collection and storage, however, there is potential for contamination. For this reason it is recommended that:

Recycled surface water is used only for flushing water closets, car washing and garden taps.	D 3.6.7 ND 3.6.7
Prior to the storage of water in a tank the rainwater should be filtered to remove leaves and other organic matter and dust or grit.	D 3.6.7 ND 3.6
Disinfection maybe required if the catchment area is likely to be contaminated with animal faeces, extensive bird droppings, oils or soil.	D 3.6.7 ND 3.6
Water storage tanks should be constructed of materials such as glass reinforced plastic (GRP), high-density polyethylene, steel or concrete, and sealed and protected against the corrosive effects of the stored water and to prevent the ingress of groundwater if located underground.	D 3.6.7 ND 3.6
Water for use in the building should be extracted from just below the water surface in the tank to provide optimum water quality.	D 3.6.7 ND 3.6
All pipework carrying rainwater for use in the building should be identified as such in accordance with the Water Regulations Advisory Scheme (WRAS) guidance notes and great care should be taken to avoid cross-connecting reclaimed water and mains water.	D 3.6.7 ND 3.6
Tanks should be accessible to allow for internal cleaning and the maintenance of inlet valves, sensors, filters or submersible pumps.	D 3.6.7 ND 3.6
An overflow should discharge to a soakaway, or to mains drainage where it is not reasonably practicable to discharge to a soakaway.	D 3.6.7 ND 3.6
Backflow prevention devices should be incorporated to prevent contaminated water from entering the system.	D 3.6.7 ND 3.6
A surface water drainage system should be tested to ensure the system is laid and is functioning correctly.	D 3.6.10 ND 3.6.10

Sanitary appliances below flood level

The basements of approximately 500 buildings in Scotland are flooded each year when the sewers surcharge (i.e. when the effluent runs back up the pipes because they are too small to take the required flow). To guard against this happening:

• wastewater from sanitary appliances and floor gullies below flood level should be drained by wastewater lifting plants or, where there is unlikely to be a risk to persons such as in a car park, via an anti-flooding device;	D 3.7.2 ND 3.7.2
• wastewater lifting plants should be constructed in accordance with the requirements of BS EN 12056-4: 2000;	
• wastewater from sanitary appliances above flood level should not be drained through anti-flooding devices and only in special case, e.g. refurbishment, by a wastewater lifting plant.	

Conversions and extensions

A careful check should be made before breaking into an existing drain to ensure it is the correct one (e.g. whether wastewater to surface water or vice versa) and a further test carried out after connection, such as a dye test, to confirm correct connection.	D 3.7.6

Ventilation of a drainage system

A wastewater drainage system serving a building should be ventilated to limit the pressure fluctuations within the system and minimize the possibility of foul air entering the building.	D 3.7.7 ND 3.7.7

Greywater recycling

Water reuse is becoming an accepted method of reducing demand on mains water and the use of greywater may be appropriate in some buildings for flushing of water closets. However, because greywater recycling systems require constant observation and maintenance they should only be used in buildings where a robust maintenance contract exists.

The approval of Scottish Water is required before any such scheme is installed.

Where a greywater system is to be used, it should be designed, installed and commissioned by a person competent and knowledgeable in the nature of the system and the regulatory requirements.	ND 3.7.9
A risk assessment on the health and safety implications should be carried out for those who will be employed to install and maintain the system.	ND 3.7.9
A comprehensive installation guide, a user's guide and an operation and maintenance manual should be handed to the occupier at the commissioning stage.	ND 3.7.9

Greywater disposal

The disposal of greywater (from baths, showers, washbasins, sinks and washing machines) may be accomplished by an infiltration field.	D 3.9.3 ND 3.9.3

The area of an infiltration field can be calculated using the following formula:

$$A = P \times V_\mathrm{p} \times 0.2$$

where:

A is the area of the sub-surface drainage trench, in m^2;
P is the number of persons served; and
Vp is the percolation value obtained, in s/mm.

Wastewater discharge

Domestic buildings

A wastewater drainage system should discharge to a public sewer or public wastewater treatment plant provided under the Sewerage (Scotland) Act 1968, where it is reasonably practicable to do so.	D 3.7.9

Non-domestic buildings

Discharge of greywater may be via a water closet into a public sewer or public wastewater treatment plant provided under the Sewerage (Scotland) Act 1968, where it is reasonably practicable to do so.	ND 3.7.11

A wastewater drainage system should discharge to a public   ND 3.7.11
sewer or public wastewater treatment plant provided under
the Sewerage (Scotland) Act 1968, where it is reasonably
practicable to do so.

Sanitary facilities

The drainage from sanitary facilities in dwellings is of prime concern.

 Note: Although not recommending that sanitary facilities on the principal living level of a dwelling be designed to an optimum standard for wheelchair users, it should be possible for **MOST** people to use these facilities unassisted and in privacy.

Every building must be designed and constructed in such   D 3.12
a way that sanitary facilities are provided for all occupants   ND 3.12
of, and visitors to, the building in a form that allows
convenience of use and that there is no threat to the health
and safety of occupants or visitors.

Every dwelling should have sanitary facilities comprising   D 3.12.1
at least one water closet (WC), or waterless closet, together
with one wash hand basin per WC, or waterless closet, one
bath or shower and one sink.

An accessible shower room should be of a size that will   D 3.12.3
accommodate both a level-access floor shower with a
drained area of not less than 1.0 m × 1.0 m (or equivalent)
or a 900 mm × 900 mm shower tray (or equivalent).

An accessible shower room should have: ND 3.12.10

- a dished floor of a gradient of not more than 1:50
  discharging into a floor drain, or a proprietary level
  access shower with a drainage area of not less than
  1.2 m by 1.2 m;
- a folding shower seat positioned 500 mm from a flanking wall and securely fixed, with a seat height that
  permits transfer to and from a wheelchair positioned
  outwith the showering area.

 Note: For further details concerning sanitary facilities, see Section 6.19, 'Water (and earth) closets,
bathrooms and showers'.

Fuel storage – containment

Oil is a common and highly visible form of water pollution. Because of the way it spreads, even a small quantity can cause a lot of harm to the aquatic environment. It can pollute rivers, lochs, groundwater and coastal waters, killing wildlife and removing vital oxygen from the water.

The UK government is required by the EEC Directive to prevent List I substances (i.e. highly dangerous substances such as mercury and cadmium) from entering groundwater and to prevent groundwater pollution by List II substances (e.g. trichloroethane, lead and zinc).

Every building must be designed and constructed in such a way that an oil storage installation, incorporating oil storage tanks used solely to serve a fixed combustion appliance installation providing space heating or cooking facilities in a building will: • reduce the risk of oil escaping from the installation; • contain any oil spillage likely to contaminate any water supply, groundwater, watercourse, drain or sewer; and permit any spill to be disposed of safely; • minimize the number of journeys by delivery vehicles.	D 3.24 ND 3.24

LPG storage – fixed tanks

A Liquefied Petroleum Gas (LPG) storage tank, together with any associated pipework, should be designed, constructed and installed in accordance with the requirements set out in the Liquefied Petroleum Gas Association (LPGA) Code of Practice 1: 'Bulk LPG Storage at Fixed Installations'.

 Note: LPG storage tanks in excess of 4 tonnes (9,000 L) capacity are uncommon in domestic applications.

Ground features such as open drains, manholes, gullies and cellar hatches should be sealed or trapped to prevent the passage of LPG vapour.	D 4.11.2 ND 4.11.2

LPG storage cylinders

Any installation should enable cylinders to stand upright, secured by straps or chains against a wall outside the building.	D 4.11.3 ND 4.11.3

Cylinders should be positioned on a firm, level base such as concrete at least 50 mm thick or paving slabs bedded on mortar, and located in a well-ventilated position at ground level, so that the cylinder valves will be:

D 4.11.3
ND 4.11.3

- at least 1 m horizontally and 300 mm vertically from openings in the buildings or from heat source such as flue terminals or tumble dryer vents;
- at least 2 m horizontally from untrapped drains, unsealed gullies or cellar hatches unless an intervening wall not less than 250 mm high is present.

Cylinders should be readily accessible, reasonably protected from physical damage and located where they do not obstruct exit routes from the building.

D 4.11.3
ND 4.11.3

Secondary containment

Externally located, above-ground, oil tanks with a capacity of not more than 2,500 L serving a building should be provided with a catchpit or be integrally bunded if the tank is:

D 3.24.3
ND 3.24.30

- located within 10 m of the water environment (i.e. rivers, lochs, coastal waters);
- located where spillage could run into an open drain or to a loose-fitting manhole cover;
- within 50 m of a borehole or spring;
- over ground where conditions are such that oil spillage could run off into a watercourse;
- located in a position where the vent pipe outlet is not visible from the fill point.

In a domestic building, secondary containment should be provided where a tank is within a building or wholly below ground.

D 3.24.3

Solid waste storage

This standard applies **only** to a dwelling.

Every building must be designed and constructed in such a way that accommodation for solid waste storage is provided which:

D 3.25

- does not threaten the health of people in and around the building;

- does not contaminate any water supply, groundwater or surface water.

Provision for washing down

Where communal solid waste storage is located within a building, such as where a refuse chute is utilized, the storage area should have provision for washing down and draining the floor into a wastewater drainage system.   D 3.25.4

Gullies should incorporate a trap that maintains a seal even during periods of disuse.   D 3.25.4

Walls and floors should be of an impervious surface that can be washed down easily and hygienically.   D 3.25.4

The enclosures should be permanently ventilated at the top and bottom of the wall.   D 3.25.4

Dungsteads and farm effluent tanks

Silage effluent is the most prevalent cause of point source water pollution from farms in Scotland and a high portion of serious pollution incidents occur each year through failure to contain or dispose of effluent satisfactorily.

Every building must be designed and constructed in such a way that there will not be a threat to the health and safety of people from a dungstead and farm effluent tank.   D 3.26
ND 3.26

Solids that are stored in dungsteads that must be properly drained and the effluent collected in a tank while liquids are stored in tanks above or below ground. The container must be impermeable.   D 3.26.0
ND 3.26.0

Construction of dungsteads and farm effluent tanks

Every dungstead or farm effluent tank, including a slurry or silage effluent tank should be constructed in such a manner so as to prevent the escape of effluent through the structure that could cause ground contamination or environmental pollution.   D 3.26.1
ND 3.26.1

| The construction should also prevent seepage and overflow that might endanger any water supply or watercourse. | D 3.26.1 ND 3.26.1 |

Location of dungsteads and farm effluent tanks

| Every dungstead or farm effluent tank, including a slurry or silage effluent tank should be located at a distance (i.e. at least 15 m) from a building so as not to prejudice the health of people in the dwelling. | D 3.26.2 ND 3.26.2 |

Safety of dungsteads and farm effluent tanks

| Where there is the possibility of injury from falls, a dungstead or farm effluent tank should be covered or fenced to prevent people from falling in. | D 3.26.3 ND 3.26.3 |
| Covers or fencing should be in accordance with the relevant recommendations of Section 8 of BS 5502: Part 50: 1993 | D 3.26.3 ND 3.26.3 |

Roof coverings and gutters

The possibility of direct flame impingement from neighbouring buildings is greater where the roof covering and gutters of the building are close to the boundary. Whilst much will depend on the fire dynamics and the velocity and direction of the wind, burning brands are also likely to be more intense. For these reasons, the vulnerability of a roof covering and guttering is determined in relation to the distance of a building to the boundary.

| The roof of a building, including any boxed gutters, soffit or barge boards, should have a low vulnerability if not more than 6 m from the boundary. | 2.8.1 |

 Note: Common materials that meet this criterion include slates, tiles, glazing, sandwich panels and certain plastic materials.

6.4 Ventilation

Ventilation is defined as 'the supply and removal of air (by natural and/or mechanical means) to and from a space or spaces in a building'.

In addition to replacing 'stale' indoor air with 'fresh' outside air, the aim of ventilation is to:

- limit the accumulation of moisture and pollutants from a building which could, otherwise, become a health hazard to people living and/or working within that building;
- dilute and remove airborne pollutants (especially odours);
- control excess humidity;
- provide air for fuel burning appliances.

In general terms, all of these aims can be met if the ventilation system:

- disperses residual pollutants and water vapour;
- extracts water vapour from wet areas where it is produced in significant quantities (e.g. kitchens, utility rooms and bathrooms);
- rapidly dilutes pollutants and water vapour produced in habitable rooms, occupiable rooms and sanitary accommodation;
- extract pollutants from areas where they are produced in significant quantities (e.g. rooms containing processes or activities which generate harmful contaminants);
- is designed, installed and commissioned so that it:
  - is not detrimental to the health of the people living and/or working in the building;
  - helps maintenance and repair;
  - is reasonably secure;
  - makes available, over long periods, a minimum supply of outdoor air for the occupants;
  - minimizes draughts;
  - provides protection against rain penetration.

Ventilation is also a means of controlling thermal comfort.

The aim of the Building (Scotland) Regulations is to suggest to the designer the level of ventilation that should be sufficient for a particular situation as opposed to how it should be achieved. The designer is, therefore, free to use whatever ventilation system he considers the most suitable for a particular building provided that it can be demonstrated that it meets the recommended performance criteria and levels concerning moisture, pollutants and air flow rate standards as shown in Table 6.15.

External pollution
In urban areas, buildings are exposed to a large number of pollution sources from varying heights and upwind distances (i.e. long-, intermediate- and short-range). Internal contamination from these pollution sources can have a

Table 6.15 Standards for performance-based ventilation

Type	Standard	Part
Intermittent extract fan	BS EN 13141-4	Clause 4
Range hood	BS EN 13141-3	Clause 4
Background ventilator (non-RH controlled)	BS EN 13141-1	Clauses 4.1 and 4.2
Background ventilator (RH controlled)	PrEN13141-9	Clauses 4.1 and 4.2
Passive stack ventilator	See Building (Scotland) Regulations	
Continuous mechanical extract ventilation (MEV system)	BS EN 13141-6	Clause 4
Continuous mechanical supply and extract with heat recovery (MVHR)	PrEN13141-7	Clauses 6.1, 6.2 and 6.2.2
Single room heat recovery ventilator	PrEN13141-8	Clauses 6.1 and 6.2

detrimental effect on the buildings' occupants and so it is very important to ensure that the ventilation system provided is sufficient and, above all, that the air intake cannot be contaminated.

Typical urban pollutants include:

- benzene (C_6H_6)
- butadiene (C_4H_6)
- carbon monoxide (CO)
- lead (Pb)
- nitrogen dioxide (NO_2)
- nitrogen oxide (NO)
- ozone (O_3)
- particles (PM_{10})
- sulphur dioxide (SO_2).

Typical emission sources include:

- building ventilation system exhaust discharges;
- combustion plant (such as heating appliances) running on conventional fuels;
- construction and demolition sites;
- discharges from industrial processes and other sources;
- other combustion type processes (e.g. waste incineration, thermal oxidation abatement schemes);
- road traffic, including traffic junctions and underground car parks;
- uncontrolled ('fugitive') discharges from industrial processes and other sources.

Indoor air pollutants

The maximum permissible level of indoor air pollutants is:

Nitrogen dioxide (NO$_2$)	Not exceeding:
	288 µg/m^3 (150 ppb) – 1 h average
	40 µg/m^3 (20 ppb) – long-term average
Carbon monoxide (CO)	Not exceeding:
	100 mg/m^3 (90 ppm) – 15 min averaging time
	60 mg/m^3 (50 ppm) – 30 min averaging time (DH, 2004)
	30 mg/m^3 (25 ppm) – 1 h averaging time (DH, 2004)
	10 mg/m^3 (10 ppm) – 8 h averaging time (DH, 2004)
Control of bio-effluents (body odours)	3.5 l/s per person
Total volatile organic compound (TVOC)*	Not exceeding: 300 µg/m^3 averaged over 8 h

Note: *TVOC is defined as any chemical compound based on carbon chains or rings (which also contain hydrogen) with a vapour pressure greater than 2 mm of mercury (0.27 kPa) at 25°C, excluding methane.

6.4.1 Requirements

Fire

Domestic and non-domestic buildings

> *Every building must be designed and constructed in such a way that:*
>
> - *in the event of an outbreak of fire within the building, the occupants, once alerted to the outbreak of the fire, are provided with the opportunity to escape from the building, before being affected by fire or smoke;*   2.92.11
> - *facilities are provided to assist fire-fighting or rescue operations.*   2.14

Non-domestic buildings

> *Every building must be designed and constructed in such a way that:*   ND 2.1
>
> - *in the event of an outbreak of fire within the building, fire and smoke are inhibited from spreading beyond the compartment of origin until any occupants have had the time to leave that compartment and any mandatory fire containment measures have been initiated.*

Every non-domestic building which is divided into more   ND 2.2
than one area of different occupation must be designed and
constructed in such a way that in the event of an outbreak of fire
within the building, fire and smoke are inhibited from spreading
beyond the area of occupation where the fire originated.

Environment

Every building must be designed and constructed in such a way
that:

- *there will not be a threat to the building or the health of*   3.10
  the occupants as a result of moisture from precipitation
  penetrating to the inner face of the building;
- *the air quality inside the building is not a threat to the health*   3.14
  of the occupants or the capability of the building to resist
  moisture, decay or infestation;
- *each fixed combustion appliance installation:*   3.17
  - *operates safely;*   3.21
  - *receives air for combustion and operation of the chimney*
    so that the health of persons within the building is not
    threatened by the build-up of dangerous gases as a result
    of incomplete combustion;
  - *receives air for cooling so that the fixed combustion*   3.22
    appliance installation will operate safely without threatening
    the health and safety of persons within the building;
- *any component part of each fixed combustion appliance*   3.18
  installation used for the removal of combustion gases will
  withstand heat generated as a result of its operation without
  any structural change that would impair the stability or
  performance of the installation;
- *the products of combustion are carried safely to the external*   3.20
  air without harm to the health of any person through
  leakage, spillage or exhaust, nor permit the re-entry of
  dangerous gases from the combustion process of fuels into
  the building.

Safety

Every building must be designed and constructed in such a way that:

- *people in and around the building are protected from injury that*
  could result from fixed glazing, projections or moving elements
  on the building;

- *fixed glazing in the building is not vulnerable to breakage where* **4.8**
  there is the possibility of impact by people in and around the
  building;
- *both faces of a window and rooflight in a building are capable of*
  being cleaned such that there will not be a threat to the cleaner
  from a fall resulting in severe injury;
- *a safe and secure means of access is provided to a roof;*
- *manual controls for ventilation and for electrical fixtures can be*
  operated safely.

This last part of the standard does not apply to domestic
buildings.

Energy

Every building must be designed and constructed in such a way **6.2**
that:

- *an insulation envelope is provided which reduces heat loss;* **6.6**
- *the form and fabric of the building minimizes the use of*
  mechanical ventilating or cooling systems for cooling purposes;
- *in non-domestic buildings, the ventilating and cooling systems*
  installed are energy efficient and are capable of being controlled
  to achieve optimum energy efficiency;
- *energy supply systems and building services which use fuel or* **6.7**
  power for heating, lighting, ventilating and cooling the internal
  environment and heating the water are commissioned to achieve
  optimum energy efficiency.

The occupiers of a building should be provided with written **6.8**
information by the owner:

- *on the operation and maintenance of the building services and*
  energy supply systems;
- *where any air-conditioning system in the building is subject to a*
  time-based interval for inspection of the system.

Every building must be designed and constructed in such a way that: **6.9**

- *an energy performance certificate for the building is affixed*
  to the building, indicating the approximate annual carbon
  dioxide emissions and energy usage of the building based
  on a standardized use of the building;

- *the energy performance for the certificate is calculated in accordance with a methodology which is asset-based, conforms with the European Directive 2002/91/EC and uses the UK climate data;*
- *the energy performance certificate is displayed in a prominent place within the building.*

6.4.2 Meeting the requirement

Ventilation

Ventilation of a building is required to prevent the accumulation of moisture that could lead to mould growth, and pollutants originating from within the building that could become a risk to the health of the occupants.

Every building must be designed and constructed in such a way that the air quality inside the building is not a threat to the health of the occupants or the capability of the building to resist moisture, decay or infestation.	D 3.14 ND 3.14
A building should have provision for ventilation by either: • natural means; or • mechanical means; or • a combination of natural and mechanical means.	D 3.14.1 ND 3.14.1
Ventilation should have the capability of: • providing outside air to maintain indoor air quality sufficient for human respiration; • removing excess water vapour from areas where it is produced in significant quantities, such as kitchens, utility rooms, bathrooms and shower rooms, to reduce the likelihood of creating conditions that support the germination and growth of mould, harmful bacteria, pathogens and allergens; • removing pollutants that are hazard to health from areas where they are produced in significant quantities, such as non-flued combustion appliances; • rapidly diluting pollutants and water vapour, where necessary, that are produced in apartments and sanitary accommodation.	D 3.14.1 ND 3.14.1
Ventilation should be to the outside (i.e. external) air.	D 3.14.1 ND 3.14.1

There is no need to ventilate:

- a room (other, that is, than a kitchen or utility room) with a floor area of not more than $4\,m^2$;
- a store room used only for storage that requires a controlled temperature.

Natural ventilation
Non-domestic
Natural ventilation of a room or building should be provided as follows:

Table 6.16   Requirements for natural ventilation

Type of construction	Requirements
For a room	A ventilator with an opening area of at least 1/30th of the floor area of the room it serves; A trickle ventilator with an opening area of at least $4{,}000\,mm^2$ if the area of the room is not more than $10\,m$; or A trickle ventilator with an opening area of $400\,mm^2$ for each square metre of room area, if the area of the room is more than $10\,m^2$.
For a room in a building constructed with an infiltration rate of not more than $10\,m^3/h/m^2$	A ventilator with an opening area of at least 1/30th of the floor area of the room it serves, and A trickle ventilator with an opening of at least $10{,}000\,mm^2$ if the room is not more than $10\,m^2$; or A trickle ventilator with an opening area of at least $12{,}000\,mm^2$ if the room is more than $10\,mm^2$.
For a toilet	Mechanical extract ventilation.
For any other building	in accordance with Section 3 of BS 5925: 1991 (1995).

Where a building is naturally ventilated, all moisture-producing areas such as bathrooms and shower rooms should have the additional facility for removing such moisture before it can damage the building. — ND 3.14.2

Where rapid ventilation is provided (e.g. an opening window), some part of the opening should be at least 1.75 m above floor level. — ND 3.14.2

Domestic

To reduce the effects of stratification of the air in a room, some part of the opening ventilator should be at least 1.75 m above floor level. — D 3.14.2

Most dwellings are naturally ventilated and ventilation should   D 3.14.2
be provided in accordance with Table 6.17.

Table 6.17   Recommended ventilation of a dwelling

	Ventilation recommendations	Trickle ventilation $>10\,m^3/h/m^2$	Trickle ventilation $<10\,m^3/h/m^2$
Apartment	A ventilator with an opening area of at least 1/30th of the floor area it serves	$8{,}000\,mm^2$	$12{,}000\,mm^2$
Kitchen	Either: • mechanical extraction capable of at least 30 L/s (intermittent) above a hob; or • mechanical extraction capable of at least 60 L/s (intermittent) if elsewhere; or • a passive stack ventilation system.	$4{,}000\,mm^2$	$10{,}000\,mm^2$
Utility room	Either: • mechanical extraction capable of at least 30 L/s (intermittent) above a hob; or • a passive stack ventilation.	$4{,}000\,mm^2$	$10{,}000\,mm^2$
Bathroom or shower room (with or without a WC)	Either: • mechanical extraction capable of at least 15 L/s (intermittent); or • a passive stack ventilation system.	$4{,}000\,mm^2$	$10{,}000\,mm^2$
Toilet	Either: • a ventilator with an opening area of at least 1/30th of the floor area it serves; or • mechanical extraction capable of at least three air changes per hour.	$4{,}000\,mm^2$	$10{,}000\,mm^2$

Trickle ventilators

A trickle ventilator (sometimes called background ventilation) is a small ventilation opening, mostly provided in the head of a window frame (but not always) with a controllable shutter. A permanent ventilator is not recommended since occupants like control over their environment, and uncontrollable ventilators are usually permanently sealed up to prevent draughts (Figure 6.21).

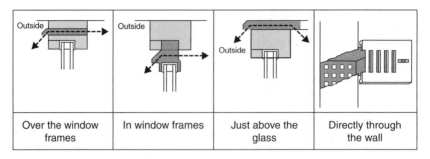

| Over the window frames | In window frames | Just above the glass | Directly through the wall |

Figure 6.21 Trickle ventilation system

To avoid cold draughts, trickle ventilators are normally positioned 1.7 m above floor level and usually include a simple control (such as a flap) to allow users to shut off the ventilation according to personal choice or external weather conditions. Nowadays, pressure-controlled trickle ventilators that reduce the air flow according to the pressure difference across the ventilator are available to reduce draught risks during windy weather.

Trickle ventilators are normally left open in occupied rooms in dwellings.

Background ventilators

The need for background ventilators will depend on the air permeability or air tightness of a building (Figure 6.22).

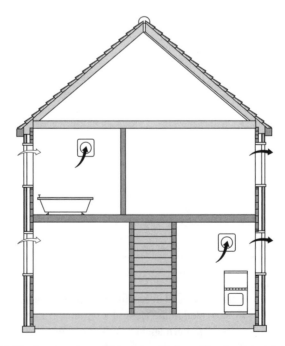

Figure 6.22 Background ventilators and intermittent extract fans

Trickle ventilators should be provided in naturally ventilated areas to allow fine control of air movement.	D 3.14.5 ND 3.14.3
Where the trickle ventilator has to be ducted (e.g. to an internal room), the geometric area of the trickle ventilator should be increased to double that is shown in Table 6.17 to compensate for the reduced air flow caused by friction.	D 3.14.5
It is recommended that trickle ventilators supply replacement air for mechanical extract and passive stack ventilation systems and routes in rooms where windows are more than likely to be left open.	D 3.14.5
A trickle ventilator should be so positioned that a part of it is at least 1.75 m above floor level.	D 3.14.5
A trickle ventilator serving a bathroom or shower room may open into an area that does not generate moisture, such as a bedroom or hallway, provided the area is fitted with a trickle ventilator.	D 3.14.5
A trickle ventilator should be provided in an area fitted with mechanical extraction to provide replacement air and ensure efficient operation when doors are closed.	D 3.14.5

Passive stack ventilation systems

Passive Stack Ventilation (PSV) is a ventilation device which uses ducts from terminals mounted in the ceiling of rooms to terminals on the roof, to extract air to the outside by a combination of the natural stack effect and the pressure effects of wind passing over the roof of the building.

The so-called 'stack effect' relies on the pressure differential between the inside and the outside of a building caused by differences in the density of the air due to an indoor/outdoor temperature difference (Table 6.18).

Table 6.18 Passive stack ventilation

Room	Internal duct diameter (mm)	Internal cross-sectional area (mm)
Kitchen	125	12,000
Utility room	100	8,000
Bathroom	100	8,000
Sanitary accommodation	80	5,000

For sanitary accommodation only, purge ventilation may be used provided that security is not an issue.

Note: Open-flued appliances may provide sufficient extract ventilation when in operation and can be arranged to provide sufficient ventilation when not firing.

The design and installation of PSV systems are crucial to their operation and Figure 6.23 shows the preferred option for kitchen and bathroom ducts with ridge terminals.

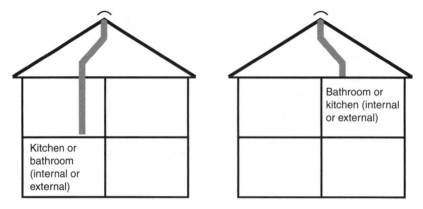

Figure 6.23 Preferred PSV system layouts

Another option (see Figure 6.24) is to have the kitchen and bathroom ducts penetrating the roof and extend its terminals to ridge height.

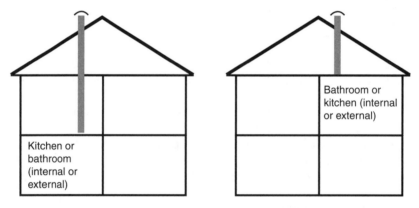

Figure 6.24 Alternative PSV system layouts

These systems are most suited for use in a building with a height of not more than four storeys as the stack effect will diminish as the air cools.

Every passive stack ventilation system should: D 3.14.6

- incorporate a ceiling-mounted automatic humidity-sensitive extract grille that will operate when the relative humidity is between 50 and 65%;
- be insulated with at least 25 mm thick material having a thermal conductivity of 0.04 W/mK where it passes through a roof space or other unheated space or where it extends above the roof level.

The flue of an open-flued combustion appliance may serve as a passive stack ventilation system provided that either: D 3.14.6

- the appliance is a solid fuel appliance and is the primary source of heating, cooking or hot water production;
- the flue has an unobstructed area equivalent to a 125 mm diameter duct and the appliance's combustion air inlet and dilution air inlet are permanently open;
- the appliance is an oil-firing appliance which is a continually burning vaporizing appliance (only) such as a cooker or room heater, and the room is fitted with a ventilator with a minimum free area of 10,000 mm^2.

A duct or casing, forming a passive stack ventilation system serving a kitchen should be non-combustible. D 3.14.6

Mechanical ventilation

Mechanical ventilation should be to the outside air, but it may be via a duct or heat exchanger. D 3.14.8

Where a mechanical ventilation system serves more than one dwelling, it should have a duplicate motor and be separate from any other ventilation system installed for any other purpose. D 3.14.8

Where the mechanical ventilation system gathers extracts into a common duct for discharge to an outlet, no connections to the system should be made between any exhaust fan and the outlet. D 3.14.8

The use of non-return valves is not recommended. D 3.14.8

A mechanical ventilation or air-conditioning system should be designed, installed and commissioned to perform in a way that is:

- not detrimental to the health of the occupants of a building; ND 3.14.5
- can be easily accessible for regular maintenance.

Mechanical extract should be provided in rooms where the cubic space per occupant is not more than 3 m and where the rooms have low ceilings and are occupied by large numbers of people. ND 3.14.5

Table 6.19   Mechanical ventilation of domestic-sized kitchens, bathrooms and toilets

Space	Ventilation provision	Trickle ventilation	
		$>10\,\text{m}^3/\text{h}/\text{m}^2$	$<10\,\text{m}^3/\text{h}/\text{m}^2$
Kitchen	Either: • mechanical extraction capable of at least 30 L/s (intermittent) above a hob; or • mechanical extraction capable of at least 60 L/s (intermittent) if elsewhere	4,000 mm^2	10,000 mm^2
Utility room or washroom	mechanical extraction capable of at least 30 L/s (intermittent)	4,000 mm^2	10,000 mm^2
Bathroom or shower room (with or without a WC)	mechanical extraction capable of at least 15 L/s (intermittent)	4,000 mm^2	10,000 mm^2
Toilet	mechanical extraction capable of at least three air changes per hour	4,000 mm^2	10,000 mm^2
Kitchen	Either: • mechanical extraction capable of at least 30 L/s (intermittent) above a hob; or • mechanical extraction capable of at least 60 L/s (intermittent) if elsewhere	4,000 mm^2	10,000 mm^2
Utility room or washroom	mechanical extraction capable of at least 30 L/s (intermittent)	4,000 mm^2	10,000 mm^2
Bathroom or shower room (with or without a WC)	mechanical extraction capable of at least 15 L/s (intermittent)	4,000 mm^2	10,000 mm^2
Toilet	mechanical extraction capable of at least three air changes per hour	4,000 mm^2	10,000 mm^2

Where a mechanical ventilation system gathers extracts into a common duct for discharge to an outlet, no connection to the system should be made between any exhaust fan and the outlet.	ND 3.14.5
Mechanical ventilation should be to the outside air (normally via duct or heat exchanger).	ND 3.14.5
An inlet to, and an outlet from, a mechanical ventilation system should be installed so as to avoid contamination of the air supply to the system.	ND 3.14.5

Continuous mechanical extract ventilation

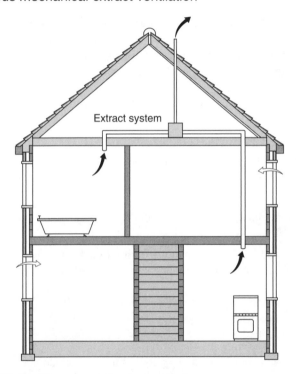

Extract system

Figure 6.25 Continuous mechanical extract

This system may consist of either a central extract system or individual room fans, or a combination of both.

Mechanical aids to ventilation

A mechanical ventilation system should be designed, installed and commissioned to perform in a way that is not detrimental to the health of the occupants of the building and when necessary, is easily accessible for regular maintenance.	D 3.14.10

The following is a list of acceptable mechanical systems that   D 3.14.10
will aid ventilation in a dwelling:

- continuously operating balanced supply and extract
  mechanical ventilation (with or without heat recovery);
- continuously operating mechanical extract ventilation;
- mechanical extract ventilation units (extract fans), either
  window or wall mounted, in rooms where there is likely
  to be high humidity such as kitchens, bathrooms and
  shower rooms;
- mechanical input air ventilation systems.

Mechanical Ventilation and Air Conditioning (MVAC)

It is not considered desirable that dwellings (or buildings consisting of dwellings) have air-conditioning systems or use mechanical ventilation systems for cooling purposes, as this leads to increased energy use and higher carbon dioxide emissions.

When considering the installation of MVAC, attention should therefore be given to:

- the form and fabric of the building;
- the energy efficiency of the equipment;
- control of the equipment.

Every building must be designed and constructed in such a way   D 6.6
that:
                                                                  ND 6.6
- the form and fabric of the building minimizes the use of
  mechanical ventilating or cooling systems for cooling
  purposes;
- in non-domestic buildings, the ventilating and cooling
  systems installed are energy efficient and are capable of
  being controlled to achieve optimum energy efficiency.

 This standard does not apply to buildings which do not
use fuel or power for ventilating or cooling the internal
environment.

Form and fabric of the building

In order to minimize any need for mechanical ventilation for cooling or air conditioning due to high internal temperatures in hot weather, the form and the fabric of a dwelling should include:

- a high proportion of translucent glazing;
- orientation of translucently glazed areas;

- solar shading or other solar control measures where areas of the external building fabric are susceptible to solar gain;
- natural ventilation (including night cooling);
- thermal mass.

A dwelling should be designed to avoid any need for cooling.	D 6.6.1
Where a dwelling has little or no cross-ventilation (e.g. flats with all external windows/rooflights on one southerly elevation or a high proportion of translucent glazing), the dwelling should be designed to avoid high internal temperature.	D 6.6.1
The form and fabric of a non-domestic building should not result in a need for excessive installed capacity of mechanical ventilation and cooling equipment.	ND 6.6.1

Control of MVAC equipment

Appropriate ways should be provided to manage, control and monitor the operation of the equipment and systems that are installed in the building.	ND 6.2.4
Temperature sensors should be provided in areas for the services being controlled.	ND 6.2.4
The temperature control should be selected for the minimum energy consumption for the given occupancy conditions.	ND 6.2.4
The control system of the air-conditioning system should be set up to avoid simultaneous heating and cooling and minimize energy consumption.	ND 6.2.4
Free cooling should be optimized in order to minimize the running costs of the MVAC system.	ND 6.2.4
Central air handling units should have damper controls to provide fresh air as the first stage of cooling.	ND 6.2.4
When the external air is higher than the space temperature the dampers should be overridden to provide a minimum level of fresh air.	ND 6.2.4
Enthalpy control should also be considered to improve free cooling.	ND 6.2.4
Night-time cooling to pre-cool the building structure overnight should be considered to limit daytime cooling demand and minimize energy consumption.	ND 6.2.4

Efficiency of MVAC equipment

Fans (other than individual fans that serve a small number of   ND 6.6.2
rooms in an otherwise naturally ventilated building), pumps,
motors, refrigeration equipment and other components should:

- have no more capacity for demand and standby than is
  needed:
- not be oversized as energy efficiency and power factor
  values will be adversely affected;
- have their characteristics matched to the volume control
  using variable speed motors and variable pitch fans to
  optimize fan performance at part load.

Where air-conditioning systems are installed to provide   ND 6.6.2
comfort cooling, the minimum energy efficiency ratios of
such systems should be in accordance with Table 6.20.

Table 6.20   Comfort cooling energy efficiency ratio

Comfort cooling systems	Required minimum energy efficiency ratio (EER)
Package air conditioners – single duct types	1.8
Package air conditioners – other types	2.2
Split and multi-split air conditioners (incl. VRF)	2.4
Vapour compression cycle chillers – water cooled	3.4
Vapour compression cycle chillers – air cooled	2.25
Water loop heat pump	3.2
Absorption chillers	0.5

Owners of buildings containing air conditioning shall ensure that:

All air-conditioning systems are inspected at regular intervals.   D 0.17.1

Appropriate advice is given to the users of the buildings on   ND 0.17.1
reducing the energy consumption of such an air-conditioning
system.

Non-domestic ventilation fans

Where fan systems are installed to provide either ventilation or   ND 6.2
air circulation:

- the total specific fan power (SFP) should not be greater than
  1.5 W/L/s;

- fan characteristics should be matched to the volume control using variable speed motors and variable pitch fans to optimize fan performance at part load;
- the individual SFP at the design-flow rate should be no worse than the values shown in Table 6.21;
- ventilation system fans rated at more than 1,100 W should be fitted with variable speed drives to ensure they operate efficiently by varying the output of the fan to match the actual demand.

 Consideration should be given to allowing greater SFP where specialist processes occur or if the external air is more heavily polluted, as better air filtration or cleaning may be appropriate.

Table 6.21   Air distribution permissible specific fan power

Air distribution systems	Maximum permissible specific fan power (Watts/(Litres/s))
Central mechanical ventilation including heating, cooling and heat recovery	2.5
Central mechanical ventilation including heating and cooling	2.0
All other central systems	1.8
Local ventilation only units remote from the area, such as ceiling void or roof mounted units serving one room or area	0.5
Local ventilation only units within local area, such as window/wall/roof units, serving one room or area	1.2
Other local units (e.g. a fan coil unit)	0.8

Extract fans

Extract fans lower the pressure in a dwelling and may cause the spillage of combustion products from open-flued appliances.

In dwellings where it is intended to install open-flued combustion appliances and extract fans, the combustion appliances should be able to operate safely whether or not the fans are running.	D 3.17.8 ND 3.17.8
The installation of extract fans should be tested to show that combustion appliances operate safely whether or not fans are running.	D 3.17.8 ND 3.17.8
For solid fuel appliances, extract ventilation should not generally be installed in the same room.	D 3.17.8 ND 3.17.8

For a gas-fired appliance, where a kitchen contains an open-flued appliance, the extract rate of the fan should not exceed 20 L/s.	D 3.17.8 ND 3.17.8

Installation of fans in dwellings

The three fan types most commonly used in domestic applications are:

- axial fans
- centrifugal fans
- in-line fans.

Axial fans

The axial fan is the most common form of fan which can be mounted on the wall, window (i.e. through a suitable glazing hole) or in the ceiling (e.g. in a bathroom).

For wall and window mounting applications up to 350 mm thick, use a short length of rigid round duct or a flexible duct pulled taut.

For bathrooms, 100 mm diameter fans can be used as an axial fan in the ceiling with a short length (1.5 m maximum) of flexible duct with (a maximum of) two 90 bends.

 Note: The duct must be pulled taut and the discharge terminal should have at least 85% free area of the duct diameter.

Centrifugal fans

Centrifugal fans (because they develop greater pressure) permit longer lengths of ducting to be used and so can be used for most wall and/or window applications in high-rise (i.e. above three storeys) buildings or in exposed locations to overcome wind pressure.

Most centrifugal fans are designed with 100 mm diameter outlets which enables them to be connected to a wide variety of duct types.

In-line fans

There are two types of in-line fan available:

- axial fans which have to be installed with the shortest possible duct length to the discharge terminal;
- mixed flow fans which have the characteristics of both axial and centrifugal fans and can, therefore, be used with longer lengths of ducting.

Both types can be used in bathrooms (110 mm diameter), utility rooms (125 mm) and kitchens (150 mm diameter).

Ventilation of conservatories

With large areas of glazing, conservatories attract large amounts of the sun's radiation that can create unacceptable heat build-up. Efficient ventilation therefore is very important to ensure a comfortable environment (Figure 6.26).

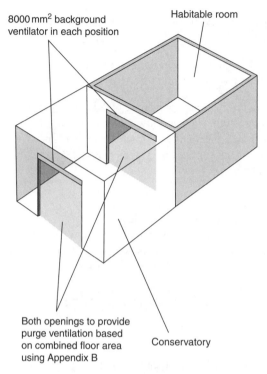

8000 mm² background ventilator in each position

Habitable room

Both openings to provide purge ventilation based on combined floor area using Appendix B

Conservatory

Figure 6.26 A habitable room ventilated through a conservatory

A conservatory should have a ventilator or ventilators with an opening area of at least 1/5th of the floor area it serves.	D 3.14.3
To minimize the effects of heat build-up, it is recommended that at least 30% of the ventilator area provision is located on the roof or at as high a level as possible.	D 3.14.3

Conservatories and extensions built over existing windows

Constructing a conservatory or extension over an existing window, or ventilator, will effectively result in an internal room, restrict air movement and could significantly reduce natural ventilation to that room.

A conservatory may be constructed over a ventilator serving a room in a dwelling provided that the ventilation of the conservatory is to the outside air and has an opening area of at least 1/30th of the total combined floor area of the internal room so formed and the conservatory.	D 3.14.7
The ventilator to the internal room should have an opening area of at least 1/30th of the floor area of the room.	D 3.14.7

Trickle ventilators should be provided relevant to the overall areas created.	D 3.14.7
If an extension is built over a ventilator, then a new ventilator should be provided to the room.	D 3.14.7
If the conservatory or extension is constructed over an area that generates moisture, such as a kitchen, bathroom, shower room or utility room, mechanical extract, via a duct if necessary or a passive stack ventilation system, should be provided direct to the outside air.	D 3.14.7

Extensions built over existing windows

Constructing an extension over an existing window, or ventilator, will effectively result in an internal room. This will restrict air movement and could significantly reduce natural ventilation to that room.

A new ventilator and trickle ventilator should be provided to the existing room.	ND 3.14.4

 BUT if this is **not** reasonably practicable (e.g. if virtually the entire external wall of the room is covered by the extension), the new extension should be treated as part of the existing room as opposed to being a separate internal room.

If the extension is constructed over an area that generates moisture, such as a kitchen, bathroom, shower room or utility room, mechanical extract, via a duct if necessary, should be provided direct to the outside air.	ND 3.14.4

Ventilation of areas designated for drying of washing

Where clothes are dried naturally indoors large quantities of moisture can be released and this will need to be removed before it can damage the building.

Where a space other than a utility room or bathroom is used, that space should be provided with either: • mechanical extraction with the fan connected through a humidistat set to activate when the relative humidity is between 50 and 65% and capable of at least 15 L/s intermittent operation; or • a passive stack ventilation system.	D 3.14.4

Ventilation of sanitary accommodation

Any area containing sanitary facilities should be well ventilated, so that offensive odours do not linger.	ND 3.14.7
Measures should be taken to prevent odours entering other rooms (e.g. by providing a ventilated area between the sanitary accommodation and the other room or by using a mechanical ventilation).	ND 3.14.7
No room containing sanitary facilities should communicate directly with a room for the preparation or consumption of food.	ND 3.14.7

Ventilation of garages

The principal reason for ventilating garages is to protect the building users from the harmful effects of toxic emissions from vehicle exhausts.

Ventilation should be provided in accordance with the following guidance:	D 3.14.11

• where the garage is naturally ventilated, by providing at least two permanent ventilators, each with an open area of at least 1/3,000th of the floor area they serve, positioned to encourage through ventilation, with one of the permanent ventilators being not more than 600 mm above floor level;
• where the garage is mechanically ventilated, by providing a system:
  – capable of continuous operation, designed to provide at least two air changes per hour;
  – independent of any other ventilation system;
  – constructed so that two-thirds of the exhaust air is extracted from outlets not more than 600 mm above floor level.
	ND 3.14.8

Domestic buildings

Where a garage is attached to a dwelling: • the separating construction should be as airtight as possible; • communicating doors should have airtight seals (in certain circumstances a lobby arrangement may be appropriate).	D 3.14.11

Garages of less than $30\,m^2$ do not require the ventilation to be designed as a degree of fortuitous ventilation is created by the imperfect fit of 'up and over' doors or pass doors.   D 3.14.11

A garage with a floor area of at least $30\,m^2$ but not more than $60\,m^2$ used for the parking of motor vehicles should have provision for natural or mechanical ventilation.   D 3.14.11

Ventilation of small garages

A garage with a floor area of at least $30\,m^2$ but not more than $60\,m^2$ used for the parking of motor vehicles should have provision for natural or mechanical ventilation as follows: ND 3.14.8

- where the garage is naturally ventilated, by providing at least two permanent ventilators, each with an open area of at least 1/3,000th of the floor area they serve, positioned to encourage through ventilation, with one of the permanent ventilators being not more than 600 mm above floor level;
- where the garage is mechanically ventilated, by providing a system:
    - capable of using any other ventilation system;
    - constructed so that 2/3rd of the exhaust air is extracted from outlets not more than 600 mm above floor level.

Ventilation of large garages

A garage with a floor area more than $60\,m^2$ for the parking of motor vehicles should have provision for natural or mechanical ventilation on every storey.

Ventilation should: ND 3.14.9

- give carbon monoxide concentrations of not more than 30 ppm averaged over an 8 h period;
- restrict peak concentrations of carbon monoxide at areas of traffic concentrations such as ramps and exits to not more than 90 ppm for periods not exceeding 15 min.

Ventilation may be achieved: ND 3.14.9

- by providing openings in the walls on every storey of at least 1/20th of the floor area of that storey with at least half of such area in opposite walls to promote extract ventilation, if the garage is naturally ventilated;

- by providing mechanical ventilation system capable of at least six air changes per hour and at least 10 air changes per hour where traffic concentrations occur;
- where it is a combined natural/mechanical ventilation system, by providing:
  - openings in the wall on every storey of at least 1/40th of the floor area of the storey with at least half of such area in opposite walls;
  - a mechanical system capable of at least three air changes per hour.

Ventilation of a drainage system

A wastewater drainage system serving a building should be ventilated to limit the pressure fluctuations within the system and minimize the possibility of foul air entering the building.	D 3.7.7 ND 3.7.7

Ventilation of wall cavities

Ventilation of external wall cavities is necessary to prevent the build-up of excessive moisture that could damage the fabric of a building.

To reduce the amount of interstitial condensation to a level that will not harm the timber frame or sheathing, a cavity of at least 50 mm wide should be provided between the sheathing and the cladding.	D 3.10.6 ND 3.10.6
Due to the air gaps inherent between the components of a timber, slate or tile clad wall, no proprietary ventilators should be necessary and a 10 mm free air space should be sufficient.	
Where the wall cavity is sub-divided into sections by the use of cavity barriers, e.g. at mid-floor level in a two-storey house, the ventilators should be provided to the top and bottom of each section of the cavity.	D 3.10.6 ND 3.10.6
Care should be taken with rendered walls to prevent blockage of the ventilators.	D 3.10.6 ND 3.10.6
Where the outer leaf is of masonry construction, venting of the cavity is normally sufficient.	D 3.10.6 ND 3.10.6
Cavities should be vented to the outside air by installing ventilators with at least 300 mm^2 free opening area at 1.2 m maximum centres.	D 3.10.6 ND 3.10.6

Additional ventilation requirements for domestic buildings

Ducted warm air heating

Where a flat or maisonette has a storey at a height of more   D 2.9.13
than 4.5 m, or a basement storey, and the flat/maisonette is
provided with a system of ducted warm air heating:

- transfer grilles should not be fitted between any room and
  the entrance hall or stair;
- supply and return grilles should be not more than 450 mm
  above floor level;
- if warm air is ducted to an entrance hall or stair, the return
  air should be ducted back to the heater;
- if a duct passes through any wall, floor or ceiling of an
  entrance hall or stair, all joints between the duct and the
  surrounding construction should be sealed;
- there should be a room thermostat in the living room, at a
  height more than 1,370 mm and not more than 1,830 mm,
  with an automatic control which will turn off the heater,
  and actuate any circulation fan should the ambient
  temperature rise to more than 35°C;
- if the system recirculates air, smoke detectors should be
  provided in every extract duct to cause the recirculation
  of air to stop and direct all extract air to the outside of the
  building in the event of fire.

Smoke control in corridors

Where a domestic building has more than one escape stair and   D 2.9.17
where a corridor, or part of a corridor, provides escape in only
one direction, automatic opening ventilators should be provided
in that part of the corridor which provides single direction escape.
Such ventilators should:

- provide for exhaust at or near ceiling level and for supply
  at or near floor level with a combined aggregate area of at
  least 1.5 m^2;
- be activated by automatic smoke detection fixed to the
  ceiling of the corridor and fitted with a manual override for
  fire service use.

Protected lobbies

A protected lobby means a lobby within a protected zone, but separated from
the remainder of the protected zone so as to resist the movement of smoke
from the adjoining accommodation to the remainder of the protected zone.

A protected lobby should be constructed within a protected zone.   D 2.9.19

Where flats or maisonettes are served by only one escape stair   D 2.9.19
and there is no alternative means of escape from the upper
storeys, there should be a protected lobby with automatic
opening ventilators at each storey within the protected zone
between the escape stair and the accommodation, including a
parking garage and any other accommodation ancillary to the
dwellings (see Figure 6.27).

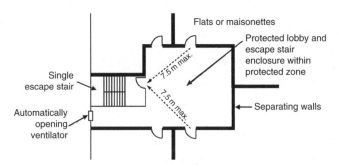

Figure 6.27 Single stair access to flats and maisonettes with any storey
at a height of more than 7.5 m, but not more than 18 m

Where automatic opening ventilators are recommended, they   D 2.9.19
should:

- provide for exhaust at or near ceiling level and for supply
  at or near floor level with a combined aggregate area of at
  least $1.5 \, \text{m}^2$;
- be activated by automatic smoke detection fixed to the
  ceiling of the protected lobby and fitted with a manual
  override for fire service use.

Detectors should be evenly spaced and: D 2.9.19

- not more than 20 m apart;
- at least 500 mm from any side of the lobby or corridor;
- with the detector-sensing element more than 35 mm and not
  more than 300 mm from the soffit of the ceiling;
- with a detector situated not more than 5 m from any change
  of direction in the lobby or corridor exceeding 45°.

Any part of a lobby or corridor divided from any other part
by a beam or other obstruction projecting more than 600 mm
below the soffit of the ceiling shall be deemed to be a separate
lobby or corridor.

External escape stairs

External escape stairs present additional hazards to people evacuating a building in the case of fire. This is because the escape stair will be exposed to the
possible effects of inclement weather.

Due to the likely smoke dissipation to atmosphere, service openings including ventilation ducts not more than 2 m from the escape stair may be protected by heat-activated sealing devices or systems.	D 2.9.22

Smoke alarms

A smoke alarm should be at least 300 mm away from any air-conditioning outlet.	D 2.11.2

Combustion appliances

Protection from combustion products

It is essential that flues continue to function effectively when in use without
allowing the products of combustion to enter the building.

Every building must be designed and constructed in such a way that any component part of each fixed combustion appliance installation used for the removal of combustion gases will withstand heat generated as a result of its operation without any structural change that would impair the stability or performance of the installation.	D 3.18 ND 3.18

Removal of products of combustion

Heating and cooking appliances fuelled by solid fuel, oil or gas all have
the potential to cause carbon monoxide (CO) poisoning if they are poorly
installed or commissioned, inadequately maintained or incorrectly used.

Every building must be designed and constructed in such a way that the products of combustion are carried safely to the external air without harm to the health of any person through leakage, spillage or exhaust, nor permit the re-entry of dangerous gases from the combustion process of fuels into the building.	D 3.20 ND 3.20

Oil-firing appliances in bathrooms and bedrooms

There is an increased risk of carbon monoxide poisoning in bathrooms, shower rooms or rooms intended for use as sleeping accommodation, such as bed-sitters. Because of this:

- open-flued oil-firing appliances should not be installed in   D 3.20.5
  these rooms or any cupboard or compartment connecting   ND 3.20.5
  directly with these rooms;
- where locating a combustion appliance in such rooms
  cannot be avoided, the installation of a room-sealed
  appliance would be appropriate.

Design of flues

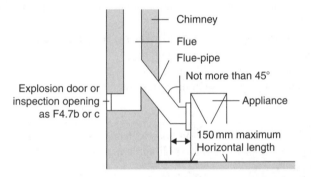

Section through appliance and flue-pipe

Figure 6.28 Design of flues

A combustion appliance should be connected to a chimney that discharges to the external air.	D 3.20.11 ND 3.20.11
Every solid fuel appliance should be connected to a separate flue.	D 3.20.11 ND 3.20.11
Every oil-firing appliance should be connected to a separate flue (unless the appliance has pressure jet burners and is connected into a shared flue).	D 3.20.11 ND 3.20.11
Every gas-fired appliance that requires a flue should connect into a separate flue.	D 3.20.11 ND 3.20.11
The flue of a natural draught appliance, such as a traditional solid fuel appliance, should offer the least resistance to the passage of combustion gases.	D 3.20.11 ND 3.20.11

Combustion appliances – air for combustion

All combustion appliances need ventilation to supply them with oxygen for combustion.

Every building must be designed and constructed in such a way that each fixed combustion appliance installation receives air for combustion and operation of the chimney so that the health of persons within the building is not threatened by the build-up of dangerous gases as a result of incomplete combustion.	D 3.21 ND 3.21

Supply of air for combustion generally

- A room containing an open-flued appliance may need permanently open air vents.
- An open-flued appliance needs to receive a certain amount of air from outside, dependent on its type and rating.
- Infiltration through the building fabric may be sufficient, but above certain appliance ratings permanent openings are necessary.

Ventilators for combustion should be located so that occupants are not provoked into sealing them against draughts and noise.	D 3.21.1 ND 3.21.1
Discomfort from draughts can be avoided by placing vents close to appliances (e.g. floor ventilators), by drawing air from intermediate spaces such as hallways or by ensuring good mixing of incoming air.	D 3.21.1 ND 3.21.1
Air vents should **not** be located within a fireplace recess except on the basis of specialist advice.	D 3.21.1 ND 3.21.1
Noise-attenuated ventilators may be required in certain circumstances.	D 3.21.1 ND 3.21.1
Appliance compartments that enclose open-flued appliances should be provided with vents large enough to admit all the air required by the appliance for combustion and proper flue operation, whether the compartment draws air from the room or directly from outside.	D 3.21.1 ND 3.21.1

Supply of air for combustion to solid fuel appliances

A solid fuel appliance installed in a room or space should have a supply of air for combustion by way of permanent ventilation, either direct to the open air or to an adjoining space (including a sub-floor space) that is itself permanently ventilated direct to the open air.	D 3.21.2 ND 3.21.2
An air supply should be provided in accordance with Table 6.22.	D 3.21.2 ND 3.21.2

Table 6.22   Supply of air for combustion

Type of appliance	Minimum ventilation opening sizes
Open appliance without a throat	A permanent air entry opening or openings with a total free area of 50% of the cross-sectional area of the flue
Open appliance with a throat	A permanent air entry opening or openings with a total free area of 50% of the throat opening area.
Any other solid fuel appliance	A permanent air entry opening or openings with a total free area of $550\,mm^2$ for each kilowatt of combustion appliance rated output more than 5 kW. (A combustion appliance with an output rating of not more than 5 kW has no minimum requirement, unless stated by the appliance manufacturer)

Where a draught stabilizer is fitted to a solid fuel appliance, or to a chimney or flue-pipe in the same room as a solid fuel appliance, additional ventilation opening should be provided with a free area of at least $300\,mm^2/kW$ of solid fuel appliance rated output.

Supply of air for combustion to oil-firing appliances

An oil-firing appliance installed in a room or space should have a supply of air for combustion by way of permanent ventilation either direct to the open air or to an adjoining space which is itself permanently ventilated direct to the open air.	D 3.21.3 ND 3.21.3
An air supply should be provided in accordance with the recommendations in BS 5410: Part 1: 1997.	D 3.21.3 ND 3.21.3

Supply of air for combustion to gas-fired appliances

A gas-fired appliance installed in a room or space should have a supply of air for combustion and an air supply should be provided in accordance with Table 6.23.	D 3.21.4 ND 3.21.4

Table 6.23   Supply of air for combustion to gas-fired appliances

For a decorative fuel-effect gas appliance	BS 5871: Part 3: 2005
For an inset live fuel-effect gas appliance	BS 5871: Part 2: 2005
For any other gas-fired appliance	BS 5440: Part 2: 2000

Combustion appliances – air for cooling

In some cases, combustion appliances may need air for cooling in addition to air for combustion. This air will keep control systems in the appliance at a safe temperature and/or ensure that casings remain safe to touch.

Every building must be designed and constructed in such a way that each fixed combustion appliance installation receives air for cooling so that the fixed combustion appliance installation will operate safely without threatening the health and safety of persons within the building.	D 3.22 ND 3.22

Supply of air for cooling to oil-firing appliances

An oil-firing appliance installed in an appliance compartment should have a supply of air for cooling by way of permanent ventilation, in addition to air for combustion, either direct to the open air or to an adjoining space.	D 3.22.2 ND 3.22.2

Supply of air for cooling to gas-fired appliances

A gas-fired appliance installed in an appliance compartment should have supply of air for cooling.	D 3.22.3 ND 3.22.3
Air for cooling should be provided in accordance with the recommendations in BS 5440: Part 2: 2000 for a gas-fired appliance located in an appliance compartment.	D 3.22.3 ND 3.22.3

Combustion appliances – safe operation

The correct installation of a heating appliance or design and installation of a flue can reduce the risk from combustion appliances and their flues:

- endangering the health and safety of persons in or around a building;
- compromising the structural stability of a building;
- causing damage by fire.

Every building must be designed and constructed in such a way that each fixed combustion appliance installation operates safely.	D 3.17 ND 3.17

Smoke and heat ventilation

Smoke clearance

Ventilation of the escape stairs, protected lobbies and common access corridors is important to assist fire service personnel during fire-fighting operations and for smoke clearance purposes after the fire has been extinguished.

Fire-fighters should be able to take control of the ventilation plant, or control plant, during fire-fighting or smoke clearing operations.	ND 2.14.1
Fire service personnel should be able to control the opening and closing of ventilators on arrival at the building.	D 2.14.3 ND 2.14.5
An escape stair within a protected zone should have either: • a ventilator of not less than 1 m² at the top of the stair; or • an opening window at each storey with an openable area of 0.5 m².	D 2.14.3 ND 2.14.5

In domestic buildings

Every access corridor or part of an access corridor, in a building containing flats or maisonettes, should be provided with openable ventilators.	D 2.14.3
The ventilators should provide exhaust at or near ceiling level and supply air at or near floor level with a combined aggregate opening area of at least 1.5 m².	D 2.14.3

Where access to the flats or maisonettes is from an open-access balcony or an access deck, openable ventilators need not be installed provided the balcony or access deck is open to the external air and the opening area extends over at least 4/5th of its length and at least 1/3rd of its height.

In non-domestic buildings

Ventilation should be provided to assist fire-fighting operations and to allow smoke clearance after the fire.	ND 2.14.7
Smoke outlets, communicating directly with the external air, should be provided from every basement storey, and where the basement storey is divided into compartments, from every compartment.	ND 2.14.7
Smoke outlets need not be provided: • in open-sided car parks; • where the floor area of the basement storey is not more than 200 m^2; • where the basement storey is at a depth of not more than 4.5 m; • where a window or windows opening direct to the external air have a total area not less than 1% of the floor area; • where the basement storey or part of the basement storey is used as a strong room; • where the basement storey has an automatic fire suppression system and is ventilated by a mechanical smoke and heat extraction system incorporating a powered smoke and heat exhaust ventilator which has a capacity of at least 10 air changes per hour.	ND 2.14.7
Smoke outlets should discharge directly to the open air at a point at least 2 m, measured horizontally, from any part of an escape route or exit.	ND 2.14.8

Smoke and heat exhaust ventilation systems
Smoke venting shafts

A Smoke and Heat Exhaust Ventilation System (SHEVS) is primarily used in shopping centres and should be installed in: • the mall of an enclosed shopping centre; • shops with a storey area more than 1,300 m^2; • large shops (other than in enclosed shopping centres), with a compartment area more than 5,600 m^2.	ND 2.1.3 ND 2.C.1

SHEVS should be designed in accordance with the following (where appropriate): ND 2.C.1

- the underside of the mall roof should be divided into smoke reservoirs, each of which should be not more than $2,000\,m^2$ in area and at least 1.5 m deep measured to the underside of the roof or to the underside of any high-level plant or ducts within the smoke reservoir or the underside of an imperforate suspended ceiling;
- smoke reservoirs should be formed by fixed or automatically descending smoke curtains which are no greater than 60 m apart, measured along the direction of the mall;
- smoke should not be allowed to descend to a height of less than 3 m above any floor level;
- each smoke reservoir should be provided with the necessary number of smoke ventilators or extract fans to extract the calculated volume of smoke produced, spaced evenly throughout the reservoir;
- where mechanical extraction is used, there should be spare fan capacity equivalent to the largest single fan in the reservoir which will operate automatically on the failure of any one of the fans, or which runs concurrently with the fans;
- any fans, ducts and reservoir screens provided should be designed to operate at the calculated maximum temperature of the smoke within the reservoir in which they are located, but rated to a minimum of 300°C for 30 min;
- structures supporting any fans, ducts or reservoir screens should have the same performance level as the component to be supported;
- the fans or ventilators within the affected smoke reservoirs should operate:
  - on the actuation of any automatic fire suppression system;
  - on the actuation of the smoke detection system within the reservoir;
  - on the operation of more than one smoke detector anywhere in the shopping centre;
  - following a delay not exceeding 4 min from initiation of the first fire alarm signal anywhere in the shopping centre;
- replacement air should be provided automatically on the operation of the ventilation or exhaust system at a level at least 0.5 m below the calculated level of the base of the smoke layer;

- any power source provided to any elements of the smoke and heat exhaust ventilation system should be connected by mineral insulated cables or by cables which are code A category specified in BS 6387: 1994 or by cables protected from damage to the same level;
- an automatically switched standby power supply provided by a generator should be connected to any fans provided as part of the smoke and heat ventilation system capable of simultaneously operating the fans in the reservoir affected and any of the two adjacent reservoirs;
- simple manual overriding controls for all smoke exhaust, ventilation and air input systems should be provided at all fire service access points and any fire control room provided;
- where outlets are provided with weather protection, they should open on the activation of the fan(s) or ventilators;
- smoke from areas adjoining the smoke reservoirs should only be able to enter one reservoir;
- where there is an openwork ceiling, the free area of the ceiling should not be less than 25% of the area of the smoke reservoir or, for natural ventilation, 1.4 times the free area of the roof-mounted fire ventilator above (3 times where the height from floor to roof ventilator is more than 12 m), whichever free area is the greater, and be evenly distributed to prevent an unbalanced air flow into the reservoir;
- when a natural ventilation system is used and the smoke reservoir includes a suspended ceiling, other than an openwork ceiling, the free area of the ventilator opening in the suspended ceiling, or any ventilator grille in the ceiling, should not be:
  - less than 1.4 times (3 times where the height from floor to roof ventilator is more than 12 m) that of the roof-mounted fire ventilator above in the case of a ventilator opening;
  - 2 times (3.5 times where the height from floor to roof ventilator is more than 12 m) for any ventilator grille.

A smoke venting shaft should be enclosed by compartment walls with medium fire resistance duration, other than at the smoke inlets and smoke outlets to the shaft.	ND 2.1.10

Environment

Every building must be designed and constructed in such a way that:	3.17
• each fixed combustion appliance installation operates safely;	3.18
• any component part of each fixed combustion appliance installation will not cause damage to the building in which it is installed by radiated, convected or conducted heat or from hot embers expelled from the appliance;	
• any component part of each fixed combustion appliance installation will not cause damage to the building in which it is installed by radiated, convected or conducted heat or from hot embers expelled from the appliance;	3.19
• the products of combustion are carried safely to the external air without harm to the health of any person through leakage, spillage or exhaust; nor permit the re-entry of dangerous gases from the combustion process of fuels into the building;	3.20
• each fixed combustion appliance installation receives air for combustion and operation of the chimney so that the health of persons within the building is not threatened by the build-up of dangerous gases as a result of incomplete combustion;	3.21
• each fixed combustion appliance installation receives air for cooling so that the fixed combustion appliance installation will operate safely without threatening the health and safety of persons within the building.	3.22

Protection from radon gas

If a dwelling is located on ground designated as a 'radon affected area' protective work should be undertaken to prevent excessive radon gas from entering the dwelling.	D 3.2.2 ND 3.2.2
Radon protective measures should be provided in accordance with the guidance contained in BRE publication BR376 – 'Radon: guidance on protective measures for new dwellings in Scotland'.	D 3.2.2 ND 3.2.2

Whilst the national reference level of $400\,\text{Bq/m}^3$ in workplaces makes it easier to stay within that level than the lower level of $200\,\text{Bq/m}^3$ for dwellings, research has shown that an impervious membrane with securely welted joints laid over the full area of the building will help to reduce radon ingress into the building.

Workplaces are less of a risk than dwellings because, generally speaking, people spend less time at work than at home and workplaces generally have better ventilation provision. In addition, large buildings tend to be mechanically ventilated which will result in the dilution of radon gas.

Control of legionellosis

An inlet to, and an outlet from, a mechanical ventilation system should be installed such that their positioning avoids the contamination of the air supply to the system.	D 3.14.9 ND 3.14.6
A mechanical ventilation system should be constructed to ensure, as far as is reasonably practicable, the avoidance of contamination by *Legionella*.	ND 3.14.6

Moisture from the ground

Permanent ventilation of the underfloor space direct to the outside air by ventilators in two external walls on opposite sides of the building is required.	D 3.4.6 ND 3.4.6

Condensation

To prevent excessive build-up of condensation in cold, pitched roof spaces, where the insulation is at ceiling level, the roof space should be cross-ventilated.	D 3.15.7
Special care should be taken with ventilation where ceilings follow the roof pitch.	D 3.15.7

Safety

Every building must be designed and constructed in such a way that all manual controls for ventilation and for electrical fixtures can be operated safely.	4.8

Emergency lighting

In non-domestic buildings, emergency lighting is designed to come into, or remain in, operation automatically in the event of a local and general power failure.

| In the case of a building with a smoke and heat exhaust ventilation system, the emergency lighting should be sited below the smoke curtains or installed so that it is not rendered ineffective by smoke-filled reservoirs. | ND 2.10.3 |

Escape lighting

| On the operation of the fire alarm in shopping centres (and subject to the 4 min grace period where appropriate) all air moving systems, mains and pilot gas outlets, combustion air blowers and gas, electrical and other heating appliances in the reservoir are shut down. | ND 2.C.5 |

Access to manual controls

| The location of any manual control device (i.e. for openable ventilators, and for controls and outlets of electrical fixtures located on a wall or other vertical surface that is intended for operation by the occupants of a building) should be:

• installed in position that allows safe and convenient use.

• Unless incorporating a restrictor or other protective device for safety reasons, controls should be operable with one hand. | D 4.8.5
ND 4.8.6 |
| An openable window or rooflight that provides natural ventilation should have controls for opening, positioned at least 350 mm from any internal corner, projecting wall or similar obstruction and at a height of:

• not more than 1.7 m above floor level, where access to controls is unobstructed;
• not more than 1.5 m above floor level, where access to controls is limited by a fixed obstruction of not more than 900 mm high which projects not more than 600 mm in front of the position of the controls, such as a kitchen base unit;
• not more than 1.2 m above floor level, in an unobstructed location, within an enhanced apartment or within accessible sanitary accommodation not provided with mechanical ventilation. | D 4.8.5
ND 4.8.6 |

Openings and service provisions

Fire and smoke can easily pass through openings and service penetrations in protected routes of escape thus preventing the occupants from escaping in the event of an outbreak of fire within the building. For this reason:

Fire-stopping may be necessary to close an imperfection of fit or design tolerance between construction elements and components, service openings and ventilation ducts.	D2.2.9

Sizes of openings and recesses

Openings in walls below ground floor should be limited to small holes for services and ventilation, etc., not more than $0.1\,m^2$ and at least 2 m apart.	1D31 1E46
Ventilation ducts not more than 2 m from the escape stair may be protected by heat-activated sealing devices or systems.	ND 2.9.24

Auditoria

In a building containing an auditorium, the ventilation system should be designed to ensure that the direction of air movement in the event of fire is from the auditorium towards the stage.	ND 2.9.27
Exhaust ventilators over an open stage should have a combined total aerodynamic free area of at least 10% of the area of the stage.	ND 2.9.27

Ductwork installation

To minimize air leakage and energy use, ventilation sheet metal ductwork should be airtight.	ND 6.2.3

Commissioning building services

Commissioning [i.e. in terms of achieving the levels of energy efficiency that the component manufacturers expect from their product(s)] should also be carried out with a view to ensuring the safe operation of the system.

Written information

Correct use and maintenance of building services equipment is essential if the benefits of enhanced energy efficiency are to be realized from such equipment. To achieve this, it is essential that user and maintenance instructions together with all other relevant documentation are available to the occupier of the building.

The occupiers of all buildings must be provided with written   6.8
information by the owner:

(a)   on the operation and maintenance of the building
       services and energy supply systems;
(b)   where any air-conditioning system in the building is
       subject to a time-based interval for inspection of the
       system.

    These requirements do not apply to:

- major power plants serving the National Grid;
- buildings which do not use fuel or power for
  heating, lighting, ventilating and cooling the internal
  environment and heating the water supply services;
- lighting, ventilation and cooling systems in a domestic
  building.

Domestic buildings

A heating, hot water service, ventilating or cooling system in a   D 6.7.1
dwelling should be inspected and commissioned in accordance
with manufacturers' instructions to ensure optimum energy
efficiency.

Non-domestic buildings
The occupiers of domestic buildings must be supplied with:

An energy performance certificate which should be   ND 6.9.2
displayed and show the following information:

- the postal address of the building for which the
  certificate is issued;
- building type;
- the name of the SBSA protocol organization issuing the
  certificate (if applicable); may include the member's
  membership number;
- the date of the certificate;
- the conditioned floor area of the dwelling;
- the main type of heating and fuel;
- the type of electricity generation;
- whether or not there is any form of building-integrated
  renewable energy generation;

- the calculation tool used for certification;
- the type of ventilation system;
- a specific indication of current CO_2 emissions and an indication of potential emissions;
- a seven-band scale in different colours representing the following bands of carbon dioxide emissions: A, B, C, D, E, F and G (where A = excellent and G = very poor);
- the approximate energy use expressed in kWh per m^2 of floor area per annum;
- a list of cost-effective improvements (lower cost measures).

6.5 Basements and cellars

In normal circumstances (i.e. unless you have the consent of the local authority) you are not allowed to construct a cellar or room in (or as part of) a house, an existing cellar, a shop, an inn, a hotel or an office if the floor level of the cellar or room is lower than the ordinary level of the sub-soil water on (under or adjacent to) the site of the house, shop, inn, hotel or office.

6.5.1 Requirements

Fire precautions

Every building must be designed and constructed in such a way that:

- facilities are provided to assist fire-fighting or rescue operations;   *2.14*
- in the event of an outbreak of fire within the building: *2.1*
  - fire and smoke are inhibited from spreading beyond the compartment of origin until any occupants have had the time to leave that compartment and any mandatory fire containment measures have been initiated.

  This standard does not apply to domestic buildings.

  - the occupants, once alerted to the outbreak of the fire, are provided with the opportunity to escape from the building, before being affected by fire or smoke;   *2.9*
  - illumination is provided to assist in escape.   *2.10*

Environment

Every building must be designed and constructed in such a way that there will not be a threat to the building or the health of the occupants as a result of moisture penetration from the ground.	*3.4*

Safety

Every building must be designed and constructed in such a way that each liquefied petroleum gas (LPG) storage installation, used solely to serve a combustion appliance providing space heating, water heating or cooking facilities, will: • be protected from fire spreading to any LPG container; • not permit the contents of any such container to form explosive gas pockets in the vicinity of any container.	4.11

6.5.2 Meeting the requirements

Loadings

The design and construction, and the loads to which an earth-retaining structure (such as a cellar or a basement to a building) will be subjected should be calculated in accordance with BS 8002: 1994.	ND 2.1.2
Structures below ground, including basements, should be constructed in accordance with the recommendation of BS 8102: 1990.	D 3.4.7 ND 3.4.7

Compartmentation

Every basement storey should form a separate compartment.	ND 2.A.1 ND 2.B.1

Maximum compartment areas

If a non-domestic building, or part of a building, has a total storey area more than the limits given in Tables 6.24 and 6.25, it should be sub-divided by compartment walls and (where appropriate) compartment floors which have a minimum fire resistance duration as shown.

Note: In most cases, a single-storey building poses less of a life risk to the occupants or to fire service personnel than a multi-storey building, therefore a greater compartment size can be constructed (Table 6.24).

Table 6.24   Single-storey buildings and compartmentation between single-storey and multi-storey buildings where appropriate

Building use	Maximum total area of any compartment (m²)	Minimum fire resistance duration for compartmentation (if any)
Assembly building	6,000	Long
Entertainment building	2,000	Medium
Factory (Class 1)	33,000	Long [3]
Factory (Class 2)	93,000	Long [3]
Office	4,000	Medium
Open sided car park	Unlimited	Not relevant
Residential care building, hospital	1,500	Medium
Residential building (other than a residential care building and hospital)	2,000	Medium
Shop	2,000 [2]	Long
Storage building (Class 1)	1,000 [1]	Long
Storage building (Class 2)	14,000 [1]	Long [3]

Notes:
1. Areas may be doubled where there is an automatic fire suppression system.
2. Unlimited provided there is an automatic fire suppression system.
3. A medium fire resistance duration compartment wall or compartment floor may be provided between the single-storey part and the multi-storey part provided the multi-storey part does not exceed the limitations for medium fire resistance duration in the following table covering multi-storey buildings.

Table 6.25   Multi-storey buildings

Building use	Maximum total area of any compartment (m²)	Minimum fire resistance duration for compartmentation in a basement or cellar
Assembly building	1,500 (1)	Medium
	3,000 (1)	Medium
	6,000 (1)	Long
Entertainment building	1,000 (1)	Medium
	2,000 (1)	Medium

(Continued)

Table 6.25 (Continued)

Building use	Maximum total area of any compartment (m²)	Minimum fire resistance duration for compartmentation in a basement or cellar
	4,000 (1)	Long
Factory (Class 1)	500 (1)	Medium
	6,000 (1)	Long
Factory (Class 2)	200 (1)	Medium (4)
	15,000 (1)	Long
Office	2,000 (1)	Medium (4)
	4,000 (1)	Medium (4)
	8,000 (1)	Long
Open sided car park	Unlimited	Medium
Residential care building	1,500	Medium
Residential building (other than a	1,000	Medium
residential care building and/or hospital)	2,000	Medium
Shop	500	Medium
	1,000	Medium (4)
	2,000	Long
Storage building (Class 1)	200 (1)	Medium
	1,000 (1)	Long
Storage building (Class 2)	500	Medium (4)
	5,000	Long

Compartment walls and compartment floors in hospitals should be constructed from materials which are non-combustible.

Note: For further details concerning compartment floors, see Section 6.6.

Basements in non-domestic buildings

If a building has a basement storey, the floor of the ground storey should be a compartment floor.	ND 2.1.7
Where a building has a basement storey at a depth of more than 10 m, every basement storey should form a separate compartment.	ND 2.1.7

In a residential building, every basement storey should form a   ND 2.1.5
separate compartment.

At least two storey exits should be provided from: ND 2.9.1

- a basement storey at a depth of more than 4.5 m;
- a basement storey at a depth of not more than 4.5 m
  where the storey is intended to be used by members
  of the general public (other than for access to sanitary
  accommodation).

 Note: Escape stair widths are calculated on the basis of the total number of available storey exits.

Escape stairs and escape routes

Where an escape stair also serves a basement storey, the   ND 2.9.26
protected zone enclosing the escape stair in the basement
storey should be separated from the protected zone
containing the escape stair serving the rest of the building,
by a wall or screen, with or without a door, at the ground
storey floor level.

 Note: The wall, screen and self-closing fire door, where provided, should have medium fire resistance duration.

Where escape routes from a storey consists of a combination   ND 2.9.10
of escape stairs and other escape routes (see Figure 6.29),
the width of the escape route from the ground storey and/or
basement storey should be increased to take account of that
proportion of the occupancy capacity from the ground storey
and/or basement storey.

Facilities on escape stairs
Facilities should be designed and installed within the building to assist the fire
service in carrying out their fire-fighting or rescue operations as efficiently as
possible.

In complex buildings and/or high-rise buildings which have   ND
two or more escape stairs, fire-fighting facilities should be   2.14.1
provided to at least two escape stairs in accordance with
Table 6.26 and positioned at least 20 m apart.

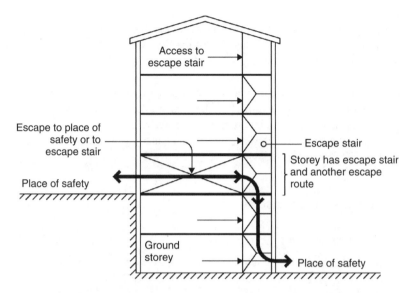

Figure 6.29 Combined escape routes

Table 6.26 Fire-fighting facilities on escape stairs

Storey height and depth of buildings	All buildings (other than those listed in column 3)	Shops, storage buildings and open sided car parks where the area of any storey is more than 900 m²
Basements at a depth more than 10 m	• Fire-fighting shaft; • Fire-fighting lift; • Dry fire main (outlet located in fire-fighting lobby)	
Basements at a depth not more than 10 m	No provision	• Fire-fighting shaft; • Dry fire main (outlet located in fire-fighting lobby)

Venting of heat and smoke from basements

Smoke outlets, communicating directly with the external air, should be provided from every basement storey, and where the basement storey is divided into compartments, from every compartment.	ND 2.14.7
Smoke outlets need not be provided where:	ND 2.14.7

• the floor area of the basement storey is not more than 200 m²;
• the basement storey is at a depth of not more than 4.5 m;

- a window or windows opening direct to the external air have a total area not less than 1% of the floor area;
- the basement storey or part of the basement storey is used as a strong room;
- the basement storey has an automatic fire suppression system and is ventilated by a mechanical smoke and heat extraction system incorporating a powered smoke and heat exhaust ventilator which has a capacity of at least 10 air changes per hour.

Smoke outlets from basements

Smoke outlets should: ND 2.14.8

- be sited at ceiling level within the room they serve;
- have an aggregate cross-sectional area of at least 2.5% of the floor area of the room they serve;
- be evenly distributed around the perimeter of the building;
- discharge directly to the open air at a point at least 2 m, measured horizontally, from any part of an escape route or exit;
- have a sign stating 'Smoke outlet from basement, do not obstruct' fixed adjacent to each external outlet point;
- (where they serve a place of special fire risk) be separate from smoke outlets from other areas;
- discharge by means of windows, panels or pavement lights which are readily accessible to fire service personnel and which can be opened or knocked out if necessary;
- where appropriate, be enclosed by a smoke venting shaft;
- (where there are smoke venting shafts from different parts of the same basement storey, or from a different basement storey) be separated;
- be covered with a metal grille or louvre.

Fire service facilities

Where a hospital with a hospital street has two or more   ND 2.B.6
escape stairs, facilities should be provided in accordance
with Table 6.27.

Table 6.27   Facilities on escape stairs in hospitals with hospital streets

Storey height and depth of hospital	Facilities on escape stairs
Basements at a depth more than 10 m	• Fire-fighting lift • Fire-fighting shaft • Dry fire main (outlet located at every departmental entrance)
Basements at a depth not more than 10 m	• Dry fire main (outlet located at every departmental entrance)

Basements in domestic buildings

In domestic buildings, there should be at least one escape route   D2.9.1
from a basement or cellar if it is:

- the main entrance door to a flat or maisonette;
- the door to a communal room;
- from a plant room.

Where an escape stair also serves a basement storey, the   D 2.9.28
protected zone enclosing the escape stair in the basement
storey should be separated from the protected zone containing
the escape stair serving the rest of the building, by a medium
fire resistance wall or screen (with or without a door) at the
ground storey floor level.

A basement storey which contains an apartment should be   D 2.9.28
provided with either:

- an alternative exit from the basement storey, which
  provides access to the external air (below the adjoining
  ground) from which there is access to a place of safety
  at ground level;
- an escape window in every basement apartment.

An apartment in a basement should not be in an inner room   D 2.9.29
unless there is an escape window or there are alternative routes
from the apartment to circulation areas or other rooms.

Where a private stair serves a basement storey, the private stair   D 2.9.29
should be in a protected enclosure.

Ducted warm air heating

Where a flat or maisonette has a basement storey and is   D 2.9.13
provided with a system of ducted warm air heating:

- transfer grilles should not be fitted between any room and
  the entrance hall or stair;

- supply and return grilles should be **not** more than 450 mm above floor level;
- where warm air is ducted to an entrance hall or stair, the return air should be ducted back to the heater;
- where a duct passes through any wall, floor or ceiling of an entrance hall or stair, all joints between the duct and the surrounding construction should be sealed;
- there should be a room thermostat in the living room, at a height more than 1,370 mm and not more than 1,830 mm, with an automatic control which will turn off the heater, and actuate any circulation fan should the ambient temperature rise to more than 35°C;
- where the system recirculates air, smoke detectors should be provided in every extract duct to cause the recirculation of air to stop and direct all extract air to the outside of the building in the event of fire.

Emergency lighting

Emergency lighting is lighting designed to come into, or remain in, operation automatically in the event of a local and general power failure.

Emergency lighting should be installed in buildings or parts of a building considered to be at higher risk such as:	D 2.10.3 ND 2.10.3

- buildings with basements;
- in rooms where the number of people is likely to exceed 60;
- in a protected zone or unprotected zone serving a basement storey;

LPG storage – fixed tanks and cylinders

An LPG storage tank, together with any associated pipework, should be designed, constructed and installed in accordance with the requirements set out in the Liquefied Petroleum Gas Association (LPGA) Code of Practice 1: 'Bulk LPG storage at fixed installations'.

Ground features such cellar hatches, within the separation distances given in column A of Table 6.28, should be sealed or trapped to prevent the passage of LPG vapour.	D 4.11.2 ND 4.11.2

Table 6.28 Separation distances for liquefied petroleum gas storage tanks

Maximum capacity (in tonnes)		Minimum separation distance for above-ground tanks (in metres)		
Of any single tank	Of any group of tanks	From a building, boundary or fixed source of ignition to the tank		
		A no fire wall	B with fire wall	Between tanks
0.25	0.8	2.5	0.3	1.0
1.1	3.5	3.0	1.5	1.0
4.0	12.5	7.5	4.0	1.0

Cylinders should be positioned on a firm, level base that is located in a well-ventilated position at ground level, at least 2 m horizontally from unsealed cellar hatches unless an intervening wall (not less that 250 mm high) is present.

D 4.11.3
ND 4.11.3

6.6 Floors

The ground floor of a building is either solid concrete or a suspended timber type. With a concrete floor, a damp proof membrane (DPM) is laid between walls. With timber floors, sleeper walls of honeycomb brickwork are built on over site-concrete between the base brickwork; a timber sleeper plate rests on each wall, and timber joists are supported on them. Their ends may be similarly supported, led into the brickwork or suspended on metal hangers. Floorboards are laid at right angles to joists. First-floor joists are supported by the masonry or hangers.

Similar to a brick-built house, the floors in a timber-framed house are either solid concrete or suspended timber. In some cases, a concrete floor may be screeded or surfaced with timber or chipboard flooring. Suspended timber floor joists are supported on wall plates and surfaced with chipboard.

6.6.1 Requirements

Fire

Every building must be designed and constructed in such a way that in the event of an outbreak of fire within the building:

2.3

- *the load-bearing capacity of the building will continue to function until all occupants have escaped, or been assisted to escape, from the building and any fire containment measures have been initiated;*

- *the unseen spread of fire and smoke within concealed spaces in*   2.4
  its structure and fabric is inhibited;
- *fire and smoke will be inhibited from spreading through the*   2.15
  building by the operation of an automatic life safety fire
  suppression system.

Non-domestic buildings

- *Every building must be designed and constructed in such a*   2.1
  way that in the event of an outbreak of fire within the building,
  fire and smoke are inhibited from spreading beyond the
  compartment of origin until any occupants have had the time
  to leave that compartment and any mandatory fire containment
  measures have been initiated.
- *Every building, which is divided into more than one area of*   2.2
  different occupation, must be designed and constructed in such
  a way that in the event of an outbreak of fire within the building,
  fire and smoke are inhibited from spreading beyond the area of
  occupation where the fire originated.

Environment

Every building must be designed and constructed in such a way   3.4
that: there will not be a threat to the building or the health of the   3.15
occupants as a result of:

- *moisture penetration from the ground;*
- *moisture caused by surface or interstitial condensation.*

Safety

Every building must be designed and constructed in such a way that: 4.2

- *in non-domestic buildings, safe, unassisted and convenient*
  means of access is provided throughout the building;
- *in residential buildings, a proportion of the rooms intended to*
  be used as bedrooms must be accessible to a wheelchair user;
- *in domestic buildings, safe and convenient means of access is*
  provided within common areas and to each dwelling;
- *in dwellings, safe and convenient means of access is provided*
  throughout the dwelling;
- *in dwellings, unassisted means of access is provided to, and*
  throughout, at least one level.

Noise

> *Every building must be designed and constructed in such a way that*   *5.1*
> *each wall and floor separating one dwelling from another, or one*
> *dwelling from another part of the building, or one dwelling from a*
> *building other than a dwelling, will limit the transmission of noise*
> *to the dwelling to a level that will not threaten the health of the*
> *occupants of the dwelling or inconvenience them in the course of*
> *normal domestic activities provided the source noise is not in excess*
> *of that from normal domestic activities.*

Energy

> *Every building must be designed and constructed in such a way that*   *6.2*
> *an insulation envelope is provided which reduces heat loss.*

6.6.2 Meeting the requirements

Floor types

Floor type 1 – concrete base with soft covering

The resistance to airborne sound transmission depends on the mass of the concrete base and on eliminating air paths. The resistance to impact sound transmission depends on the soft covering.

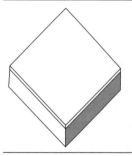

Points to watch:

- Fill all joints between parts of the floor to avoid air paths.
- Limit pathways around the floor to reduce flanking transmission.
- Workmanship and detailing should be given special attention at the perimeter and wherever the floor is penetrated by a pipe or duct, to reduce flanking transmission and to avoid air paths.

Constructions – floor type 1

There are four recommended (soft covered) concrete base constructions (A–D) as shown in Table 6.29.

Table 6.29   Constructions – floor type 1

Floor base A Floor type 1	Section	• Solid concrete slab, cast *in situ*. • Floor screed and/or ceiling finish optional. Mass (including any screed and/or ceiling finish) 365 kg/m².
Floor base B Floor type 1	Section	• Solid concrete slab, cast *in situ*, with permanent shuttering. • Floor screed and/or ceiling finish optional. • Mass 365 kg/m² including shuttering only if it is solid concrete or metal, and including any screed and/or ceiling finish.
Floor base C Floor type 1		• Concrete beams with infilling blocks. • Floor screed and/or structural topping should be used. • Ceiling finish optional. • Mass 365 kg/m² including beams, blocks, any structural topping, screed, and any ceiling finish.
Floor base D Floor type 1		• Concrete planks (solid or hollow). • Floor screed and/or structural topping should be used. • Ceiling finish optional. • Mass 365 kg/m², including planks, any structural topping and screed, including any ceiling finish.

Soft covering
Floor type 1

Soft covering, fully bonded to the floor base:
- a resilient material, or material with a resilient base, with an overall uncompressed thickness of at least 4.5 mm;
- a material with a weighted reduction in impact sound pressure level (ΔLw) of at least 17.

It is **NOT** suitable as a means to limit impact transmission to a dwelling below a walkway or a roof that acts as a floor.

Junctions at walls at external or cavity separating walls – floor type 1

Floor type 1

Soft covering →
Screed if used →

Floor base →

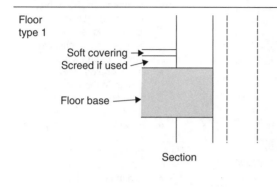

Section

- The mass of the wall leaf adjoining the floor should be 120 kg/m², including any plaster. This is not necessary where the area of openings in the external wall exceeds 20% of its area: there is no recommendation for the minimum mass of such a wall.
- The floor base, excluding any screed, should pass through the leaf whether spanning parallel to, or at right angles to, the wall.
- The cavity should not be bridged.

Junctions at walls at internal or solid separating wall – floor type 1

Floor type 1

Soft covering →
Screed if used →

Floor base →

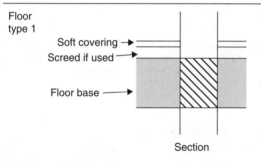

Section

- If the wall mass is less than 355 kg/m² including any plaster then the floor base excluding any screed should pass through.
- If the wall mass is more than 355 kg/m² including any plaster, either the wall or the floor base excluding any screed may pass through.

 Where the wall does pass through, tying the floor base to the wall and grouting the joint is recommended.

Floor penetrations – floor type 1

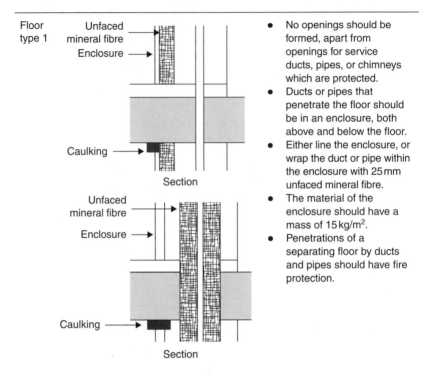

Floor type 1

Unfaced mineral fibre

Enclosure

Caulking

Section

Unfaced mineral fibre

Enclosure

Caulking

Section

- No openings should be formed, apart from openings for service ducts, pipes, or chimneys which are protected.
- Ducts or pipes that penetrate the floor should be in an enclosure, both above and below the floor.
- Either line the enclosure, or wrap the duct or pipe within the enclosure with 25 mm unfaced mineral fibre.
- The material of the enclosure should have a mass of 15 kg/m^2.
- Penetrations of a separating floor by ducts and pipes should have fire protection.

 A flue-pipe may penetrate the floor, provided that it discharges into either a masonry chimney carried by the floor or any other type of chimney enclosed within a non-combustible duct that is lined with absorbent mineral fibre.

Floor type 2 – concrete base with floating layer

The resistance to airborne sound transmission depends mainly on the mass of the concrete base and partly on the mass of the floating layer. The resistance to impact sound depends on the resilient layer to isolate the floating layer from the base and from the surrounding construction.

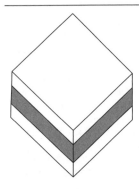

Points to watch:

- Fill all joints between parts of the floor base to avoid air paths.
- Limit the pathways around the floor to reduce flanking transmission.
- Workmanship and detailing should be given special attention at the perimeter and wherever the floor is penetrated, to reduce flanking transmission and to avoid air paths.
- Take care not to create a bridge between the floating layer and the base, surrounding walls, or adjacent screeds.
- With bases C and D a screed is used to accommodate surface irregularities and prevent reduced resistance to noise transmission at joints.

Constructions – floor type 2

There are four recommended (floating layer) concrete base constructions (A–D) together with two floating layer constructions (F1 and F2) as shown in Table 6.30, and any of these can be used in combination.

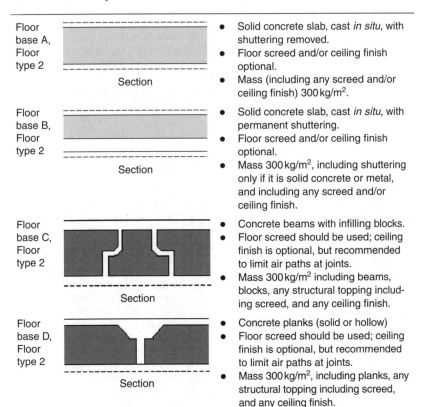

Floor base A, Floor type 2	• Solid concrete slab, cast *in situ*, with shuttering removed.
Section	• Floor screed and/or ceiling finish optional.
	• Mass (including any screed and/or ceiling finish) 300 kg/m².
Floor base B, Floor type 2	• Solid concrete slab, cast *in situ*, with permanent shuttering.
Section	• Floor screed and/or ceiling finish optional.
	• Mass 300 kg/m², including shuttering only if it is solid concrete or metal, and including any screed and/or ceiling finish.
Floor base C, Floor type 2	• Concrete beams with infilling blocks.
Section	• Floor screed should be used; ceiling finish is optional, but recommended to limit air paths at joints.
	• Mass 300 kg/m² including beams, blocks, any structural topping including screed, and any ceiling finish.
Floor base D, Floor type 2	• Concrete planks (solid or hollow)
Section	• Floor screed should be used; ceiling finish is optional, but recommended to limit air paths at joints.
	• Mass 300 kg/m², including planks, any structural topping including screed, and any ceiling finish.

Floating layer constructions

Table 6.30   Constructions – floor type 1

Floating layer F1 (Timber raft) Floor type 2 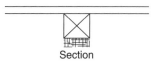 Section	• Timber boarding or wood-based board, minimum 18 mm thick, with tongued and grooved edges, fixed to minimum 45 × 45 mm (nominal) timber battens with a bonded integral resilient polymer-based layer. • Polymer-based layers include foams, man-made fibres and elastomers. • Resilient flanking strips at least 5 mm thick should be fitted between floor edge and wall/ skirting junction. • Floating floor treatment to demonstrate a weighted reduction in impact sound pressure level (Δw) of at least 25 dB.
Floating layer F2 (Screed over resilient layer) Floor type 2 Section	• Cement sand screed, 65 mm thick with mesh underlay to protect the resilient layer while the screed is being laid. • Resilient layer of extruded closed cell polyethylene foam, 12.5 mm thick, density 30–45 kg/m³. • To protect the material from puncture it should be laid over a levelling screed. • Lay with taped joints. • The resilient layer should be faced with a membrane to prevent screed entering the layer. • Lay the material tightly butted and turned up at the edges of the floating layer.

Junctions at walls at external or cavity separating walls – floor type 2

Floor type 2 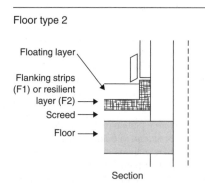 Floating layer Flanking strips (F1) or resilient layer (F2) → Screed → Floor → Section	• The mass of the leaf adjoining the floor should be 120 kg/m², including any plaster. • The floor base, excluding any screed should pass through the wall whether spanning parallel to, or at right angles to, the wall. • The cavity should not be bridged. • Carry the resilient layer up at all edges to isolate the floating layer. • Leave a 5 mm gap between skirting and floating layer or turn resilient layer under skirting. • Where a seal is necessary, it should be flexible.

Junctions at walls at internal or solid separating wall – floor type 2

Floor type 2

Floating layer

Flanking strips
(F1) or resilient
Layer (F2) →

Screed →

Floor →

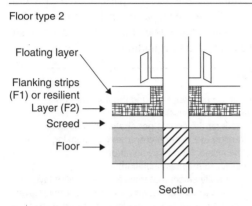

Section

- If the wall mass is less than 355 kg/m² including any plaster, then the floor base excluding any screed should pass through.
- If the wall mass is more than 355 kg/m² including any plaster, either the wall or the floor base excluding any screed may pass through.

 Where the wall does pass through, tying the floor base to the wall and grouting the joint is recommended.

Floor penetrations – floor type 2

Floor type 2

Unfaced
Mineral fibre →
Enclosure →
Caulking →

Caulking →

Unfaced
Mineral fibre →
Enclosure →
Caulking →

Caulking →

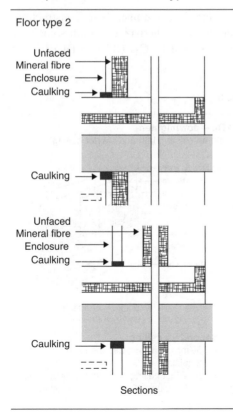

Sections

- No openings should be formed, apart from openings for service ducts, pipes or chimneys which are protected.
- Ducts or pipes that penetrate the floor should be in an enclosure, both above and below the floor.
- Either line the enclosure, or wrap the duct or pipe within the enclosure with 25 mm unfaced mineral fibre.
- The material of the enclosure should have a mass of 15 kg/m².

- Penetrations of a separating floor by ducts and pipes should have fire protection.
- Leave a 5 mm gap between enclosure and floating layer, and seal with acrylic caulking or neoprene.
- A flue-pipe may penetrate the floor, provided that it discharges into either a masonry chimney carried by the floor or any other type of chimney enclosed within a non-combustible duct that is lined with absorbent mineral fibre.

Floor type 3 – timber base with floating layer

The resistance to airborne sound transmission depends partly on the structural floor plus absorbent blanket or deafening, and partly on the floating layer. Resistance to impact sound transmission depends on the resilient layer to isolate the floating layer from the base and the surrounding construction.

Points to watch:

- Limit the pathways around the floor to reduce flanking transmission.
- Wherever the floor is penetrated, pay special attention at the perimeter to reduce flanking transmission and to avoid air paths.
- In order to maintain isolation:
  - carefully select materials for the resilient layer;
  - take care not to bridge between the floating layer and the base or surrounding walls (e.g. with services or fixings which penetrate the resilient layer);
  - allow for movement of materials, e.g. expansion of chipboard after laying (to maintain isolation).

Construction – floor type 3

There are three recommended (floating layer) timber base constructions (A–C) as shown in Table 6.31. There are also alternatives within some constructions and four constructions (A-DL, B-DL, C-a-DL, C-b-DL) for use with downlighters.

Table 6.31   Constructions – floor type 3

Floor type 3C-DL – downlighters
(This construction could be used with an existing lath and plaster ceiling)

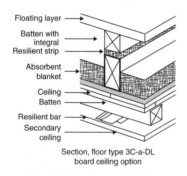

Section, floor type 3C-a-DL
board ceiling option

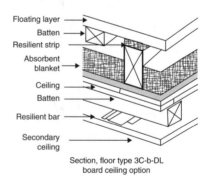

Section, floor type 3C-b-DL
board ceiling option

If downlighters are to be installed in a separating floor, the lights should be fitted within the depth of a secondary ceiling, to avoid the creation of air paths by penetration of the main ceiling layers, and the ceiling layers should be fixed directly to the joists.

Secondary ceiling:
- 50 mm × 50 mm battens fixed through to joists.
- Resilient ceiling bars perpendicular to battens; and 12.5 mm gypsum-based board.

**Floor type 3A – platform floor with absorbent blanket
(for use in conversions only)**

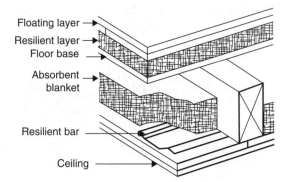

Floating layer →
Resilient layer →
Floor base →
Absorbent blanket →
Resilient bar
Ceiling →

Floating layer

Two types of floating layer may be used: timber or wood-based board, 18 mm thick with tongued and grooved edges and all joints glued, spot bonded to a substrate of gypsum-based board with a minimum mass of 13.5 kg/m^2.
A floating layer of 2 thicknesses of cement bonded particleboard with joints staggered, glued and screwed together, total thickness 24 mm.

Ceiling

- Resilient ceiling bars, fixed perpendicular to joist direction at 400 mm centres.
- Absorbent blanket of 100 mm mineral fibre, density 10–33 kg/m^3, laid on ceiling between joists.
- Two or more layers of gypsum-based board with joints staggered, overall minimum mass 24 kg/m^2, or total thickness 30 mm.

Resilient layer

- Resilient layer of a material with a weighted reduction in impact sound pressure level (Δw) of at least 14 dB when measured in combination with the floating layer;
- Resilient flanking strips at least 5 mm thick should be fitted between floor edge and wall/skirting junction.

Floor base

- Floor base of 12 mm timber boarding or wood-based board nailed to timber joists.

Floor type 3A-DL – downlighters

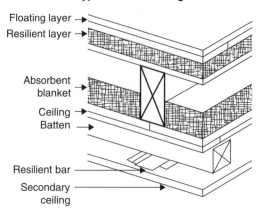

Floating layer →
Resilient layer →
Absorbent blanket →
Ceiling →
Batten →
Resilient bar
Secondary ceiling

Secondary ceiling
- A secondary ceiling should be fitted if downlighters are to be installed in a separating floor, to avoid penetration of the main ceiling layers.
- The ceiling layers should be fixed directly to the joists.
- 50 × 50 mm battens, resilient ceiling bars perpendicular to battens, and 12.5 mm gypsum-based board.

Floor type 3B – ribbed floor with absorbent blanket

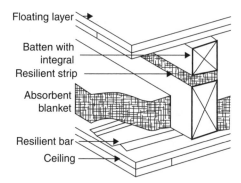

Floating layer →
Batten with integral
Resilient strip
Absorbent blanket
Resilient bar
Ceiling

Floating layer
- Floating layer of timber or wood-based board, minimum 18 mm thick with tongued and grooved edges and all joints glued, spot bonded to, and fixed through, a substrate of gypsum-based board (minimum mass 13.5 kg/m^2) to minimum

Floor base
Floor base of 45 mm wide timber joists.
- Structural bracing is not shown.
- Ribbed floors are routinely built with an additional sub-deck board (not shown) over the joists to provide safe access before fixing of the floating layer.

45 × 45mm nominal timber battens with a bonded integral resilient polymer-based layer. Polymer-based layers include foams, man-made fibres and elastomers.

- Resilient flanking strips at least 5mm thick should be fitted between floor edge and wall/ skirting junction.
- Floating floor treatment to demonstrate a weighted reduction in impact sound pressure level (ΔLw) of at least 14dB when measured in accordance with Annex 5.B.
- Follow manufacturer's instructions for installation of proprietary systems.

- Such boarding should not introduce noise problems, but does not add to the sound insulation.
- The sub-deck board should be level and should not sag between joists.
- When such boarding is used, the battens may be laid either in line with, or at 90° to the joists.

Ceiling
- Resilient ceiling bars fixed perpendicular to joist direction at 400mm centres.
- Absorbent blanket of 100mm mineral fibre, density 10–33kg/m^3, laid on ceiling between joists.
- Ceiling of two or more layers of gypsum-based board with joints staggered, overall minimum mass 24kg/m^2, or total thickness 30mm.

Floor type 3B-DL – downlighters

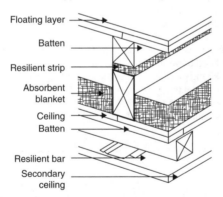

Floating layer
Batten
Resilient strip
Absorbent blanket
Ceiling
Batten
Resilient bar
Secondary ceiling

- If downlighters are to be installed in a separating floor, the lights should be fitted within the depth of a secondary ceiling to avoid the creation of air paths by penetration the main ceiling layers.
- Ceiling layers should be fixed directly to the joists.

Secondary ceiling:
50 × 50mm battens fixed through to joist.
- Resilient ceiling bars perpendicular to battens; 12.5mm gypsum-based board. See also notes above on floor base.

Floor type 3C-a – battens along top of joists

Floating layer ⟶
Batten with integral resilient strip ⟶
Deafening ⟶
Liner ⟶
Plywood ceiling ⟶

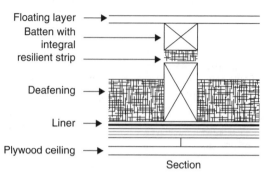

Section

Board ceiling option

Floating layer

- Floating layer of timber- or wood-based board, minimum 18 mm thick with tongued and grooved edges, and all joints glued, fixed to minimum 45 × 45 mm (nominal) timber battens with a bonded integral resilient polymer-based layer. Polymer-based layers include foams, man-made fibres and elastomers.
- Battens placed on top of the joists in the same direction as the joists.
- Resilient flanking strips at least 5 mm thick should be fitted between floor edge and wall/skirting junction.
- Floating floor treatment to demonstrate a weighted reduction in impact sound pressure level (Δw) of at least 14 dB.

Floor base

- Floor base of 45 mm wide timber joists.
- Ribbed floors.
- Battens may be laid either in line with, or at 90° to the joists.

Floor type 3C-b – battens between joists (only for use in conversions)

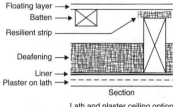

Floating layer →
Batten →
Resilient strip
Deafening →
Liner →
Plaster on lath →

Section

Lath and plaster ceiling option

Floating layer →
Batten →
Resilient strip
Deafening →
Liner →
Plywood ceiling →

Section

Board ceiling option

- Floating layer of timber or wood-based board, 18 mm thick with tongued and grooved edges and all joints glued, nailed to 45 × 45 mm timber battens; floating layer placed onto resilient strip on top of joists, laid along their length.
- Resilient strips of a material with a weighted reduction in impact sound pressure level (Δw) of at least 17 dB.

Ceiling
- 19 mm dense plaster on expanded metal lath;
- 6 mm plywood fixed under the joists plus two layers of gypsum-based board with joints staggered, total thickness 25 mm.

Deafening
- Deafening (pugging) of mass 80 kg/m² laid on a polyethylene layer.

Floor type 3 – junctions at timber-frame wall

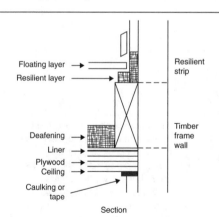

Floating layer →
Resilient layer

Resilient strip

Deafening →
Liner →
Plywood →
Ceiling →
Caulking or tape

Timber frame wall

Section

- Seal the gap between wall and floating layer with a resilient strip glued to the wall.
- Leave a 5 mm gap between skirting and floating layer.
- Block air paths between the floor base and the wall, including the space between joists when joists are at right angles to the wall.
- Seal the junction of ceiling and wall with tape or caulking.

Floor type 3 – junctions at heavy masonry leaf

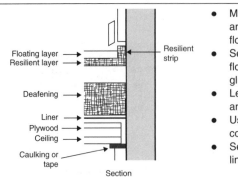

Floating layer
Resilient layer
Resilient strip
Deafening
Liner
Plywood
Ceiling
Caulking or tape

Section

- Mass of leaf 355 kg/m^2 including any plaster, both above and below floor.
- Seal the gap between wall and floating layer with a resilient strip glued to the wall.
- Leave a 5 mm gap between skirting and floating layer.
- Use any normal method of connecting floor base to wall.
- Seal the junction of ceiling and wall lining with tape or caulking.

Floor type 3 – junctions at light masonry leaf

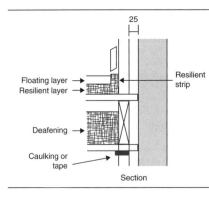

25

Floating layer
Resilient layer
Resilient strip
Deafening
Caulking or tape

Section

- Seal the gap between wall and floating layer with a resilient strip glued to the free-standing panel.
- Leave a 5 mm gap between skirting and floating layer.
- Take ceiling through to masonry, seal junction with free-standing panel with tape or caulking.

Floor type 3 – floor penetrations

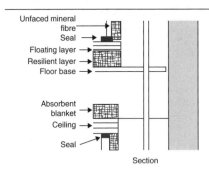

Unfaced mineral fibre
Seal
Floating layer
Resilient layer
Floor base
Absorbent blanket
Ceiling
Seal

Section

- No openings should be formed, apart from openings for service ducts, pipes or chimneys.
- Ducts or pipes that penetrate the floor should be in an enclosure, both above and below the floor.
- Either line the enclosure, or wrap the duct or pipe within the enclosure, with 25 mm unfaced mineral fibre.

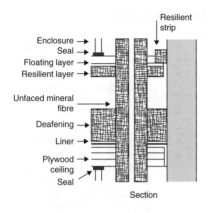

Enclosure →
Seal →
Floating layer →
Resilient layer →

Resilient
strip

Unfaced mineral
fibre →

Deafening →

Liner →

Plywood →
ceiling
Seal

Section

- The material of the enclosure should have mass of 15 kg/m².
- Leave a 5 mm gap between enclosure and floating layer, seal with acrylic caulking or neoprene.
- A flue-pipe may penetrate the floor, provided that it discharges into either a masonry chimney carried by the floor or any other type of chimney enclosed within a duct that is lined with absorbent mineral fibre.
- Seal the junction of ceiling and enclosure with tape or caulking.
- Penetrations of a separating wall by ducts and pipes should have fire protection.

Floor type 4 – timber base with independent ceiling

The resistance to airborne and impact sound depends mainly on the mass and isolation of the independent ceiling, and partly on the mass of the floor base.

 This type of floor should only be used with heavy masonry walls.

Points to watch:

- Limit the pathways around the floor, especially at the edges of the independent ceiling, to reduce flanking transmission and to avoid air paths.
- Take care not to create bridges between the floor base and the independent ceiling.

Floor type 4 – timber floor, incorporating deafening

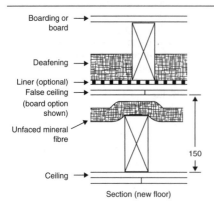

Boarding or board →
Deafening →
Liner (optional) →
False ceiling (board option shown) →
Unfaced mineral fibre →
Ceiling →

150

Section (new floor)

- Timber boarding or wood-based boarding,
- 18 mm thick with tongued and grooved edges or 3.2 mm hardboard over the whole floor to seal gaps.
- 45 mm thick joists.
- Deafening of mass 80 kg/m^2
- Intermediate ceiling of either:
  - 19 mm dense plaster on lath;
  - or more layers of gypsum-based board with joints staggered, overall minimum mass 24 kg/m^2, or total thickness 30 mm.

 Note: In existing floors deafening may be on boards between joists; in new separating floors use 6 mm plywood fixed to underside of joists. A polyethylene liner may be used if desired.

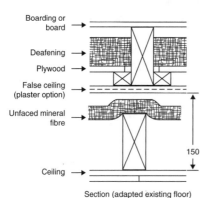

Boarding or board →
Deafening →
Plywood →
False ceiling (plaster option) →
Unfaced mineral fibre →
Ceiling →

150

Section (adapted existing floor)

Independent ceiling:

- Absorbent blanket of 25 mm with unfaced mineral fibre, density 12–36 kg/m^3, draped over 45 mm thick joists supported independently of the floor.
- Ceiling of two layers of gypsum-based board with joints staggered, total thickness 30 mm.

 Keep ceiling 150 mm away from the underside of the intermediate ceiling.

Floor type 4 – junctions at wall

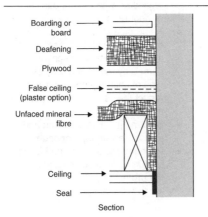

Boarding or board
Deafening
Plywood
False ceiling (plaster option)
Unfaced mineral fibre
Ceiling
Seal

Section

External or cavity separating walls

- Mass of leaf should be 355 kg/m^2, including any plaster, both above and below the floor, on at least three sides.
- Leaf on fourth side should be at least 180 kg/m^2.
- Use bearers on walls to support the edges of the ceiling and to block air paths.
- Seal the junction of ceiling and wall with tape or caulking.

Internal wall

- If masonry, mass should be 180 kg/m^2. There is no recommendation for the mass of stud partitions.
- Support and seal as for external walls.

Floor type 4 – floor penetrations

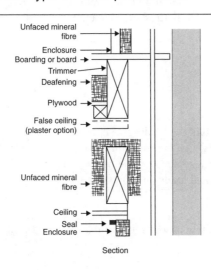

Unfaced mineral fibre
Enclosure
Boarding or board
Trimmer
Deafening
Plywood
False ceiling (plaster option)
Unfaced mineral fibre
Ceiling
Seal
Enclosure

Section

- Ducts or pipes that penetrate the floor should be in an enclosure, both above and below the floor.
- Either line the enclosure or wrap the duct or pipe within the enclosure, with 25 mm unfaced mineral fibre.
- The material of the enclosure should have a mass of 15 kg/m^2.
- Penetrations of a separating floor by ducts and pipes should have fire protection.
- A flue-pipe may penetrate the floor, provided that it discharges into either a masonry chimney carried by the floor or any other type of chimney enclosed within a duct that is lined with absorbent mineral fibre.

- Seal the junction of ceiling and enclosure with tape or caulking or neoprene.
- A flue-pipe may penetrate the floor, provided that it discharges into either a masonry chimney carried by the floor or any other type of chimney enclosed within a duct that is lined with absorbent mineral fibre.
- Seal the junction of ceiling and enclosure with tape or caulking.
- Penetrations of a separating wall by ducts and pipes should have fire protection.

Structure

Buildings should be stable whatever the combinations of dead load, imposed load and wind loading in terms of their individual structural elements, their interaction together and the overall stability as a structure.

Intermediate floors should be capable of providing local support to the walls as well as acting as horizontal diaphragms capable of transferring the wind loads to buttressing elements of the building.	D 1B1c D 1B1d

Foundations

Foundations should be designed so that the loadings (i.e. dead loads, imposed loads and wind loads) that are transmitted from the building to the subsoil will not cause undue settlement. (See Table 6.32 for examples of typical loads on foundations.)	D 1C0

The imposed loads on floors and ceilings should be less than $2.00\,\text{kN/m}^2$.	D 1D4

Table 6.32   Loads on foundations

Number of storeys	Wall type	Roof span (m)	Floor span (m)	Loading (kN/m)
3	Masonry cavity	12	6	80
3	Masonry cavity	7.5	6	70
2	Masonry cavity	12	6	60
2	Masonry cavity	7.5	6	50
2	Timber frame	7.5	6	40
1	Masonry cavity	7.5	6	30
1	Timber frame	7.5	6	30
1	Single-leaf masonry	5	5	20

Maximum floor area

The maximum floor area that may be enclosed by: D 1D3
- structural walls on all sides shall not be greater than $70 \, m^2$;   D 1E4
- a structural wall on one side shall not be greater than $36 \, m^2$.

The number of areas of floors which can be connected should not be more than four.

The imposed loads on floors should be less than $2.00 \, kN/m^2$.   D 1D4

Maximum span of floors

The maximum span for any floor supported by a wall is 6 m,   D 1D25
where the span is measured centre to centre of bearing as
shown in Figure 6.31.

Lateral support by floors

Floors should: D 1D35

- transfer lateral forces from walls to buttressing walls, piers or chimneys;
- be secured to the supported wall.

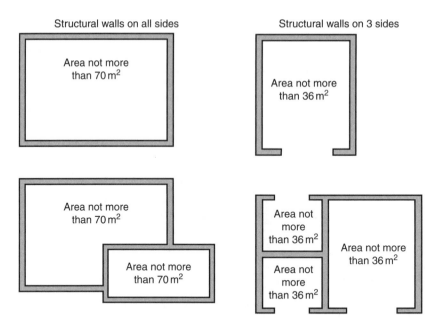

Figure 6.30 Maximum floor area

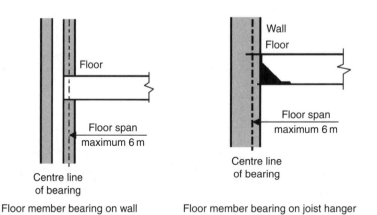

Figure 6.31 Maximum floor span

Timber-framed buildings

Timber-framed buildings typically consist of full-height timber wall panels for each storey built on to the floor below and with intermediate floors built on top of the wall panels. The roof is constructed on top of the top storey wall panels with the masonry cladding connected to the timber panels by wall ties (Figure 6.32).

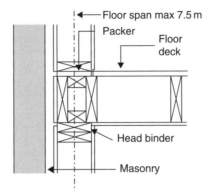

Reproduced by permission of TRADA

Figure 6.32 Maximum floor span

The maximum span for any floor supported by a wall should be 7.5 m.	D 1E41

The floor deck of intermediate floors should be fixed directly to the top faces of the joists.	D 1E49a
Notches and holes in simply supported floor joists should only have: • their holes drilled at the neutral axis; • notches and holes that are at least 100 mm apart horizontally; • notches at the top or bottom of a joist (but not coinciding).	D 1F4
Solid timber strutting should be at least 38 mm thick extending at least 3/4 depth of joist.	D 1F5

Dead floor loads

Table 6.33 provides an indication of typical dead loads on floors and is based on 600 mm joist centres and excluding the weight of the joists, partitioning and rafters.

Floor joist

Floor joists spanning more than 2.5 m should be strutted by one or more rows of solid timber strutting as listed in Table 6.34.	D 1F5

Table 6.33   Dead loads on floors

Construction	Dead load (kN/m^2)
Floors	
Floorboards, 12.5 mm plasterboard	0.22
Floorboards, 19 mm plasterboard	0.27
Floating floor, 18 mm plywood deck, 100 mm quilt insulation, 12.5 mm plasterboard and 19 mm plasterboard	0.66

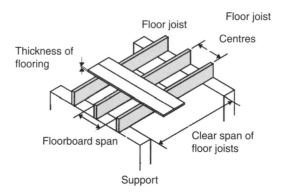

Figure 6.33 Floor joists

Table 6.34   Joist strutting

Joist span (m)	Number of rows of strutting	Position
Less than 2.5	None	N/A
2.5 to 4.5	1	At mid span
More than 4.5	2	At 1/3 span

Centres and spans for floor joists should support the dead loads and an imposed load of not more than 1.5 kN/m^2.   D 1F9

Softwood tongued and grooved floorboards: D 1F9b

- if supported at joist centres of up to 450 mm, should be at least 16 mm thick;
- if supported at wider centres not more than 600 mm, should be at least 19 mm thick.

Wood chipboard, type P5:   D 1F9c

- if supported at joist centres of not more than 450 mm, should be at least 18 mm thick;
- if supported at wider centres up to 600 mm, should be at least 22 mm thick.

Tongue and grooved chipboard flooring should be: D 1F9e

- fixed by 3.35 × 65 mm angular ring shank nails at 200 mm centres around the perimeter and 300 mm centres intermediately;
- glued with PVA adhesive between boards and joists to boards to prevent creaking.

Joists that have been designed only to give access for maintenance   D 1F10
and repair purposes should support the dead loads and an imposed
load not more than 0.75 kN/m² or an imposed concentrated load
of 0.9 kN.

Joists that have been designed to give access for maintenance,   D 1F11
repair and other purposes should support the dead loads and
an imposed load not more than 0.75 kN/m² or an imposed
concentrated load of 0.9 kN.

Openings for stairs

Where openings in floors are required for stairs: D 1F15

- the perimeter of the opening should be supported on all sides by load-bearing walls;
- the floor joists should be strengthened by means of additional joists and trimmers as follows:
  - doubling up the trimming joists on either side of the openings in floors parallel to the floor joists and connecting them together;
  - installing trimmer joists perpendicular to floor joists.

The plan size of openings for stairs should be not more than   D 1F15
2.70 m parallel to the floor joists by 1.15 m perpendicular to the
floor joists.

Supports to non-load-bearing partitions

Provided that timber studs are lightweight lined on each side   D 1F16
and 12.5 mm plasterboard is used:

- where the partition is parallel to the floor joists and directly above a floor joist, an additional joist should be used alongside the main joist;
- where the partition is parallel to the floor joists, but not directly above a floor joist, an extra joist should be used below the partition;
- where the partition is perpendicular to the floor joists and the joists are at not more than 600 mm centres, no additional supports are required.

Places with a special fire risk

Where a place of special fire risk contains any appliance or equipment using hazardous liquid, any opening in a floor dividing it from the remainder of the building should be constructed in such a manner that, in the event of any liquid spillage, the room will contain all the liquid in the appliance or equipment, plus 10%.	ND 2.1.8

Compartment floors

Buildings with different uses should be divided by compartment walls and compartment floors.	ND 2.1.4
In high-rise buildings, every floor at a storey height of more than 18 m above the ground should be a compartment floor.	ND 2.1.6
If a building has a basement storey, the floor of the ground storey should be a compartment floor.	ND 2.1.7
A compartment floor can be constructed of combustible materials having a low, medium, high or very high risk provided the compartment floor has the appropriate fire resistance duration.	ND 2.1.12
Compartment floors in hospitals should be constructed from materials which are non-combustible.	
Where an element of structure provides support to a compartment floor which attracts higher fire resistance duration, the supporting element of structure should have at least the same period of fire resistance.	ND 2.1.13

Compartment floors are intended to prevent fire passing from one compartment to another. Openings and service penetrations through these floors can compromise their effectiveness and should be kept to a minimum.

Openings and service penetrations should be carefully detailed and constructed to resist fire.	ND 2.1.14
A metal chimney should not pass through a compartment floor or separating floor.	D 3.18.4 ND 3.18.4

| There should be no joints within any floor that make accessing the chimney for maintenance purposes difficult. | D 3.18.4 ND 3.18.4 |
| Compartment floors in a hospital should be constructed of non-combustible material. | ND 2.B.1 |

Fire shutters

| Where an opening in a compartment floor contains a stair (but not an escape stair) and not more than two escalators: | ND 2.1.14 |

- a horizontal fire shutter may be installed which maintains the fire resistance duration of the compartment floor and is activated by a fusible link or other heat-sensitive device positioned to detect fire in the lower compartment;
- vertical fire shutters may be installed at each floor level (other than the topmost storey) which maintains the fire resistance duration of the compartment floor and are activated by smoke detection positioned to detect smoke in the lower compartment.

| Where an escalator passes through the opening, it should come to a controlled halt before the fire shutter is activated. | ND 2.1.14 |

Service openings

| A service opening (other than a ventilating duct) which penetrates a compartment wall or compartment floor should be fire-stopped providing at least the appropriate fire resistance duration for the wall or floor. | ND 2.1.14 |
| Where a pipe connects to another pipe which attracts more demanding fire resistance duration, and is within 1 m from the compartment wall or compartment floor, the pipe should be fire-stopped to the more demanding guidance. | ND 2.1.14 |

Ventilating ducts

| A ventilating duct passing through a compartment wall or compartment floor should be fire-stopped in accordance with BS 5588: Part 9: 1999. | ND 2.1.14 |

Junctions

The basic principle is that junctions between compartment floors and other parts of the building should be designed and constructed to prevent fire in one compartment flanking the floor and entering another compartment at the junctions including any solum space or roof space.

Where a compartment floor forms a junction with an external wall, a separating wall, another compartment wall or a wall or screen used to protect routes of escape, the junction should maintain the fire resistance of the compartment floor.	ND 2.1.15
Where a compartment wall (or sub-compartment wall) does not extend to the full height of the building, the wall should form a junction with a compartment floor.	ND 2.1.15

Separating floors

In dwellings

In dwellings, a separating floor:	D 2.2.2
• with medium fire resistance duration should be provided between a dwelling and any other part of the building in common occupation;	
• with a medium fire resistance duration provided between a domestic and non-domestic building;	D 2.2.3
• with a short fire resistance duration provided between solid waste storage accommodation and the rest of the building;	D 2.2.5
• with a short fire resistance duration provided between an integral or attached garage and a dwelling in the same occupation.	D 2.2.6
At dwellings in different occupation, a separating floor with medium fire resistance duration should be provided between adjoining dwellings.	D 2.2.1

In domestic buildings

Where the lift well does not extend to the full height of the building, the lift well should form a junction with a separating floor with medium fire resistance duration.	D 2.2.6
In any separating floor being built to one of the specified constructions, no openings should be formed, apart from openings for service ducts, pipes or chimneys which are protected and are enclosed above and below the floor.	D 5.1.3

In non-domestic buildings

Separating floors should have a medium fire resistance duration.	ND 2.2.1

For buildings in different occupation: ND 2.2.2

- a separating floor should be provided between parts of a building where they are in different occupation;
- if each unit is under the control of an individual tenant, employer or self-employed person, then separating floors should be provided between the areas intended for different occupation;
- if this is not possible, then the building should have a common fire alarm system/evacuation strategy and the same occupancy profile.

A separating floor with medium fire resistance duration should be provided between parts of a building where one part is in single occupation and the other is in communal occupation. ND 2.2.3

Combustibility
In domestic buildings

Where a domestic building also contains non-domestic accommodation, every part of a separating floor (other than a floor finish such as laminate flooring) should be constructed from non-combustible material. D 2.2.7

In non-domestic buildings

Every part of a separating wall or separating floor (other than a floor finish such as laminate flooring) should be of materials that are non-combustible; and to reduce the risk of a fire starting within a combustible separating wall or a fire spreading rapidly on or within the wall construction: ND 2.2.4

- insulation material exposed in a cavity should be of low-risk or non-combustible materials;
- the internal wall lining should be constructed from material which is low risk or non-combustible;
- the wall should contain no pipes, wires or other services.

 This is not necessary for a floor:

- between a shop or office and a dwelling above the shop or office in the same occupation where there is no other dwelling above the shop or

office, and the area of the shop or office is not more than 1½ times the area of the separating floor;
- above an open-ended passageway through a building (i.e. a pend) where the floor has at least medium fire resistance duration and the ceiling of the pend is constructed of non-combustible material;
- between a domestic building and a unit of shared residential accommodation.

Supporting structure

Separating floors are intended to prevent fire passing from one part of the building to another part under different occupation.

In domestic buildings, where an element of structure provides support to:

> - a non-combustible separating floor, the supporting element of structure should also be constructed from materials which are non-combustible;   2.2.8
> - a separating floor which attracts a higher fire resistance duration, the supporting element of structure should have at least the same fire resistance duration.

In non-domestic buildings, where an element of structure provides support to a non-combustible separating floor, the supporting element of structure should:

> - also be constructed from non-combustible materials;   ND 2.2.5
> - have at least the same period of fire resistance as the separating floor it supports.

Openings and service provisions

Openings and service penetrations through separating floors can compromise their effectiveness and should be kept to a minimum. The solum and roof space should not be forgotten. Openings and service penetrations should be carefully detailed and constructed to resist fire.

In domestic buildings, this can be achieved by following the guidance below.

> A service opening (other than a ventilating duct) which   D2.2.9
> penetrates a separating floor should be fire-stopped providing at least the appropriate fire resistance duration for the wall or floor.

> Where a pipe connects to another pipe which attracts more   D2.2.9
> demanding fire resistance duration, and is within 1 m from the separating floor, the pipe should be fire-stopped to the more demanding guidance.

A ventilating duct passing through a separating floor should be fire-stopped in accordance with section 6 of BS 5588: Part 9: 1999.   D2.2.9

Fire-stopping may be necessary to close an imperfection of fit or design tolerance between construction elements and components, service openings and ventilation ducts.   D2.2.9

Proprietary fire-stopping products, including intumescing products, should be tested to demonstrate their ability to maintain the appropriate fire resistance duration under the conditions appropriate to their end use.   D2.2.9

Where minimal differential movement is anticipated, either in normal use or during fire exposure, proprietary fire-stopping products and materials such as cement mortar, gypsum-based plaster, cement or gypsum-based vermiculite/perlite mixes, mineral fibre, crushed rock and blast furnace slag or ceramic-based products (with or without resin binders) may be used.

In non-domestic buildings,

a fire shutter should **NOT** be installed in a separating floor.   ND 2.2.6

Junctions
In domestic buildings

Junctions between separating floors and other parts of the building should be designed and constructed in such a way to prevent a fire in one part of the building flanking the separating floor and entering another part of the building under different occupation, including any solum space or roof space.   D 2.2.10

Where a separating floor forms a junction with an external wall, another separating wall, or a wall or screen used to protect routes of escape, the junction should maintain the fire resistance of the separating floor.   D 2.2.10

In non-domestic buildings

Where a separating floor meets an external wall, another separating wall, a compartment wall or any other wall or screen used to protect routes of escape, the junction should maintain the fire resistance duration of the separating wall or separating floor.   ND 2.2.7

Structural protection

During a fire the elements of structure should continue to function and remain capable of supporting and retaining the fire protection to floors, escape routes and fire access routes, until all occupants have escaped (assisted to escape by staff or been rescued by the fire service).	D 2.3.0 ND 2.3.0

Cavity barriers

Every cavity should be divided by cavity barriers so that the maximum distance between cavity barriers is: • not more than 20 m where the cavity has surfaces which are non-combustible or low-risk materials; • 10 m where the cavity has surfaces which are medium-, high- or very high-risk materials.	D 2.4.2 ND 2.4.2
Cavity barriers are not necessary to divide a cavity in a ceiling void between a floor and a fire-resisting ceiling; • below a floor next to the ground where the cavity is either inaccessible or is not more than 1 m high;	D 2.4.2 ND 2.4.2
In non-domestic buildings, cavity barriers are not necessary to divide a cavity between a floor which is an element of structure and a raised floor consisting of removable panels.	ND 2.4.2f
All cavity barriers should be tightly fitted to a rigid construction.	D 2.4.7 ND 2.4.9
At junctions with slates, tiles, corrugated sheeting or similar materials, the junction should be fire-stopped.	D 2.4.7 ND 2.4.9
A cavity barrier should be installed in all floors or other parts of a building which have a fire resistance duration and which are adjacent to a structure containing a cavity unless the cavity is: • formed by two leaves of masonry or concrete at least 75 mm thick; • formed by external wall or roof cladding, where the surfaces of the cladding are non-combustible or low-risk materials and attached to a masonry or concrete external wall or a concrete roof, and where the cavity contains only non-combustible or low-risk material; • in a wall which has a fire resistance duration for load-bearing capacity only.	D 2.4.7 ND 2.4.9

Galleries

A gallery is a raised floor or platform, including a raised storage floor, which is open to the room or space into which it projects and which:

- has every part of its upper surface not less than 1.8 m above the surface of the main floor of the said room or space;
- occupies not more than one-half of the floor area of the said room or space.

The gallery may be wholly or partly enclosed below, where: D 2.9.18

- the floor of the gallery has a short fire resistance duration;
- at least one route of escape from the gallery is by way of a protected zone.

Openings in floors

In the event of a fire, smoke and possibly flames rising through an opening in a floor may impede evacuees from leaving the building (Figure 6.34).

In a building where there is an opening in any floor, not   ND 2.9.19
being a compartment floor or separating floor:

- escape from any point on the floor, not more than 4.5 m from the opening, should be directly away from the opening (see Figure 6.34 Route A);
- the route from any point on the storey, more than 4.5 m from the opening, should pass no closer to the opening than 4.5 m (see Figure 6.34 Route B).

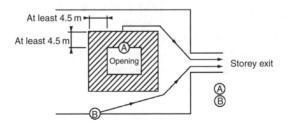

Figure 6.34 Openings in floors

Means of escape

The normal means of escape from a dwelling in the event of a fire will be by way of the internal stairs or other circulation areas.

The bottom of an escape window or the openable area should not be more than 1.1 m above the floor.
D 2.9.4

An escape stair from an openwork floor should have:
ND 2.9.22

- an occupancy capacity of not more than 60;
- an occupancy capacity of 61–100 and at least one route of escape by way of a protected zone, an external escape stair or to another compartment.

If an openwork floor serves more than one level within the room, an unenclosed escape stair may be provided between the floor of the room and the openwork floor only.
ND 2.9.22

Every part of an escape stair and the floor of a protected zone or protected lobby should be constructed of non-combustible material, unless it is:
ND 2.9.29

- an escape stair in shared residential accommodation;
- a handrail, balustrade or protective barrier on an escape stair;
- an escape stair which connects two or more levels within a single storey where the difference in height between the highest and lowest level is not more than 1.8 m;
- an escape stair from a gallery, catwalk or openwork floor;
- a floor finish (e.g. laminate flooring) applied to the escape stair (including landings) or to the floor of a protected zone or protected lobby.

Ducted warm air heating

Where a flat or maisonette has a storey at a height of more than 4.5 m, or a basement storey and is provided with a system of ducted warm air heating where a duct passes through a floor of an entrance hall or stair, all joints between the duct and the surrounding construction should be sealed.
D 2.9.13

Smoke clearance

Ventilation of the escape stairs, protected lobbies and common access corridors is important to assist fire service personnel during fire-fighting operations and for smoke clearance purposes after the fire has been extinguished.

In domestic buildings,

Every access corridor or part of an access corridor, in a building containing flats or maisonettes, should be provided with openable ventilators.	D 2.14.3
The ventilators should provide exhaust at or near ceiling level and supply air at or near floor level with a combined aggregate opening area of at least $1.5\,m^2$.	D 2.14.3
Smoke outlets need not be provided:	ND 2.14.7

- where the floor area of the basement storey is not more than $200\,m^2$;
- where a window or windows opening direct to the external air, have a total area not less than 1% of the floor area.

Resistance to fire

The recommended fire resistance duration can be attained where the construction follows the guidance in the Columns 3, 4 and 5 of Table 6.35.

Treatment of building elements adjacent to the ground

Floors adjacent to the ground should be constructed:	D 3.4.1 ND 3.4.1
- so as to prevent moisture from the ground reaching the inner surface of any part of a building that it could damage;	
- as shown in Table 6.36.	D 3.4.1–D 3.4.4 ND 3.4.1–ND 3.4.4

Floors at or near the ground level should be constructed in accordance with the recommendations in Clause 11 of CP 102: 1973.	D 3.4.6 ND 3.4.6
Permanent ventilation of the underfloor space direct to the outside air should be provided by ventilators in two external walls on opposite sides of the building.	D 3.4.6 ND 3.4.6

Sanitary appliances below flood level

The basements of approximately 500 buildings in Scotland are flooded each year when the sewers surcharge (the effluent runs back up the pipes because they are too small to take the required flow) (Figure 6.35).

Table 6.35 Recommended fire resistance duration

Column 1	Column 2	Column 3	Column 4	Column 5	Column 6	Column 7
Construction	Fire resistance duration	British standards			European standards	Test exposure
		Load-bearing capacity (min)	Integrity (min)	Insulation (min)		
Compartment floor, separating floor, or any other floor, flat roof or access deck used as a protected route of escape	Short Medium Long	30 60 120	30 60 120	30 60 120	REI 30 RE1 60 REI 120	From the underside
Other than a floor shown above or an intermediate floor within a flat or maisonette	Short Medium Long	30 60 120	None None None	None None None	R 30 R 60 R 120	From the underside

Table 6.36   Construction of concrete and timber floors

Ground supported concrete floors

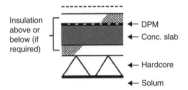

- The solum is brought to a level surface.
- Hardcore bed 100 mm thick of clean broken brick or similar inert material.
- Concrete slab 100 mm thick with insulation, if any, laid above or below the slab; with or without a screed or floor finish.
- Damp-proof membrane above or below the slab or as a sandwich; jointed and sealed to the damp-proof course or damp-proof structure in walls, columns and other adjacent elements.

Suspended concrete floors

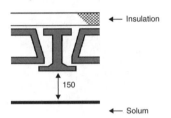

- The solum is brought to an even surface.
- Any upfilling to be of hard, inert material.
- Suspended concrete floor of *in situ* or precast concrete slabs or beams.
- Permanent ventilation of the underfloor space direct to the outside air by ventilators in two external walls on opposite sides of the building.
- The ventilated space to be 150 mm to the underside of the floor slab or beams.

Suspended timber floors

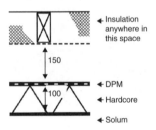

- The solum is brought to an even surface.
- Any up filling to be of hard, inert material.
- Hardcore bed 100 mm thick of clean broken brick; or concrete 50 mm thick laid on 0.25 mm (1,000 gauge) polyethylene sheet; or concrete 100 mm thick; so that in any case the top surface is not below that of the adjacent ground.
- Suspended timber floor with or without insulation.
- Floor joists carried on wall-plates supported as necessary by sleeper walls with a dip under the wall-plates.
- Permanent ventilation of the underfloor space direct to the outside air by ventilators in two external walls on opposite sides of the building.

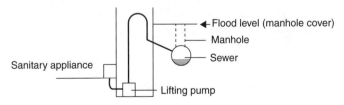

Figure 6.35 Sanitary discharge stack systems

Wastewater from floor gullies below flood level should
be drained by wastewater lifting plants or, where there is
unlikely to be a risk to persons such as in a car park, via an
anti-flooding device.

D 3.7.2
ND 3.7.2

Enhanced apartment

Smaller apartments or those with an unusual shape may limit how space
within can be used. At least one apartment on the principal living level of a
dwelling should be of a size and form that allow greater flexibility of use.
These are called enhanced apartments.

Enhanced apartments should:

D 3.11.2
D 3.11.5

- have a floor area of at least 12 m² and a length and width
  of at least 3.0 m;
- exclude any space less than 1.8 m in height and any
  portion of the room designated as a kitchen;
- contain an unobstructed manoeuvring space of at least
  1.5 by 1.5 m square or an ellipse of at least 1.4 by 1.8 m,
  which may overlap with activity spaces.

 Note: A door may open over this space; and have
unobstructed access, at least 800 mm wide, to the
controls of any openable window or any heating
appliance and between doors within the apartment.

Accessible bathrooms and shower rooms

An accessible shower room should have a dished floor of
a gradient of not more than 1:50 discharging into a floor
drain, or a proprietary level access shower with a drainage
area of not less than 1.2 by 1.2 m.

ND 3.12.10

Every bathroom or shower room should have a floor
surface that minimizes the risk of slipping when wet.

ND 3.12.6

Surface condensation – thermal bridging

Thermal bridging occurs when the continuity of the building fabric is broken
by the penetration of an element allowing a significantly higher heat loss than
its surroundings.

| To minimize the risk of condensation on any inner surface, cold bridging at a floor, wall, roof or other building element should be avoided. | D 3.15.4 |
| A floor should minimize the risk of interstitial condensation in any part of a dwelling that it could damage. | D 3.15.5 |

Provision for washing down

Where communal solid waste storage is located within a building, such as where a refuse chute is utilized, the storage area should have provision for washing down and draining the floor into a wastewater drainage system.	D 3.25.4
Floor gullies should incorporate a trap that maintains a seal even during periods of disuse.	D 3.25.4
Floors should be of an impervious surface that can be washed down easily and hygienically.	D 3.25.4

Floor surfaces in common areas of buildings

| Floor surfaces within common areas of a domestic building and/or corridors, and circulation areas within a non-domestic building should be uniform, permit ease in manoeuvring and be of a material and finish that, when clean and dry, will provide a level of traction that will minimize the possibility of slipping. | D 4.2.2
ND 4.2.3 |
| Where there is a change in the characteristics of materials on a circulation route, such as from a tile to carpet finish, transition should be level and, where reasonably practicable, differing surfaces should contrast visually to identify the change in material and reduce the potential for trips. | D 4.2.2
ND 4.2.3 |

Noise

There are currently *NO* Building Standards to protect the occupants or users of a non-domestic building from noise, but the need may well arise for such standards at a later date.

Resisting sound transmission to dwellings

Measures to reduce the transmission of sound vary according to the type of construction and its reaction to sound energy. The most important factors which affect the behaviour of separating floors are mass, cavities, isolation and absorption.

Note: Section 5 of the Technical Handbooks applies to dwellings other than those that are totally detached.

> Every building must be designed and constructed in such a   ND 5.1
> way that each floor separating one dwelling from another,
> or one dwelling from another part of the building, or one
> dwelling from a building other than a dwelling, will limit
> the transmission of noise to the dwelling to a level that will
> not threaten the health of the occupants of the dwelling
> or inconvenience them in the course of normal domestic
> activities, provided the source noise is not in excess of that
> from normal domestic activities.
>
>   This standard does not apply to:
>
> • fully detached houses;
> • roofs or walkways with access solely for maintenance, or
>   solely for the use of the residents of the dwelling below.

General application to dwellings

The following requirements should be met in all dwellings other than those that are totally detached.

> • Airborne sound resisting separating floors should be   D 5.1.1
>   provided between dwellings so that each dwelling is
>   protected from noise emanating from the other one and
>   from other parts, such as common stair enclosures and
>   passages, solid waste disposal chutes, lift shafts, plant
>   rooms, communal lounges and car parking garages.
> • Impact sound resisting separating floors should be provided
>   between dwellings so that the lower dwelling is protected
>   from sound emanating from the upper dwelling and from
>   other parts of the building above.
> • Impact sound resisting construction should be provided
>   between a dwelling and a roof that acts as a floor or a
>   walkway directly above the dwelling so that the dwelling
>   below is protected from sound emanating from the roof
>   or walkway above (e.g. roofs that act as floors are access
>   decks, car parking, escape routes and roof gardens).

Except where:

- where two houses are linked only by an imperforate separating wall between their ancillary garages, it is not necessary for the wall to be airborne sound resisting;
- where the wall between a dwelling and another part of the building is substantially open to the external air, it is not necessary for the wall to resist airborne sound transmission (e.g. a wall between a dwelling and an access deck);
- where the wall between a dwelling and another part of the building incorporates a fire door, it is not necessary for the door to be airborne sound resisting;
- when a roof or walkway is providing access solely for the purpose of maintenance or is solely for the use of the residents of the dwelling directly below, it is not necessary to provide impact sound resisting construction;
- in the case of a separating floor between a dwelling and a private garage or a private waste storage area which is ancillary to the same dwelling, it is not necessary for the wall or floor to be airborne or impact sound resisting.

Performance testing

Use of the performance testing approach is particularly useful where the separating or flanking construction is of innovative design and for conversions where flanking transmission may be significant. Performance values are given in terms of two acoustic parameters, one related to airborne sound, the other related to impact sound.

The airborne sound insulation characteristic of a floor is the sound pressure level difference between the room containing the noise source and the receiving room. The larger the difference, the higher the level of airborne sound insulation. Impact sound insulation is the sound pressure level in the receiving room.

Recommended performance values for separating walls and separating floors are given in Table 6.37.

Note:

- Annex 5.A to the Non-Domestic Technical Handbooks describes methods for calculating the mass of masonry wall leafs (mortar joints, *in situ* concrete, screeds, slabs and composite floor bases).
- Annex 5.B describes methods for the selection of resilient materials used for soft coverings.
- Annex 5.C describes methods of measurement and test procedures.

The scheme operated by Robust Details Ltd for the English and Welch Building Regulations is also a useful measurement method, but at the time of

Table 6.37   Recommended performance values for separating walls and separating floors

Airborne sound (minimum values)

Minimum values of weighted standardized level difference as defined in BS EN ISO 717-1: 1997:

	Mean value	Individual value
Floors	52 dB	48 dB

Impact sound (minimum values)

Maximum values of weighted standardized level difference as defined in BS EN ISO 717-2: 1997:

	Mean value	Individual value
Floors	62 dB	65 dB

writing this book, this scheme has not yet been fully reviewed in relation to construction practice in Scotland.

Energy

The maximum U-values for floors within the insulation envelope are shown in Table 6.38.

Table 6.38   Maximum U-values for floors within elements of the insulation envelope

Type of element	(a) Area-weighted average U-value (W/m^{2}K) for all elements of the same type	(b) Individual element U-value (W/m^{2}K)
Floor (normal)[1]	0.25	0.70
Floor (shell and fit-out buildings)	0.22	0.70
Floor (where conversion of a heated building is to be carried out)	0.70	
Floor (when constructing an extension)	0.22	0.70

 Note: [1] Excluding separating floors, where thermal transmittance should be ignored.

Alterations to the insulation envelope

Alterations that involve increasing the floor area and/or bringing parts of the existing building that were previously outwith the insulation envelope into the heated part of the dwelling are considered as extensions and/or conversions.

The infill of an existing opening of approximately $4\,m^2$ D 6.2.11
or less in the building fabric should have a U-value which   ND 6.2.12
matches at least that of the remainder of the surrounding
element and in the case of a floor it should not be worse
than $0.70\,W/m^2K$ (for a roof, not worse than $0.35\,W/m^2K$).

Where additional windows, doors and rooflights are being   D 6.2.10
created, the overall total area (including existing) should
not exceed 25% of the total dwelling floor area and in the
case of a heated communal room or other area (exclusively
associated with the dwellings), it should not exceed 25% of
the total floor area of these rooms/areas.

A building that was in a ruinous state should, after   D 6.2.10
renovation, be able to achieve almost the level expected of
new construction and after an alteration of this nature to the
insulation envelope, a roof should be able to achieve at least
an average U-value of 0.35 and in the case of a wall or floor,
$0.70\,W/m^2K$.

When alterations are carried out, attention should still be   D 6.2.11
paid to limiting infiltration thermal bridging at junctions   ND 6.2.12
and around windows, doors and rooflights and limiting air
infiltration.

Conservatories

Although conservatories that are attached to dwellings are normally considered as stand-alone buildings:

the dividing floor element should have U-values equal or   D 6.2.12
better than the corresponding exposed elements in the rest
of the dwelling.

Heat pump systems efficiency (warm and hot water)

All heat pumps are at their most efficient when the source temperature is as high as possible, the heat distribution temperature is as low as possible and pressure losses are kept to a minimum.

Supply water temperatures should be in the range of 30–40°C to   D 6.3.4
an underfloor heating system.

Hot water underfloor heating

The following controls should be fitted to ensure safe system operating temperatures: D 6.3.8   • a separate flow temperature high limit thermostat for warm water systems connected to any high water temperature heat supply;   • a separate means of reducing the water temperature to the underfloor heating system.

Controls for dry space heating and hot water systems

Zone controls are not considered necessary for single apartment dwellings.

For large dwellings with a floor area over $150\,m^2$, independent time and temperature control of multiple space heating zones is recommended.   D 6.3.9
Each zone (not exceeding $150\,m^2$) should have a room thermo-stat, and a single multi-channel programmer or multiple heating zone programmers.   D 6.3.9

Electric underfloor heating

For electric storage, direct acting systems and undertile systems programmable room timer/thermostats with manual override feature room controls are recommended for all heating zones, with air and floor (or floor void) temperature sensing capabilities to be used individually or combined.

A storage system should have: D 6.3.9   • anticipatory controllers for controlling low-tariff input charge with external temperature sensing and floor temperature sensing;   • a manual override facility, for better user control.
Controls for storage systems with room timer/thermostats should take advantage of low-tariff electricity.   D 6.3.9

Information to be provided for dwellings

The Energy Performance Certificate should display the following information:	D 6.9.2
• the conditioned floor area of the dwelling; • the approximate energy use expressed in kWh/m² of floor area per annum.	ND 6.9.2

Location of an Energy Performance Certificate
Domestic buildings

The Energy Performance Certificate should be indelibly marked and located in a position that is readily accessible (such as a cupboard containing the gas or electricity meter or the water supply stopcock), and it should be protected from weather and not easily obscured.	D 6.9.3
For conservatories and for other ancillary stand-alone buildings of less than 50 m² floor area, an energy performance certificate need not be provided.	D 6.9.4
For those buildings of a floor area of 50 m² or more, the guidance in the Non-Domestic Technical Handbook should be followed and an additional certificate supplementing the one for the dwelling should be provided.	D 6.9.4
For stand-alone ancillary buildings (such as a kiosk for a petrol filling station) of less than 50 m² floor area, an Energy Performance Certificate need not be provided.	ND 6.9.4
For stand-alone buildings of a floor area of 50 m² or more that are heated or cooled which are ancillary or subsidiary to the main building, a certificate should be provided, in addition to the one for the main building.	ND 6.9.4

Loudspeakers

The development of flat panel loudspeakers and loudspeakers integrated within floor constructions has introduced an additional neighbour noise concern.

NO loudspeaker should be fitted within a separating floor.	D 5.1.3

6.7 Walls

In a brick-built house:

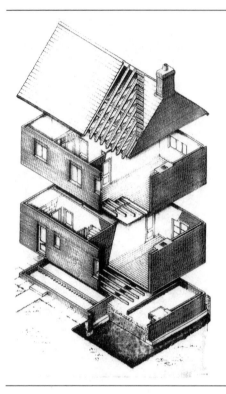

- The external walls are load-bearing elements that support the roof, floors and internal walls.
- These walls are normally cavity walls comprising two leaves braced with metal ties, but older houses will have solid walls, at least 225 mm (9″) thick.
- Bricks are laid with mortar in overlapping bonding patterns to give the wall rigidity, and a damp-proof course (DPC) is laid just above ground level to prevent the moisture rising.
- Window and door openings are spanned above with rigid supporting beams called lintels.
- The internal walls are either non-load-bearing divisions (made from lightweight blocks, manufactured boards or timber studding) or load-bearing structures made of brick or block.

In modern timber-framed houses:

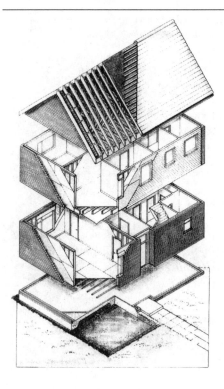

- Walls are constructed of vertical timber studs with horizontal top and bottom plates nailed to them.
- The frames, which are erected on a concrete slab or a suspended timber platform supported by cavity brick walls, are faced on the outside with plywood sheathing to stiffen the structure.
- Breather paper is fixed over the top to act as a moisture barrier. Insulation quilt is used between studs.
- Rigid timber lintels at openings carry the weight of the upper floor and roof.
- Brick cladding is typically used to cover the exterior of the frame and is attached to the frame with metal ties.
- Weatherboarding often replaces the brick cladding on upper floors.

 Note: Timber-frame walls should commence above ground level and are, therefore, not subject to lateral loads other than from wind.

 Note: When reading this particular section, you will probably notice that a number of the requirements have already been covered in Section 6.6 *Floors* and Section 6.8 *Ceilings*. This has been done in order to save the reader having to constantly turn back and re-read a previous page.

6.7.1 Requirements

Load-bearing wall

 Note: Load-bearing wall construction includes masonry cross-wall construction, and walls comprising close centred timber or lightweight steel section studs.

> *The nominal length of load-bearing wall should be taken as:*   *1.2.0*
>
> - *in the case of a reinforced concrete wall, the distance between lateral supports subject to a length not more than 2.25 times storey height;*
> - *in the case of an external masonry wall, or timber or steel stud wall, the length is measured between vertical lateral supports;*
> - *in the case of an internal masonry wall, or timber or steel stud wall, a length is not more than 2.25 times storey height.*

Fire

Non-domestic buildings

> *Every building must be designed and constructed in such a*   *ND 2.1*
> *way that in the event of an outbreak of fire within the building,*
> *fire and smoke are inhibited from spreading beyond the*
> *compartment of origin until any occupants have had the time*
> *to leave that compartment and any mandatory fire containment*
> *measures have been initiated.*
>
> **Note:** This standard does not apply to domestic buildings.

Domestic and non-domestic buildings

> *Every building, which is divided into more than one area of different*   *2.2*
> *occupation, must be designed and constructed in such a way that in*
> *the event of an outbreak of fire within the building, fire and smoke*
> *are inhibited from spreading beyond the area of occupation where*
> *the fire originated.*

Fire safety

> *Every building must be designed and constructed in such a way that*   *2.4*
> *in the event of an outbreak of fire within the building (or from an*
> *external source):*
>
> - *the unseen spread of fire and smoke within concealed spaces in its structure and fabric is inhibited;*
> - *the development of fire and smoke from the surfaces of walls*   *2.5*
>   *within the area of origin is inhibited;*
> - *the spread of fire to neighbouring buildings is inhibited;*   *2.6*
> - *the spread of fire on the external walls of the building is inhibited.*   *2.7*

Fire service facilities

Every building must be designed and constructed in such a way that facilities are provided to assist fire-fighting or rescue operations.	*2.14*

Environment

There will not be a threat to the building or the health of the occupants as a result of moisture from precipitation penetrating to the inner face of the building.

Every building must be designed and constructed in such a way that there will not be a threat to the building or the health of the occupants or people in or around the building:	*3.1*
• *due to the presence of harmful or dangerous substances;*	
• *as a result of moisture penetration from the ground.*	*3.4 and 3.10*
Every building must be designed and constructed in such a way that:	*3.12*
• *sanitary facilities are provided that allow convenience of use and without threat to the health and safety of occupants or visitors;*	
• *the air quality inside the building is not a threat to the capability of the building to resist moisture, decay or infestation;*	*3.14*
• *any component part of each fixed combustion appliance installation used for the removal of combustion gases will withstand heat generated as a result of its operation without any structural change that would impair the stability or performance of the installation;*	*3.18*
• *any component part of each fixed combustion appliance installation will not cause damage to the building in which it is installed by radiated, connected or conducted heat or from hot embers expelled from the appliance;*	*3.19*
• *an oil storage installation, incorporating oil storage tanks used solely to serve a fixed combustion appliance installation providing space heating or cooking facilities in a building, will inhibit fire from spreading to the tank and its contents from within, or beyond the boundary;*	*3.23*
• *a container for the storage of woody biomass fuel will inhibit fire from spreading to its contents from within, or beyond the boundary;*	

- *accommodation for solid waste storage is provided which:*   *3.25*
  - *permits access for storage and for the removal of its contents;*
  - *does **NOT** threaten the health of people in and around the building;*
  - *does **NOT** contaminate any water supply, ground water or surface water.*

For dwellings

Every building must be designed and constructed in such a way   *3.15*
that there will not be a threat to the building or the health of the
occupants as a result of moisture caused by surface or interstitial
condensation.

Noise

Every building must be designed and constructed in such a way that   *5.1*
each wall separating one dwelling from another, or one dwelling from
another part of the building, or one dwelling from a building other
than a dwelling, will limit the transmission of noise to the dwelling
to a level that will not threaten the health of the occupants of the
dwelling or inconvenience them in the course of normal domestic
activities provided the source noise is not in excess of that from
normal domestic activities.

Energy

Every building must be designed and constructed in such a way that   *6.2*
an insulation envelope is provided which reduces heat loss.

6.7.2 Meeting the requirements

Basic requirements for stability

Buildings should be stable whatever the combinations of dead load, imposed load and wind loading in terms of their individual structural elements, their interaction together and the overall stability as a structure.

| A layout of internal and external walls should form a robust three-dimensional box structure in plan. | D 1B1b ND 1B1b |
| Internal and external walls should be connected either by masonry bonding or via mechanical connections. | D 1B1c ND 1B1c |

Foundations

| The foundations of a building should be situated centrally under the wall and (vice versa) walls should be central on the foundation. | D 1C3a and g ND C3a and g |

Wall heights

Building heights are measured from the lowest finished ground level to the highest point of the roof as shown in Figure 6.36.

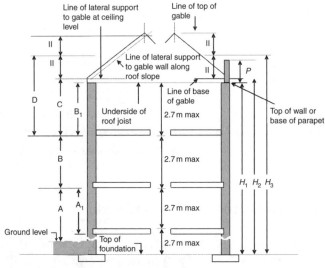

Where:
H_1 is the height of a wall that does not include a gable
H_2 is the height of a separating wall which may extend to the underside of the roof
H_3 is the height for a wall (except a separating wall) which includes a gable
P is the height of the parapet. If the parapet height more than 1.2 m add height to H_1

Figure 6.36 Rules for measuring the heights of storeys and walls

| Differences in level of ground or other solid construction between one side of the wall and the other should not be more than four times the thickness of the wall. | D 1D26 ND 1D26 |
| The combined dead load and imposed load should be not more than 70 kN/m at the base of the wall. | D 1D26 ND 1D26 |

Masonry walls

The thickness of the wall depends on the general conditions relating to the building of which the wall forms a part (e.g. floor area, roof loading, wind speed) and the design conditions relating to the wall (e.g. type of materials, loading, end restraints, openings, recesses, overhangs and lateral floor support requirements).

All small traditional masonry wall buildings should be designed to take into account of loading conditions and limitations on dimensions, openings, etc.

Brick and block construction

Walls should be bonded and put together with mortar.   D 1D18

Materials should be chosen from Table 6.39 based on their
intended use and for the exposure conditions likely to prevail.

Table 6.39   Bricks used for masonry walls

Type of brick	Relevant standard
Clay bricks or blocks	BS 3921: 1985 or BS EN 771-1: 2003
Calcium silicate bricks	BS 187: 1978 or BS 6649: 1985; or BS EN 771-2: 2003
Concrete bricks or blocks	BS EN 771-3: 2003 or BS EN 771-4: 2003
Square dressed natural stone	BS 5628-3: 2006 or BS EN 771-6: 2005
Manufactured stone	BS 6457: 1984 or BS EN 771-5: 2003

Solid walls in coursed brickwork or blockwork

Solid external walls, compartment walls and separating walls   D 1D6
that are constructed in coursed brickwork or blockwork   ND 1D6
should be at least as thick as 1/16 of the storey height and in
accordance with Table 6.40.

Solid walls in uncoursed stone and flints, etc.

The thickness of walls constructed in uncoursed stone or   D 1D7
bricks or other burnt or vitrified material should be at least   ND 1D7
1.33 times the thickness shown in Table 6.40.

Table 6.40   Wall thicknesses

Height of wall	Length of wall	Minimum thickness of wall For solid walls in coursed brickwork	Minimum thickness of wall For internal load-bearing walls in brickwork or blockwork
Not more than 3.5 m	Not more than 12 m	190 mm for the whole of its height	90 mm for the whole of its height (see note)
More than 3.5 m but not more than 9 m	Not more than 9 m	190 mm for the whole of its height	90 mm for the whole of its height (see note)
	More than 9 m but not more than 12 m	290 mm from the base for the height of one storey and 190 mm for the rest of its height	140 mm from the base for the height of one storey and 190 mm for the rest of its height
More than 9 m but not more than 12 m	Not more than 9 m	290 mm from the base for the height of one storey and 190 mm for the rest of its height	140 mm from the base for the height of one storey and 190 mm for the rest of its height
	More than 9 m but not more than 12 m	290 mm from the base for the height of two storeys and 190 mm for the rest of its height	140 mm from the base for the height of two storeys and 190 mm for the rest of its height

Cavity walls in coursed brickwork and blockwork

All cavity walls should have leaves at least 90 mm thick and structural cavities at least 50 mm wide.   D 1D8
                                                                        ND 1D8

For external walls, compartment walls and separating walls in cavity construction, the combined thickness of the two leaves (plus 10 mm) should be at least 1.33 times the thickness shown in Table 6.40 for a solid wall of the same height and length.

Walls providing vertical support to other walls

The thickness of the wall should be at least that of any part of the wall to which it gives vertical support – regardless of the materials used in the construction.   D 1D9
                                                                        ND 1D9

Internal load-bearing walls in brickwork or blockwork

> With the exception of separating walls, internal load-bearing walls in brickwork or blockwork should have a thickness at least that shown in Table 6.40.
>
> D 1D10
> ND 1D10
>
> **Note:** If the wall is the lowest wall in a three-storey building and it is carrying load from both of the upper storeys, the thickness should be a minimum of 140 mm.

Parapet walls

> The minimum thickness and maximum height of parapet walls should be as shown in Figure 6.37.
>
> D 1D11
> ND 1D11

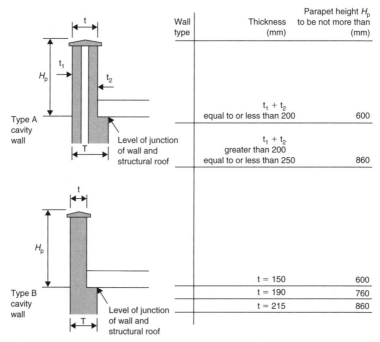

Figure 6.37 Height of parapet walls

Single-leaf external walls

> The single leaf of an external wall to a single-storey, single-leaf building (and/or extension) should be at least 90 mm thick.
>
> D 1D12
> ND 1D12

Compressive strength of masonry units

The minimum compressive strengths of masonry units should be derived by obtaining the Condition (A, B or C) from Figure 6.38 and then reading the compressive strength from Table 6.41.

D 1D19-22
ND 1D19-22

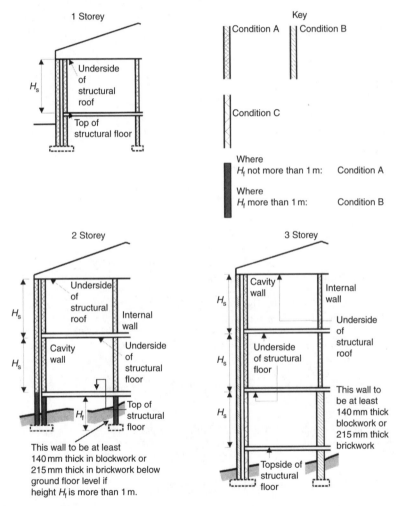

Notes:
1. If H_s is not more than 2.7 m, the compressive strength of bricks or blocks should be used in walls as indicated by the key.
2. If H_s is more than 2.7 m, the compressive strength of bricks or blocks used in the wall should be at least Condition B, or as indicated by the key, whichever is the greater.
3. If the *external* wall is solid *construction*, the masonry units should have a compressive strength of at least that shown for the internal leaf of a cavity wall in the same position.

Figure 6.38 Compressive strengths of masonry units in walls

Table 6.41 Compressive strength of masonry units (N/mm²)

Masonry unit	Clay masonry units		Calcium silicate masonry units		Aggregate concrete masonry units	Autoclaved aerated concrete masonry units	Manufactured stone masonry units
	Group 1	Group 2	Group 1	Group 2			
Condition A							Any unit following the guidance of BSEN 771-5
Brick	6.0	9.0	6.0	9.0	6.0	x	
Block	5.0	8.0	5.0	8.0	2.9	2.9	
Condition B							
Brick	9.0	13.0	9.0	13.0	9.0	x	
Block	7.5	11.0	7.5	11.0	7.3	7.3	
Condition C							
Brick	18.0	25.0	18.0	25.0	18.0	x	
Block	15.0	21.0	15.0	21.0	7.3	7.3	

Lateral support of walls

Walls should be strapped to floors above ground level by tension straps whose centres are not more than: 2 m for ground and first floors; 1.25 m above first floor level.	D 1D33 ND 1D33
Tension straps should be made of galvanised steel or austenitic stainless steel.	D 1D33 ND 1D33
The tensile strength of tension straps should not be less than 8 kN.	D 1D33 ND 1D33
The lateral support of walls shall be in accordance with Table 6.42.	D 1D33 ND 1D33

Note: Tension straps need **NOT** be provided:

- in the longitudinal direction of joists in houses of not more than two storeys if the joists:
  - are at not more than 1.2 m centres;
  - have at least 90 mm bearing on the supported walls or 75 mm bearing on a timber wall-plate at each end;
  - are carried on the supported wall by joist hangers (see Figure 6.39); and
  - are incorporated at not more than 2 m centres;

- when a concrete floor has at least 90 mm bearing on the supported wall (see Figure 6.40);
- where floors are at or about the same level on each side of a supported wall, and contact between the floors and wall is either continuous or at intervals not exceeding 2 m. Where contact is intermittent, the points of contact should be in line or nearly in line on plan (see Figure 6.41).

Table 6.42 Lateral support of walls

Wall type	Wall length	Lateral support required
Solid or cavity external wall	Any length	Roof lateral support by every roof forming a junction with a supporting wall
Solid or cavity separating wall or compartment wall	More than 3 m	Floor lateral support by every floor forming a junction with a supporting wall
Internal load-bearing wall (that is not a separating wall or compartment wall)	Any length	Roof or floor lateral support at the top of each storey

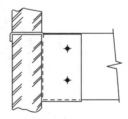

Figure 6.39 Restraint type joist hanger

X to be not less than 90 mm

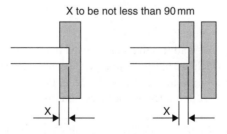

Figure 6.40 Restraint on concrete floor or roof

Vertical lateral restraint to walls

The ends of every wall should be bonded or otherwise securely tied throughout their full height to a buttressing wall, pier or chimney.	D 1D27 ND 1D27
Long walls may be provided with intermediate support dividing the wall into distinct lengths.	D 1D27 ND 1D27
The buttressing wall, pier or chimney should provide support from the base to the full height of the wall.	D 1D27 ND 1D27

Lateral support by roofs and floors

The walls in each storey of a building should: • extend to the full height of that storey; • have horizontal lateral supports to restrict movement of the wall at right angles to its plane.	D 1D35 ND 1D35

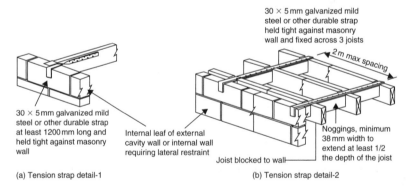

30 × 5 mm galvanized mild steel or other durable strap held tight against masonry wall and fixed across 3 joists

2 m max spacing

30 × 5 mm galvanized mild steel or other durable strap at least 1200 mm long and held tight against masonry wall

Internal leaf of external cavity wall or internal wall requiring lateral restraint

Joist blocked to wall

Noggings, minimum 38 mm width to extend at least 1/2 the depth of the joist

(a) Tension strap detail-1   (b) Tension strap detail-2

Figure 6.41 Lateral support by floors

Wall ties

There are two types of wall ties that can be used in masonry cavity walls: type A (butterfly ties) which are normal and type B (double triangle ties) which are used only in external masonry cavity walls where tie type A fails to satisfy the requirements of the Building (Scotland) Regulations.

Wall ties should be selected in accordance with Table 6.43 and in line with the guidance provided in BS EN 845-1: 2003.	D 1D17 ND 1D17

Table 6.43 Permissible type of cavity wall tie

Normal cavity width (mm)	Tie length (mm)
50–75	220
76–90	225
91–100	225
101–125	250
126–150	275
151–175	300
176–300	The embedment depth of the tie should be at least 50 mm in both leaves

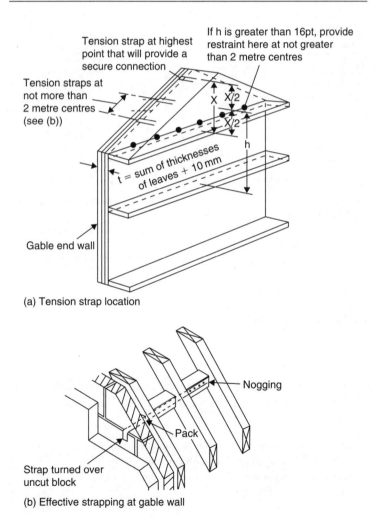

(a) Tension strap location

(b) Effective strapping at gable wall

Figure 6.42 Lateral support at roof level

Openings, recesses, chases and overhangs

The number, size and position of openings and recesses should not interfere with: • the stability of a wall; • the lateral support provided by a buttressing wall to a supported wall. Construction over openings and recesses should be adequately supported.	1D30 1D30

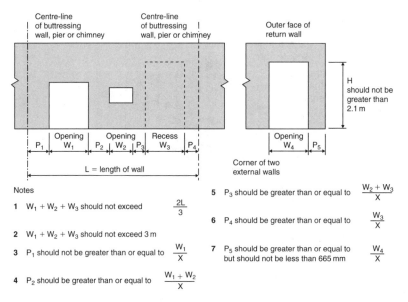

Figure 6.43 Openings and recesses

Notes

1   $W_1 + W_2 + W_3$ should not exceed $\dfrac{2L}{3}$

2   $W_1 + W_2 + W_3$ should not exceed 3 m

3   P_1 should not be greater than or equal to $\dfrac{W_1}{X}$

4   P_2 should be greater than or equal to $\dfrac{W_1 + W_2}{X}$

5   P_3 should be greater than or equal to $\dfrac{W_2 + W_3}{X}$

6   P_4 should be greater than or equal to $\dfrac{W_3}{X}$

7   P_5 should be greater than or equal to $\dfrac{W_4}{X}$ but should not be less than 665 mm

Openings in walls below ground floor should be limited to small holes for services and ventilation, etc., not more than $0.1\,m^2$ and at least 2 m apart.	1D31
The value of Factor X should be taken from Table 6.44 or can be given the value 6, provided the compressive strength of the bricks or blocks (in the case of a cavity wall, in the inner leaf) is not less than $7\,N/mm^2$.	1D32

Table 6.44 Value of factor X

Nature of roof span	Maximum roof span (mm)	Minimum thickness of inner leaf of wall (m)	Nature of floor span		
			Parallel to wall	Perpendicular to wall (max 4.5 m)	Perpendicular to wall (max 6.0 m)
Parallel to the wall	Non-applicable	100	6	6	6
		90	6	6	6
Perpendicular to the wall	9	100	6	6	5
		90	6	4	4

Lintels for openings

Proprietary steel or concrete lintels used with masonry cavity wall construction should be tested by a notified body or justified by calculations.	D 1D24 ND 1D24

Buttressing walls

Buttressing walls are external masonry return walls or internal walls that are perpendicular to the supported wall.

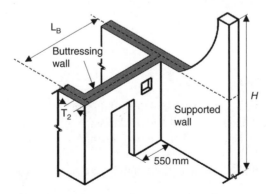

Figure 6.44 Openings in a buttressed wall

If the buttressing wall is not itself a supported wall, then its thickness T2 should be at least:	D 1D28 ND 1D28

- half the thickness normally required by these Regulations for an external wall or separating wall of similar height and length – less than 5 mm;
- 75 mm if the wall forms part of a dwelling and is not more than 6 m in total height and 10 m in length;
- 90 mm in all other cases.

The buttressing wall should be bonded or securely tied to the supported wall, another buttressing wall, pier or chimney.	D 1D28 ND 1D28
The length, L_B, of the buttressing wall should be at least 1/6th of the overall height, H, of the supported wall.	D 1D28 ND 1D28
The position and shape of the openings should not impair the lateral support being given by the buttressing wall.	D 1D28 ND 1D28
Openings or recesses in the buttressing wall that are more than $0.1\,m^2$ should be at least 550 mm from the supported wall.	D 1D28 ND 1D28
Only one opening or recess of not more than $0.1\,m^2$ is permitted within 550 mm of the supported wall.	D 1D28 ND 1D28
The opening height in a buttressing wall:	D 1D28 ND 1D28

- should be not more than 0.9 times the floor to ceiling height.

The depth of lintel including any masonry over the opening should be not less than 150 mm.

Note: See Figure 6.36 for details on how to measure the height of supported walls, etc.

Chases

Chases should not:	D 1D33 ND 1D33

- if vertical, be deeper than 1/3rd of wall thickness or in cavity walls and be 1/3rd of leaf thickness;
- if horizontal, be deeper than 1/6th of the thickness of the leaf or wall;
- be so positioned as to impair the stability of the wall.

If hollow blocks are used, then at least 15 mm thickness of block should be retained.	D 1D33 ND 1D33

Overhangs

The amount of any projection should not impair the stability of the wall.	D 1D34 ND 1D34

Gable wall strapping

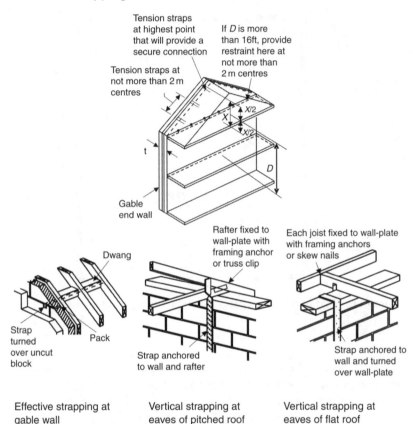

Figure 6.45 Gable wall strapping

Vertical strapping at least 1 m long should be provided at eaves level at intervals not more than 2 m if the roof:	D 1D36 ND 1D36

- has a pitch of not more than 150;
- is not tiled or slated;
- is not of a type known by local experience to be resistant to wind gusts;
- does not have main timber members spanning onto the supported wall at not more than 1.2 m centres.

Size and proportions of openings

Other than windows and a single-leaf door that meet the requirements shown in Figure 6.46:

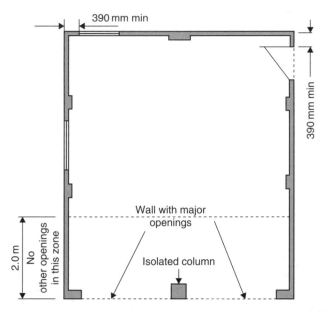

Figure 6.46 Size of openings in small single-storey, single-leaf buildings

No more than two major openings (maximum height 2.1 m, maximum width 5.0 m) are allowed in any one wall of the building or extension, either a single opening or the combined width of two openings.	D 1D40 ND 1D40
No other openings shall be within 2.0 m of a wall containing a major opening.	D 1D36 ND 1D36
The total size of openings in a wall not containing a major opening should not exceed 2.4 m².	D 1D36 ND 1D36
There should not be more than one opening between piers.	D 1D36 ND 1D36
The distance from a window or a door to a corner should be at least 390 mm unless there is a corner pier.	D 1D36 ND 1D36

Wall thicknesses and recommendations for piers

Walls without a major opening

- the walls should be at least 90 mm thick;   D 1D41
- pier sizes (Ap × Bp) should be at least 390 × 190 mm or
  327 × 215 mm depending on the size of the masonry units;
- isolated columns should be at least 325 × 325 mm
  (CC × CC).

Walls which do not contain a major opening but are more than   D 1D41
2.5 m long or wide should be bonded or tied to piers for their full
height at not more than 3 m centres as shown in Figure 6.47.

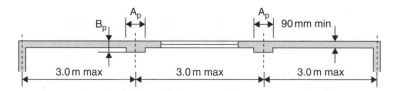

Figure 6.47 Walls without a major opening

Walls with one or two major openings

Walls with one or two major openings should, in addition,   D 1D41
have piers oriented as shown in Figure 6.48.

Ties used to connect piers to walls should be: D 1D41

- flat, 20 × 3 mm in cross-section;
- placed in pairs and spaced no more than 300 mm centres
  vertically.

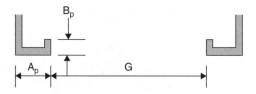

Figure 6.48 Walls with one or two major openings

Timber-frame walls

In Scotland, more and more use is being made of small, timber-frame buildings with external masonry cladding as they can more easily be designed to

take into account loading conditions and/or restrictions on dimensions and openings.

These types of buildings typically consist of full height timber wall panels for each storey built on to the floor below and with intermediate floors built on top of the wall panels. The roof is constructed on top of the top-storey wall panels with the masonry cladding connected to the timber panels by wall ties.

The following guidance is for timber-framed wall constructions for domestic buildings of not more than two storeys.

General

The timber-frame walls are the actual load-bearing parts of the walls whilst the individual parts act as:

- wind walls that resist the wind loads acting which act on the walls facing the wind;
- racking walls that support these walls by means of the racking resistance of the sheathing;
- walls studs that support the vertical loads (snow, floor loads and self-weight of the roof, floors and upper timber floors) and wind loads;
- cripple studs and lintels frame openings that carry the loads to the floor below;
- floors and roofs that support the walls horizontally top and bottom;
- walls, roofs and floors that act as a robust three-dimensional structural box giving overall stability to the building;
- masonry cladding that provides secondary support to the sheathing;
- internal walls that can provide additional racking resistance.

 Note: A procedure which can be used to determine the member sizing for timber-frame wall construction is summarized in the flow chart shown on 1E9 of the Domestic Technical Handbook.

Internal, party and external wall dead loads should not exceed than 1.5 kN/m excluding masonry cladding.	D 1E2i

Minimum thicknesses of external cavity walls

Masonry-clad, timber-frame walls should comprise:	D 1E7

- masonry cladding at least 100 mm thick;
- 50 mm nominal cavity width;
- 9 mm nominal sheathing thickness
- timber studs at least 89 mm depth;
- inner wall lining.

Walls providing vertical support to other walls

The thickness of the wall should be at least that of any part of the wall to which it gives vertical support – regardless of the materials used in the construction.   D 1D9   D 1E8

Wall sheathing

The wind loads are resisted and transferred to the base of the walls by the resistance of the racking (or wind) wall panels. This is achieved by the sheathing to the external wall panels although internal walls can also be used in certain circumstances.

Walls should only be considered to be racking walls (and hence able to provide resistance to wind loads) if they are designed in accordance with the Regulations and have at least one layer of plywood sheathing or oriented strand board (OSB) secured to the timber studs.   D 1E10

External walls which are perpendicular to the walls that are subject to the wind load should be designed as racking walls.   D 1E43

 Note: Internal walls can be used to provide additional racking resistance.

Racking resistance

The wind load on an external wall is supported by the racking resistance of external walls that are perpendicular to that wall.

The racking resistance of the perpendicular walls is 50% of the wind load (as shown in Figure 6.49) and is dependent on:

- the overall building height from ground level to ridge;
- the roof shape;
- the length/width ratio;
- the wind speed;
- the altitude/distance category.

 Guidance on the racking procedure is provided in the Domestic Technical Handbook clauses 1E11–1E21 and this is shown in outline in Figure 6.50.

Wall stud sizing

The wall studs carry the vertical and wind loads imposed on the timber-frame panels and the sizes of studs should be selected using the tables shown in Clause 1E22 of the Domestic Technical Handbook.   1E20–22

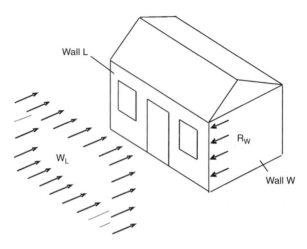

Key:
WL: = wind load on to long wall L
RW = racking resistance from wall W, supporting 50% of the total wind load, WL, on wall L

Figure 6.49 Racking loads

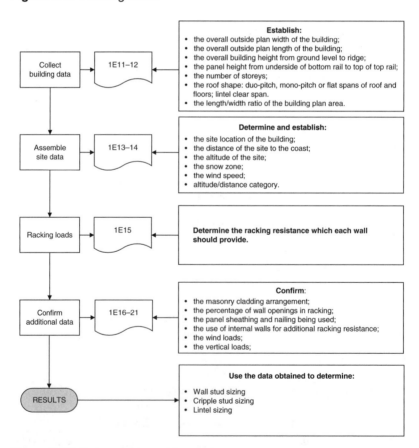

Figure 6.50 Determining racking resistance

Construction materials

Construction materials and methods are restricted to those materials, timber strength classes, specifications and dimensions which are most commonly used in Scotland for simple platform timber-frame buildings (see Figure 6.51).

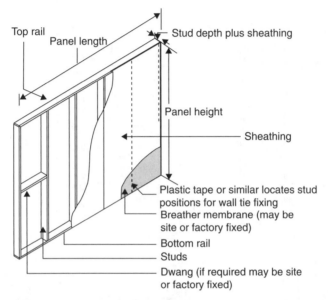

Figure 6.51 Typical timber-frame wall panel

Wall panels

Panel heights are measured from the underside of the bottom rail to the top of the top rail as shown in Figure 6.52.

Sandwich panels

A sandwich panel used for internal walls or linings in a domestic building should be fully filled with a core of non-combustible material.	D 2.5.8
A sandwich panel used for internal walls or linings in a non-residential building should be designed and installed in accordance with Chapter 8 of 'Design, Construction, Specification and Fire Management of Insulated Envelopes for Temperature Controlled Environments', International Association of Cold Storage Contractors (European Division), 1999.	ND 2.5.8

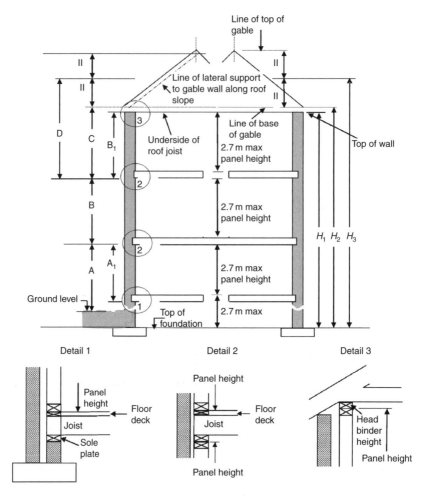

Figure 6.52 Panel heights

 Note: A sandwich panel is a factory-made non-load-bearing component of a wall, ceiling or roof consisting of a panel having an insulated core filling the entire area between sheet metal outer facings, which may or may not have decorative and/or weatherproof coatings.

Wall ties

Wall ties are stainless steel nails and brackets installed to tie back and brace the external brick cladding to the timber frame. Wall ties provide lateral restraint, but do not carry the weight of the wall which is self-supporting.

Cavity walls

Wall ties should be: D 1D8

- spaced not more than 300 mm apart vertically;
- with horizontal centres of 900 mm and vertical centres of
  450 mm (i.e. equivalent to 2.5 ties per square metre);
- within 225 mm from the vertical edges of all openings,
  movement joints and roof verges.

 For a selection of wall ties that can be used in a range
of cavity widths, see Section D 1D17 of the Domestic
Technical Handbook.

Masonry-clad, timber-frame walls

Wall ties should be nailed to the vertical studs (and not to the   D 1E7
sheathing) at the following centres:

- not more than 300 mm centres vertically, within a distance
  of 225 mm from the vertical edges of all openings,
  movement joints and roof verges;
- brickwork cladding: horizontal centres of 600 mm and
  vertical centres of 375 mm;
- blockwork cladding: horizontal centres of 400 or 600 mm
  and vertical centres of 450 mm.

Masonry cladding

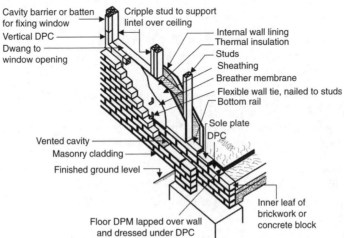

Figure 6.53 Typical masonry-clad timber-frame wall

Brick and block used as masonry cladding should be at least 100 mm thick with a minimum density of 7.36 kN/m³.   D 1E31

Proprietary steel or concrete lintels used with masonry cladding to timber-frame construction should be tested by a notified body or justified by calculations.   D 1E33

Under **NO** circumstances should any part of the masonry walls be supported by the timber frame.

All structural timber members should be dry graded and marked with the timber species and grade combinations.   D 1E34

The minimum thickness of plywood used as wall sheathing to timber frames should be 9.5 mm.   D 1E34

Oriented strand board used as wall sheathing to timber frame should be at least 9.0 mm thick.   D 1E34

Plasterboard used as wall linings should be 12.5 mm minimum thickness for stud centres not more than 600 mm.   D 1E34

Fasteners: D 1E36

- all structural fasteners should be corrosion resistant and checked for compatibility with preservative treatments used and any other metalwork with which they are in contact;
- nails should be manufactured from mild or stainless steel and be of round head or 'D' head configuration;
- ground floor fasteners should be stainless steel or galvanized.

Timber members in wall panels should be at least 38 × 89 mm rectangular section with linings fixed to the narrower face, with ends cut square.   D 1E37

Wall studs should be spaced at not more than 600 mm centres and should be vertically aligned to coincide with the floor joists and roof trusses.   D 1E37

Masonry cladding should be constructed on to the building foundation and tied back to the timber-frame structure with a cavity width of 50 mm between the inside face of the masonry cladding and the outer face of the timber-frame wall.   D 1E37

Composite action: D 1E38
Sheathings and linings shall be nailed to all perimeter and intermediate timber members as shown on Figure 6.54 as follows:

- sheathing edges should be backed by and nailed to timber framing at all edges;   D 1E38a

- where sheathing is nailed to studs, the nails should be at least 7 mm from the edge of the board or the face of the stud;   D 1E38b

- for plasterboard linings, nails should be at least 10 mm from formed board edges and at least 13 mm from ends of the board at centres not more than 150 mm;   D 1E38c

- internal walls which are lined with plasterboard should be connected to the wall studs at the same perimeter nail centres as for external sheathing material;   1D E38d

- fixing of perimeter studs to sheathing should be at the centres (see Figure 6.19);   D 1E38e

- fixing of intermediate studs to sheathing should be at not more than twice the centres of the perimeter nailing (see Figure 6.19).   D 1E38f

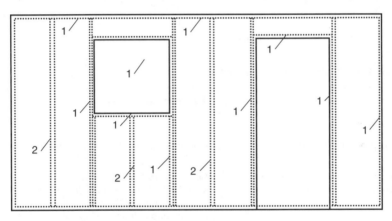

Figure 6.54 Perimeter and intermediate nailing diagram

Wall panel connections

To ensure that wall panels are able to resist overturning forces, they should be combined to form lengths of wall as follows:

- tops and bottoms of individual wall panels should be linked by head binders and sole plates, respectively, that are continuous across panel joints including at junctions of the same dimensions as the top and bottom rails;   D 1E39a

- sole plates should be secured to either the concrete floor slab or the header joists in the case of a timber ground floor or the header joists of the intermediate floor;   D 1E39b

- header plates should be secured to the header joists of the intermediate floor or the roof trusses;   D 1E39c
- faces of end studs of contiguous panels should be fixed such that any vertical shear is transferred;   D 1E39d
- all edges (including those to openings for windows, doors, etc., but other than at the bases of door openings and small openings) should be supported by timber members having a thickness not less than the thickness of the studs;   D 1E39e
- nail lengths should be increased to take account of the additional thickness where a secondary board is fixed on the same side of a wall as the primary sheathing;   D 1E39f
- panels above and below openings should be fixed so that the horizontal forces are transferred in the plane of the panel above and below openings by 3.35 mm nails of length 75 mm at 300 mm centres.   D 1E39g

Nailing and fixing schedule

Table 6.45   Nailing and fixing schedule

Item	Recommended fixing
Foundations	
Sole plate to under building	Mechanical fixings at 600 mm centres rated at 4.7 kN shear resistance
Holding down straps providing at least 3.5 kN of resistance	Stainless steel strap 30 mm × 2.5 mm attached to stud by 6 no. 3.36 mm × 65 mm ring shank nails at 2.4 m centres, at every opening and at the end studs of a wall attaching the strap to the stud and placing the L-shaped end of the strap under the masonry cladding creating the holding down resistance
Wall panels	
Top rail of panels to head binders	Tops of individual wall panel members linked by member continuous across panel joints secured with 4.0 × 90 mm galvanized wire nails, two nails between stud centres
Sole plate to ring beam/joist	4.0 × 90 mm galvanized wire nails, two nails between stud centres
Bottom rail to sole plate	4.0 × 90 mm galvanized wire nails, two nails between stud centres
Wall panel stud to wall panel stud	4.0 × 90 mm galvanized wire nails at 600 mm centres each side staggered

(Continued)

Table 6.45 (Continued)

Item	Recommended fixing
Wall panels	
Header plate to intermediate floor	4.0 × 90 mm galvanized wire nails at 300 mm centres. Nails skewed externally through rimboard into headbinder and internally skewed through the headbinder into the joists
Sheathing to perimeter studs	3.1 × 50 mm wire nails at 100 or 150 mm centres as calculated
Intermediate studs to sheathing	3.1 × 50 mm wire nails at twice perimeter centres
Studs to plasterboard	2.65 × 40 mm smooth shanked galvanized flat round headed nails at 150 mm centres
Top and bottom rails to studs	4.0 × 90 mm nails end fixed
Spandrel panels to wall panel centres	4.0 mm × 90 mm galvanized wire nails, two nails between stud head

Other loading conditions

Combined dead load and imposed load should be not more than 70 kN/m at the base of the wall.   D E42

Vertical loading on walls from timber floors and flat roofs should be designed in accordance with Annex 1.F of the Domestic Technical Handbook.   D 1E42

Note: Timber-frame walls should commence above ground level and are, therefore, not subject to lateral loads other than from wind.

End restraint

The ends of every wall should be securely tied to the walls which are providing the racking resistance.   D 1E43

Openings, notching and drilling

The number, size and position of openings should not impair the stability of a wall or the lateral support afforded to a supported wall.   D 1E44

Construction over openings should be supported.   D 1E44

Framing of openings

Loads over openings in timber frame wall panels are carried   D 1E45
independently by timber lintels which should be supported by
cripple studs as shown in Figure 6.55.

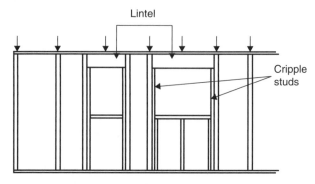

Figure 6.55 Openings in timber-frame wall panels. Reproduced by
permission of TRADA

Openings in walls below ground floor should be limited to   D 1E46
small holes for services and ventilation, etc., which are not
more than $0.1\,\text{m}^2$ and at least $2\,\text{m}$ apart.

Lateral support by roofs and floors

The wall panels in each storey of a building should: D 1E49

- extend to the full height of that storey;
- be connected to the floors and roofs to provide diaphragm
  action;
- transfer lateral forces from the walls to the racking walls.

The floor deck of intermediate floors should be fixed directly   D 1E49a
to the top faces of the joists.

The plasterboard ceiling of the top storey of pitched roofs   D 1E49b
should be:

- fixed directly under the roof;
- secured to the supported wall.   D 1E49c

Spandrel panels should be tied into roof bracing with dwangs   D 1E39d
placed between vertical elements of the spandrel and fixed by
at least $3.1 \times 75\,\text{mm}$ screws as shown in Figure 6.56.

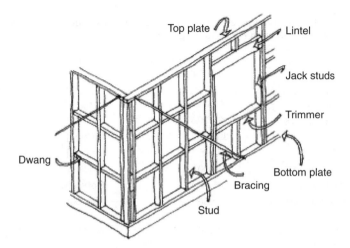

Figure 6.56 Roof bracing with dwangs

Compartmentation

 The requirements for compartmentation given below do *not* apply to domestic buildings.

The aim of compartmentation is to inhibit rapid fire spread within the building by reducing the fuel available in the initial stages of a fire. The intention is to limit the severity of the fire which in turn should help the occupants to evacuate the building and assist fire service personnel with fire-fighting and rescue operations. This is achieved by dividing the building into a series of fire-tight boxes, termed compartments, which will form a barrier to the products of combustion: smoke, heat and toxic gases.

A number of proprietary fire-stopping and sealing systems (including those designed for service penetrations) which have been shown by test to maintain the fire resistance of a wall are available, as well as other fire-stopping materials such as:

- cement mortar;
- gypsum-based plaster;
- cement or gypsum-based vermiculite/perlite mixes;
- glass fibre, crushed rock, blast furnace slag or ceramic-based products (with or without resin binders);
- intumescent mastics.

Maximum compartment areas

In non-domestic buildings (or parts of a building) with a total storey area greater than the limits given in the Tables 6.46 and 6.47, the building should be subdivided by compartment walls and, where appropriate, compartment floors.

The minimum fire resistance duration can also be obtained from these tables.

Table 6.46   Single-storey buildings and compartmentation between single-storey and multi-storey buildings where appropriate

Building use	Maximum total area of any compartment (m^2)	Minimum fire resistance duration for compartmentation (if any)
Assembly building	6,000	Long
Entertainment building	2,000	Medium
Factory (Class 1)	33,000	Long [3]
Factory (Class 2)	93,000	Long [3]
Office	4,000	Medium
Open-sided car park	Unlimited	Not relevant
Residential care building, hospital	1,500	Medium
Residential building (other than a residential care building and hospital)	2,000	Medium
Shop	2,000 [2]	Long
Storage building (Class 1)	1,000 [1]	Long
Storage building (Class 2)	14,000 [1]	Long [3]

Notes:
1. Areas may be doubled where there is an automatic fire suppression system.
2. Unlimited provided there is an automatic fire suppression system.
3. A medium fire resistance duration compartment wall or compartment floor may be provided between the single-storey part and the multi-storey part provided the multi-storey part does not exceed the limitations for medium fire resistance duration in the following table covering multi-storey buildings.

Note: In most cases, a single-storey building poses less of a life risk to the occupants or to fire service personnel than a multi-storey building, therefore a greater compartment size can be constructed.

Buildings with different uses

Buildings with different uses should be divided by compartment walls.	ND 2.1.4

Places of special fire risk

A place of special fire risk should be enclosed by compartment walls with medium fire resistance duration.	ND 2.1.8

Table 6.47 Multi-storey buildings

Building use	Maximum total area of any compartment (m²)	Minimum fire resistance duration for compartmentation (if any)			
		Basements	The topmost storey of a building is at a height of not more than 7.5 m above ground	The topmost storey of a building is at a height of not more than 18 m above ground	The topmost storey of a building is at a height of more than 18 m above ground
Assembly building	1,500 (1)	Medium	Short	Medium	Long [2]
	3,000 (1)	Medium	Medium	Medium	Long (2)
	6,000 (1)	Long	Long	Long	Long
Entertainment building	1,000 (1)	Medium	Short	Medium	Long (2)
	2,000 (1)	Medium	Medium	Medium	Long (2)
	4,000 (1)	Long	Long	Long	Long
Factory (Class 1)	500 (1)	Medium	Medium	Medium	Long (2)
	6,000 (1)	Long	Long	Long	Long
Factory (Class 2)	200 (1)	Medium (4)	Medium (4)	Medium (4)	Medium (4)
	15,000 (1)	Long	Long	Long	Long
Office	2,000 (1)	Medium (4)	Short	Medium (4)	Long (2)
	4,000 (1)	Medium (4)	Medium (4)	Medium (4)	Long (2)
	8,000 (1)	Long	Long	Long	Long
Open-sided car park	Unlimited	Medium	Short	Short	Medium
Residential care building	1,500	Medium	Medium	Medium	Long (2)

(Continued)

Table 6.47 (Continued)

Building use	Maximum total area of any compartment (m²)	Minimum fire resistance duration for compartmentation (if any)			
		Basements	The topmost storey of a building is at a height of not more than 7.5m above ground	The topmost storey of a building is at a height of not more than 18m above ground	The topmost storey of a building is at a height of more than 18m above ground
Residential building (other than a residential care building and/or hospital)	1,000	Medium	Short	Medium	Long (2)
Shop	2,000	Medium	Medium	Medium	Long (2)
	500	Medium	Short	Medium (4)	Long (2)
	1,000	Medium (4)	Medium (4)	Medium (4)	Long (2)
	2,000	Long	Long	Long	Long
Storage building (Class 1)	200 (1)	Medium	Medium	Medium	Long (2)
	1,000 (1)	Long	Long	Long	Long
Storage building (Class 2)	500	Medium (4)	Medium (4)	Medium (4)	Medium (4)
	5,000	Long	Long	Long	Long

Where a place of special fire risk contains any appliance or   ND 2.1.8
equipment using hazardous liquid, any opening in a wall
dividing it from the remainder of the building should be
constructed so that, in the event of any liquid spillage,
the room will contain all the liquid in the appliance or
equipment, plus 10%.

Smoke venting shafts

A smoke venting shaft should be enclosed by compartment   ND 2.1.10
walls with medium fire resistance duration, other than at the
smoke inlets and smoke outlets to the shaft.

Lift wells

A lift well should be enclosed by compartment walls with   ND 2.1.11
medium fire resistance duration.

Combustibility

A compartment wall can be constructed of combustible   ND 2.1.12
materials having a low, medium, high or very high risk
provided the compartment wall has the appropriate fire
resistance duration.

 However, compartment walls in hospitals should be
constructed from materials which are non-combustible.

Supporting structure

Where an element of structure provides support to a   ND 2.1.13
compartment which attracts higher fire resistance duration,
the supporting element of structure should have at least the
same period of fire resistance.

Openings and service penetrations

Compartment walls are intended to prevent fire passing from one compartment to another. Openings and service penetrations through these walls can
compromise their effectiveness and should be kept to a minimum.

Openings and service penetrations should be carefully detailed and constructed to resist fire.	ND 2.1.14
A self-closing fire door with the same fire resistance duration as the compartment wall should be installed in accordance with the recommendations in the Building Hardware Industry Federation, Code of Practice, 'Hardware for Timber Fire and Escape Doors' Issue 1, November 2000.	ND 2.1.14

> For metal doorsets, reference should be made to the 'Code of Practice for fire resisting metal doorsets' published by the Door and Shutter Manufacturers' Association, 1999.

A lockable door to a cupboard or service duct where the cupboard or the service duct has a floor area not more than $3\,m^2$ need not be self-closing.	ND 2.1.14
Self-closing fire doors can be fitted with hold-open devices as specified in BS 5839: Part 3: 1988 **provided** the door is not:	ND 2.1.14

- an emergency door;
- a protected door serving the only escape stair in the building (or the only non-escape stair serving part of the building);
- a protected door serving a fire-fighting shaft.

Electrically operated hold-open devices should deactivate:	ND 2.1.14

- on operation of an automatic fire alarm system;
- for any loss of power to the hold open device, apparatus or switch;
- via a manually operated switch fitted in a position at the door.

All fire shutters in compartment walls should be capable of being opened and closed manually by fire service personnel.	ND 2.14.6

Fire shutters

A fire shutter which is not motorized may be fitted in a compartment provided the shutter has the same fire resistance duration as the compartment wall.	ND 2.1.14

Service opening

A service opening (other than a ventilating duct) which penetrates a compartment wall or compartment floor should be fire-stopped providing at least the appropriate fire resistance duration for the wall or floor.	ND 2.1.14

Where a pipe connects to another pipe which attracts more | ND 2.1.14
demanding fire resistance duration, and is within 1 m from
the compartment wall, the pipe should be fire-stopped to the
more demanding guidance.

Ventilating ducts

A ventilating duct passing through a compartment wall | ND 2.1.14
should be fire protected.

Junctions

The basic principle is that junctions between compartment walls and compartment floors (including fire resisting ceilings) and other parts of the building should be designed and constructed to prevent a fire in one compartment flanking the wall, floor or ceiling and entering an other compartment at the junctions including any solum space or roof space.

Building elements, materials or components should not be | ND 2.1.15
built into, or carried through or across the ends of, or over
the top of a compartment wall in such a way as to impair the
fire resistance between the relevant parts of the building.

Where a compartment wall forms a junction with an external | ND 2.1.15
wall, a separating wall, another compartment wall or a wall
or screen used to protect routes of escape, the junction should
maintain the fire resistance of the compartment wall.

If a compartment wall forms a junction with a roof, the junction should maintain the fire resistance duration of the compartment wall in accordance with the following:

- where the roof has a combustible substrate, the wall | ND 2.1.15
  should project through the roof to a distance of at least
  375 mm above the top surface of the roof;
- where the wall is taken to the underside of a non-
  combustible roof substrate, the junction should be
  fire-stopped and the roof covering should be low
  vulnerability for a distance of at least 1.7 m to each
  side of the centre-line of the wall;

- in the case of a pitched roof covered by slates nailed directly to sarking and underlay, the junction between the sarking and wall-head should be fire-stopped as described in BRE Housing Defects Prevention Unit Defect Action Sheet (Design) February 1985 (DAS 8);
- in the case of a pitched roof covered by slates or tiles fixed to tiling battens and any counter-battens, the junction between the tiles or slates and the underlay should be fully bedded in cement mortar (or other fire-stopping material) at the wall-head.

Where a compartment wall (or sub-compartment wall) does not extend to the full height of the building, the wall should form a junction with a compartment floor.　ND 2.1.15

If a localized capping system is constructed at the head of a protected zone or a lift shaft, then the system should be fire protected on both sides and be sufficiently robust to protect against premature collapse of the roof structure.　ND 2.1.15

Separating walls

In dwellings

Fire separation should be provided between dwellings, and between dwellings and any common spaces.　D 2.2.0

Fire separation should form a complete barrier to the products of combustion: smoke, heat and toxic gases.　D 2.2.0

In semi-detached or terraced houses, or between flats and maisonettes, the barrier will normally be in the form of fire-resisting walls and floors where appropriate.　D 2.2.0

At dwellings in different occupation, a separating wall or separating floor with medium fire resistance duration should be provided between adjoining dwellings.　D 2.2.1

A separating wall or separating floor:　D 2.2.2　D 2.2.3

- with medium fire resistance duration should be provided between a dwelling and any other part of the building in common occupation;
- with a medium fire resistance duration provided between a domestic and non-domestic building;

- with a short fire resistance duration provided between solid waste storage accommodation and the rest of the building;   D 2.2.5
- with a short fire resistance duration provided between an integral or attached garage and a dwelling in the same occupation.   D 2.2.6

Every lift well should be enclosed by separating walls with medium fire resistance duration.   D 2.2.6

Where the lift well does not extend the full height of the building, the lift well should form a junction with a separating floor with medium fire resistance duration.   D 2.2.6

Where a lift is installed, the landing controls and lift car controls should be of a type that do not operate on heat or pressure resulting from a fire.   D 2.2.6

In a building with no storey at a height above 18 m, separating walls may be constructed from combustible materials provided the appropriate fire resistance duration is maintained.   D2.2.7

To reduce the risk of a fire starting within a combustible separating wall or a fire spreading rapidly on or within the wall construction: D 2.2.7

- insulation material exposed in a cavity should be constructed from materials which are non-combustible or of a low risk classification;
- the internal wall linings should be constructed from materials which are non-combustible or of a low risk classification;
- the wall should contain no pipes, wires or other services.

Where a domestic building also contains non-domestic accommodation, every part of a separating floor (other than floor finish, e.g. laminate flooring) should be constructed from non-combustible material.   D 2.2.7

In non-domestic buildings

Separating walls and separating floors should have medium fire resistance duration.   ND 2.2.1

For buildings in different occupation: ND 2.2.2

- a separating wall or separating floor should be provided between parts of a building where they are in different occupation;

- if each unit is under the control of an individual tenant, employer or self-employed person, then separating walls and separating floors should be provided between the areas intended for different occupation;
- if this is not possible, then the building should have a common fire alarm system/evacuation strategy and the same occupancy profile.

A separating wall or separating floor with medium fire resistance duration should be provided between parts of a building where one part is in single occupation and the other is in communal occupation.   ND 2.2.3

Every part of a separating wall or separating floor (other than a floor finish such as laminate flooring) should be of materials that are non-combustible and to reduce the risk of a fire starting within a combustible separating wall or a fire spreading rapidly on or within the wall construction: ND 2.2.4

- insulation material exposed in a cavity should be of low risk or non-combustible materials;
- the internal wall lining should be constructed from material which is low risk or non-combustible;
- the wall should contain no pipes, wires or other services.

Supporting structure

Separating walls and separating floors are intended to prevent fire passing from one part of the building to another part under different occupation.

In domestic buildings

Where an element of structure provides support to: D 2.2.8

- a non-combustible separating wall, the supporting element of structure should also be constructed from materials which are non-combustible;
- a separating wall which attracts a higher fire resistance duration, the supporting element of structure should have at least the same fire resistance duration.

In non-domestic buildings

Where an element of structure provides support to a non-combustible separating wall or separating floor, the supporting element of structure should: ND 2.2.5

- also be constructed from non-combustible materials;
- have at least the same period of fire resistance as the separating wall/floor it supports.

Openings and service provisions

Openings and service penetrations through separating walls can compromise their effectiveness and should be kept to a minimum. Openings and service penetrations should be carefully detailed and constructed to resist fire as shown below:

 The solum and roof space should not be overlooked!

In domestic buildings

A self-closing fire door with the same fire resistance duration as the separating wall should be installed.	D 2.2.9
A self-closing fire door should not be fitted in a separating wall between two dwellings in different occupation.	D 2.2.9
A lockable door to a cupboard or service duct with a floor area not more than $3\,m^2$ need not be self-closing.	D 2.2.9
Self-closing fire doors can be fitted with hold-open devices as specified in BS 5839: Part 3: 1988 provided the door is not an emergency door, a protected door serving the only escape stair in the building (or the only escape stair serving part of the building) or a protected door serving a fire-fighting shaft.	D 2.2.9
Hold-open devices shall deactivate on operation of the fire alarm.	D 2.2.9
Electrically operated hold-open devices should deactivate on operation of: • an automatic fire alarm system • any loss of power to the hold-open device, apparatus or switch; • a manually operated switch fitted in a position at the door.	D 2.2.9
A chimney or flue-pipe should be constructed so that, in the event of a fire, the fire resistance duration of the separating wall is maintained.	D 2.2.9
A service opening (other than a ventilating duct) which penetrates a separating wall should be fire-stopped providing at least the appropriate fire resistance duration for the wall or floor.	D 2.2.9

Where a pipe connects to another pipe which attracts more demanding fire resistance duration, and is within 1 m from the separating wall, the pipe should be fire-stopped to the more demanding guidance.	D 2.2.9
A ventilating duct passing through a separating wall should be fire-stopped in accordance with section 6 of BS 5588: Part 9: 1999.	D 2.2.9

In non-domestic buildings

Self-closing fire doors should not be installed: • in separating walls other than in buildings with common occupation; • where the building is in the same occupation, but in different use.	ND 2.2.6
A fire shutter should not be installed in a separating wall.	ND 2.2.6

Junctions
In domestic buildings

Junctions between separating walls and other parts of the building should be designed and constructed in such a way to prevent a fire in one part of the building flanking the separating wall and entering another part of the building under different occupation, including any solum space or roof space.	D 2.2.10
Where a separating wall forms a junction with an external wall, another separating wall, or a wall or screen used to protect routes of escape, the junction should maintain the fire resistance of the separating wall.	D 2.2.10
Where a separating wall forms a junction with a roof, the junction should maintain the fire resistance duration of the separating wall in accordance with the following: • where the roof has a combustible substrate, the wall should project through the roof to a distance of at least 375 mm above the top surface of the roof; • where the wall is taken to the underside of a non-combustible roof substrate, the junction should be fire-stopped and the roof covering should be at low vulnerability for a distance of at least 1.7 m to each side of the centre-line of the wall;	D 2.2.10

- in the case of a pitched roof covered by slates, it should be nailed directly to sarking and underlay:
  - the junction between the sarking and wall-head should be fire-stopped;
- in the case of a pitched roof covered by slates or tiles fixed to tiling battens and any counter-battens, the junction between the tiles or slates and the underlay should be fully bedded in cement mortar at the wall-head.

In non-domestic buildings

Where a separating wall or separating floor meets an external wall, another separating wall, a compartment wall or any other wall or screen used to protect routes of escape, the junction should maintain the fire resistance duration of the separating wall or separating floor.   ND 2.2.7

Cavities

Fire and smoke in concealed spaces are particularly hazardous as fire can spread quickly throughout a building and remain largely undetected by the occupants of the building or by fire service personnel. Ventilated cavities generally encourage the rapid spread of fire around the building more than unventilated cavities owing to the plentiful supply of replacement air. Buildings containing sleeping accommodation pose an even greater risk to life safety and demand a higher level of fire precautions. For these reasons, it is important to control the size of cavities and the type of material in the cavity.

Cavity barriers

In order to inhibit fire spread in a cavity: D 2.4.1
                                                                            ND 2.4.1
- every cavity within a building should have cavity barriers with at least short fire resistance duration installed around the edges of the cavity (e.g. around the head, jambs and sill of an external door or window opening);
- cavity barriers should be installed between a roof space and any other roof space; and between a cavity and any other cavity (e.g. at the wall-head between a wall cavity and a roof space cavity).

A cavity barrier is any construction provided to seal a cavity against the penetration of fire and smoke or to restrict its movement within the cavity.

Dividing up cavity barriers

Every cavity should be divided by cavity barriers so that the maximum distance between cavity barriers is:	D 2.4.2 ND 2.4.2

- not more than 20 m where the cavity has surfaces which are non-combustible or low-risk materials;
- 10 m where the cavity has surfaces which are medium-, high- or very high-risk materials.

Cavity barriers are not necessary to divide a cavity:	D 2.4.2 ND 2.4.2

- formed by two leaves of masonry or concrete at least 75 mm thick;
- in a ceiling void between a floor and a fire-resisting ceiling;
- between a roof and a fire-resisting ceiling;
- below a floor next to the ground where the cavity is either inaccessible or is not more than 1 m high;
- formed by external wall or roof cladding, where the exposed surfaces of the cladding are low-risk materials or non-combustible materials attached to a masonry or concrete external wall or a concrete roof, and where the cavity contains only non-combustible material.

In non-domestic buildings, cavity barriers are not necessary to divide a cavity between a floor which is an element of structure (see Clause 2.3.0) and a raised floor consisting of removable panels.	ND 2.4.2f

Cavities above ceilings in residential buildings (other than residential care buildings and hospitals)

In non-domestic buildings, where a roof space cavity or a ceiling void cavity extends over a room intended for sleeping, or over such a room and any other part of the building, cavity barriers should be installed on the same plane as on the wall.	ND 2.4.4

Junctions

All cavity barriers should be tightly fitted to a rigid construction.	D 2.4.7 ND 2.4.9
At junctions with slates, tiles, corrugated sheeting or similar materials, the junction should be fire-stopped.	D 2.4.7 ND 2.4.9

A cavity barrier should be installed in all walls, floors
or other parts of a building which have a fire resistance
duration and which are adjacent to a structure containing
a cavity unless the cavity is:

D 2.4.7
ND 2.4.9

- formed by two leaves of masonry or concrete at least
  75 mm thick;
- formed by external wall or roof cladding, where the
  surfaces of the cladding are non-combustible or low-risk
  materials and attached to a masonry or concrete external
  wall or a concrete roof, and where the cavity contains
  only non-combustible or low-risk material;
- in a wall which has a fire resistance duration for load-
  bearing capacity only.

Internal linings

Materials used in walls can significantly affect the spread of fire and its rate
of growth. Fire spread on internal linings in escape routes is particularly
important because rapid fire spread in protected zones and unprotected zones
could prevent the occupants from escaping.

Every room, fire-fighting shaft, protected zone or unprotected
zone should have wall surfaces whose reaction to fire is in
accordance with Table 6.48.

ND 2.5.1

Table 6.48   Reaction to fire of wall and ceiling surfaces

Building	Residential care buildings and hospitals	Shops	All other buildings
Room not more than 30 m²	Medium risk [2]	High risk	High risk
Room more than 30 m²	Low risk [3]	Medium risk [4]	Medium risk
Unprotected zone	Low risk	Low risk	Medium risk
Protected zone and fire-fighting shaft [1]	Low risk	Low risk	Low risk

Notes:
1. Including any toilet or washroom within a protected zone.
2. High risk in a room not greater than 4 m².
3. Ceilings may be medium risk.
4. Low risk in storage buildings (Class 1).

Fire resistance of external walls

In order to reduce the danger to the occupants of other buildings, one building should be isolated from another by either construction or distance – the distance between a building and its relevant boundary being dictated by the amount of heat that is likely to be generated in the event of fire, influenced by the extent of openings or other unprotected areas in the external wall of the building.

The installation of an automatic fire suppression system greatly reduces the amount of radiant heat flux from a fire through an unprotected opening.

External walls should have: 2.6.1

- short fire resistance duration, if more than 1 m from the boundary;
- medium fire resistance duration, if less than 1 m from the boundary except for:
  - a detached building such as a garden hut or store;
  - a building for keeping animals, birds or other livestock for domestic purposes;
  - a conservatory or porch attached to a dwelling;
  - a garage wall.

Fire resistance duration need not be provided for an   2.6.1
additional building to a dwelling (e.g. such as a carport, covered area, greenhouse, summerhouse or swimming pool enclosure), unless the building contains oil or liquefied petroleum gas fuel storage.

In non-domestic buildings

External walls should have at least the fire resistance   ND 2.6.1
duration as shown in Table 6.49.

Unprotected area

Where the external wall of a building is not more than 500 mm   D 2.6.2
from the boundary there should be no unprotected area, other than any wallhead fascia, soffit or barge board, or any cavity vents or solum vents.

Where the external wall of a building is more than 500 mm but   D 2.6.2
not more than 1 m from the boundary, the level of unprotected area is limited to:

- the external wall of a protected zone;
- an area of not more than 0.1 m², which is at least 1.5 m from any other unprotected area in the same wall;

- an area of not more than $1\,m^2$, which is at least $4\,m$ from any other unprotected area in the same wall (this $1\,m^2$ unprotected area may consist of two or more smaller areas which when combined do not exceed an aggregate area of $1\,m^2$).

In non-domestic buildings, where the external wall of a building is more than 1 m from the boundary, the minimum distance to the boundary may vary with the amount of unprotected area.

Table 6.49   Recommended fire resistance duration of external walls

Use of building	Not more than 1 m from the boundary		More than one 1 m from the boundary	
	No fire suppression system	Fire suppression system	No fire suppression system	Fire suppression system
Assembly building	Medium	Medium [2c]	Medium [1, 2c]	None
Entertainment building	Medium	Medium [2b]	Medium [2b]	Medium [1]
Factory and storage building	Medium	Medium	Medium [3]	Medium [1, 4]
Residential building (other than a residential care building or hospital)	Medium	Medium [2b]	Medium [1, 2b]	None
Residential care building and hospital	Medium	Medium	Medium	None
Shop	Medium	Medium [2a]	Medium [2a]	Medium [1]
Office	Medium	Medium [2d]	Medium [1, 2d]	None
Open-sided car park	Short	Short	None	None

Notes:
1. Short fire resistance duration is sufficient where the building is single-storey.
2. Short fire resistance duration is sufficient where the building is a multi-storey building and the area of any compartment does not exceed: (a) $500\,m^2$ (b) $1,000\,m^2$ (c) $1,500\,m^2$ (d) $2,000\,m^2$.
3. Short fire resistance duration is sufficient where the building is a factory (Class 2) and is single-storey.
4. No fire resistance duration is necessary where the building is a factory (Class 2).

External wall cladding

External wall cladding includes all non-load-bearing external wall cladding systems attached to the structure, for example, clay or concrete tiles, slates, pre-cast concrete panels, stone panels, masonry, profiled metal sheeting including sandwich panels, weather boarding, thermally insulated external wall rendered systems, glazing systems and all other ventilated cladding systems.

External wall cladding not more than 1 m from a boundary should be constructed of non-combustible material.	D 2.6.4

Combustibility

Every part of an external wall (including external wall cladding not more than 1 m from a boundary) should be constructed of non-combustible material.	D 2.6.5 ND 2.6.6

Supporting structure

The fire resistance duration of supporting elements of a structure should have the same fire resistance duration as the external wall structure supports.	D 2.6.6 ND 2.6.7

External walls

Although there is a risk of fire spread on the external walls of a building, for most buildings it is only necessary to consider this if the external wall is in close proximity to the boundary. Having said that:

- entertainment and assembly buildings should always be given special consideration owing to the high risk of wilful fire-raising against the external walls;
- residential care buildings and hospitals also present a greater risk because the mobility, awareness and understanding of the occupants could be impaired;
- in high-rise buildings, there is a need to take further precautions as external fire spread could involve a large number of floors thus presenting greater risk both to the occupants of the building and to fire-fighters;
- in domestic buildings, although there is a small risk of fire spread on the external walls of a building, for most buildings it is only necessary to consider this if the external wall is in close proximity to the boundary.

The reaction to fire characteristics of cladding materials are therefore more demanding the higher the building.

External wall cladding

External wall cladding includes all non-load-bearing external wall cladding systems attached to the structure, for example, clay or concrete tiles, slates, pre-cast concrete panels, stone panels, masonry, profiled metal sheeting including sandwich panels, weather boarding, thermally insulated external wall rendered systems, glazing systems and all other ventilated cladding systems.

External wall cladding (including any insulation core) not more than 1 m from a boundary should be constructed of non-combustible material.	D 2.7.1
External wall cladding constructed from combustible material more than 1 mm thick which is low, medium, high or very high risk and attached to the outside face of an external wall may be used provided the external wall is more than 1 m from the boundary.	D 2.7.1

In non-domestic buildings

If the cladding is more than 1 m from the boundary and is constructed from combustible material more than 1 mm thick, that has a low, medium, high or very high risk, the cladding should be constructed from materials with a reaction to fire in accordance with Table 6.50.	ND 2.7.1

Table 6.50   Reaction to fire of external walls more than 1 m from boundary

Building height	Building type	Location	Maximum level of risk
Not more than 18 m above the ground	Entertainment and assembly buildings	Not more than 10 m above the ground (or above a roof or any part of the building to which the general public have access)	Low risk
		10–18 m above the ground	Very high risk
	Residential care buildings and hospitals	Any	Low risk
	All other buildings	Any	Very high risk
More than 18 m above the ground	Any	Any	Low risk

A cavity formed by external wall cladding should be protected.   ND 2.7.1

If a building has storeys more than 18 m above the ground,   ND 2.7.2
then the insulation material situated or exposed in a cavity
formed by external wall cladding should be non-combustible.

Means of escape

Life safety is the paramount objective of fire safety. Everyone within a building should be provided with at least one means of escape from fire that offers a safe passage to a place of safety outside the building. This should be short enough for them to escape from the building before being affected by fire or smoke. In certain circumstances, however, a second route of escape will be necessary to provide the occupants an alternative means of escape from the building should be the first fire escape become impassable. This will allow the occupants to turn away from the fire and make their escape in the other direction.

Fire and smoke control in corridors

For purposes of smoke control: ND 2.9.16

- a corridor should be sub-divided with a wall or screen with a short fire resistance duration;
- any door in the wall or screen should be a self-closing fire door.

Residential building

In a residential building (other than a residential care   ND 2.9.16
building and/or hospital) rooms intended for use as sleeping
accommodation (including any en-suite sanitary
accommodation where provided) should be:

- separated from an escape route by a wall providing short fire resistance duration.

Any door in the wall should be a suitable self-closing fire door with a short fire resistance duration.

Flat roofs and access decks

Escape across flat roofs and access decks can be hazardous because the surface can be exposed to adverse weather conditions and in the case of flat roofs, may also have obstructions or no edge protection. Therefore in order to protect evacuees from fire:

> - every wall not more than 2 m from either side of the escape route should have a short fire resistance duration up to a height of at least 1.1 m measured from the level of the escape route;   ND 2.9.17
> - a wall or protective barrier at least 1.1 m high should be provided on each side of the escape route or along the edge of the access deck as appropriate.

Protected lobbies

A protected lobby means a lobby within a protected zone, but separated from the remainder of the protected zone so as to resist the movement of smoke from the adjoining accommodation to the remainder of the protected zone.

> The wall dividing a protected lobby from the remainder of the protected zone may have short fire resistance duration and any door in the wall should be a self-closing fire door with short fire resistance duration.   ND 2.9.21

Rooms, toilets and washrooms in protected zones

> The wall separating the rooms/cupboard from a protected zone should have short fire resistance duration and any door in the wall should be a self-closing fire door.   ND 2.9.23

> The walls separating the toilets or washrooms from a protected zone need not have a fire resistance duration.   ND 2.9.23

External escape stairs

External escape stairs present additional hazards to people evacuating a building in the event of fire because the escape stair will be exposed to the possible effects of inclement weather and people who are unfamiliar with the escape routes can feel less confident using an unenclosed stair high above the ground. For these reasons:

> Every part of an external wall (including a door, window or other opening) not more than 2 m from the escape stair should have short fire resistance duration.   ND 2.9.24

Escape stairs in basements

Where an escape stair also serves a basement storey, the protected zone enclosing the escape stair in the basement storey should be separated from the protected zone containing the escape stair serving the rest of the building, by a wall or screen, with or without a door, at the ground-storey floor level.

ND 2.9.26

 Note: The wall, screen and self-closing fire door where provided should have medium fire resistance duration.

Auditoriums

In a building containing an auditorium having an occupancy capacity of more than 500, any separated stage and stage area should be separated from the remainder of the building by a wall with medium fire resistance duration.

ND 2.9.27

Flat roofs and access decks

Where an access deck, open access balcony or flat roof forms part of an escape route; every wall not more than 2 m from either side of the escape route should have short fire resistance duration up to a height of at least 1.1 m measured from the level of the escape route.

D 2.9.6

If it is a flat roof there should also be a wall or protective barrier of at least 1.1 m high provided on each side of the escape route.

D 2.9.6

Ducted warm air heating

Where a flat or maisonette has a storey at a height of more than 4.5 m, or a basement storey, and is provided with a system of ducted warm air heating, where a duct passes through any wall of an entrance hall or stair, all joints between the duct and the surrounding construction should be sealed.

D 2.9.13

Protected lobbies
A protected lobby means a lobby within a protected zone, but separated from the remainder of the protected zone so as to resist the movement of smoke from the adjoining accommodation to the remainder of the protected zone.

A protected lobby should be constructed within a protected zone.

D 2.9.19

The wall dividing the protected lobby from the rest of the protected zone should have:

D 2.9.19

- at least a short fire resistance duration for integrity only.

Any door in the wall should be a self-closing fire door with a short fire resistance duration.

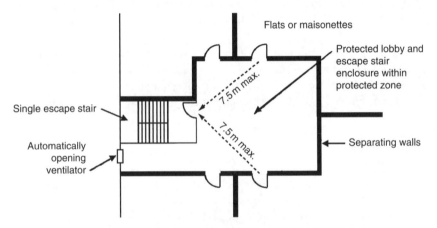

Figure 6.57 Single stair access to flats and maisonettes with any storey at a height of more than 7.5 m but not more than 18 m

Protected zones

A protected zone and which is enclosed by separating walls should have at least medium fire resistance duration.

D 2.9.20

Where any part of an external wall of a protected zone is not more than 2 m from, and makes an angle of not more than 135° with any part of an external wall of another part of the building, the escape stair should be protected by construction for a distance of 2 m with a:

D 2.9.20

- short fire resistance duration where every storey in the building is at a height of not more than 7.5 m above the ground;
- medium fire resistance duration where any storey is at a height of more than 7.5 m.

Protected enclosures

> If a protected enclosure is constructed within a dwelling: D 2.9.21
>
> - the walls should have a short fire resistance duration;
> - any door in the wall should be a self-closing fire door with a short fire resistance duration.

External escape stairs

External escape stairs present additional hazards to people evacuating a building in the case of fire. This is because the escape stair will be exposed to the possible effects of inclement weather.

> Every part of an external wall (including a door, window or   D 2.9.22
> other opening) within 2 m from the escape stair, should have
> short fire resistance duration other than a door opening from
> the top storey to the external escape stair.

Basements

> Where an escape stair also serves a basement storey, the   D 2.9.28
> protected zone enclosing the escape stair in the basement
> storey should be separated from the protected zone containing
> the escape stair serving the rest of the building, by a wall or
> screen (with or without a door) at the ground-storey floor
> level.
>
> The wall, screen and self-closing fire door should have   D 2.9.28
> medium fire resistance duration.

Fire service access

Vehicle access to the exterior of a building is required to allow high-reach appliances, such as turntable ladders and hydraulic platforms, to be used and to enable pumping appliances to supply water and equipment for fire-fighting and rescue activities. The access arrangements will obviously increase in line with the building size and height.

Fire shutters

> All fire shutters in compartment walls should be capable   ND 2.14.6
> of being opened and closed manually by fire service
> personnel.

Additional guidance for high-rise domestic buildings

As occupants of high-rise domestic buildings will probably not be aware that there is a fire in progress, or if there is a fire somewhere else in the building, they may not immediately perceive themselves to be at risk. Additional recommendations for these types of buildings have, therefore, been provided for separation, cavities, spread to neighbouring buildings, spread on external walls, escape lighting and fire service facilities.

Due to the increased hazards associated with fires in high-rise domestic buildings both to the occupants of the building and to fire-fighters, additional active and passive fire protection should be provided.	D 2.A.0
Additional structural fire protection is necessary with increased height and automatic life safety fire suppression systems should be installed within every dwelling.	D 2.A.0

Separation
In all high-rise domestic buildings:

• separating walls should be constructed of non-combustible materials; • separating walls should have at least medium fire resistance duration; • any door in the separating wall should be a self-closing fire door with medium fire resistance duration.	D 2A1

Cavities

Material situated or exposed within a cavity or a cavity formed by external wall cladding, including thermal insulation material, should be constructed of non-combustible materials.	D 2A3

Spread to neighbouring buildings

Every part of an external wall (including thermal insulation or external wall cladding) should be constructed of non-combustible material.	D 2A4

Spread on external walls

External wall cladding should be constructed of non-combustible materials.	D 2A5
Any wall insulation material situated or exposed within a cavity formed by external wall cladding should also be constructed of non-combustible materials.	D 2A5

Fire service facilities

A fire-fighting shaft should be enclosed by walls forming the enclosing structure of a protected zone with long fire resistance duration.	D 2A7

Resistance to fire

The recommended fire resistance duration can be attained where the construction follows the guidance in Columns 3, 4 and 5 of Table 6.51.

Reaction to fire

The guidance in Table 6.52 will be sufficient to attain the appropriate levels of performance (in terms of risk) identified in the Building (Scotland) Regulations.

Table 6.51   Recommended fire resistance duration

Column 1	Column 2	Column 3	Column 4	Column 5	Column 6	Column 7
Construction	Fire resistance duration	British standards			European standards	Test exposure
		Load-bearing capacity (min)	Integrity (min)	Insulation (min)		
Compartment wall, sub-compartment wall, separating wall or an internal wall or screen used as a protected route of escape	Short Medium Long	30 60 120	30 60 120	30 60 120	REI 30 RE1 60 REI 120	Each side separately

(*Continued*)

Table 6.51   (continued)

Column 1	Column 2	Column 3	Column 4	Column 5	Column 6	Column 7
Construction	Fire resistance duration	British standards			European standards	Test exposure
		Load-bearing capacity (min)	Integrity (min)	Insulation (min)		
Load-bearing wall, other than a compartment wall, sub-compartment wall, separating wall or an internal wall or screen protecting routes of escape	Short Medium Long	30 60 120	None None None	None None None	R 30 R 60 R 120	
Fire door in a compartment wall, sub-compartment wall, separating wall or an internal wall or screen protecting routes of escape	Short Medium Long	None None None	30 60 120	None None None	E 30 Sa E 60 Sa E 120 Sa	Each side separately when fitted in frame
Fire shutter in a compartment wall, or in a wall or screen protecting routes of escape in a non-domestic building	Short Medium Long	None None None	30 60 120	30 60 120	EI 30 Sa EI 60 Sa EI 120 Sa	Each side separately when fitted in frame
External wall more than 1 m from a boundary	Short Medium	30 60	30 60	None 30	RE 30 RE 60	From the inside only
External wall not more than 1 m from a boundary	Short Medium	30 60	30 60	None 30	REI 30 REI 60	From the inside only
Roof against an external wall of a non-domestic building	Medium	None	60	60	EI60	From the inside

Table 6.52 Reaction to fire

Column 1 Risk	Column 2 British standards	Column 3 European standards
Medium risk	The material when tested to BS 476: Part 7: 1987 (1993), attains a Class 1 surface spread of flame	The material has achieved a classification of C-s3, d2 or better when tested in accordance with BS EN: 13823 and BS EN ISO: 11925-2
High risk	The material when tested to BS 476: Part 7: 1987 (1993), attains a Class 2 or Class 3 surface spread of flame	The material has achieved a classification of D-s3, d2 or better when tested in accordance with BS EN: 13823 and BS EN ISO: 11925-2
Very high risk	A material which does not attain the recommended performance for high risk	

The designer, however, is free to choose materials or products which satisfy either the British Standard Tests or the Harmonized European Tests.

Additional guidance for residential care buildings

All residential buildings pose special problems because the occupants may be asleep when a fire starts. In residential care buildings the problems are greater as the mobility, awareness and understanding of the occupants may also be impaired.

Compartmentation

To prevent rapid fire spread and to reduce the chances of fire becoming large, the spread of fire within a building can be restricted by sub-dividing that building into compartments that are separated from one another by walls and/ or floors of fire-resisting construction.

The appropriate degree of sub-division depends on:

- the use of and fire load in the building;
- the height to the floor of the top storey in the building;
- the availability of a sprinkler system.

Fire hazard rooms

In order to contain a fire in its early stages, the following rooms are considered to be hazardous and should be enclosed by walls with short fire resistance duration:

- chemical stores;
- cleaners' rooms;
- clothes storage;
- day rooms with a floor area greater than $20\,m^2$;

- smoking rooms;
- disposal rooms;
- lift motor rooms;
- linen stores;
- bedrooms;
- kitchens;
- laundry rooms;
- staff changing and locker rooms;
- store rooms.

Every compartment in a residential care building should be divided into at least two sub-compartments by a sub-compartment wall with short fire resistance duration, so that each sub-compartment is not greater than 750 m².	ND 2.A.1
A sub-compartment wall can be constructed with combustible material (i.e. material that is low, medium, high or very high risk) provided the wall has short fire resistance duration.	ND 2.A.1
Where a lower roof abuts an external wall, the roof should provide medium fire resistance duration for a distance of at least 3 m from the wall.	ND 2.A.1

Additional guidance for hospitals

Hospitals also pose special problems because not only might occupants be asleep when a fire starts, but lack of mobility, awareness and understanding might hinder their escape and (in certain circumstances) it may be harmful to move occupants (i.e. patients) within the building.

Compartmentation

To assist in the safe horizontal evacuation of the occupants in a hospital, every compartment should be divided into at least two sub-compartments by a sub-compartment wall with short fire resistance duration, so that no sub-compartment is more than 750 m².	ND 2.B.1
A compartment wall with medium fire resistance duration should be provided between different hospital departments and between a hospital department and a protected zone.	ND 2.B.1
Intensive therapy units should be divided into at least two sub-compartments by sub-compartment walls with short fire resistance duration.	ND 2.B.1

Rooms such as chemical stores, laboratories, laundry rooms   ND 2.B.1
and X-ray film, record stores, etc., are considered to be
hazardous and should be enclosed by walls providing short
fire resistance duration.

Compartment walls or compartment floors in a hospital   ND 2.B.1
should be constructed of non-combustible material.

Where a compartment wall or sub-compartment wall meets   ND 2.B.1
an external wall, there should be a 1 m wide strip of the
external wall which has the same level of fire resistance
duration as the compartment wall or sub-compartment wall,
to prevent lateral fire spread.

Where a lower roof abuts an external wall, the roof should   ND 2.B.1
provide medium fire resistance duration for a distance of at
least 3 m from the wall.

Additional guidance for enclosed shopping centres

Enclosed shopping centres can be extremely complex to design. There are large fire loads and large numbers of people, all within a complicated series of spaces – where most people only know one way in or out.

A separating wall is not necessary between a shop and a mall   ND 2.C.2
except to shops having mall-level storey areas more than
2,000 m^2 that are located opposite each other.

Environment

Site preparation
Surface soil and vegetable matter can be detrimental to a building's structure if left undisturbed within the building footprint.

The solum (prepared area within the containing walls of a   D 3.1.1
building) should be treated to prevent vegetable growth and   ND 3.1.1
reduce the evaporation of moisture from the ground to the
inner surface of any part of a dwelling that it could damage.

Protection from radon gas
Radon is a naturally occurring, radioactive, colourless and odourless gas that is formed where uranium and radium are present. It can move through cracks and fissures in the subsoil, and so into buildings. Where this gas occurs under

a dwelling, the external walls contain it and the containment of radon can build up inside the dwelling over the long term posing a risk to health.

To reduce the risk, all new dwellings, extensions and alterations built in areas where there might be radon concentration may need to incorporate precautions against radon gas.

Every building must be designed and constructed in such a way that there will not be a threat to the health of people in or around the building due to the emission and containment of radon gas.	D 3.2 ND 3.2

Treatment of building elements adjacent to the ground

Walls adjacent the ground should be constructed so as to prevent moisture from the ground reaching the inner surface of any part of a building that it could damage.	D 3.4.1 ND 3.4.1

Precipitation

Rain penetration occurs most often through walls exposed to the prevailing wet winds (usually south-westerly or southerly) and unless there are adequate damp-proof courses and flashings, etc., materials in parapets and chimneys can collect rainwater and deliver it to other parts of the dwelling below roof level.

Walls exposed to precipitation or wind-driven moisture should prevent penetration of moisture to the inner surface of any part of a building so as to protect the occupants and to ensure that the building is not damaged.	D 3.10.1 ND 3.10.1

Table 6.53   Precipitation – general provisions

Element	Type of construction
Masonry walls of bricks and/or blocks incorporating damp-proof courses, flashings and other materials and components	Constructed to suit the degree of exposure to wind and rain
Masonry walls incorporating external rendering	Constructed to suit the degree of exposure and the type of masonry
Masonry walls of natural stone or cast stone blocks	Constructed to suit the degree of exposure to wind and rain
Masonry cavity walls incorporating insulation material, either as a complete or partial cavity fill, where the insulating material is the subject of a current certificate issued under the relevant conditions of an independent testing body	Constructed in accordance with the terms of the certificate

The following general recommendations should be followed for walls, as appropriate.

Wall constructions

How to prevent rain penetrating to the inner surfaces of the building depends largely on the type of construction being considered. There are two main categories, namely:

- solid masonry and cavity masonry (solid masonry and cavity masonry)
- framed (solid masonry)

Note: The thickness and other dimensions quoted are the minimum recommended unless otherwise stated.

Solid masonry and cavity masonry (see Table 6.54)

Framed (see Table 6.55)

Extensions and conservatories

The outer leaf of a wall that was previously an external wall will become an internal wall of the extension/conservatory and any moisture that enters the cavity could collect and cause serious damage to the building.

Where the building is located in an exposed location or where the existing construction might allow the passage of rain either through facing brick or a poorly rendered masonry wall, the use of a cavity tray along the line of the roof of the conservatory or extension may be appropriate.	D 3.10.4 ND 3.10.4
In sheltered situations or where the detailing can prevent damage to the building as a result of rain penetration a raggled flashing (chased into the wall) may be sufficient.	D 3.10.4 ND 3.10.4

Ventilation of wall cavities

Ventilation of external wall cavities is necessary to prevent the build-up of excessive moisture that could damage the fabric of a building.

To reduce the amount of interstitial condensation to a level that will not harm the timber frame or sheathing, a cavity of at least 50 mm wide should be provided between the sheathing and the cladding.	D 3.10.6 ND 3.10.6
Where the outer leaf is of timber, slate or tile clad construction, a vented cavity should be provided.	D 3.10.6 ND 3.10.6
A ventilated cavity should be provided for extra protection in severely exposed areas.	D 3.10.6 ND 3.10.6

Table 6.54 Wall constructions: solid masonry and cavity masonry

Type	Sub-type		Common requirements	System specific requirements
Solid, masonry	Wall type A (solid wall with internal insulation)	Plasterboard, Insulation, Cavity, 25	• Solid wall, 200 mm thick of bricks • Blocks or slabs of clay • Calcium silicate • Concrete or cast stone	• Wall rendered or unrendered externally • Insulation and plasterboard internally, with a cavity 25 mm wide
	Wall type B (solid wall with external insulation)	Insulation, External protection		• Insulation applied to the external surface of the wall and protected either by cladding (of sheets, tiles or boarding) with permanent ventilation, or by rendering • Wall with or without an internal surface finish of plaster or plasterboard
Cavity, masonry	Wall type A (cavity wall with internal insulation)	Plasterboard, Insulation, Cavity, 100 50 100	• Cavity wall of two leaves of masonry separated by a 5 mm cavity • Each leaf, 100 mm thick, of either bricks or blocks of clay, calcium silicate or concrete • Wall rendered or unrendered externally	• Insulation applied as a lining to the internal surface of the wall and plasterboard

(Continued)

Table 6.54 (Continued)

Type	Sub-type	Common requirements	System specific requirements
	Wall type B (cavity wall with cavity fill insulation)		• Insulation applied as a cavity fill • Wall with or without an internal surface finish of plaster or plasterboard This construction is only recommended for sheltered conditions.
	Wall type C (cavity wall with partial fill insulation)		• Insulation applied to either leaf as a partial cavity fill so as to preserve a residual space of 50 mm wide • Wall with or without an internal surface finish of plaster or plasterboard

Table 6.55 Wall constructions: framed

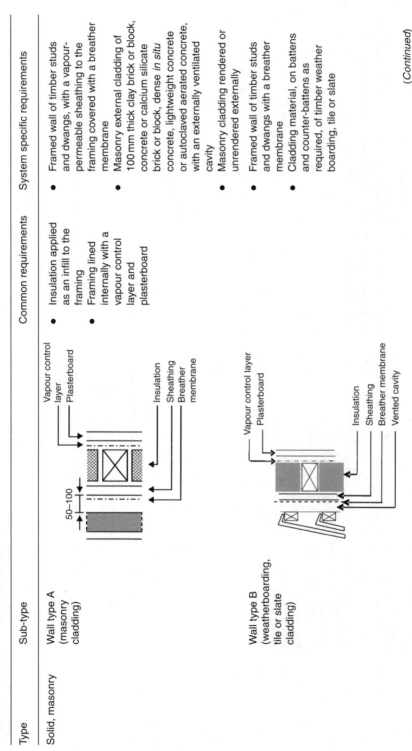

Type	Sub-type	Common requirements	System specific requirements
Solid, masonry	Wall type A (masonry cladding)	• Insulation applied as an infill to the framing • Framing lined internally with a vapour control layer and plasterboard	• Framed wall of timber studs and dwangs, with a vapour-permeable sheathing to the framing covered with a breather membrane • Masonry external cladding of 100 mm thick clay brick or block, concrete or calcium silicate brick or block, dense *in situ* concrete, lightweight concrete or autoclaved aerated concrete, with an externally ventilated cavity • Masonry cladding rendered or unrendered externally
	Wall type B (weatherboarding, tile or slate cladding)		• Framed wall of timber studs and dwangs with a breather membrane • Cladding material, on battens and counter-battens as required, of timber weather boarding, tile or slate

(Continued)

Table 6.55 (Continued)

Type	Sub-type	Common requirements	System specific requirements
	Wall type C (sheet or panel cladding with/without ventilated cavity)	Lining, Vapour control layer, Insulation Lining, Vapour control layer, Insulation, Vented cavity	• Framed wall of timber or metal studs and dwangs • Sheet or panel cladding material of fibre cement, plastic, metal, GRP or GRC • Insulation applied either to the internal face of the framing with permanent ventilation behind any impervious cladding, or as an infill to the framing • Wall lined internally with a vapour control layer and a lining

Due to the air gaps inherent between the components of a timber, slate or tile-clad wall, no proprietary ventilators should be necessary and a 10 mm free air space should be sufficient.

Where the wall cavity is sub-divided into sections by the use of cavity barriers, e.g. at mid-floor level in a two-storey house, the ventilators should be provided to the top and bottom of each section of the cavity.	D 3.10.6 ND 3.10.6
Care should be taken with rendered walls to prevent blockage of the ventilators.	D 3.10.6 ND 3.10.6
Where the outer leaf is of masonry construction, venting of the cavity is normally sufficient.	D 3.10.6 ND 3.10.6
Cavities should be vented to the outside air by installing ventilators with at least $300\,mm^2$ free opening area at $1.2\,m$ maximum centres.	D 3.10.6 ND 3.10.6

Sanitary facilities

The walls of sanitary accommodation that is adjacent to any sanitary facility should be of robust construction that will permit secure fixing of grab rails or other aids; and where incorporating a WC, space for at least one recognized form of unassisted transfer from a wheelchair to the WC.	D 3.12.3

Ventilation of large garages

A garage with a floor area more than $60\,m^2$ for the parking of motor vehicles should be provided with natural or mechanical ventilation on every storey.

Ventilation may be achieved:	ND 3.14.9

- by providing openings in the walls on every storey of at least 1/20th of the floor area of that storey with at least half of such area in opposite walls to promote extract ventilation, if the garage is naturally ventilated;
- where it is a combined natural/mechanical ventilation system, by providing:
  - openings in the wall on every storey of at least 1/40th of the floor area of the storey with at least half of such area in opposite walls;
  - a mechanical system capable of at least three air changes per hour.

Condensation

Condensation can occur in dwellings when water vapour, usually produced by the occupants and their activities, condenses on exposed building surfaces (surface condensation) where it supports mould growth, or within building elements (interstitial condensation).

Thermal bridging occurs when the continuity of the building fabric is broken by the penetration of an element allowing a significantly higher heat loss than its surroundings.

To minimize the risk of condensation on any inner surface, cold bridging at a wall should be avoided.	D 3.15.4
To maintain an adequate internal surface temperature and thus minimize the risk of surface condensation, it is recommended that the thermal transmittance (U-value) of any part and at any point of the external fabric does not exceed $1.2 \, W/m^2K$.	D 3.15.4

Interstitial condensation

A wall should minimize the risk of interstitial condensation in any part of a dwelling that it could damage.	D 3.15.5

Natural lighting provision

Every apartment should have a translucent glazed opening, or openings, of an aggregate glazed area equal to at least 1/15th of the floor area of the apartment and located in an external wall or roof or in a wall between the apartment and a conservatory.	D 3.16.1

Chimneys

Masonry chimneys

A new masonry chimney, usually custom-built on site, and normally with an outer wall of brick, block or stone, should be well constructed and incorporate a flue liner, or flue-blocks, of either clay material or pre-cast concrete.	D 3.18.3 ND 3.18.3

Metal chimneys

Metal chimneys may be either single-walled or double-walled.	D 3.18.4 ND 3.18.4
A metal chimney should not pass through a compartment wall, compartment floor, separating wall or separating floor.	D 3.18.4 ND 3.18.4
There should be no joints within any wall that make accessing the chimney for maintenance purposes difficult.	D 3.18.4 ND 3.18.4

Flue-pipes

A flue-pipe connecting a solid fuel appliance to a chimney should **not** pass through an internal wall (although it is acceptable to discharge a flue-pipe into a flue in a chimney formed wholly or partly by a non-combustible wall), a ceiling or a floor.	D 3.18.5 ND 3.18.5

Flue liners

A flue liner is the wall of the chimney that is in contact with the products of combustion. It can generally be of concrete, clay, metal or plastic depending on the designation of the application.

Existing chimneys for solid fuel applications may also be relined using approved rigid metal liners or single-walled chimney products, an approved cast *in situ* technique or an approved spray- or brush-on coating. Approved products are listed in the HETAS Guide.	D 3.18.6 ND 3.18.6

Combustion appliances – relationship to combustible materials

Combustion appliances and their component parts, particularly solid fuel appliance installations, generate or dissipate considerable temperatures. Certain precautions need to be taken to ensure that any high temperatures are not sufficient to cause a risk to people and the building.

Combustible material should not be located where the heat dissipating through the walls of fireplaces or flues could ignite it.	D 3.19.1 ND 3.19.1

Relationship of metal chimneys to combustible material

There should be a separation distance where a metal chimney passes through combustible material. This is specified, as part of the designation string for a system chimney when used for oil or gas, as (Gxx) where xx is the distance in millimetres.

There is no need for a separation distance if the flue gases are not likely to exceed 100°C.

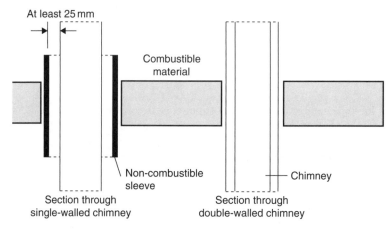

Figure 6.58 Relationship of metal chimneys to combustible material

Where no data is available, the separation distance for oil or gas applications with a flue gas temperature limit of T250 or less should be 25 mm from the outer surface of a single-walled chimney to combustible material measured from the surface of the inner wall of a double-walled chimney.	D 3.19.3 ND 3.19.3
There should be a separation distance of 25 mm from the outer surface of a single-walled chimney to combustible material (measured from the surface of the inner wall of a double-walled chimney) where the metal chimney runs in close proximity to combustible material.	D 3.19.3 ND 3.19.3

Relationship of hearths to combustible materials

Walls that are not part of a fireplace recess or a prefabricated appliance chamber but are adjacent to hearths or appliances should also protect the dwelling from catching fire.	D 3.19.8 ND 3.19.8

This is particularly relevant to timber-framed buildings.

Bulk storage of woody biomass fuel

By its very nature, woody biomass fuel is highly combustible and precautions need to be taken to reduce the possibility of the stored fuel igniting.

Wood pellets can be damaged during delivery thus producing dust that can cause an explosion and precautions need to be taken to reduce this risk.

Storage containers for wood pellets, where they are to be pumped from a transporter to the container, should include a protective rubber mat over the wall to reduce damage to the pellets when they hit the wall.	D 3.23.4 ND 3.23.4

Fuel storage – containment

Oil is a common and highly visible form of water pollution. Because of the way it spreads, even a small quantity can cause a lot of harm to the aquatic environment as oil can pollute rivers, lochs, groundwater and coastal waters, killing wildlife and removing vital oxygen from the water.

The UK government is required by the EC Groundwater Directive (80/68/ EEC) directive to prevent List I substances from entering groundwater and to prevent groundwater pollution by List II substances.

Provision for washing down

Walls and floors should be of an impervious surface that can be washed down easily and hygienically.	D 3.25.4
The enclosures should be permanently ventilated at the top and bottom of the wall.	D 3.25.4

Noise

There are currently no Building Standards to protect the occupants or users of a non-domestic building from noise, but the need may well arise for such standards at a later date.

Measures to reduce the transmission of sound vary according to the type of construction and its reaction to sound energy. The most important factors which affect the behaviour of separating walls and separating floors are mass, cavities, isolation and absorption.

Every building must be designed and constructed in such a way that each wall separating one dwelling from another, or one dwelling from another part of the building, or one dwelling from a building other than a dwelling, will limit the transmission of noise to the dwelling to a level that will not threaten the health of the occupants of the dwelling or inconvenience them in the course of normal domestic activities provided the source noise is not in excess of that from normal domestic activities.	5.1

General application to dwellings

Section 5 of the Technical Handbook applies to dwellings other than those that are totally detached.

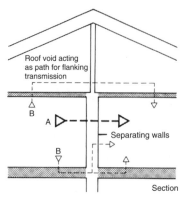

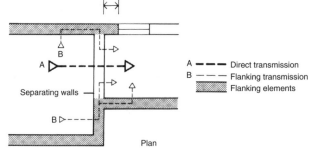

Figure 6.59 Direct and flanking transmission

The following requirements should be met in all dwellings other than those that are totally detached.

- airborne sound-resisting separating walls should be provided between dwellings so that each dwelling is protected from noise emanating from the other one and from other parts, such as common stair enclosures and passages, solid waste disposal chutes, lift shafts, plant rooms, communal lounges and car parking garages.   D 5.1.1

- impact sound-resisting walls are not necessary: D 5.1.1
  - where two houses are linked only by an imperforate separating wall between their ancillary garages;
  - where the wall between a dwelling and another part of the building is substantially open to the external air (e.g. a wall between a dwelling and an access deck);

> – where the wall between a dwelling and another part of
>   the building incorporates a fire door;
> – in the case of a separating wall or separating floor
>   between a dwelling and a private garage or a private waste
>   storage area which is ancillary to the same dwelling.

Figure 6.60 illustrates the parts of a building that should be protected from airborne and impact sound.

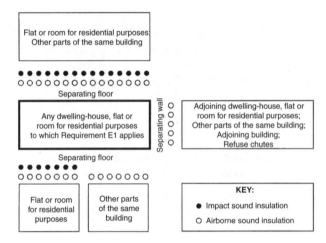

Figure 6.60 Resistance to sound

Flanking transmission

Flanking transmission occurs when there is a path for sound to travel along elements adjacent to separating walls. If the flanking construction and its connections with the separating structure are not correctly detailed, flanking transmission can equal or even exceed sound levels perceived as a result of direct transmission. Figure 6.61 shows typical routes (viewed either in plan or section).

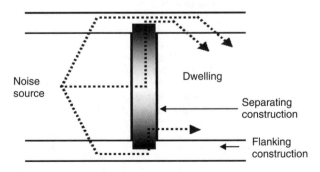

Figure 6.61 Flanking transmission

Note: Specified constructions

One of the possible approaches to conforming to this standard (i.e. Standard 5.1) is to use specified constructions. The specified constructions use common building techniques and materials and the thickness, mass and other dimensions shown in the following clauses are suggested minimum values. Timber sizes refer to actual sizes. Where the mass per unit area (kg/m^2) is given, it refers to the wall surface area (see Annex. 5.A to the Technical Handbook for method of calculating mass in relation to the specified constructions).

Chimneys

Custom-built and system chimneys should not be built into timber-frame separating walls, including wall type 4. Only masonry chimneys (including chimneys built of pre-cast concrete flue blocks) should be built into other types of separating wall.	D 5.1.3

Loudspeakers

The development of flat panel loudspeakers and loudspeakers integrated within floor constructions has introduced an additional neighbour noise concern.

NO loudspeaker should be fitted within a separating wall.	D 5.1.3

Wall type 1 – solid masonry

The resistance to airborne sound transmission depends mainly on the mass of the wall.

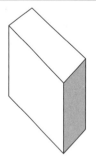

Points to watch:

- Fill masonry joints with mortar in order to achieve the mass and avoid air paths.
- Limit the pathways around the wall (to reduce flanking transmission).

Note: Chases for services may be provided if:

- the depth of any horizontal chase does not exceed 1/6th of the thickness of the leaf;
- the depth of any vertical chase does not exceed one third of the thickness;
- chases are not back to back.

Constructions – wall type 1

There are five recommended solid masonry wall constructions (A–E) as shown in Table 6.56.

Table 6.56   Constructions – wall type 1

Wall type 1A

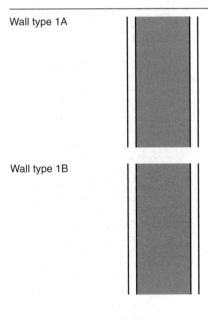

- Brick, plastered both sides.
- Mass including plaster 375 kg/m².
- 13 mm plaster each side.
- Lay bricks in a bond which includes headers and with frogs uppermost.

Example: 215 mm brick, 75 mm coursing, brick density 1,610 kg/m³; lightweight plaster.

Wall type 1B

- Concrete block, plastered both sides.
- Mass including plaster 415 kg/m².
- 13 mm plaster each side.
- Use blocks which extend to the full thickness of the wall.
- Two leaves of block side by side are not recommended.

Example: 215 mm block, 110 mm coursing, block density of 1,840 kg/m³; lightweight plaster.

Wall type 1C

Render/parge coat

Gypsum-based board

Section

- Concrete block, parged both sides, gypsum-based board both sides.
- Mass of masonry alone 415 kg/m².
- 13 mm internal render (parge coat) both sides, should not be smoothed or float finished.
- Minimum mass per unit area of internal render 18 kg/m², both sides.
- Typical internal render mix: cement:lime:sand 1:½:4, by dry volume, in accordance with BS 5492: 1990.
- 12.5 mm gypsum-based board each side, minimum mass per unit area 8.5 kg/m², both sides, fixed with plaster dabs, not battens. Use blocks which extend to the full thickness.
- Lay bricks in a bond which includes headers and frogs uppermost.

Example: 215 mm brick, 75 mm coursing, brick density 1,610 kg/m³.

(Continued)

Table 6.56   (Continued)

Wall type 1D

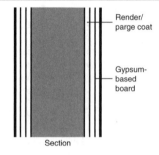

Render/parge coat

Gypsum-based board

Section

- Concrete block, parged both sides, gypsum-based board both sides.
- Mass of masonry alone 415 kg /m^2.
- 13 mm internal render (parge coat) both sides, should not be smoothed or float finished.
- Minimum mass per unit area of internal render 18 kg/m^2, both sides.
- Typical internal render mix: cement:lime:sand 1:½:4, by dry volume, in accordance with BS 5492: 1990.
- 12.5 mm gypsum-based board each side, minimum mass per unit area 8.5 kg/m^2, both sides, fixed with plaster dabs, not battens. Use blocks which extend to the full thickness of the wall.
- Two leaves of block side by side are not recommended.

Example: 215 mm block, 150 mm coursing, block density 1,840 kg/m^3.

Wall type 1E

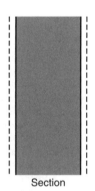

Section

- *In situ* concrete or large concrete panel. Minimum density 1,500 kg /m^3, plaster optional.
- Mass (including plaster if used) 415 kg /m^2. Fill joints between panels with mortar.

Example: 190 mm thick unplastered wall, density 2,200 kg /m^3.

Junctions at roof, ceilings, floors – wall type 1

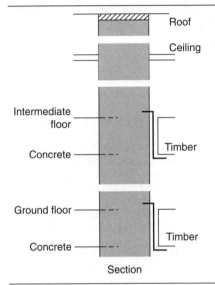

Section

- Fire-stop the joint between wall and roof.
- Where there is a heavy ceiling with sealed joints (12.5 mm gypsum-based board or board material of equivalent mass), the mass of the wall above the ceiling may be reduced to 150 kg/m^2.
- If lightweight aggregate blocks are used to reduce mass, seal one side with cement paint or plaster skim.
- With a timber floor, use joist hangers instead of building joists into separating walls.
- With a concrete floor the wall should be carried through, unless the concrete floor has a mass of 365 kg/m^2 or more.

Junctions at external walls – wall type 1

The outer leaf of a cavity wall adjacent to a type 1 wall may be of any construction.

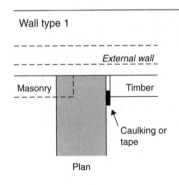

Plan

Where a cavity wall has an inner leaf of masonry, or where the external wall is of solid masonry:
- the masonry of the separating wall should:
  - be bonded together with the masonry of the inner leaf or the solid external wall;
  - abut the masonry of the external wall;
  - be tied to it with ties at not more than 300 mm centres vertically;
- the masonry should have a mass of 120 kg/m^2 unless the length of the external wall is limited by openings:
  - of 1 m high;
  - on both sides of the separating wall at every storey;
  - within 700 mm of the face of the separating wall on both sides (a short length of wall does not vibrate excessively at low frequencies to give flanking transmissions).

Where a cavity wall has an inner leaf of timber construction it should:
- abut the separating wall;
- be tied to it with ties at not more than 300 mm centres vertically;
- have the joints sealed with tape or caulking.

Wall type 2 – cavity masonry

The resistance to airborne sound transmission depends on the mass of the leaves and on the degree of isolation achieved.

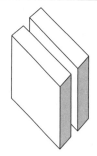

Points to watch:

- Fill masonry joints with mortar in order to achieve the mass and avoid air paths. Maintain the cavity up to the underside of the roof.
- Connect the leaves only where necessary for structural reasons.
- Use only butterfly pattern ties, spaced no further apart than 900 mm horizontally and 450 mm vertically.
- Cavities should be kept clear of mortar droppings, which can reduce acoustic performance by creating a bridge between the two leaves.
- 50 mm cavities are acceptable for wall types 2A, 2B and 2C, but 75 mm cavities make it is easier to avoid this problem.
- If external walls are to be filled with an insulating material, other than loose fibre, the insulating material should be prevented from entering the cavity in the separating wall.

Chases for services can be provided if:

- the depth of any horizontal chase does not exceed 1/6th of the thickness of the leaf;
- the depth of any vertical chase does not exceed 1/3rd of the thickness;
- chases are not back to back.

Constructions – wall type 2

There are four recommended solid masonry wall constructions (A–D) as shown in Table 6.57.

Two of these specified constructions are only intended for use between houses with a step in elevation and/or a stagger in plan at the separating wall (i.e. C and D).

Table 6.57   Constructions – wall type 2

Wall type 2A

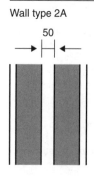

- Two leaves of brick with 50 mm cavity, plastered on both room faces.
- Mass including plaster 415 kg/m². 13 mm plaster each face.

Example: 102 mm leaves laid frogs uppermost, 75 mm coursing, brick density 1,970 kg /m³; lightweight plaster.

(Continued)

Table 6.57   (Continued)

Wall type 2B

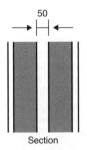

Section

- Two leaves of concrete block with 50 mm cavity, plastered on both room faces.
- Mass including plaster 415 kg/m². 13 mm plaster each face.

Example: 100 mm leaves, 225 mm coursing, block density 1,990 kg/m³; lightweight plaster.

Wall type 2C

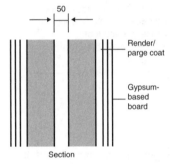

Render/
parge coat

Gypsum-
based
board

Section

- Two leaves of concrete block with 50 mm cavity, both leaves parged, gypsum-based board on both room faces.
- Mass of masonry alone 415 kg/m².
- 12.5 mm gypsum-based board each side, minimum mass per unit area 8.5 kg/m², both sides, fixed with plaster dabs.
- 13 mm internal render (parge coat) both leaves; should not be smoothed or float finished.
- Minimum mass per unit area of internal render 18 kg/m², both leaves.
- Typical internal render mix: cement:lime:sand 1:½:4, by dry volume.

Example: 100 mm leaves, 225 mm coursing; block density of 1,990 kg/m² gives the required mass.

 Note: Adjacent dwellings using wall type 2C should be stepped and/or staggered by at least 300 mm.

Wall type 2D

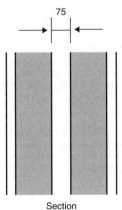

Section

- Two leaves of lightweight aggregate concrete block with 75 mm cavity, plastered on both room faces.
- Maximum block density 1,500 kg/m³.
- Mass including plaster 250 kg/m².
- 13 mm plaster each face.
- Seal the face of the blockwork, with cement paint or plaster, through the full width/depth of any intermediate floor.

Example: 100 mm leaves, 225 mm coursing, block density 1,105 kg/m³; lightweight plaster.

 Note: Adjacent dwellings using wall type 2D should be stepped and/or staggered by at least 300 mm.

Junctions at roof, ceilings, floors – wall type 2

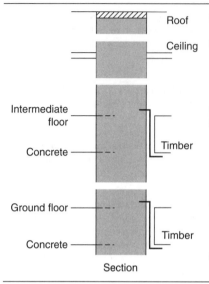

Section

- Fire-stop the joint between wall and roof.
- Where there is a heavy ceiling with sealed joints (12.5 mm gypsum-based board or board material of equivalent mass), the mass of the wall above the ceiling may be reduced to 150 kg/m^2.
- The cavity should still be maintained.
- If lightweight aggregate blocks are used to reduce mass, one face of the wall should be sealed with cement paint or plaster skim. With a timber floor, use joist hangers for any joists supported on the wall.
- With a concrete intermediate or suspended ground floor the floor may be carried through, only to the cavity face of each leaf.
- A concrete slab on the ground may be continuous.

Junctions at external walls – wall type 2

The outer leaf of a cavity wall adjacent to a type 2 wall may be of any construction.

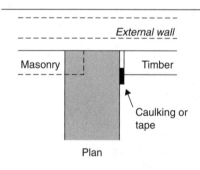

Plan

Where a cavity wall has an inner leaf of masonry:

- the masonry of the separating wall should:
  - be bonded together with the masonry of the inner leaf of the external wall to create a homogeneous unit;
  - abut the masonry of the external wall and be tied to it with ties at no more than 300 mm centres vertically;
- the masonry should have a mass of 120 kg/m^2 except where separating wall type 2B is used, when there is no minimum appropriate mass.

(*Continued*)

Where a cavity wall has an inner leaf of timber construction, it should:

● abut the separating wall;
● be tied to it with ties at not more than 300 mm centres vertically;
● have the joints sealed with tape or caulking.

Wall type 3 – solid masonry between isolated panels

The resistance to airborne sound transmission depends on the mass and type of core, and on the isolation and mass of the panels.

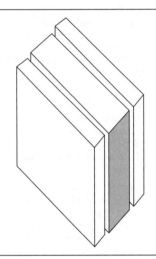

Points to watch:

● Fill masonry joints with mortar in order to achieve the mass and avoid air paths.
● To achieve isolation, support the panels only from floor and ceiling without fixing or tying to the core.
● Cavity barriers between the masonry core and isolated panels should be the minimum necessary, and should be of a flexible type, to maintain the isolation.
● Services may penetrate the free-standing panels, but any gaps should be sealed with tape or caulking.

Constructions – wall type 3

There are four recommended solid masonry wall constructions (A–D) as shown in Table 6.58 and two panels (P1 and P2).

Any of the masonry cores may be used in combination with either of the panels.

Table 6.58   Constructions – wall type 3

Basic construction		Masonry cores
A masonry core, with an isolated panel on each side.		**Core A** – Brick. Mass 300 kg/m². Example: 215 mm core, laid with frogs uppermost, 75 mm coursing; brick density 1,290 kg/m³.
Minimum air space between panels and core 25 mm. Framing should be kept clear of the masonry core by at least 10 mm.	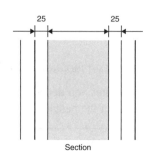 Section	**Core B** – Concrete block. Mass 300 kg/m². Example: 140 mm core, 110 mm coursing, block density 2,200 kg/m³.
		Core C – Lightweight aggregate concrete block. Mass 200 kg/m². Maximum density 1,500 kg/m³. Examples: 140 mm core, 225 mm coursing; block density 1,405 kg/m³. 215 mm core, 150 mm coursing; block density 855 kg/m³.
		Core D – Autoclaved aerated concrete block. Mass 160 kg/m². Examples: 200 mm core, 225 mm coursing; block density 730 kg/m³. 215 mm core, 150 mm coursing; block density 855 kg/m³.
Isolated panels	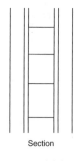 Section	**Panel P1** – Two sheets of gypsum-based board joined by cellular core. ● Mass (including plaster finish if used) 18 kg/m² ● Fit to ceiling and floor only. ● Tape joints between panels.
	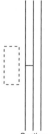 Section	**Panel P2** – Two sheets of gypsum-based board with joints staggered. ● Mass (including plaster finish if used) 18 kg/m². ● Thickness of each sheet 12.5 mm if a supporting framework is used or total thickness of 30 mm if no framework is used.

Junctions at roof, ceilings, floors – wall type 3

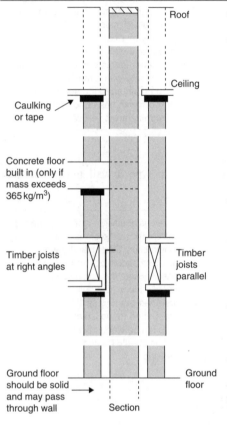

Roof

Ceiling

Caulking or tape

Concrete floor built in (only if mass exceeds 365 kg/m³)

Timber joists at right angles

Timber joists parallel

Ground floor should be solid and may pass through wall

Ground floor

Section

- Fire-stop the joint between masonry cores.
- Where there is a heavy ceiling with sealed joints (12.5 mm gypsum-based board or board material of equivalent mass), the free-standing panels may be omitted in the roof space, and mass of the core above the ceiling may be reduced to 150 kg/m². If lightweight aggregate blocks are used to reduce mass, seal one side with cement paint or plaster skim.
- Seal the junction between ceiling and free-standing panels with tape or caulking. With a timber intermediate floor use joist hangers for any joists supported on the wall and seal the spaces between joists with full depth timber dwangs.
- With a concrete intermediate floor, the floor base may only be carried through where it has a mass of 365 kg/m².
- Seal the junction between ceiling and panel with tape or caulking.

Note: The ground floor should be a solid slab laid on the ground to prevent air paths.

Junctions at external walls – wall type 3

The outer leaf of a cavity wall adjacent to a type 3 wall may be of any construction.

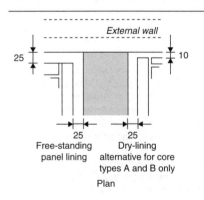

External wall

25

10

25
Free-standing panel lining

25
Dry-lining alternative for core types A and B only

Plan

- The inner leaf of a cavity wall should have an internal finish of isolated panels as specified for the separating wall.
- This is not necessary where the separating wall has core A or B, in which case plaster or dry-lining with joints sealed with tape or caulking may be used.

(Continued)

- A layer of insulation may be added to such internal finish provided the 25 mm and 10 mm gaps shown in the diagram are maintained. The inner leaf may be of any construction if it is lined with isolated panels.
- If the inner leaf is dry-lined, it should be masonry with a mass of 120 kg/m^2, butt jointed to the separating wall core with ties at no more than 300 mm centres, vertically.

Junctions at partitions – wall type 3

The outer leaf of a cavity wall adjacent to a type 3 wall may be of any construction.

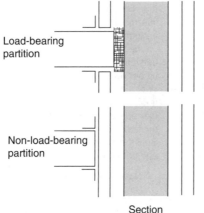

Load-bearing partition

Non-load-bearing partition

Section

- Partitions abutting a type 3 separating wall should not be of masonry construction. Other load-bearing partitions should be fixed to the masonry core through a continuous pad of mineral fibre quilt.
- Non-load-bearing partitions should be tight butted to the isolated panels.
- All joints between partitions and panels should be sealed with tape or caulking.

Wall type 4 – timber frames with absorbent curtain

The resistance to airborne sound transmission depends on the isolation of the frames plus absorption in the air space between them.

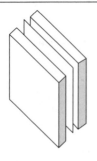

Points to watch:

- Only connect frames if necessary for structural reasons, and then use as few ties as possible. These should not be more than 14–16 gauge (40 × 3 mm) metal straps fixed at or just below ceiling level, 1.2 m apart.
- Where cavity barriers are needed in the cavity between frames, they should be either flexible or fixed to only one frame.
- Services should not be contained in the wall. This is a structural fire precaution, but also limits the creation of air paths through the lining.

Constructions – wall type 4

Two recommended timber-frame constructions (A and B) are as shown in Table 6.59.

Table 6.59   Constructions – wall type 4

Wall type 4A

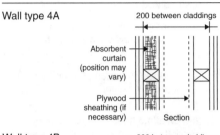

- Timber frames plus absorbent curtain in cavity.
- 200 mm between claddings.
- Plywood sheathing may be used in the cavity as necessary for structural reasons.

Wall type 4B

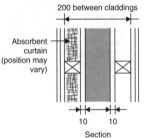

- Timber frames, masonry core, plus absorbent curtain in a cavity.
- 200 mm between claddings (ignore core). Framing should be clear of core by 10 mm. The masonry core is not considered as part of the means of providing sound resistance, but it may be useful for structural support and/or easing the transition to external masonry cladding in stepped or staggered situations.
- There are no restrictions on the type of masonry but the core may be connected to only one of the frames.

Cladding

On each side: two or more layers of gypsum-based board, combined thickness 30 mm, joints staggered to avoid air paths.

Absorbent curtain

- Unfaced mineral fibre quilt (which may be wire reinforced), density 12–36 kg/m^3.
- Thickness 25 mm if suspended in the cavity between frames, 50 mm if fixed to one frame, or 25 mm per quilt if one fixed to each frame.

Junctions at roof, ceilings, floors – wall type 4

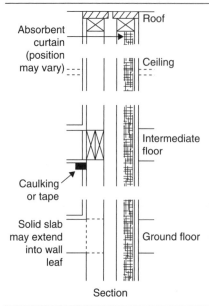

Absorbent curtain (position may vary)

Roof

Ceiling

Caulking or tape

Intermediate floor

Solid slab may extend into wall leaf

Ground floor

Section

Roof

- Fire-stop the joint between masonry core and roof (see Section 2, Fire).

Ceiling and roof space

- Carry the complete construction through to the underside of the roof.

 Note: Provision of a ceiling of any type is optional.

Intermediate floor and ground floor
Block the air path to the wall cavity either by carrying the cladding through the floor or by using a solid timber edge to the floor. Where the joists are at right angles to the wall, seal spaces between joists with full depth timber dwangs.

Junctions at external walls - wall type 4
The outer leaf of a cavity wall adjacent to a type 3 wall may be of any construction.

There are no restrictions on a traditional timber-framed wall, but if the wall is of cavity construction, the cavity should be sealed between the ends of the separating wall and the outer leaf to prevent air paths.

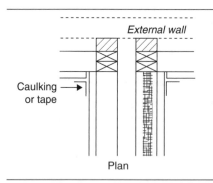

External wall

Caulking or tape

Plan

- The internal finish of the external wall should be 12.5 mm gypsum-based board or other equally heavy material having a mass of at least 10 kg/m^2 (thermal insulation may be incorporated within the framing).
- Where there is an adjacent separating floor, an additional layer of 12.5 mm gypsum-based board should be mounted on the inner leaf of the external wall.

Performance testing

Use of a performance testing approach is particularly useful where the separating or flanking construction is of innovative design and for conversions where flanking transmission may be significant.

Performance values are given in terms of two acoustic parameters, one related to airborne sound, the other related to impact sound.

The airborne sound insulation characteristic of a wall is the sound pressure level difference between the room containing the noise source and the receiving room. The larger the difference the higher the level of airborne sound insulation. Impact sound insulation is the sound pressure level in the receiving room.

Recommended performance values for separating walls are given in Table 6.37.

Airborne sound (minimum values)
The minimum values of airborne sound for walls (as defined in BS EN ISO 717-1: 1997) are:

Mean value	Individual value
53 dB	49 dB

Note:
- Annex 5.A to the Domestic Technical Handbook describes methods for calculating the mass of masonry wall leafs (mortar joints, *in situ* concrete, screeds, slabs and composite floor bases).
- Annex 5.B describes methods for the selection of resilient materials used for soft coverings.
- Annex 5.C describes methods of measurement and test procedures.

The scheme operated by Robust Details Ltd for the English and Welsh Building Regulations is also a useful measurement method, but at the time of writing this book, this scheme has not yet been fully reviewed in relation to construction practice in Scotland.

Energy

The current edition of the Building (Scotland) Regulations focuses on the reduction of carbon dioxide emissions arising from the use of heating, hot water and lighting in buildings, and its guidance sets an overall level for maximum carbon dioxide emissions in buildings incorporating a range of parameters which will influence energy use.

This means that for **all** new buildings, designers are now obliged to consider energy as a complete package rather than only looking at individual elements such as insulation or boiler efficiency favouring, and localized or building-integrated Low and Zero Carbon Technologies (LZCT) (e.g. photovoltaics, active solar water heating, combined heat and power and heat pumps) can be used as a contribution towards meeting this standard.

Maximum *U*-values

The design of a dwelling should involve extensive use of building-integrated or LZCT levels of thermal insulation.	D 6.2.1 LD 6.2.1
The maximum *U*-values for building elements of the insulation envelope should not be greater than those shown in Table 6.60.	D 6.2.1 LD 6.2.1

Table 6.60   Maximum *U*-values for building elements of the insulation envelope

Type of element	(a) Area-weighted average *U*-value (W/m²K) for all elements of the same type	(b) Individual element *U*-value (W/m²K)
Wall [1]	0.30	0.70
Floor [1]	0.25	0.70
Roof	0.20 (0.25 for non-domestic buildings)	0.35
Windows, doors, rooflights	2.2	3.3

Note: 1. Excluding separating walls and separating floors where thermal transmittance should be ignored.

Shell and fit-out buildings

New non-domestic buildings which have been constructed as a shell under one Building Warrant for later fit-out under a separate warrant should meet the maximum *U*-values for building elements of the insulation envelope as given in column (a) of Table 6.61.	ND 6.2.3
Localized areas of the same type of element may be designed to give poorer performance, but nevertheless these areas should be better than the figures given in column (b) of Table 6.61.	ND 6.2.3

The opening areas in the building 'shell' should be designed in accordance with Table 6.62.	ND 6.2.3

Table 6.61 Maximum U-values for building elements of the insulation envelope

Type of element	(a) Area-weighted average U-value (W/m²K) for all elements of the same type	(b) Individual element U-value (W/m²K)
Wall [1]	0.25	0.70
Floor [1]	0.22	0.70
Roof	0.16	0.35
Windows, doors, rooflights	1.8	3.3

Note: 1. Excluding separating walls and separating floors where thermal transmittance should be ignored.

Table 6.62 Maximum windows, doors and rooflight areas

Type of building	Windows and doors as a percentage of the area of the exposed wall	Rooflights as a percentage of the area of the roof
Residential (non-domestic)	30	20
Offices, shops and buildings for entertainment and assembly	40	20
Industrial and storage buildings	15	20
High-usage entrance doors and display windows, and similar glazing	As required	As required

Conversion of heated buildings

Converting a building that was previously designed to be heated (such as building an extension or conservatory or making alterations to the building fabric) will normally present an opportunity for improving the energy efficiency of that building.

Where conversion of a heated building is to be carried out, the insulation envelope should be examined and upgraded (if necessary) in accordance with Table 6.63.	D 6.2.7 ND 6.2.8

Table 6.63 Maximum U-values for building elements of the insulation envelope

Type of element	Area-weighted average U-value (W/m²K) for all elements of the same type
Wall	0.70
Floor	0.70
Roof	0.35
New and replacement windows, doors, rooflights	1.8

Extensions to the insulation envelope

When constructing an extension, the majority of the work will be new and seldom will there be the need to construct to a lesser specification as is sometimes the case for alteration work.

Where the insulation envelope of a non-domestic building is extended, the new building fabric should be designed in accordance with Table 6.64.	ND 6.2.10
Where the insulation envelope of a domestic building is extended, the area of windows, doors, rooflights and roof windows should be limited to 25% of the floor area of the extension.	D 6.2.9
Where the insulation envelope of a dwelling or a building consisting of dwellings is extended the new building fabric should be designed in accordance with Table 6.64.	D 6.2.9
For ease of understanding, the *U*-values (area weighted average *U*-values) in Table 6.64 are summarized in Figure 6.62.	

Table 6.64   Maximum *U*-values for building elements of the insulation envelope

Type of element	(a)Area-weighted average *U*-value (W/m²K) for all elements of the same type	(b) Individual element *U*-value (W/m²K)
Wall	0.27	0.70
Floor	0.22	0.70
Pitched roof (insulation between ceiling ties or collars)	0.16	0.35
Flat roof or pitched roof (insulation between rafters or roof with integral insulation)	0.20	0.35
Windows, doors, rooflights	1.8	3.3

 Note: In Figure 6.62, the extension is the shaded portion; the existing dwelling is in elevation behind.

The *U*-values for the elements involved in the work may be varied provided that the area-weighted overall *U*-value of all the elements in the extension is not greater than that of a 'notional' extension. (see Annex 6.B of the Technical Handbook for an example of this approach).	D 6.2.9

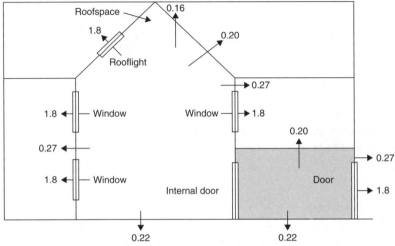

Figure 6.62 Maximum *U*-values for building elements of the insulation envelope

For non-domestic buildings, where the insulation envelope is extended, the new opening areas should be designed in accordance with Table 6.62.	ND 6.2.10

Alterations to the insulation envelope

For alterations, it is more than likely that the existing construction will be from a different era, in Building Regulation terms, and in certain cases (e.g. historic buildings) it may be necessary to adopt alternative energy efficiency measures which relate to the amount of alteration work being undertaken.

Alterations that involve increasing the floor area and/or bringing parts of the existing building that were previously outwith the insulation envelope into the heated part of the dwelling are considered as extensions and/or conversions (regulation 4, schedule 2).

The infill of an existing opening of approximately $4\,m^2$ or less in the building fabric should have a *U*-value which matches at least that of the remainder of the surrounding element and in the case of a wall it should not be worse than $0.70\,W/m^2K$ (for a roof, not worse than $0.35\,W/m^2K$).	D 6.2.11 ND 6.2.12
The infill of an existing opening of greater area (than approximately $4\,m^2$) in the building fabric should have *U*-values which achieve those in column (a) of Table 6.64.	D 6.2.11 ND 6.2.12

Where the alteration causes an existing internal part or other
element of a building to form the insulation envelope, that part
of the building (including any infill construction) should have
U-values which achieve those in column (a) of Table 6.64.

D 6.2.11
ND 6.2.12

A building that was in a ruinous state should, after
renovation, be able to achieve almost the level expected
of new construction and after an alteration of this nature to
the insulation envelope, a roof should be able to achieve at
least an average U-value of 0.35 and in the case of a wall or
floor, 0.70 W/m²K.

D 6.2.10

Conservatories

Although conservatories are normally attached to dwellings they are, neverthe-less, for the purpose of the Building (Scotland) Regulations, treated as stand-alone buildings.

The dividing elements (wall, door, window and on the rare
occasion floor) should have U-values equal to or better than
the corresponding exposed elements in the rest of the dwelling.

D 6.2.12

6.8 Ceilings

For details concerning wall and floor junctions with ceilings, see Sections 6.6 and 6.7.

6.8.1 Requirements

Fire

*Every building must be designed and constructed in such a way
that in the event of an outbreak of fire within the building:*

- *the unseen spread of fire and smoke within concealed spaces
  in its structure and fabric is inhibited;*
- *the development of fire and smoke from the surfaces ceilings
  within the area of origin is inhibited;*
- *the occupants, once alerted to the outbreak of the fire, are
  provided with the opportunity to escape from the building,
  before being affected by fire or smoke;*
- *fire and smoke are inhibited from spreading beyond the
  compartment of origin until any occupants have had the
  time to leave that compartment and any mandatory fire
  containment measures have been initiated.*

2.4

2.5

2.9
2.11

2.1

This standard (i.e. 2.1) does not apply to domestic buildings

*Every building must be designed and constructed in such a way   2.8
that:*

- *in the event of an outbreak of fire in a neighbouring building,
  the spread of fire to the building is inhibited.*
- *facilities are provided to assist fire-fighting or rescue   2.14
  operations.*

Environment

*Every building must be designed and constructed in such a way   3.14
that:*

- *the air quality inside the building is not a threat to the
  capability of the building to resist moisture, decay or
  infestation;   3.18*
- *any component part of each fixed combustion appliance
  installation used for the removal of combustion gases will
  withstand heat generated as a result of its operation without
  any structural change that would impair the stability or
  performance of the installation.   3.11*
- *there will not be a threat to the building or the health of
  the occupants as a result of moisture caused by surface
  or interstitial condensation.*

Safety

*Every building must be designed and constructed in such a   4.8
way that:*

- (a) *people in and around the building are protected from
  injury that could result from fixed glazing, projections
  or moving elements on the building;*
- (b) *fixed glazing in the building is not vulnerable to breakage
  where there is the possibility of impact by people in and
  around the building;*
- (c) *both faces of a window and rooflight in a building are
  capable of being cleaned such that there will not be
  a threat to the cleaner from a fall resulting in severe
  injury;*

(d) *a safe and secure means of access is provided to a roof;
and manual controls for ventilation and for electrical
fixtures can be operated safely.*

  Standard 4.8(d) does not apply to domestic buildings.

Energy

*Every building must be designed and constructed in such a
way that:*   6.2

- *an insulation envelope is provided which reduces heat
  loss;*
- *the form and fabric of the building minimizes the use of*   6.6
  *mechanical ventilating or cooling systems for cooling
  purposes;*
- *in non-domestic buildings, the ventilating and cooling
  systems installed are energy efficient and are capable of
  being controlled to achieve optimum energy efficiency;*
- *energy supply systems and building services which use*   6.7
  *fuel or power for heating, lighting, ventilating and cooling
  the internal environment and heating the water, are
  commissioned to achieve optimum energy efficiency.*

The occupiers of a building must be provided with written   6.8
*information by the owner on the operation and maintenance
of the building services and energy supply systems.*

6.8.2 Meeting the requirements

Structure
Imposed loads on roof, floors and ceilings

The imposed loads on ceilings should not be more than 0.25 kN/m². 1D4

Lateral support by roofs and floors

The plasterboard ceiling of the top storey of pitched roofs   1E49b
should be fixed directly under the roof.

Openings and service penetrations

Fire-resisting ceilings are intended to prevent fire passing from one compartment to another. Openings and service penetrations through these walls or floors can compromise their effectiveness and should be kept to a minimum.

Openings and service penetrations should be carefully detailed and constructed to resist fire.	ND 2.1.14

Junctions

Junctions between fire-resisting ceilings and other parts of the building should be designed and constructed to prevent a fire in one compartment flanking the wall, floor or ceiling from entering an other compartment at the junctions including any solum space or roof space.

Where a fire-resisting ceiling forms a junction with an external wall, a separating wall, another compartment wall or a wall or screen used to protect routes of escape, the junction should maintain the fire resistance of the compartment wall or compartment floor.	ND 2.1.15

Cavity barriers

Where cavity barriers are installed between a roof and a ceiling above an undivided space, cavity barriers should be installed not more than 20 m apart.	ND 2.4.2 ND 2.B.2
Cavity barriers are not necessary to divide a cavity: • in a ceiling void between a floor and a fire-resisting ceiling; • between a roof and a fire-resisting ceiling.	ND 2.4.2
In non-domestic buildings, where a ceiling void cavity extends over a room intended for sleeping, or over such a room and any other part of the building, cavity barriers should be installed on the same plane as the wall.	ND 2.4.4

Roof constructions

Special care should be taken with ventilation where ceilings follow the roof pitch.	D 3.15.7

To prevent excessive build-up of condensation in cold, pitched   D 3.15.7
roof spaces, where the insulation is at ceiling level the roof
space should be cross-ventilated.

Chimneys

There should be no joints within a ceiling that make accessing   D 3.18.4
a metal chimney for maintenance purposes difficult.   ND 3.18.4

A flue-pipe connecting a solid fuel appliance to a chimney   D 3.18.5
should **not** pass through a ceiling.   ND 3.18.5

Height of activity spaces

Reduced headroom, such as beneath a sloping ceiling, can cause problems
in use of both facilities and furniture, particularly if a person has difficulty in
bending or has a visual impairment.

Activity spaces within the enhanced apartment or kitchen   D 3.11.4
should have an unobstructed height of at least 1.8 m.

Fire precautions

Resistance to fire

The recommended fire resistance duration can be attained where the construction follows the guidance in Columns 3, 4 and 5 of Table 6.65.

Table 6.65   Recommended fire resistance duration

Column 1	Column 2	Column 3	Column 4	Column 5	Column 6	Column 7
Construction	Fire resistance duration	British standards			European standards	Test exposure
		Load-bearing capacity (min)	Integrity (min)	Insulation (min)		
Ceiling in place of a cavity barrier	Short	None	30	30	EI 30	From the underside

Reaction to fire

The guidance in Table 6.66 will be sufficient to attain the appropriate levels of performance (in terms of risk) identified in the Building (Scotland) Regulations.

The designer is free to choose materials or products which satisfy either the British Standard Tests or the Harmonized European Tests.

Table 6.66   Reaction to fire

Risk	British standards	European standards
Medium risk	The material of the wall or ceiling when tested to BS 476: Part 7: 1987 (1993), attains a Class 1 surface spread of flame	The material has achieved a classification of C-s3, d2 or better when tested in accordance with BS EN: 13823 and BS EN ISO: 11925-2
High risk	The material of the wall or ceiling when tested to BS 476: Part 7: 1987 (1993) attains a Class 2 or Class 3 surface spread of flame.	The material has achieved a classification of D-s3, d2 or better when tested in accordance with BS EN: 13823 and BS EN ISO: 11925-2
Very high risk	A material which does not attain the recommended performance for high risk	

Fire-resistant ceilings

Buildings must be designed and constructed so that in the event of a fire within the building, the development of fire and smoke from the surfaces of ceilings within the area of origin is inhibited.	D 2.5 ND 2.5

Where a fire-resisting ceiling, including a suspended ceiling, contributes to the fire resistance duration of a compartment:

the ceiling should not be easily demountable;	ND 2.1.15
openings and service penetrations in the ceiling should be protected;	ND 2.1.15
the ceiling should form a junction with the compartment wall (or sub-compartment wall).	ND 2.1.15

Fire-resisting ceilings as an alternative to cavity barriers

Fire-resistant ceilings provided as an alternative to cavity barriers should:

D 2.4.3

ND 2.4.5

- have a short fire resistance duration;
- not be easily demountable;
- have protected openings and service penetrations in the ceiling;
- contain an access hatch which, when closed, will maintain the fire resistance duration of the ceiling.

Smoke control

Where a building has more than one escape stair and where a corridor, or part of a corridor, provides escape in only one direction, automatic opening ventilators should be provided in that part of the corridor which provides single direction escape; such ventilators should be activated by automatic smoke detection fixed to the ceiling of the corridor and fitted with a manual override for fire service use.

D 2.9.17

Smoke alarms

A smoke alarm should be ceiling mounted and located:

D 2.11.2

- where the circulation area is more than 15 m long;
- in a circulation area which will be used as a route along which to escape;
- not more than 7 m from the door to a living room or kitchen; and
- not more than 3 m from the door to a room intended to be used as sleeping accommodation;
- not more than 7.5 m from another smoke alarm on the same storey;
- at least 300 mm away from any wall or light fitting, heater or air-conditioning outlet;
- on a surface which is normally at the ambient temperature of the rest of the room or circulation area in which the smoke alarm is situated.

Smoke outlets from basements

Smoke outlets should be sited at ceiling level within the
room they serve.

ND 2.14.8

Internal linings

Materials used in ceilings can significantly affect the spread of fire and its
rate of growth. Fire spread on internal linings in escape routes is particularly
important because rapid fire spread in protected zones and unprotected zones
could prevent the occupants from escaping.

Every room, fire-fighting shaft, protected zone or
unprotected zone should have ceiling surfaces with a
reaction to fire which follows the guidance in Table 6.67.

ND 2.5.1

Table 6.67   Reaction to fire of wall and ceiling surfaces

Building	Residential care buildings and hospitals	Shops	All other buildings
Room not more than 30 m²	Medium risk [2]	High risk	High risk
Room more than 30 m²	Low risk [3]	Medium risk [4]	Medium risk
Unprotected zone	Low risk	Low risk	Medium risk
Protected zone and fire-fighting shaft [1]	Low risk	Low risk	Low risk

Notes:
1. Including any toilet or washroom within a protected zone.
2. High risk in a room not greater than 4 m².
3. Ceilings may be medium risk.
4. Low risk in storage buildings (Class 1).

Thermoplastic materials in ceilings

Thermoplastic materials in ceilings, rooflights and lighting diffusers provide
a significant hazard in a fire. Burning droplets can rapidly increase the fire
growth rate and the smoke produced is normally dense and toxic which com-
bine to produce extremely hazardous conditions. For these reasons:

Thermoplastic material should **not** be used in protected
zones or fire-fighting shafts.

D 2.5.4
ND 2.5.4

A ceiling constructed from thermoplastic materials, either as a suspended or stretched skin membrane with a TP(a) flexible classification should be supported on all its sides and not exceed 5 m².	D 2.5.5 ND 2.5.5
A ceiling with a TP(a) flexible classification should not be installed in the ceiling of a protected zone or fire-fighting shaft.	D 2.5.5 ND 2.5.5

Thermoplastic materials in rooflights

Thermoplastic materials (other than TP(a) flexible) may be used in rooflights subject to the recommendations in Table 6.68 and Figure 6.63.	D 2.5.6 ND 2.5.6

Table 6.68   Thermoplastic rooflights and light fittings with diffusers

Classification of lower surface	Protected zone or fire-fighting shaft	Unprotected zone or protected enclosure		Room	
	Any thermoplastic	TP(a) rigid	TP(a) flexible and TP(b)	TP(a) rigid	TP(a) and flexible TP(b)
Maximum area of each diffuser panel or rooflight (m²)	Not advised	No limit	5 m²	No limit	5 m²
Maximum total area of diffuser panels or rooflights as a percentage of the floor area of the space in which the ceiling is located (%)	Not advised	No limit	15%	No limit	50%
Minimum separation distance between diffuser panels or rooflights (m)	Not advised	No limit	3 m	No limit	3 m

Notes:
1. Smaller panels can be grouped together provided that the overall size of the group and the space between any others, satisfies the dimensions shown in the diagram opposite.
2. The minimum 3 m separation in the diagram opposite should be maintained between each 5 m² panel. In some cases therefore, it may not be possible to use the maximum percentage quoted.
3. TP(a) flexible is not recommended in rooflights.

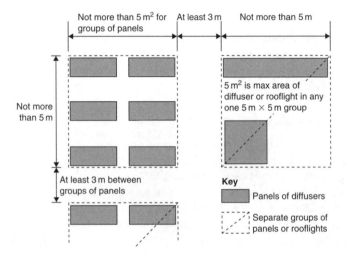

Figure 6.63 Layout restrictions on thermoplastic rooflights and light fittings with diffusers

Thermoplastic materials in light fittings with diffusers

Thermoplastic materials may be used in light fittings with diffusers.	D 2.5.7 ND 2.5.7
Where the lighting diffuser forms an integral part of the ceiling, the size and disposition of the lighting diffusers should be installed subject to the recommendations in Table 6.68 and Figure 6.63.	D 2.5.7 ND 2.5.7
Where the lighting diffusers form an integral part of a fire-resisting ceiling which has been satisfactorily tested, the amount of thermoplastic material is unlimited.	D 2.5.7 ND 2.5.7
Where light fittings with thermoplastic diffusers do not form an integral part of the ceiling, the amount of thermo-plastic material is unlimited provided the lighting diffuser is designed to fall out of its mounting when softened by heat.	D 2.5.7 ND 2.5.7

Ducted warm air heating

Where a flat or maisonette has a storey at a height of more than 4.5 m, or a basement storey (and is provided with a system of ducted warm air heating) and where a duct passes through a ceiling of an entrance hall or stair, all joints between the duct and the surrounding construction should be sealed.	D 2.9.13

Ventilation

Passive stack ventilation systems

A passive stack ventilation system uses a duct running from a ceiling (normally in a kitchen or shower room) to a terminal on the roof to remove any moisture-laden air.

Every passive stack ventilation system should:	D 3.14.6

- incorporate a ceiling mounted automatic humidity-sensitive extract grille that will operate when the relative humidity is between 50 and 65%;
- be insulated with at least 25 mm thick material having a thermal conductivity of 0.04 W/mK where it passes through a roof space or other unheated space or where it extends above the roof level.

Mechanical ventilation

Mechanical extract should be provided in rooms where the cubic space per occupant is not more than 3 m and where the rooms have low ceilings and are occupied by large numbers of people.	ND 3.14.5

Rooflights

Annex 6A of the Technical Handbook offers advice about compensating U-values for rooflights.

Maximum U-values

The design of a dwelling should involve extensive use of building-integrated or localized Low and Zero Carbon Technologies (LZCT) levels of thermal insulation.	D 6.2.1 LD 6.2.1
The maximum U-values for building elements of the insulation envelope should not be greater than those shown in Table 6.69.	D 6.2.1 LD 6.2.1

Note: Excluding separating walls and separating floors where thermal transmittance should be ignored.

The opening areas in the building 'shell' should be designed in accordance with Table 6.70.	ND 6.2.3

Table 6.69   Maximum U-values for building elements of the insulation envelope

Situation	(a) Area-weighted average U-value (W/m²K) for all elements of the same type	(b) Individual element U-value (W/m²K)
Normal	2.2	3.3
Shell and fit-out buildings	1.8	3.3
Conversion of heated buildings	1.8	

Table 6.70   Maximum rooflight areas

Type of building	Rooflights as a percentage of the area of the roof
Residential (non-domestic)	20
Offices, shops and buildings for entertainment and assembly	20
Industrial and storage buildings	20
High-usage entrance doors and display windows, and similar glazing	As required

Converting a building that was previously designed to be heated (such as forming an extension or conservatory or making alterations to the building fabric) will normally present an opportunity for improving the energy efficiency of that building.

Extensions to the insulation envelope

When constructing an extension, the majority of the work will be new and seldom will there be the need to construct to a lesser specification as is sometimes the case for alteration work.

Where the insulation envelope of a domestic building is extended, the area of windows, doors, rooflights and roof windows should be limited to 25% of the floor area of the extension.   D 6.2.9

Where the insulation envelope of a dwelling or a building consisting of dwellings is extended the new building fabric should be designed in accordance with Table 6.70.   D 6.2.9

For ease of understanding, the U-values (area-weighted average U-values) in Table 6.70 are summarized in Table 6.70a.

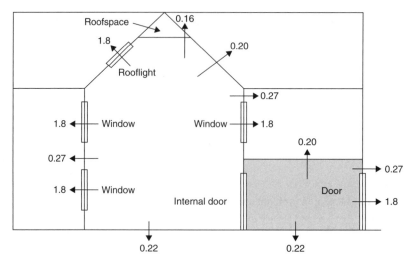

Figure 6.65   Maximum *U*-values for building elements of the insulation envelope

Note: The extension is the shaded portion; the existing dwelling is in elevation behind.

Table 6.70a Maximum *U*-values for for ceilings and rooflights

Type of element	(a) Area-weighted average *U*-value (W/m²K) for all elements of the same type	(b) Individual element *U*-value (W/m²K)
Pitched roof (insulation between ceiling ties or collars)	0.16	0.35
Windows, doors, rooflights	1.8	3.3

Roof coverings

The roof of a building, including any rooflights, but excluding   2.8.1
any wallhead fascia, flashing or trim, boxed gutters, soffit or
barge boards:

- should have a low vulnerability if not more than 6 m from
  the boundary;
- should have a boundary low or a medium vulnerability if
  more than 6 m but not more than 24 m from the boundary.

Alterations to the insulation envelope

Where windows, doors and rooflights are being created or replaced, they should achieve the U-values recommended in Column (a) of Table 6.70a (an example of a compensatory approach is shown in Annex 6C to the Technical Handbooks).	D 6.2.11 ND 6.2.12
Where additional windows, doors and rooflights are being created, the overall total area (including existing) should not exceed 25% of the total dwelling floor area and in the case of a heated communal room or other area (exclusively associated with the dwellings), it should not exceed 25% of the total floor area of these rooms/areas.	D 6.2.10
When alterations are carried out, attention should still be paid to limiting infiltration thermal bridging at junctions and around windows, doors and rooflights and limiting air infiltration.	D 6.2.11 ND 6.2.12

Stand-alone buildings

In these rooms or accommodation, the area of windows, doors, rooflights and roof windows should be limited to 25% of the total floor area of these parts.	D 6.2.13

Heating system

Where a dwelling has little or no cross-ventilation (e.g. flats with all external windows/rooflights on one southerly elevation or a high proportion of translucent glazing), the dwelling should be designed to avoid high internal temperature.	D 6.6.1

Thermoplastic materials in rooflights

Thermoplastic materials (other than TP(a) flexible) may be used in rooflights subject to the recommendations in Table 6.68 and Figure 6.63.	D 2.5.6 ND 2.5.6

Danger from accidents

The majority of accidents occur during normal use involving building features such as doors, windows and areas of fixed glazing. Collisions with glazing are very common as it can, if transparent, be difficult to see and may create confusing lighting effects, presenting particular difficulties for a person with a visual or cognitive impairment.

Every building must be designed and constructed in such a   D 4.8
way that both faces of a window and rooflight in a building   ND 4.8
are capable of being cleaned such that there will not be a
threat to the cleaner from a fall resulting in severe injury.

Any rooflight, all or part of which is more than 4 m above   *D 4.8.3*
adjacent ground or internal floor level, should be constructed   *ND 4.8.3*
so that any external and internal glazed surfaces can be
cleaned safely from:

- inside the building in accordance with the recommendations of Clause 8 of BS 8213: Part 1: 2004;
- a load-bearing surface, such as a balcony or catwalk,
  large enough to prevent a person falling further;
- a window access system, such as a cradle or travelling
  ladder, mounted on the building, as described in Annex
  C3 of BS 8213: Part 1: 2004;
- (for non-domestic buildings) a ladder sited on adjacent
  ground or from an adjacent load-bearing surface.

6.9 Roofs

Requirements for rooflights can be found in Section 6.8.

6.9.1 Requirements

Structure

Every building must be designed and constructed in such a way   *1.1*
that the loadings that are liable to act on it, taking into account
the nature of the ground, will not lead to:

- *the collapse of the whole or part of the building;*
- *deformations which would make the building unfit for its*
  intended use, unsafe, or cause damage to other parts of the
  building or to fittings or to installed equipment;
- *impairment of the stability of any part of another building.*

Every building must be designed and constructed in such a way   *1.2*
that in the event of damage occurring to any part of the structure
of the building the extent of any resultant collapse will not be
disproportionate to the original cause.

Fire

Every building, which is divided into more than one area of different occupation, must be designed and constructed in such a way that in the event of an outbreak of fire within the building, fire and smoke are inhibited from spreading beyond the area of occupation where the fire originated.	*2.2*
Every building must be designed and constructed in such a way that in the event of an outbreak of fire within the building:	*2.4*
• *the unseen spread of fire and smoke within concealed spaces in its structure and fabric is inhibited;*	
• *the occupants, once alerted to the outbreak of the fire, are provided with the opportunity to escape from the building, before being affected by fire or smoke.*	*2.9*
Every building must be designed and constructed in such a way that in the event of an outbreak of fire in a neighbouring building, the spread of fire to the building is inhibited.	*2.8*

Domestic buildings

Every building must be designed and constructed in such a way that in the event of an outbreak of fire within the building, fire and smoke are inhibited from spreading beyond the compartment of origin until any occupants have had the time to leave that compartment and any mandatory fire containment measures have been initiated.	*2.1*

Environment

Every building must be designed and constructed with a surface water drainage system that will:	*3.6*
• *ensure the disposal of surface water without threatening the building and the health and safety of the people in and around the building;*	
• *have facilities for the separation and removal of silt, grit and pollutants.*	

Every building must be designed and constructed so that:	*3.10*

- *there will be no threat to the building or the health of the occupants as a result of moisture from precipitation penetrating to the inner face of the building;*
- *the air quality inside the building is not a threat to the health of the occupants or the capability of the building to resist moisture, decay or infestation;*   *3.14*
- *any component part of each fixed combustion appliance installation used for the removal of combustion gases will withstand heat generated as a result of its operation without any structural change that would impair the stability or performance of the installation.*   *3.18*

Every building must be designed and constructed in such a way that there will not be a threat to the building or the health of the occupants as a result of moisture caused by surface or interstitial condensation.   *D 3.15*

Standard D 3.15 applies only to a dwelling.

Safety

Every building must be designed and constructed in such a way that: *4.8*

(a) *people in and around the building are protected from injury that could result from fixed glazing, projections or moving elements on the building;*

(b) *fixed glazing in the building is not vulnerable to breakage where there is the possibility of impact by people in and around the building;*

(c) *both faces of a window and rooflight in a building are capable of being cleaned such that there will not be a threat to the cleaner from a fall resulting in severe injury;*

(d) *a safe and secure means of access is provided to a roof; and manual controls for ventilation and for electrical fixtures can be operated safely.*

Standard 4.8(d) does not apply to domestic buildings.

Noise

Every building must be designed and constructed in such a way that   5.1
will limit the transmission of noise from one dwelling to another
or another part of the building to a level that will not threaten the
health of the occupants of the dwelling or inconvenience them in the
course of normal domestic activities provided the source noise is
not in excess of that from normal domestic activities.

Energy

Every building must be designed and constructed in such a   6.2
way that an insulation envelope is provided which reduces
heat loss.

6.9.2 Meeting the requirements

Roof construction

There is evidence to suggest that condensation in cold deck flat roofs is a
problem and should be avoided. Both the warm deck and warm deck inverted
roof constructions, where the insulation is placed above the roof deck, are
considered preferable.

Loadings

Buildings should be stable whatever the combinations of dead load, imposed
load and wind loading in terms of their individual structural elements, their
interaction together and the overall stability as a structure.

The building shall be designed to transmit loads safely to the ground.	1.1.0
The roof should be capable of providing local support to the walls as well as acting as horizontal diaphragms capable of transferring the wind loads to buttressing elements of the building.	D 1B1c D 1B1d
Trussed rafter roofs should be braced in accordance with the recommendations of BS 5268-3: 2006.	D 1B2

Flat roof constructions

Table 6.71 Roof constructions (flat)

Type	Sub-type	Construction	Common requirements	System-specific requirements
Flat roof constructions	Roof type A (concrete – warm roof)	Weatherproof covering / Insulation / Vapour control layer / Screed, if required	Flat roof structure of *in situ* or precast concrete with or without a screed; with or without a ceiling or soffit	External weatherproof covering insulation laid on a vapour control layer between the roof structure and the weatherproof covering [1]
	Roof type B (concrete – inverted roof)	Protective covering / Insulation / Waterproof membrane / Screed, if required		External protective covering; with low permeability insulation laid on a waterproof membrane between the roof structure and the external covering
	Roof type C (timber or metal frame – warm roof)	Weatherproof covering / Insulation / Vapour control layer	Flat roof structure of timber or metal-framed construction with a board decking 19 mm thick; with or without a ceiling or soffit	External weatherproof covering insulation laid on a vapour control layer between the roof structure and the weatherproof covering [1]

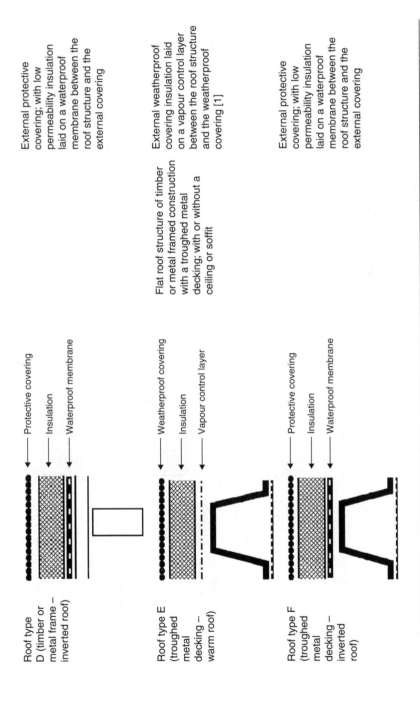

Roof type D (timber or metal frame – inverted roof)

Protective covering
Insulation
Waterproof membrane

External protective covering; with low permeability insulation laid on a waterproof membrane between the roof structure and the external covering

Roof type E (troughed metal decking – warm roof)

Weatherproof covering
Insulation
Vapour control layer

Flat roof structure of timber or metal framed construction with a troughed metal decking; with or without a ceiling or soffit

External weatherproof covering insulation laid on a vapour control layer between the roof structure and the weatherproof covering [1]

Roof type F (troughed metal decking – inverted roof)

Protective covering
Insulation
Waterproof membrane

External protective covering; with low permeability insulation laid on a waterproof membrane between the roof structure and the external covering

Note: 1. Roof types A, C and E are not suitable for sheet metal coverings that require joints to allow for thermal movement.

Pitched roof constructions

Table 6.72 Roof constructions (pitched)

Type	Sub-type		Common requirements	System-specific requirements
Pitched roof constructions	Roof type A (slates or tiles – insulation on a level ceiling)	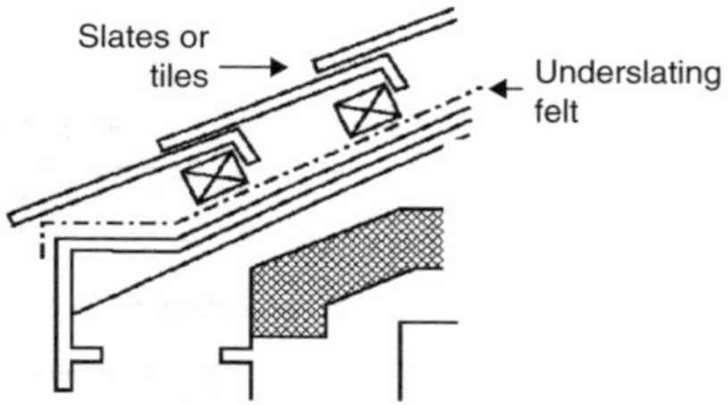	Pitched roof structure of timber or metal framed construction	• External weatherproof covering of slates or tiles on under slating felt with or without boards or battens
	Roof type B (slates or tiles – insulation on a sloping ceiling)	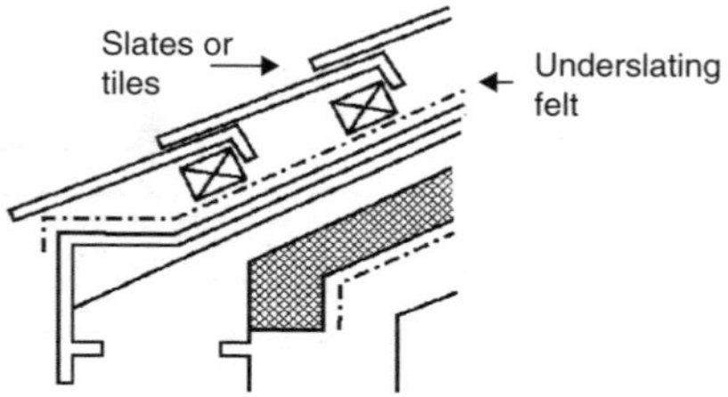		• External weatherproof covering of slates or tiles on under slating felt with or without boards or battens
	Roof type C (slates or tiles – insulation as decking)	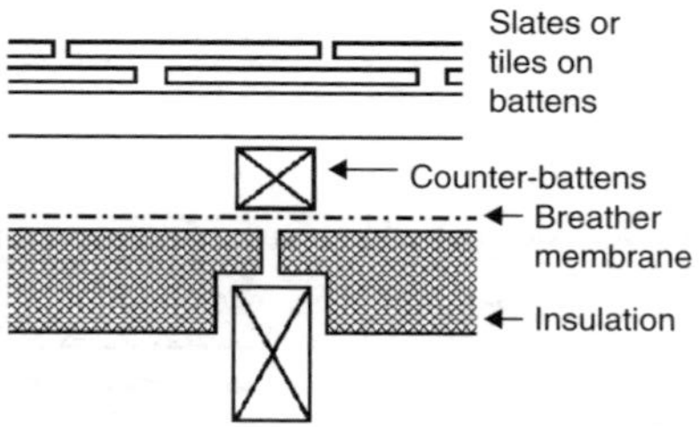		• A decking of low permeability insulation fitted to and between the roof framing

- External weatherproof covering of slates or tiles, with tiling battens and counter-battens (located over roof framing), and a breather membrane laid on the insulation decking with a sloping ceiling

- External weatherproof covering of metal or fibre cement sheet sandwich construction laid on purlins
- Insulation sandwiched between the external and soffit sheeting with or without a ceiling

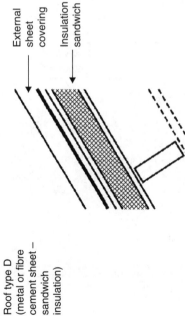

External sheet covering

Insulation sandwich

Roof type D
(metal or fibre
cement sheet –
sandwich
insulation)

Note: Roof type D is not suitable for sheet metal coverings that require joints to allow for thermal movement.

Note: A traditional cut timber roof (i.e. using rafters, purlins and ceiling joists) generally has sufficient built-in resistance to instability and wind loads (e.g. from hipped ends, tiling battens, rigid sarking, etc.). However, diagonal rafter bracing, equivalent to that recommended in BS 5268-3: 2006 or Annex H of BS 8103-3: 1996, should be provided particularly for single-hipped and non-hipped roofs of more than 40° pitch to detached houses.

Extensions to existing buildings

Where a new roof connects to an existing roof, the loads from the new roof should be carried down to the new foundation and no new loads should be carried by the existing structure.	D 1C5

Imposed loads on roof

The imposed loads on roof should not be more than those shown in Table 6.73.	D 1D4

Table 6.73   Imposed loading

Element	Loading
Spans of at least 12 m	$1.00\,kN/m^2$
Spans at least 6 m	$1.50\,kN/m^2$

Lateral support by roofs

Roofs should: • transfer lateral forces from walls to buttressing walls, piers or chimneys; • be secured to the supported wall.	D 1D35

Lateral support of walls

The lateral support of walls shall be in accordance with Table 6.74.	D 1D35

Where a roof is alongside a supported wall and interrupts the continuity of lateral support: • the length of the opening should be not more than 3 m; • if connection is via by anchors, then these should be spaced less than 2 m on each side of the opening;	D 1D37

- if connection is not via anchors, then this should be provided throughout the length of each portion of the wall, and on each side of the opening;

 There should be **NO** other interruption of lateral support.

Table 6.74   Lateral support of walls

Wall type	Wall length	Lateral support required
Solid or cavity external wall	Any length	Roof lateral support by every roof forming a junction with a supporting wall
Solid or cavity separating wall or compartment wall	More than 3 m	Floor lateral support by every floor forming a junction with a supporting wall
Internal load-bearing wall (that is not a separating wall or compartment wall)	Any length	Roof or floor lateral support at the top of each storey

Gable wall strapping

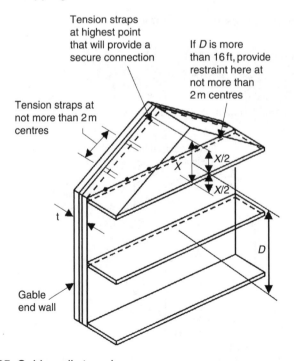

Figure 6.65 Gable wall strapping

Vertical strapping at least 1 m long should be provided at   D 1D36
eaves level at intervals not more than 2 m if the roof:

- has a pitch of not more than 15°;
- is not tiled or slated;
- is not of a type known by local experience to be resistant
  to wind gusts;
- does not have main timber members spanning onto the
  supported wall at not more than 1.2 m centres.

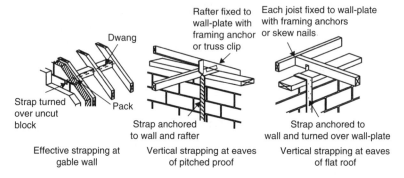

Figure 6.66 Types of strapping

Horizontal lateral restraint at roof level

In small single-storey, single-leaf buildings where straps   D 1D42
cannot pass through a wall, they:

- should be secured to the masonry;
- should be tied to the roof structure as shown in Figure 6.67.

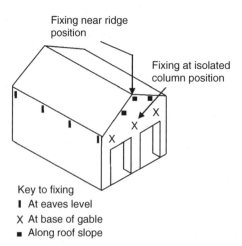

Figure 6.67 Horizontal lateral restraint at roof level

Timber-frame walls

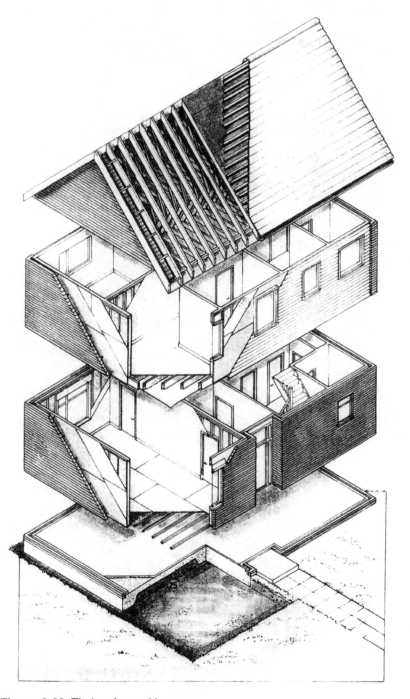

Figure 6.68 Timber-framed house

This type of building typically consists of full height timber wall panels for each storey built on to the floor below and with intermediate floors built on top of the wall panels. The roof is constructed on top of the top storey wall panels with the masonry cladding connected to the timber panels by wall ties.

The following guidance is for timber-framed wall constructions for domestic buildings of not more than two storeys.

General

The roof construction should be:	D 1E2 g
• a flat, raised tie or collared roof; • duo or mono pitch trussed rafters with 15–45° pitch and dead load not more than 1.04 kN/m² on the slope.	

Lateral support by roofs

The wall panels in each storey of a building should be connected to the floors and roofs to provide diaphragm action.	D 1E49
The plasterboard ceiling of the top storey of pitched roofs should: • be fixed directly under the roof; • be secured to the supported wall.	D 1E49b D 1E49c
Spandrel panels should be tied into roof bracing with dwangs placed between vertical elements of the spandrel and fixed by at least 3.1 mm × 75 mm screws as shown in Figure 6.69.	D 1E39d

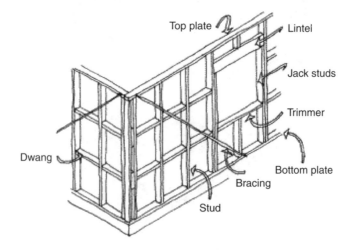

Figure 6.69 Roof bracing with dwangs

Differential movement

The allowances shown in Figure 6.70 should be made for differential movement between timber and masonry construction.   D 1E50

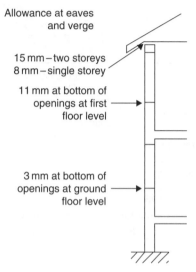

Allowance at eaves and verge

15 mm – two storeys
8 mm – single storey

11 mm at bottom of openings at first floor level →

3 mm at bottom of openings at ground floor level →

Figure 6.70 Allowances for differential movement

The roof of any timber-frame extension that is connected to an existing traditional masonry wall should be supported on a timber bearer connected to the existing wall.   D 1E50

Notches and holes

Notches and holes in flat roof joists should only have: D 1F4

- their holes drilled at the neutral axis;
- notches and holes that are at least 100 mm apart horizontally;
- notches at the top or bottom of a joist (but not coinciding).

Notches and holes should not be cut in rafters, ties, collars or hangers.   D 1F4

Members of trussed rafters should not be cut, trimmed, notched or otherwise altered.   D 1F4

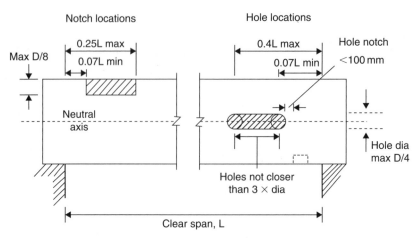

Figure 6.71 Notches and holes

Dead floor and roof loads

Table 6.75 provides an indication of typical dead loads on floors and roofs and is based on 600 mm joist centres and excluding the weight of the joists, partitioning and rafters.

Table 6.75   Dead roof loads

Construction	Dead load (kN/m^2)
Flat roofs	
3 layer felt, 120 mm rigid insulation, vapour layer, 18 mm plywood decking and 12.5 mm plasterboard	0.38
13 mm chippings, 3 layer felt, 120 mm rigid insulation, vapour layer, 22 mm plywood decking and 12.5 mm plasterboard	0.63
Pitched roof	
Concrete tiles, battens and sarking	0.75

Imposed roof loads

Depending on geographical location and altitude (see Figure 6.72) snow loading should be not more than the values shown in Table 6.76.   D 1F7

Figure 6.72 Scottish snow zones

Table 6.76 Imposed roof loads (kN/m²)

Zone	Altitude		
	Below 100 m	Between 100 m and 200 m	Between 200 and 260 m
A	0.75	1.00	See BS 6399-3:1988
B	1.00	1.50	1.50

Raised tie roof

Raised tie roofs designed for access (and limited to mainte- nance or repair purposes) should be capable of supporting a dead load of not more than $0.75\,kN/m^2$ and an imposed load not more than $1.5\,kN/m^2$ for truss centres of not more than 600 mm and a span not more than 5 m.   D 1F12

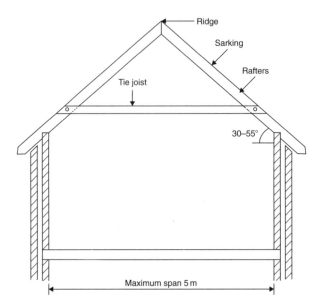

Figure 6.73 Raised roof tie

Collared roof

The roof space is for access only with an imposed load on to the ceiling ties of not more than $0.25\,kN/m^2$ together with a concentrated load of $0.9\,kN$, and does not include for water tanks (Figure 6.74).

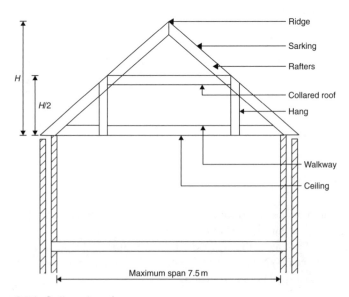

Figure 6.74 Collared roof

Collared roofs that have been designed access limited to maintenance or repair purposes should support dead loads not more than 0.75 kN/m^2 and imposed load not more than 1.5 kN/m^2 for truss centres not more than 600 mm and a span not more than 7.5 m.

D 1F13

Connection details for raised tied and collared roofs

Connections for raised and collared roofs should be:

D 1F14

- 450 mm rafter centres: 38 mm diameter double-sided toothed connector and M10, grade 4.6 bolts should be used; or
- 600 mm rafter centres: 51 mm diameter double-sided toothed connector and M12, grade 4.6 bolts should be used.

Fire precautions

Fire spread from junctions with roofs

Where a compartment wall or separating wall forms a junction with a roof, the junction should maintain the fire resistance duration of the compartment wall in accordance with the following:

- where the roof has a combustible substrate, the wall should project through the roof to a distance of at least 375 mm above the top surface of the roof;

D 2.2.10

- where the wall is taken to the underside of a non-combustible roof substrate, the junction should be fire-stopped and the roof covering should be of low vulnerability for a distance of at least 1.7 m to each side of the centre-line of the wall;

ND 2.1.15

- in the case of a pitched roof covered by slates nailed directly to sarking and underlay, the junction between the sarking and wall-head should be fire-stopped as described in BRE Housing Defects Prevention Unit Defect Action Sheet (Design) February 1985 (DAS 8);
- in the case of a pitched roof covered by slates or tiles fixed to tiling battens and any counter-battens, the junction between the tiles or slates and the underlay should be fully bedded in cement mortar (or other fire-stopping material) at the wall-head.

Fire spread from neighbouring buildings

Buildings are at risk from fires starting beyond their boundaries. The area of greatest vulnerability is the roof and there may be a risk of ignition or penetration by burning brands, flames or heat. The degree of protection for roof coverings (i.e. one or more layers of material such as felt, tiles, slates, sandwich panels, etc.) is dependent on the distance to the boundary.

Solar roof panels should be regarded as forming part of the roof covering and as such should be able to resist ignition from an external source.	D 2.8 ND 2.8

Roof coverings

The possibility of direct flame impingement from neighbouring buildings is greater where the roof covering of the building is close to the boundary. Whilst much will depend on the fire dynamics and the velocity and direction of the wind, burning brands are also likely to be more intense. For these reasons, the vulnerability of a roof covering is determined in relation to the distance of a building to the boundary.

The roof of a building, including any rooflights, but excluding any wall-head fascia, flashing or trim, boxed gutters, soffit or barge boards, should have a low vulnerability if not more than 6 m from the boundary.	D 2.8.1 ND 2.8.1

 Note: Common materials that meet these criteria include slates, tiles, glazing, sandwich panels and certain plastic materials.

The roof of a building, including any rooflights, but excluding any wall-head fascia, flashing or trim, boxed gutters, soffit or barge boards, should have a boundary low or a medium vulnerability if more than 6 m but not more than 24 m from the boundary.	D 2.8.1 ND 2.8.1

 Note: Common materials that meet these criteria are felts and certain plastic materials.

Flat roofs and access decks

Escape across flat roofs can be hazardous because the surface can be exposed to adverse weather conditions and may also have obstructions or no edge protection. Therefore:

Escape routes over flat roofs should only be used where: • the building or part of the building is inaccessible to the general public; • there is more than one escape route from the room or storey leading to the flat roof.	ND 2.9.17

In order to protect evacuees from fire:

> • a flat roof forming part of an escape route should have   ND 2.9.17
>   a medium fire resistance duration for the width of the
>   escape route and for a further 3 m on either side of
>   the escape route.

Escape windows

The normal means of escape from a dwelling in the event of a fire will be by
way of the internal stairs or other circulation areas; however; in some circumstances an escape window may be necessary.

> Escape windows should: D 2.9.4
>   ND 2.9.4
> • be a window, or door (french window) situated in a roof;
> • have an unobstructed openable area that is at least $0.33\,m^2$;
> • be at least 450 mm high and 450 mm.
>
> The bottom of the openable area should not be more than   D 2.9.4
> 1.1 m above the floor.   ND 2.9.4
>
> If a conservatory is located below an escape window   D 2.9.4
> consideration should be given to the design of the conservatory   ND 2.9.4
> roof to withstand the loads exerted from occupants lowering
> themselves onto the roof in the event of a fire.

Destination of escape routes

> An escape route from a flat or maisonette should lead to a   D 2.9.5
> place of safety or an access deck directly or by way of a flat   ND 2.9.6
> roof (but only where there is more than one escape route from
> the storey).

Flat roofs and access decks

> Where a flat roof forms part of an escape route, it should have: D 2.9.6
>
> • a medium fire resistance duration for the width of the escape
>   route and for a further 3 m on either side of the escape route
>   where appropriate;
> • no exhausts of any kind less than 2 m from the escape route;
> • a wall or protective barrier at least 1.1 m high provided on
>   each side of the escape route.

Resistance to fire

The recommended fire resistance duration can be attained where the construction follows the guidance in Columns 3, 4 and 5 of Table 6.77.

Table 6.77   Recommended fire resistance duration

Column 1	Column 2	Column 3	Column 4	Column 5	Column 6	Column 7
Construction	Fire resistance duration	British standards			European standards	Test exposure
		Load-bearing capacity (min)	Integrity (min)	Insulation (min)		
Flat roof or access deck used as a protected route of escape	Short Medium Long	30 60 120	30 60 120	30 60 120	REI 30 RE1 60 REI 120	From the underside
Roof against an external wall of a non-domestic building	Medium	None	60	60	EI60	From the inside

Compartmentation

Where a lower roof abuts an external wall, the roof should provide medium fire resistance duration for a distance of at least 3 m from the wall.

ND 2.A.1
ND 2.B.1

Fire detection and alarm system

In residential care buildings and hospitals, an automatic fire detection and alarm system need not be provided in a roof space which contains only mineral insulated wiring, or wiring laid on metal trays, or in metal conduits, or metal/plastic pipes used for water supply, drainage or ventilating ducting.

ND 2.A.5
ND 2.B.5

Smoke and heat exhaust ventilation systems

A smoke and heat exhaust ventilation system (SHEVS) should be installed in the mall of an enclosed shopping centre and in shops with a storey area more than 1,300 m².

ND 2.C.1

SHEVS should be designed in accordance with the following where appropriate: ND 2.C.1

- the underside of the mall roof should be divided into smoke reservoirs, each of which should be not more than 2,000 m² in area and at least 1.5 m deep measured to the underside of the roof or to the underside of any high-level plant or ducts within the smoke reservoir or the underside of an imperforate suspended ceiling;
- where there is an openwork ceiling, the free area of the ceiling should not be less than 25% of the area of the smoke reservoir or, for natural ventilation, 1.4 times the free area of the roof mounted fire ventilator above (3 times where the height from floor to roof ventilator is more than 12 m), whichever free area is the greater, and be evenly distributed to prevent an unbalanced air flow into the reservoir;
- when a natural ventilation system is used and the smoke reservoir includes a suspended ceiling, other than an openwork ceiling, the free area of the ventilator opening in the suspended ceiling, or any ventilator grille in the ceiling, should not be:
  - less than 1.4 times (3 times where the height from floor to roof ventilator is more than 12 m) that of the roof mounted fire ventilator above in the case of a ventilator opening;
  - 2 times (3.5 times where the height from floor to roof ventilator is more than 12 m) for any ventilator grille.

Cavities

Where cavity barriers are installed between a roof and a ceiling above an undivided space, cavity barriers should be installed not more than 20 m apart.   ND 2.4.2
ND 2.B.2

 Cavity barriers need not be provided to divide a cavity above an operating theatre and its ancillary rooms.

Cavity barriers

In order to inhibit fire spread in a cavity barriers should be installed: 2.4.1

- between a roof space and any other roof space;

- between a cavity and any other cavity (e.g. at the wall-head between a wall cavity and a roof space cavity).

 A cavity barrier is any construction provided to seal a cavity against the penetration of fire and smoke or to restrict its movement within the cavity.

Cavity barriers are not necessary to divide a cavity: 2.4.2

- between a roof and a fire-resisting ceiling;
- formed by external wall or roof cladding, where the exposed surfaces of the cladding are low-risk materials or non-combustible materials attached to a masonry or concrete external wall or a concrete roof, and where the cavity contains only non-combustible material.

Roof space cavities above undivided spaces

The need to provide cavity barriers in a roof space above undivided (or open plan) spaces is less important because roof spaces are regarded as having a low fire risk and the occupants in an undivided (or open plan) space should be aware of any fire developing. In such cases, the occupants should be able to make their escape in the early stages of the fire growth. However, where there is sleeping accommodation, the material exposed in the cavity and the size of a cavity should be controlled due to the nature of the risk. In such cases, the limits set in Table 6.78 should not be exceeded (ND 2.4.3).

Table 6.78 Recommended distance between cavity barriers in roof spaces above undivided spaces (m)

	Where surfaces are non-combustible or low-risk materials (m)	Where surfaces are medium, high- or very high-risk materials (m)
Intended for sleeping	20	15
Not intended for sleeping	No limit (unless it is a residential care building or hospital, in which case the limit is 20 m)	20

Cavities above ceilings in residential buildings (other than residential care buildings and hospitals)

In non-domestic buildings, where a roof space cavity or a   ND 2.4.4
ceiling void cavity extends over a room intended for sleeping, or over such a room and any other part of the building, cavity barriers should be installed on the same plane as the wall.

A cavity barrier installed in a roof space does not need to protect roof members that support the cavity barrier.

Precipitation

Rain penetration occurs most often through walls exposed to the prevailing wet winds (usually south-westerly or southerly) and unless there are adequate damp-proof courses and flashings, etc., materials in parapets and chimneys can collect rainwater and deliver it to other parts of the dwelling below roof level.

A roof exposed to precipitation, or wind-driven moisture, should prevent penetration of moisture to the inner surface of any part of a building so as to protect the occupants and to ensure that the building is not damaged.	D 3.10.1 ND 3.10.1

 Note: Roofs with copper, lead, zinc and other sheet metal roof coverings require provision for expansion and contraction of the sheet material. In 'warm deck' roofs, in order to reduce the risk of condensation and corrosion, it may be necessary to provide a ventilated air space on the cold side of the insulation and a high-performance vapour control layer between the insulation and the roof structure. It may also be helpful to consult the relevant trade association.

Surface water drainage from dwellings

Every building should be provided with a drainage system to remove rainwater from the roof, or other areas where rain water might accumulate, without causing damage to the structure or endangering the health and safety of people in and around the building.	D 3.6.1 ND 3.6.1

 Gutters and rainwater pipes may be omitted from a roof at any height provided it has an area of not more than 8 m^2 and no other area drains onto it.

Treatment of building elements adjacent to the ground

Floors, walls or other building elements adjacent to the ground should be constructed so as to prevent moisture from the ground reaching the inner surface of any part of a building that it could damage.	D 3.4.1 ND 3.4.1
Floors, walls or other building elements adjacent to the ground should be constructed as shown in Table 6.54.	D 3.4.1 to D 3.4.4 ND 3.4.1 to ND 3.4.4

Control of condensation in roofs

Cold, level deck roofs should be avoided because interstitial condensation is likely and its effect on the structure and insulation can be severe. Warm deck and warm deck inverted roof constructions (i.e. where the insulation is placed above the roof deck) are considered preferable.

Fully supported metal roof finishes (including aluminium, copper, lead stainless steel and zinc) should have a ventilated airspace on the cold side of the insulation in addition to a high performance vapour control layer near the inner surface.	D 3.15.3

Surface condensation – thermal bridging

Thermal bridging occurs when the continuity of the building fabric is broken by the penetration of an element allowing a significantly higher heat loss than its surroundings.

To minimize the risk of condensation on any inner surface, cold bridging at a roof should be avoided.	D 3.15.4

Interstitial condensation

A roof should minimize the risk of interstitial condensation in any part of a dwelling that it could damage.	D 3.15.5

Flat roof construction

For the control of condensation in roofs, including cold deck roofs, the recommendations and requirements of BS 5250: 2002 should be followed.	D 3.15.6

Pitched roof construction

To prevent excessive build-up of condensation in cold, pitched roof spaces, where the insulation is at ceiling level the roof space should be cross-ventilated.	D 3.15.7

Special care should be taken with ventilation where ceilings   D 3.15.7
follow the roof pitch.

Ventilation

Passive stack ventilation systems

A passive stack ventilation system uses a duct running from a ceiling (normally in a kitchen or shower room) to a terminal on the roof to remove any moisture-laden air. It operates by a combination of natural stack effect, that is, the movement of air due to the difference in temperature between inside and outside temperatures and the effect of wind passing over the roof of the building. These systems are most suited for use in a building with a height of not more than four storeys as the stack effect will diminish as the air cools.

Every passive stack ventilation system should: D 3.14.6

- be insulated with at least 25 mm thick material having
  a thermal conductivity of 0.04 W/mK where it passes
  through a roof space or other unheated space or where
  it extends above the roof level.

Ventilation of conservatories

With large areas of glazing, conservatories attract large amounts of the sun's radiation that can create unacceptable heat build-up. Efficient ventilation therefore is very important to ensure a comfortable environment.

To minimize the effects of heat build-up, it is recommended   D 3.14.3
that at least 30% of the ventilator area provision is located on
the roof or at as high a level as possible.

Chimneys

Flue terminals in close proximity to roof coverings that   D 3.20.17
are easily ignitable, such as thatch or shingles, should be   ND 3.20.17
located outside Zones A and B in Figure 6.75.

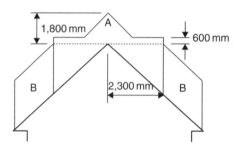

Figure 6.75 Location of flue terminals relative to easily ignitable roof coverings

Where:
Zone A is at least 1.8 m vertically above the weather skin, and at least 600 mm above the ridge; and Zone B is at least 1.8 m vertically above the weather skin, and at least 2.3 m horizontally from the weather skin.

Metal chimneys

A metal chimney should **ONLY** pass through a storage space, cupboard or roof space provided any flammable material is shielded from the chimney by a movable, imperforate casing.	D 3.18.4 ND 3.18.4
Where the chimney passes through the roof space, such as an attic, it should be surrounded be a rigid mesh prevent vermin from building a nest beside the warm chimney. Mesh should prevent an 8 mm diameter sphere from passing.	D 3.18.4 ND 3.18.4

Flue-pipes

A flue-pipe connecting a solid fuel appliance to a chimney should not pass through a roof space.	D 3.18.5 ND 3.18.5

Roof access

Working on roofs can be dangerous. Indeed, falls account for more deaths and serious injuries in the construction industry than any other cause.

A means of safe and secure access should be provided to any roof of building unless it has eaves that, at any part, are at a height of less than 4.5 m above the adjacent ground.	ND 4.8.7
Where fixed ladders are provided and could be accessible to the public, the lowest section of the ladder (i.e. up to 4.5 m) should be demountable to prevent unauthorized access.	ND 4.8.7

Where access to roofs is provided, precautions should be taken to limit the hazards presented by fragile roof surfaces.

There should be: ND 4.8.8

- a clear visible warning identifying any part of a roof that is not capable of bearing a concentrated load of 0.9 kN on a 130 mm by 130 mm^2;
- the relevant hazard sign from BS 5499: Part 5: 2002.

Enclosed storage

Communal enclosures with a roof that are also accessible to   D 3.25.2
people should be at least 2 m high while individual enclosures of wheeled bins only need to be high enough to allow the lid to open.

Noise

There are currently no Building Standards to protect the occupants or users of a non-domestic building from noise but the need may well arise for such standards at a later date.

Resisting sound transmission to dwellings

Measures to reduce the transmission of sound vary according to the type of construction and its reaction to sound energy. The most important factors which affect the behaviour of separating walls and separating floors are mass, cavities, isolation and absorption.

Dwellings directly below a walkway or a roof that is   D 5.1.0
accessible other than for maintenance should be protected by sound-resisting construction.

General application to dwellings

Section 5 of the Domestic Technical Handbook applies to dwellings other than those that are totally detached.

The following requirements should be met in all dwellings other than those that are totally detached.

Impact sound-resisting construction should be provided   D 5.1.1
between a dwelling and a roof that acts as a floor or a walkway
directly above the dwelling so that the dwelling below is
protected from sound emanating from the roof or walkway
above (e.g. roofs that act as floors are access decks, car
parking, escape routes and roof gardens).

Except:

- where two houses are linked only by an imperforate
  separating wall between their ancillary garages, it is not
  necessary for the wall to be airborne sound resisting;
- where the wall between a dwelling and another part of the
  building is substantially open to the external air, it is not
  necessary for the wall to resist airborne sound transmission
  (e.g. a wall between a dwelling and an access deck);
- where the wall between a dwelling and another part of the
  building incorporates a fire door, it is not necessary for the
  door to be airborne sound resisting;
- when a roof or walkway is providing access solely for
  the purpose of maintenance or is solely for the use of the
  residents of the dwelling directly below, it is not necessary
  to provide impact sound-resisting construction;
- in the case of a separating wall or separating floor between
  a dwelling and a private garage or a private waste storage
  area which is ancillary to the same dwelling, it is not
  necessary for the wall or floor to be airborne or impact
  sound resisting.

   Requirements for roof junctions with walls are
contained in Section 6.7.

Maximum *U*-values

The maximum *U*-values for building elements of the   D 6.2.1
insulation envelope should not be greater than those shown   LD 6.2.1
in Table 6.79.

Stand-alone buildings

Where the area of a communal room or other heated   D 6.2.13
accommodation associated with a block of dwellings is
less than $50\,m^2$, these rooms or accommodation may be
treated as a stand-alone building.

Table 6.79   Maximum _U_-values for building elements of the insulation envelope

Type of element	(a) Area-weighted average _U_-value (W/m²K) for all elements of the same type	(b) Individual element _U_-value (W/m²K)
Normal roof	0.20 (0.25 for non-domestic buildings)	0.35
Shell and fit-out buildings	0.16	0.35
Conversion of heated buildings	0.35	
Conversion of a historic building with a pitched roof (insulation between ceiling ties or collars)	0.16	0.35
Conversion of a historic building with a flat roof or pitched roof (insulation between rafters or roof with integral insulation)	0.20	0.35
Conversion of a historic building with a pitched roof (insulation between ceiling ties or collars)	0.16	0.35

Elements (including dividing elements) should have _U_-values equal to or better than those chosen for the rest of the building.   D 6.2.13

In these rooms or accommodation, the area of windows, doors, rooflights and roof windows should be limited to 25% of the total floor area of these parts.   D 6.2.13

Alterations to the insulation envelope

Alterations that involve increasing the floor area and/or bringing parts of the existing building that were previously outwith the insulation envelope into the heated part of the dwelling are considered as extensions and/or conversions (regulation 4, schedule 2).

The infill of an existing roof void of approximately 4 m² or less in the building fabric should have a _U_-value which matches at least that of the remainder of the surrounding element and in the case of a wall or floor it should not be worse than 0.35 W/m²K for a roof.   D 6.2.11   ND 6.2.12

For a building that was in a ruinous state the roof should, after   D 6.2.10
renovation, be able to achieve at least an average U-value of
0.35 and in the case of a wall or floor, 0.70 W/m^{2}K.

Conversion of unheated buildings

Converting an unheated building (such as the roof space of a dwelling or
a garage, etc.) for future domestic use can be quite a challenge in terms of
achieving energy efficiency and because of this, the most demanding of measures are recommended when conversion occurs. Table 6.80 shows some examples of the work involved.

Table 6.80   Conversion of unheated buildings

Location	Work involved
Conversion of a roof space in a dwelling	This will usually involve extending the insulation envelope to include, the gables, the collars, a part of the rafters and the oxters, as well as any new or existing dormer construction.
	The opportunity should also be taken at this time to upgrade any remaining poorly performing parts of the roof which are immediately adjacent to the conversion, for example, insulation to parts of the ceiling ties at the eaves.
Conversion of an unheated garage	This will usually involve extending the insulation envelope to include the existing floor, perimeter walls and the roof/ceiling to the new habitable part.
Conversion of a deep solum space into an apartment	This will usually involve extending the insulation envelope to include the solum/existing floor and perimeter walls to the new habitable part.

Vehicle protective barriers

Where vehicles are introduced into a building, measures should be taken to
protect people from any additional risks presented.

If vehicles have access to a roof, a vehicle protective barrier   D 4.12.1
(400 mm minimum height) should be provided to the edge of   ND 4.12.1
any such area that is above the level of any adjoining floor,
ground or any other route for vehicles.

6.10 Chimneys and fireplaces

If a person erects or raises a building that is (or is going to be) taller than the chimneys and/or flues from an adjoining building that is either joined by a party wall or less than six feet away from the taller building, then the local authority may, if reasonably practical:

- require that person to build up those chimneys and flues, so that their top is of the same height as the top of the chimneys of the taller building or the top of the taller building, whichever is the higher;
- require the owner or occupier of the adjoining building to allow the person erecting or raising the building, access to the adjacent building so that he can carry out such work as may be necessary to comply with the notice served on him.

6.10.1 Requirements

Design and construction

Every building must be designed and constructed in such a way that in the event of damage occurring to any part of the structure of the building the extent of any resultant collapse will not be disproportionate to the original cause.	*1.2*

Fire

Every building, which is divided into more than one area of different occupation, must be designed and constructed in such a way that in the event of an outbreak of fire within the building, fire and smoke are inhibited from spreading beyond the area of occupation where the fire originated.	*2.2*
Every building must be designed and constructed in such a way that in the event of an outbreak of fire within the building, fire and smoke are inhibited from spreading beyond the compartment of origin until any occupants have had the time to leave that compartment and any mandatory fire containment measures have been initiated.	*2.1*

 Standard 2.1 does not apply to a domestic building.

Escape

Every building must be designed and constructed in such a way that in the event of an outbreak of fire within the building, the occupants, once alerted to the outbreak of the fire, are provided with the opportunity to escape from the building, before being affected by fire or smoke.   2.9

Environment

Every building must be designed and constructed in such a way that: 3.10

- *there will not be a threat to the building or the health of the occupants as a result of moisture from precipitation penetrating to the inner face of the building;*
- *the products of combustion are carried safely to the external air without harm to the health of any person through leakage, spillage, or exhaust nor permit the re-entry of dangerous gases from the combustion process of fuels into the building;*   3.20
- *each fixed combustion appliance installation operates safely;*   3.17
- *each fixed combustion appliance installation receives air for combustion and operation of the chimney so that the health of persons within the building is not threatened by the build-up of dangerous gases as a result of incomplete combustion;*   3.21
- *each fixed combustion appliance installation receives air for cooling so that the fixed combustion appliance installation will operate safely without threatening the health and safety of persons within the building;*   3.22
- *component parts of a fixed combustion appliance installation will not cause damage to the building in which it is installed by radiated, connected or conducted heat or from hot embers expelled from the appliance;*   3.19
- *component parts of a fixed combustion appliance installation used for the removal of combustion gases will withstand heat generated as a result of its operation without any structural change that would impair the stability or performance of the installation.*   3.18

6.10.2 Meeting the requirements

Structure

 Note: Also see Section 6.20 of this book 'Combustion appliances'.

Vertical lateral restraint to walls

The ends of every wall should be bonded or otherwise securely tied throughout their full height to a buttressing wall, pier or chimney.	D 1D27
Long walls may be provided with intermediate support dividing the wall into distinct lengths.	D 1D27
The buttressing wall, pier or chimney should provide support from the base to the full height of the wall.	D 1D27

Buttressing walls

Buttressing walls are external masonry return walls or internal walls that are perpendicular to the supported wall (Figure 6.76).

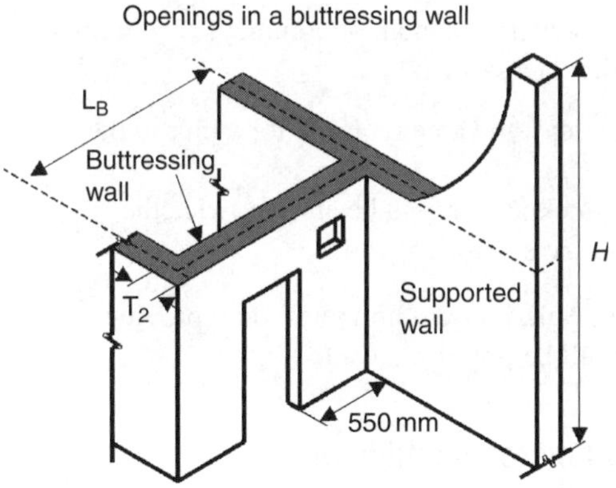

Openings in a buttressing wall

Figure 6.76 Openings in a buttressed wall

The buttressing wall should be bonded or securely tied to the supported wall, a buttressing wall, pier or chimney.	D 1D28

Chimneys providing restraint

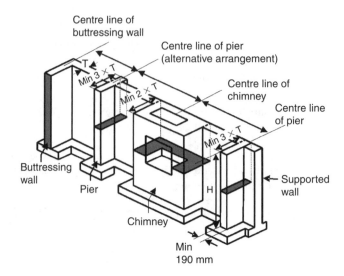

Figure 6.77 Supporting chimneys

Chimneys should be twice the thickness of the supported wall, measured at right angles to the wall.	D 1D29
The cross-sectional area on plan of chimneys (excluding openings for fireplaces): • should be at least the area required for a pier in the same wall; • the overall thickness should be at least twice the thickness of the supported wall.	D 1D29
The buttressing wall, pier or chimney should provide support to the full height of the wall.	D 1D29

Lateral support by roofs and floors

Floors and roofs should: • transfer lateral forces from walls to buttressing walls, piers or chimneys; • be secured to the supported wall.	D 1D35

Masonry chimneys

If a chimney is not supported by ties or securely restrained then its height should be no greater than 4.5 × W.	D 1D43

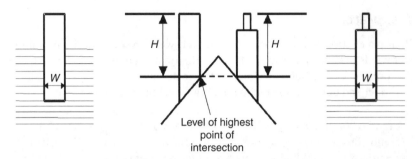

Figure 6.78 Masonry chimneys

Chimneys and flue-pipes

A chimney or flue-pipe should be constructed so that, in the event of a fire, the fire resistance duration of the compartment wall or compartment floor is maintained.	ND 2.1.14

Openings and service provisions

Openings and service penetrations through walls or floors can compromise their effectiveness and should be kept to a minimum and should be carefully detailed and constructed to resist fire and the spread of fire.

A chimney or flue-pipe should be constructed so that, in the event of a fire, the fire resistance duration of the separating wall or separating floor is maintained.	D 2.2.9
Where greater differential movement is anticipated, either in normal use or during fire exposure, proprietary fire-stopping products should be used.	D 2.2.9

 Proprietary fire-stopping products and materials such as cement mortar; gypsum-based plaster; cement or gypsum-based vermiculite/perlite mixes; mineral fibre; crushed rock and blast furnace slag or ceramic-based products (with or without resin binders) may be used.

Openings and service penetrations

Fire and smoke can easily pass through openings and service penetrations in protected routes of escape thus preventing the occupants from escaping in the event of an outbreak of fire within the building. For this reason:

• any openings or service penetrations in protected routes of escape should be limited to chimneys, flue-pipes, self-closing fire doors and service openings; and • openings should be fire-stopped.	ND 2.9.31

Precipitation

Rain penetration occurs most often through walls exposed to the prevailing wet winds (usually south-westerly or southerly) and unless there are adequate damp-proof courses and flashings, etc., materials in chimneys can collect rainwater and deliver it to other parts of the dwelling below roof level.

A roof exposed to precipitation, or wind-driven moisture, should prevent penetration of moisture to the inner surface of any part of a building so as to protect the occupants and to ensure that the building is not damaged.	D 3.10.1 ND 3.10.1

The following general recommendations should be followed for walls or roofs, as appropriate (Table 6.81):

Table 6.81   Precipitation – general provisions

Element	Type of construction
Masonry walls of bricks and/or blocks incorporating damp-proof courses, flashings and other materials and components	Constructed to suit the degree of exposure to wind and rain
Masonry walls incorporating external rendering	Constructed to suit the degree of exposure and the type of masonry
Masonry walls of natural stone or cast stone blocks	Constructed to suit the degree of exposure to wind and rain

Note: Roofs with copper, lead, zinc and other sheet metal roof coverings require provision for expansion and contraction of the sheet material.

Labelling

Where a hearth, fireplace (including a flue box) or system chimney is provided, extended or altered (or a flue liner is provided as part of refurbishment work) information essential to the correct application and use of these facilities should be permanently posted in the dwelling to alert future workmen to the specification of the installed system.

The labels should be indelibly marked and contain the following information: • the location of the hearth, fireplace (or flue box), or the location of the beginning of the flue; • a chimney designation string in accordance with BS EN 1443: 2003; • the category of the flue and generic types of appliance that can safely be accommodated;	D 3.17.7 ND 3.17.8

- the type and size of the flue (or its liner);
- the installation date.

Labels should be located in a position that will not easily be obscured such as adjacent to:	D 3.17.7 ND 3.17.8

- the gas or electricity meter;
- the water supply stopcock;
- the chimney or hearth described.

A label should be provided similar to that shown in Figure 6.79.	D 3.17.7 ND 3.17.8

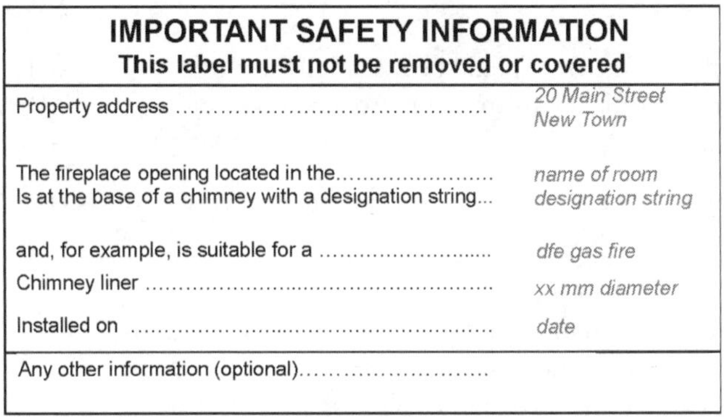

IMPORTANT SAFETY INFORMATION
This label must not be removed or covered

Property address .... *20 Main Street*
                                                          New Town

The fireplace opening located in the........................ *name of room*
Is at the base of a chimney with a designation string... *designation string*

and, for example, is suitable for a .... *dfe gas fire*

Chimney liner .... *xx mm diameter*

Installed on .... *date*

Any other information (optional)..........................

Figure 6.79 Labelling of combustion appliances

Construction

Protection from combustion products

It is essential that flues continue to function effectively when in use without allowing the products of combustion to enter the building.

Every building must be designed and constructed in such a way that any component part of each fixed combustion appliance installation used for the removal of combustion gases will withstand heat generated as a result of its operation without any structural change that would impair the stability or performance of the installation.	D 3.18 ND 3.18

Combustion appliances, other than flueless appliances such as gas cookers, should incorporate, or be connected to, a flue-pipe and/or a chimney that will withstand the heat generated by the normal operation of the appliance.	D 3.18.1 ND 3.18.1

Chimneys and flue-pipes should be swept **at least annually** if smokeless solid fuel is burnt and more often if wood, peat and/or other high volatile solid fuel such as bituminous coal is burnt.

D 3.18.1
ND 3.18.1

Mechanical sweeping with a brush is the recommended method of cleaning.

Every chimney should have such capacity, be of a height and location and with an outlet so located that the products of combustion are discharged freely and will not present a fire hazard.

D 3.18.1
ND 3.18.1

A flue should be free from obstructions.

D 3.18.1
ND 3.18.1

The surface of the flue should be essentially uniform, gas-tight and resistant to corrosion from combustion products.

D 3.18.1
ND 3.18.1

Chimneys should be constructed in accordance with:

D 3.18.1

- the recommendations of BS 6461: Part 1: 1984 for masonry chimneys; or

ND 3.18.1

- the recommendations of BS 7566: Parts 1–4: 1992 for metal system chimneys;
- BS 5410: Part 1: 1997 and OFTEC Technical Information Sheets TI/129, TI/132 and TI/135, where serving an oil-firing appliance;
- BS 5440: Part 1: 2000, where serving a gas-fired appliance.

Table 6.82   Recommended designation for chimneys and flue-pipes for use with oil-firing appliances with a flue gas temperature not more than 250°C

Appliance type	Fuel oil	Designation
Boiler including combination boiler – pressure jet burner	Class C2	T250 N2 D 1 Oxx
Cooker – pressure jet burner	Class C2	T250 N2 D 1 Oxx
Cooker and room heater – vaporizing burner	Class C2	T250 N2 D 1 Oxx
Cooker and room heater – vaporizing burner	Class D	T250 N2 D 2 Oxx
Condensing pressure jet burner appliances	Class C2	T160 N2 W 1 Oxx
Cooker – vaporizing burner appliances	Class D	T160 N2 W 2 Oxx

Table 6.83   Recommended designation for chimneys and flue-pipes for use with gas appliances

Appliance type	Fuel oil	Designation
Boiler – open-flued	Natural draught	T250 N2 D 1 Oxx
	Fanned draught	T250 P2 D 1 Oxx
	Condensing	T160 P2 W 1 Oxx
Boiler – room-sealed	Natural draught	T250 N2 D 1 Oxx
	Fanned draught	T250 P2 D 1 Oxx
Gas fire	Radiant/convector, ILFE or DFE	T250 N2 D 1 Oxx
Air heater	Natural draught	T250 N2 D 1 Oxx
	Fanned draught	T200 P2 D 1 Oxx
	SE duct	T450 N2 D 1 Oxx

Masonry chimneys

A new masonry chimney, usually custom-built on site and normally with an outer wall of brick, block or stone, should be:	D 3.18.3
• well constructed and incorporate a flue liner, or flue-blocks, of either clay material or pre-cast concrete;	ND 3.18.3
• constructed in accordance with the recommendations in BS 6461: Part 1: 1984.	D 3.18.3 ND 3.18.3

> If an outer wall is constructed of concrete it should be constructed in accordance with BS EN 12446: 2003.

Flue-blocks should be constructed and installed in accordance with recommendations in:	D 3.18.3 ND 3.18.3
• BS EN 1858: 2003, for a precast concrete flue-block chimney;	
• BS EN 1806: 2000, for a clay flue-block chimney.	

Note: Chimneys can also be constructed of prefabricated block components, designed for quick construction.

Metal chimneys

Metal chimneys may be either single-walled or double-walled.	D 3.18.4 ND 3.18.4
Some metal chimneys are specifically designed for use with gas-fired appliances and should not be used for solid fuel appliances because of the higher temperatures and greater corrosion risk.	D 3.18.4 ND 3.18.4

A metal chimney should:	D 3.18.4
• not pass through a compartment wall, compartment floor, separating wall or separating floor;	ND 3.18.4 D 3.18.4
• only pass through a storage space, cupboard or roof space if all flammable material is shielded from the chimney by movable, imperforate casing.	ND 3.18.4
Where the chimney passes through the roof space, such as an attic, it should be surrounded by a rigid mesh that will prevent vermin from building a nest beside the warm chimney.	D 3.18.4 ND 3.18.4
The mesh should prevent an 8 mm diameter sphere from passing.	
There should be no joints within any wall, floor or ceiling that make accessing the chimney for maintenance purposes difficult.	D 3.18.4 ND 3.18.4

Location of metal chimneys

To minimize the possibility of condensation in a metal chimney, it should not be fixed externally to a building, but should be routed inside the building.	D 3.20.14 ND 3.20.14

Protection of metal chimneys

Metal chimneys should be guarded if there could be a risk of damage or if they present a risk to people that is not immediately apparent (e.g. when they traverse intermediate floors out of sight of the appliance).	D 3.20.7 ND 3.20.7
Where the metal chimney passes through a room or accessible space such as a walk-in cupboard it should be protected in accordance with Table 6.83.	D 3.20.7 ND 3.20.7

Table 6.84   Protection of metal chimneys

For solid fuel appliances	BS EN 12391-1: 2003
For oil-firing appliances	BS 5410: Part 1: 1997
For gas appliances	BS 5440: Part 1: 2000

It is not necessary to provide protection where a system chimney runs within the same space as the appliance served.

Flues

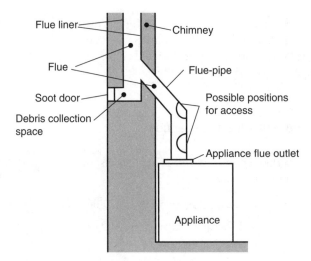

Flue liner

Chimney

Flue

Flue-pipe

Soot door

Possible positions for access

Debris collection space

Appliance flue outlet

Appliance

Figure 6.80 Chimneys and flues

Size of flues – solid fuel appliances

The size of a flue serving a solid fuel appliance should be at least the size shown in Table 6.85.   D 3.20.8
ND 3.20.8

Table 6.85   Minimum areas of flues

Appliance	Minimum flue size
Fireplaces with an opening more than 500 mm × 550 mm, or a fireplace exposed on 2 or more sides	15% of the total face area of the fireplace opening
Fireplaces with an opening not more than 500 mm × 550 mm	200 mm diameter or rectangular/square flues having the same cross-sectional area and a minimum diameter not less than 175 mm
Closed appliance with rated output more than 30 kW but not more than 50 kW, burning any fuel	175 mm diameter or rectangular/square flues having the same cross-sectional area and a minimum diameter not less than 150 mm
Closed appliance with rated output not more than 30 kW burning any fuel	150 mm diameter or rectangular/square flues having the same cross-sectional area and a minimum diameter not less than 125 mm
Closed appliance with rated output not more than 20 kW that burns smokeless or low volatiles fuel	125 mm diameter or rectangular/square flues having the same cross-sectional area and a minimum diameter not less than 100 mm for straight flues or 125 mm for flues with bends or offsets

Size of flues – oil-firing appliances

The cross-sectional area of a flue serving an oil-firing appliance should be in accordance with the recommendations in BS 5410: Part 1: 1997 and should be the same size as the appliance flue spigot.	D 3.20.9 ND 3.20.9

Size of flues – gas-fired appliances

The area of a flue serving a gas-fired appliance should have a size to ensure safe operation.	D 3.20.10 ND 3.20.10
A flue should be provided in accordance with Table 6.86.	D 3.20.10 ND 3.20.10

Table 6.86   Size of flues – oil-firing appliances

For a decorative fuel-effect gas appliance	Clause 9 of BS 5871: Part 3: 2005
For an inset live fuel-effect gas appliance	BS 5871: Part 2: 2005
For any other gas-fired appliance	BS 5440: Part 1: 2000

Design of flues

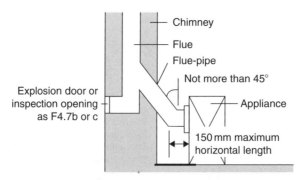

Section through appliance and flue-pipe

Figure 6.81 Design of flues

A combustion appliance should be connected to a chimney that discharges to the external air.	D 3.20.11 ND 3.20.11
Every solid fuel appliance should be connected to a separate flue.	D 3.20.11 ND 3.20.11

Every oil-firing appliance should be connected to a separate flue (unless the appliance has pressure jet burners and is connected into a shared flue).	D 3.20.11 ND 3.20.11
Every gas-fired appliance that requires a flue should connect into a separate flue.	D 3.20.11 ND 3.20.11
The flue of a natural draught appliance, such as a traditional solid fuel appliance, should offer the least resistance to the passage of combustion gases.	D 3.20.11 ND 3.20.11
The horizontal length of the back-entry flue-pipe at the point of discharge from the appliance should be not more than 150 mm.	D 3.20.11 ND 3.20.11
Where bends are essential, they should be angled at not more than 45° to the vertical.	D 3.20.11 ND 3.20.11

Openings in flues

A flue-pipe may penetrate the floor, provided that it discharges into either a masonry chimney carried by the floor or any other type of chimney enclosed within a non-combustible duct that is lined with absorbent mineral fibre.

The flue should have no intermediate openings.	D 3.20.12 ND 3.20.12

Access to flues

Access should be provided for inspection and cleaning of the flue and the appliance and an opening that is fitted with a non-combustible, rigid, gas-tight cover would be acceptable.	D 3.20.13 ND 3.20.13

Flue-pipes

A flue-pipe should be of a material that will safely discharge the products of combustion into the flue under all conditions that will be encountered.	D 3.18.5 ND 3.18.5
A flue-pipe serving a solid fuel appliance should be non-combustible and of a material and construction capable of withstanding the effects of a chimney fire without any structural change that would impair the stability and performance of the flue-pipe.	D 3.18.5 ND 3.18.5

Flue-pipes should be manufactured from the following materials:	D 3.18.5 ND 3.18.5

- cast iron pipe;
- mild steel at least 3 mm thick;
- vitreous enamelled steel;
- stainless steel designated Vm-L50100; or
- any other material approved and tested under the relevant conditions of a Notified Body.

Flue-pipes should have the same diameter or equivalent cross-sectional area as that of the appliance flue outlet and should be to the size recommended by the appliance manufacturer.	D 3.18.5 ND 3.18.5
A flue-pipe connecting a solid fuel appliance to a chimney should not pass through:	D 3.18.5 ND 3.18.5

- a roof space;
- an internal wall, although it is acceptable to discharge a flue-pipe into a flue in a chimney formed wholly or partly by a non-combustible wall;
- a ceiling or floor.

Flue liners

A flue liner is the wall of the chimney that is in contact with the products of combustion. It can generally be of concrete, clay, metal or plastic depending on the designation of the application.

All new chimneys will have flue liners installed and there are several types, as follows:	D 3.18.6 ND 3.18.6

- rigid sections of clay or refactory liner;
- rigid sections of concrete liner;
- rigid metal pipes.

Stainless steel flexible flue liners meeting BS EN 1856-2: 2005 may be used for lining or relining flues for oil and gas appliances, and for lining flues for solid fuel applications provided that the designation is in accordance with the intended application.	D 3.18.6 ND 3.18.6
Single-skin, stainless steel flexible flue liners may be used for lining flues for gas and oil appliances.	D 3.18.6 ND 3.18.6
Double-skin, stainless steel flexible flue liners for multi-fuel use.	D 3.18.6 ND 3.18.6

Existing custom-built masonry chimneys may be lined or re-lined by one of the following flue liners:

D 3.18.6
ND 3.18.6

- flexible, continuous length, single-skin stainless steel for lining or re-lining chimney flues for C2 oil and gas installations designated T250;
- flexible, continuous length, double-skin stainless steel for lining or re-lining systems designated T400 for multi-fuel installations;
- insulating concrete pumped in around an inflatable former;
- spray-on or brush-on coating by specialist.

Existing chimneys for solid fuel applications may also be re-lined using approved rigid metal liners or single-walled chimney products, an approved cast-*in situ* technique or an approved spray-on or brush-on coating. Approved products are listed in the Heating Equipment Testing Approval Scheme's (HETAS) Guide.

D 3.18.6
ND 3.18.6

Masonry liners for use in existing chimneys should meet the following requirements:

D 3.18.6
ND 3.18.6

- be selected to form the flue without cutting and to keep joints to a minimum;
- bends and offsets should only be formed with factory-made components;
- liners should be placed with the sockets or rebate ends uppermost to contain moisture and other condensates in the flue.

Solid fuel appliance flue outlets

The outlet from a flue should be located externally at a safe distance from any opening, obstruction or flammable or vulnerable materials and in accordance with Figure 6.82.

D 3.20.17
ND 3.20.17

Flue terminals in close proximity to roof coverings that are easily ignitable, such as thatch or shingles, should be located outside Zones A and B in Figure 6.95:

D 3.20.17
ND 3.20.17

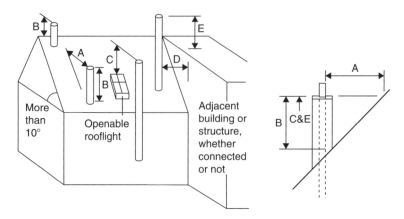

Figure 6.82 Solid fuel appliance flue outlets. Where: A is 2.3 m horizontally clear of the weather skin; B is 1.0 m provided A is satisfied or 600 mm where above the ridge; C is 1.0 m above the top of any flat roof; and 1.0 m above any openable rooflight, dormer or ventilator, etc. within 2.3 m measured horizontally; and E must be at least 600 mm where D is not more than 2.3 m.

Notes: Horizontal dimensions are to the surface surrounding the flue. Vertical dimensions are to the top of the chimney terminal.

Terminal discharges at low level

Flues discharging at low level where they may be within reach of people should be protected with a terminal guard.	D 3.20.15 ND 3.20.15
A flue terminal should be protected with a guard if a person could come into contact with it or if it could be damaged.	D 3.20.15 ND 3.20.15
If the flue outlet is in a vulnerable position (such as where the flue discharges within reach of the ground, or a balcony, veranda or window) it should be designed to prevent the entry of matter that could obstruct the flow of gases.	D 3.20.15 ND 3.20.15

Air for combustion

All combustion appliances need ventilation to supply them with oxygen for combustion.

Every building must be designed and constructed in such a way that each fixed combustion appliance installation receives air for combustion and operation of the chimney so that the health of persons within the building is not threatened by the build-up of dangerous gases as a result of incomplete combustion.	D 3.21 ND 3.21

Supply of air for combustion generally

- A room containing an open-flued appliance may need permanently open air vents.
- An open-flued appliance needs to receive a certain amount of air from outside dependent upon its type and rating.
- Infiltration through the building fabric may be sufficient but above certain appliance ratings permanent openings are necessary.

Ventilators for combustion should be located so that occupants are not provoked into sealing them against draughts and noise.	D 3.21.1 ND 3.21.1
Discomfort from draughts can be avoided by placing vents close to appliances, e.g. floor ventilators, by drawing air from intermediate spaces such as hallways or by ensuring good mixing of incoming air.	D 3.21.1 ND 3.21.1
Air vents should not be located within a fireplace recess except on the basis of specialist advice.	D 3.21.1 ND 3.21.1
Noise-attenuated ventilators may be needed in certain circumstances.	D 3.21.1 ND 3.21.1
Appliance compartments that enclose open-flued appliances should be provided with vents large enough to admit all the air required by the appliance for combustion and proper flue operation, whether the compartment draws air from the room or directly from outside.	D 3.21.1 ND 3.21.1

Supply of air for combustion to solid fuel appliances

A solid fuel appliance installed in a room or space should have a supply of air for combustion by way of permanent ventilation either direct to the open air or to an adjoining space (including a sub-floor space) that is itself permanently ventilated direct to the open air.	D 3.21.2 ND 3.21.2
An air supply should be provided in accordance with Table 6.87.	D 3.21.2 ND 3.21.2

Combustion appliances and combustible material
Relationship to combustible materials

Combustion appliances and their component parts, particularly solid fuel appliance installations, generate or dissipate considerable temperatures. Certain precautions need to be taken to ensure that any high temperatures are not sufficient to cause a risk to people and the building.

Table 6.87 Supply of air for combustion

Type of appliance	Minimum ventilation opening sizes [2]
Open appliance without a throat [1]	A permanent air entry opening or openings with a total free area of 50% of the cross-sectional area of the flue
Open appliance with a throat [1]	A permanent air entry opening or openings with a total free area of 50% of the throat opening area
Any other solid fuel appliance	A permanent air entry opening or openings with a total free area of 550 mm^2 for each kW of combustion appliance rated output more than 5 kW. (A combustion appliance with an output rating of not more than 5 kW has no minimum requirement, unless stated by the appliance manufacturer)

Notes:

1. Throat means the contracted part of the flue lying between the fireplace opening and the main flue.
2. Where a draught stabilizer is fitted to a solid fuel appliance, or to a chimney or flue-pipe in the same room as a solid fuel appliance, additional ventilation opening should be provided with a free area of at least 300 mm^2/kW of solid fuel appliance rated output.
3. Nominal fire size is related to the free opening width at the front of the fireplace opening.

Every building must be designed and constructed in such a way that any component part of each fixed combustion appliance installation will not cause damage to the building in which it is installed by radiated, convected or conducted heat or from hot embers expelled from the appliance.	D 3.19 ND 3.19

Relationship of masonry chimneys to combustible material

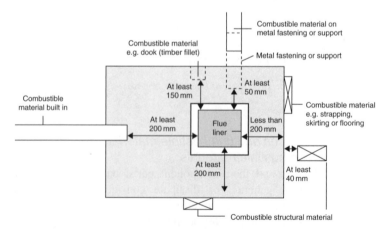

Figure 6.83 Plan view of a masonry chimney

Combustible material should not be located where the heat dissipating through the walls of fireplaces or flues could ignite it.

D 3.19.1
ND 3.19.1

All combustible materials should be located at least 200 mm from the surface surrounding a flue in a masonry chimney.

D 3.19.1
ND 3.19.1

Unless they are:

- a damp-proof course(s) firmly bedded in mortar;
- small combustible fixings may be located not less than 150 mm from the surface of the flue;
- combustible structural material may be located not less than 40 mm from the outer face of a masonry chimney;
- flooring, strapping, sarking, or similar combustible material may be located on the outer face of a masonry chimney.

Any metal fastening in contact with combustible material, such as a joist hanger, should be at least 50 mm from the surface surrounding a flue to avoid the possibility of the combustible material catching fire due to conduction.

D 3.19.1
ND 3.19.1

Relationship of system chimneys to combustible material

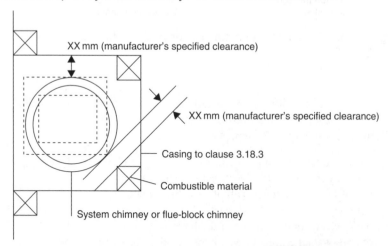

Figure 6.84 Plan of casing round a factory-made chimney

System chimneys do not necessarily require to be located at such a distance from combustible material. It is the responsibility of the chimney manufacturer to declare a distance 'XX', as stipulated in BS EN 1856-1: 2003 and BS EN 1858: 2003, as being a safe distance from the chimney to combustible material.

D 3.19.2
ND 3.19.2

Relationship of metal chimneys to combustible material

There should be a separation distance where a metal chimney passes through combustible material. This is specified, as part of the designation string for a system chimney when used for oil or gas, as (Gxx), where xx is the distance in mm.

 There is no need for a separation distance if the flue gases are not likely to exceed 100°C (Figure 6.85).

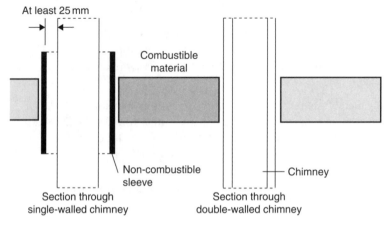

Figure 6.85 Relationship of metal chimneys to combustible material

Where no data is available, the separation distance for oil or gas applications with a flue gas temperature limit of T250 or less should be 25 mm:	D 3.19.3
• measured from the outer surface of a single-walled chimney to combustible material measured from the surface of the inner wall of a double-walled chimney;	ND 3.19.3
• measured from the surface of the inner wall of a double-walled chimney) where the metal chimney runs in close proximity to combustible material.	D 3.19.3 ND 3.19.3

Relationship of flue-pipes to combustible material

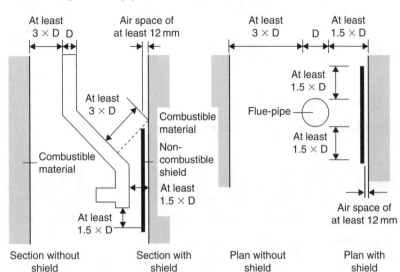

Figure 6.86 Relationship of flue-pipes to combustible material

To prevent the possibility of radiated heat starting a fire, a flue-pipe should be separated from combustible material by:

D 3.19.4
ND 3.19.4

- a distance according to the designation of the connecting flue-pipe in accordance with BS EN 1856-2: 2005; or
- a distance equivalent to at least 3 times the diameter of the flue-pipe. However, this distance may be reduced:

  - to 1.5 times the diameter of the flue-pipe, if there is a non-combustible shield provided in accordance with the following sketch; or
  - to 0.75 times the diameter of the flue-pipe, if the flue-pipe is totally enclosed in non-combustible material at least 12 mm thick with a thermal conductivity of not more than 0.065 W/mK.

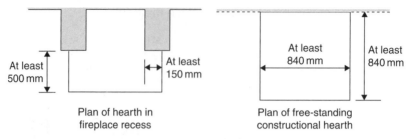

Plan of hearth in
fireplace recess

Plan of free-standing
constructional hearth

Figure 6.87 Relationship of solid fuel appliance to combustible material

A solid fuel appliance should be provided with a solid, non-combustible hearth that will prevent the heat of the appliance from igniting combustible materials.	D 3.19.5 ND 3.19.5
A hearth should be provided to the following dimensions: • a constructional hearth at least 125 mm thick and with plan dimensions in accordance with Figures 6.88 and 6.89; or • a free-standing, solid, non-combustible hearth at least 840 × 840 mm minimum plan area and at least 12 mm thick, provided the appliance will not cause the temperature of the top surface of the hearth on which it stands to be more than 100°C.	D 3.19.5 ND 3.19.5
A solid fuel appliance should: • sit on a hearth; • be located on the hearth such that protection will be offered from the risk of ignition of the floor by direct radiation, conduction or falling embers; • be in accordance with Figure 6.88.	D 3.19.5 ND 3.19.5

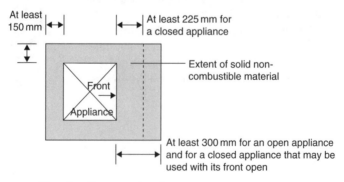

Plan of appliance on a hearth

Figure 6.88 Locating a solid fuel appliance

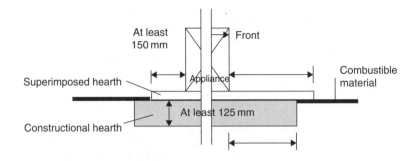

Section through superimposed hearth

Figure 6.89 Solid fuel appliance – superimposed hearth

Unless the solid fuel appliance is located in a fireplace recess or where the back or sides of the hearth either abut or are carried into a solid, non-combustible wall, in which case it may sit on a superimposed hearth provided that the hearth is of solid, non-combustible material and at least 50 mm thick in accordance with Figure 6.89.	D 3.19.5 ND 3.19.5

Relationship of oil-firing appliances to combustible material

A hearth is not required beneath an oil-firing appliance if it incorporates a full-sized, rigid non-combustible base and does not raise the temperature of the floor beneath it to more than 100°C under normal working conditions.	D 3.19.6 ND 3.19.6
A floor-standing, oil-firing appliance should be positioned on the hearth in accordance with Figure 6.90 and in such a way as to minimize the risk of ignition of any part of the floor by direct radiation or conduction.	D 3.19.6 ND 3.19.6

An oil-firing appliance should be separated by: • a shield of non-combustible material at least 25 mm thick; or • an air space of at least 75 mm from any combustible material if the temperature of the back, sides or top of the appliance is more than 100°C under normal working conditions.	D 3.19.6 ND 3.19.6

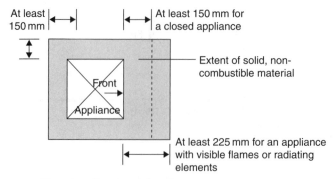

Plan of appliance on a hearth

Figure 6.90 Location of oil-fired appliance

 Note: The 150 mm does not apply where the appliance is located in a fire-place recess, nor does it apply where the back or sides of the hearth either abut or are carried into a solid, non-combustible wall.

Relationship of gas-fired appliances to combustible material

A gas-fired appliance should be provided with a hearth as shown in Table 6.88:

Table 6.88 Gas-fired appliance – alternative hearth

Gas fire, convector heater and fire/back boiler	Clause 12 of BS 5871: Part 1: 2005
Inset live fuel-effect gas appliance	Clause 12 of BS 5871: Part 2: 2005
Decorative fuel-effect gas appliance	Clause 11 of BS 5871: Part 3: 2005
All other gas-fired appliances	The hearth should be a solid, heat-resistant, non-combustible, non-friable material at least 12 mm thick and at least the plan dimension shown in Fig. 6.91

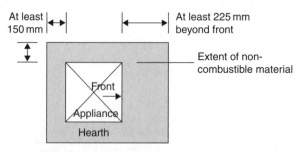

Plan of appliance on a hearth

Figure 6.91 Gas-fired appliances

A hearth need not be provided:	D 3.19.7
• where every part of any flame or incandescent material in the appliance is at least 225 mm above the floor; or	ND 3.19.7
• where the appliance is designed not to stand on a hearth, such as a wall-mounted appliance or a gas cooker.	
A gas-fired appliance should be separated from any combustible material if the temperature of the back, sides or top of the appliance is more than 100°C under normal working conditions. Separation (see Figure 6.92) may be by:	D 3.19.7 ND 3.19.7
• a shield of non-combustible material at least 25 mm thick; or	
• an air space of at least 75 mm.	

Note: A gas-fired appliance with a CE marking and installed in accordance with the manufacturer's written instructions may not require this separation.

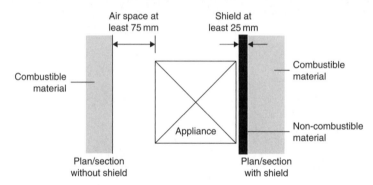

Figure 6.92 Gas-fired appliance separation from combustible material

Relationship of hearths to combustible materials

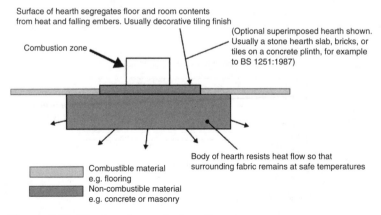

Figure 6.93 The functions of a hearth

Walls that are not part of a fireplace recess or a prefabricated appliance chamber but are adjacent to hearths or appliances should also protect the dwelling from catching fire.

D 3.19.8
ND 3.19.8

 This is particularly relevant to timber-framed buildings.

Any part of a building that abuts or is adjacent to a hearth, should be constructed in such a way as to minimize the risk of ignition by direct radiation or conduction from a solid fuel appliance located upon the hearth.

D 3.19.8
ND 3.19.8

The hearth should be located:

D 3.19.8
ND 3.19.8

- in a fireplace recess in accordance with BS 8303: Part 1: 1994; or
- (in any part of the dwelling, other than the floor) not more than 150 mm from the hearth, constructed of solid, non-combustible material in accordance with Figure 6.94 and Table 6.89.

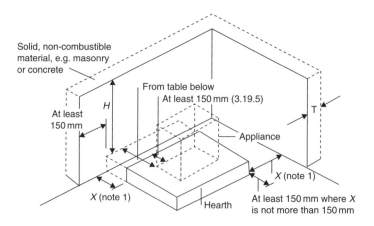

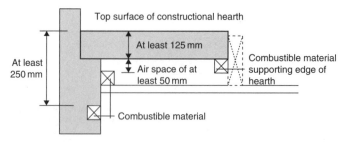

Figure 6.94 Location of hearths

Table 6.89 Hearth and appliance adjacent to any part of a building

Location of hearth or appliance	Thickness (T) of solid non-combustible material	Height (H) of solid non-combustible material
Where the hearth abuts a wall and the appliance is not more than 50 mm from the wall	200 mm	At least 300 mm above the appliance or 1.2 m above the hearth whichever is the greater
Where the hearth abuts a wall and the appliance is more than 50 mm but not more than 300 mm from the wall	75 mm	At least 300 mm above the appliance or 1.2 m above the hearth whichever is the greater
Where the hearth does not abut a wall and is not more than 150 mm from the wall	75 mm	At least 1.2 m above the hearth

Notes:
1. There is no requirement for protection of the wall where X is more than 150 mm.
2. All combustible material under a constructional hearth should be separated from the hearth by an air space of at least 50 mm. However, an air space is not necessary where:
   - the combustible material is separated from the top surface of the hearth by solid, non-combustible material of at least 250 mm; or
   - the combustible material supports the front and side edges of the hearth.

Fireplace recesses

A fireplace recess should be constructed of solid, non-combustible material and should incorporate a constructional hearth.	D 3.19.9 ND 3.19.9

Combustion appliances – removal of products of combustion

Heating and cooking appliances fuelled by solid fuel, oil or gas all have the potential to cause carbon monoxide (CO) poisoning if they are poorly installed or commissioned, inadequately maintained or incorrectly used.

Every building must be designed and constructed in such a way that the products of combustion are carried safely to the external air without harm to the health of any person through leakage, spillage or exhaust, nor permit the re-entry of dangerous gases from the combustion process of fuels into the building.	D 3.20 ND 3.20

Chimneys and flue-pipes serving appliances burning any fuel

A chimney or flue-pipe serving any appliance should be suitable for use with the type of appliance served.	D 3.20.1 ND 3.20.1
A chimney should be manufactured using products in accordance with the standards shown in Table 6.90.	D 3.20.1 ND 3.20.1

Table 6.90   Chimney manufacturing standards

For concrete chimney blocks	BS EN 1858: 2003
For clay chimney blocks	BS EN 1806: 2000
For purpose-made concrete flue linings	BS EN 1857: 2003
For purpose-made clay flue linings	BS EN 1457: 1999
For a factory-made metal chimney	BS EN 1856-1: 2003

Chimneys and flue-pipes serving solid fuel appliances

A flue in a chimney should be separated from every other flue and extend from the appliance to the top of the chimney.	D 3.20.1 ND 3.20.1
Every flue should be surrounded by non-combustible material that is capable of withstanding the effects of a chimney fire, without any structural change that would impair the stability or performance of the chimney.	D 3.20.1 ND 3.20.1

Chimneys and flue-pipes serving oil-firing appliances

A chimney or flue-pipe serving an oil-firing appliance should be constructed to the recommendations of BS 5410: Part 1: 1997.	D 3.20.3 ND 3.20.3
Where the flue gas temperatures are more than 250°C, under normal working conditions, custom-built chimneys, system chimneys and flue-pipes should be designed and constructed for use with a solid fuel appliance.	D 3.20.3 ND 3.20.3

Chimneys and flue-pipes serving gas-fired appliances

A chimney or flue-pipe should be constructed and installed in accordance with BS 5440-1: 2000 and 'GE/UP/7: Edition 2, 'Gas installations in timber framed and light steelframed buildings'.	D 3.20.4 ND 3.20.4

Oil-firing appliances in bathrooms and bedrooms

There is an increased risk of carbon monoxide poisoning in bathrooms, shower rooms or rooms intended for use as sleeping accommodation, such as bed-sitters. Because of this:

• open-flued oil-firing appliances should **not** be installed in these rooms or any cupboard or compartment connecting directly with these rooms;	D 3.20.5 ND 3.20.5
• where locating a combustion appliance in such rooms cannot be avoided, the installation of a room-sealed appliance would be appropriate.	

Gas-fired appliances in bathrooms and bedrooms

Regulation 30 of the Gas Safety (Installations and Use) Regulations 1998 has specific requirements for room-sealed appliances in these locations.

Terminal discharge from condensing boilers

The condensate plume from a condensing boiler can cause damage to external surfaces of a building if the terminal location is not carefully considered. The manufacturer's instructions should be followed (Figure 6.95).

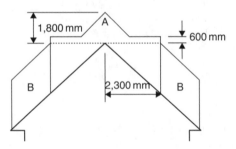

Figure 6.95 Location of flue terminals relative to easily ignitable roof coverings. Where: Zone A is at least 1.8 m vertically above the weather skin, and at least 600 mm above the ridge; and Zone B is at least 1.8 m vertically above the weather skin, and at least 2.3 m horizontally from the weather skin

6.11 Windows

Basically, a window is an opening in an otherwise solid and opaque surface that lets light and possibly air into the room and allows occupants to see out. Windows are usually glazed or covered in some other transparent or translucent material.

6.11.1 Requirements

Fire

Every building must be designed and constructed in such a way that:	*2.4*
• *in the event of an outbreak of fire within the building, the unseen spread of fire and smoke within concealed spaces in its structure and fabric is inhibited;*	
• *in the event of an outbreak of fire within the building, the occupants, once alerted to the outbreak of the fire, are provided with the opportunity to escape from the building, before being affected by fire or smoke;*	*2.9*
• *facilities are provided to assist fire-fighting or rescue operations.*	*2.14*

Environment

Every building, and hard surface within the curtilage of a building, must be designed and constructed with a surface water drainage system that will:	*D 3.6*
• *ensure the disposal of surface water without threatening the building and the health and safety of the people in and around the building; and*	
• *have facilities for the separation and removal of silt, grit and pollutants.*	
Every building must be designed and constructed in such a way that:	*D 3.14*
• *the air quality inside the building is not a threat to the health of the occupants or the capability of the building to resist moisture, decay or infestation.*	
• *the products of combustion are carried safely to the external air without harm to the health of any person through leakage, spillage, or exhaust nor permit the re-entry of dangerous gases from the combustion process of fuels into the building.*	*D 3.20*

Safety

Every building must be designed and constructed in such a way that every sudden change of level that is accessible in, or around, the building is guarded by the provision of pedestrian protective barriers.	*D 4.4*

Standard D 4.4 does not apply where the provision
of pedestrian protective barriers would obstruct the
use of areas so guarded.

Every building must be designed and constructed in such a   D 4.8
way that:
- *people in and around the building are protected from injury that could result from fixed glazing, projections or moving elements on the building;*
- *fixed glazing in the building is not vulnerable to breakage where there is the possibility of impact by people in and around the building;*
- *both faces of a window and rooflight in a building are capable of being cleaned such that there will not be a threat to the cleaner from a fall resulting in severe injury;*
- *a safe and secure means of access is provided to a roof; and manual controls for ventilation and for electrical fixtures can be operated safely.*

 The last part of the Standard does not apply to
domestic buildings.

Energy

Every building must be designed and constructed in such a   D 6.2
way that:

- *an insulation envelope is provided which reduces heat loss;*
- *the form and fabric of the building minimizes the use of mechanical ventilating or cooling systems for cooling*   D 6.6
  purposes; and
- *in non-domestic buildings, the ventilating and cooling systems installed are energy efficient and are capable of being controlled to achieve optimum energy efficiency.*

6.11.2 Meeting the requirements

 See Annex 6A of the Technical Handbooks for guidance concerning the
calculation of average *U*-values for windows.

Structure

To prevent the collapse, excessive deformation or the disproportionate collapse
of buildings, the climatic conditions in Scotland should be carefully considered in the assessment of loadings and in the structural design of buildings.

Form and fabric of the building

In order to minimize any need for mechanical ventilation for cooling or air-conditioning due to high internal temperatures in hot weather the form and the fabric of a dwelling should include:

- a high proportion of translucent glazing;
- orientation of translucently glazed areas;
- solar shading or other solar control measures where areas of the external building fabric are susceptible to solar gain;
- natural ventilation (including night cooling);
- thermal mass.

A dwelling should be designed to avoid any need for a cooling system	D 6.6.1
Where a dwelling has little or no cross-ventilation (e.g. flats with all external windows/rooflights on one southerly elevation or a high proportion of translucent glazing, the dwelling should be designed to avoid high internal temperature.	D 6.6.1

Note: The current EU Directive allows energy performance to be evaluated in a number of different ways. Within Scotland, the method chosen is carbon dioxide using the Directive compliant methodology.

Size and proportions of openings

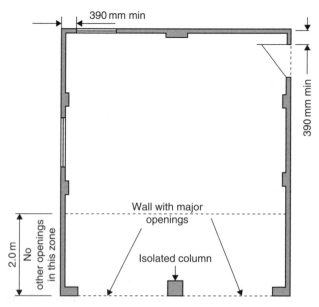

Figure 6.96 Size of openings in small single-storey, single-leaf buildings

No more than two major openings (maximum height 2.1 m, maximum width 5.0 m) are allowed in any one wall of the building or extension, either a single opening or the combined width of two openings.	D 1D40
No other openings shall be within 2.0 m of a wall containing a major opening.	D 1D40
The total size of openings in a wall not containing a major opening should not exceed 2.4 m².	D 1D40
There should not be more than one opening between piers.	D 1D40
The distance from a window or a door to a corner should be at least 390 mm unless there is a corner pier.	D 1D40
In a domestic building, the total area of windows, doors and rooflights, should not exceed 25% of the floor area of the dwelling created by the conversion of a heated building.	D 6.2.7
Where the insulation envelope of a domestic building is extended, the area of rooflights windows, doors, rooflights and roof windows should be limited to 25% of the floor area of the extension.	D 6.2.9

Timber-frame wall

Openings for doors and windows are formed by additional vertical cripple studs at the open edges supporting timber lintels.

Wall panel connections

To ensure that wall panels are able to resist overturning forces:

All edges (including those to openings for windows, doors, etc.) should be supported by timber members having a thickness not less than the thickness of the studs.	D 1E39e

Cavity barriers

In order to inhibit fire spread in a cavity (e.g. around a window opening) all cavities within a building should have cavity barriers with at least short fire resistance duration installed around the edges of the cavity.	D 2.4.1 ND 2.4.1

 Note: A cavity barrier is any construction provided to seal a cavity against the penetration of fire and smoke or to restrict its movement within the cavity.

Escape

Everyone within a dwelling should be provided with at least one means of escape from fire that offers a safe passage to a place of safety outside the building. This should be short enough for them to escape from the dwelling before being affected by fire or smoke.

Escape windows

Although the normal means of escape from a dwelling in the event of a fire will be by way of the internal stairs or other circulation areas, in certain circumstances (e.g. on ground and first floor stories) windows may be (or may form part of) an escape route.

 The use of an escape window will be the last resort for the occupants and inevitably involves some risk.

Openable windows large enough to escape through should be provided from every apartment from which the occupants could make their escape by lowering themselves from the window.	D 2.9.4
An escape window should be provided in every apartment on an upper storey (or an inner room on any storey):	D 2.9.4
• at a height of not more than 4.5 m; and	D 2.9.30
• in every apartment which is an inner room in a storey at a height of not more than 4.5 m.	D 2.9.30
Escape windows should:	D 2.9.4
• be a window, or door (e.g. french window) situated in an external wall or roof; • have an unobstructed openable area that is at least 0.33 m²; and • be at least 450 mm high and 450 mm wide.	
The bottom of the openable area should not be more than 1.1 m above the floor.	D 2.9.4
If a conservatory is located below an escape window consideration should be given to the design of the conservatory roof to withstand the loads exerted from occupants lowering themselves onto the roof in the event of a fire.	D 2.9.4

External escape stairs

External escape stairs present additional hazards to people evacuating a building in the case of fire as the escape stair will be exposed to the possible effects of inclement weather.

Every part of an external wall that includes a window or other opening within 2 m from the escape stair should have short fire resistance duration other than a door opening from the top storey to the external escape stair.   D 2.9.22   ND 2.9.24

Basements

A basement storey which contains an apartment should be provided with either: D 2.9.28
• an escape window in every basement apartment; or
• an alternative exit from the basement storey, which provides access to the external air (below the adjoining ground) from which there is access to a place of safety at ground level.

An apartment in a basement should not be in an inner room **unless** there is an escape window or there are alternative routes from the apartment to circulation areas or other rooms.   D 2.9.29

Ventilation of smoke from a basement is normally provided to assist fire-fighting operations and to allow smoke clearance after the fire, but:

• smoke outlets need not be provided where a window or windows opening direct to the external air have a total area not less than 1% of the floor area.   ND 2.14.7

• in basements, smoke outlets should discharge by means of windows (panels or pavement lights) which are readily accessible to fire service personnel and which can be opened or knocked out if necessary.   ND 2.14.8

Environment

Every building should be provided with a drainage system to remove rainwater from the roof, or other areas where rainwater might accumulate, without causing damage to the structure or endangering the health and safety of people in and around the building

Methods other than gutters and rainwater pipes may be utilized to remove rainwater from roofs but, if an eaves drop system to allow rainwater to drop freely to the ground is used, it should be designed taking into account the need to prevent water from entering windows.   D 3.6.1   ND 3.6.1

Ventilation and smoke clearance

Ventilation of the escape stairs, protected lobbies and common access corridors is important to assist fire service personnel during fire-fighting operations and for smoke clearance purposes after the fire has been extinguished.

> An escape stair within a protected zone should have either: D 2.14.3
>
> - an opening window at each storey with an openable   ND 2.14.5
>   area of 0.5 m²; or
> - a ventilator of not less than 1 m² at the top of the stair.

Natural ventilation

Where a building is naturally ventilated, all moisture producing areas such as bathrooms and shower rooms should have the additional facility for removing such moisture before it can damage the building.

> Where rapid ventilation is provided by a opening window,   ND 3.14.2
> some part of the opening should be at least 1.75 m above
> floor level.

Trickle ventilators

A trickle ventilator, sometimes called background ventilation, is a small ventilation opening, normally provided in the head of a window frame with a controllable shutter.

> In rooms where windows are more than likely to be left   D 3.14.5
> open, it is recommended that trickle ventilators supply
> replacement air for mechanical extract systems.

Mechanical aids to ventilation

A mechanical ventilation system should be designed, installed and commissioned to perform in a way that is not detrimental to the health of the occupants of the building and when necessary, is easily accessible for regular maintenance.

> Mechanical extract ventilation units (extract fans) may be   D 3.14.10
> either window or wall mounted, in rooms where there is
> likely to be high humidity such as kitchens, bathrooms and
> shower rooms.

Terminal discharges at low level

> If the flue outlet is in a vulnerable position so that the flue   D 3.20.15
> discharges within reach of a window, then it should be   ND 3.20.15
> designed to prevent the entry of matter that could obstruct
> the flow of gases.

Conservatories and extensions built over existing windows

Constructing a conservatory or an extension over an existing window (or ventilator) will effectively result in an internal room, restrict air movement and could significantly reduce natural ventilation to that room.

Conservatories built over existing windows

A conservatory may be constructed over a ventilator serving a room in a dwelling provided that the ventilation of the conservatory is to the outside air and has an opening area of at least 1/30th of the total combined floor area of the internal room so formed and the conservatory.	D 3.14.7
The ventilator to the internal room should have an opening area of at least 1/30th of the floor area of the room.	D 3.14.7
Trickle ventilators should be provided relevant to the overall areas created.	D 3.14.7
If an extension is built over a ventilator, then a new ventilator should be provided to the room.	D 3.14.7
If the conservatory or extension is constructed over an area that generates moisture, such as a kitchen, bathroom, shower room or utility room, mechanical extract, via a duct if necessary or a passive stack ventilation system should be provided direct to the outside air.	D 3.14.7

Extensions built over existing windows

A new ventilator and trickle ventilator should be provided to the existing room.	ND 3.14.4

 BUT if this is not reasonably practicable (e.g. if virtually the entire external wall of the room is covered by the extension) the new extension should be treated as part of the existing room as opposed to being a separate internal room.

Because an extension will be relatively airtight, the opening area between the two parts of the room should be not less than 1/15th of the total combined area of the existing room plus the extension.	ND 3.14.4
If the extension is constructed over an area that generates moisture, such as a kitchen, bathroom, shower room or utility room, mechanical extract, via a duct if necessary, should be provided direct to the outside air.	ND 3.14.4

Safety (danger from accidents)

The majority of accidents occur during normal use involving building features such as windows and areas of fixed glazing. Collisions with glazing are very common as it can, if transparent, be difficult to see and may create confusing lighting effects, presenting particular difficulties for a person with a visual or cognitive impairment.

Every building must be designed and constructed in such a   4.8
way that:

- people in and around the building are protected from injury that could result from fixed glazing, projections or moving elements on the building;
- fixed glazing in the building is not vulnerable to breakage where there is the possibility of impact by people in and around the building;
- both faces of a window and rooflight in a building are capable of being cleaned such that there will not be a threat to the cleaner from a fall resulting in severe injury;
- a safe and secure means of access is provided to a roof; and manual controls for ventilation and for electrical fixtures can be operated safely.

 The last part of this Standard does not apply to domestic buildings.

Whether windows are cleaned professionally or by the   D 4.8.0
building owner, provision should be made to permit glazing   ND 4.8.0
to be cleaned safely.

Design of pedestrian protective barriers

In and around domestic buildings, gaps in any protective barrier should not be large enough to permit a child to pass through. To ensure this:

- openings in a protective barrier should prevent the passage   D 4.4.2
  of a 100 mm diameter sphere;   ND 4.4.2
- a protective barrier (and any wall, partition or fixed glazing used instead of a barrier) should be secure, capable of resisting loads calculated in accordance with BS 6399: Part 1: 1996 and be at least 800 mm high:
- protective barriers should be installed where the opening window has:
  - a sill that is less than 800 mm above finished floor level;
  - an operation that will allow the possibility of falling out;
  - a difference in level between the floor level and the ground level of more than 600 mm.

Collision with projections

Fixtures that project into, or open onto any place to which people have access can be a hazard and should be positioned, secured or guarded so that they do not present a risk to building users.

Where a building element projects into a circulation route or space, and any part of the obstruction is less than 2.0 m above the ground, guarding should be provided to both highlight the hazard and prevent collision with the building element to:

D 4.8.1
ND 4.8.1

- any movable projection, such as a window frame, that opens across a circulation route or into a circulation space;
- any permanent projection of more than 100 mm into a circulation route or space that begins at a height of more than 300 mm above the ground, or the projection of which increases with height by more than 100 mm;
- any accessible area where headroom reduces to less than 2.0 m, such as beneath a stair flight.

Cleaning of windows and rooflights

Falls account for most window cleaning accidents, and generally occur from loss of balance through over-extension of reach or due to breakage of part of the building fabric through improper use or access. It is therefore important that transparent or translucent glazing should be designed so that it may be cleaned safely.

Any window or rooflight, all or part of which is more than 4 m above adjacent ground or internal floor level, should be constructed so that any external and internal glazed surfaces can be cleaned safely from:

D 4.8.3
ND 4.8.3

- inside the building in accordance with the recommendations of Clause 8 of BS 8213: Part 1: 2004;
- a load-bearing surface, such as a balcony or catwalk, large enough to prevent a person falling further;
- a window access system, such as a cradle or travelling ladder, mounted on the building, as described in Annex C3 of BS 8213: Part 1: 2004;
- (for non-domestic buildings) a ladder sited on adjacent ground or from an adjacent load-bearing surface.

 Note: Within a dwelling, any rooflight, all of which is more than 1.8 m above both adjacent ground and internal floor level, need not be constructed so that it may be safely cleaned.

A ladder should not be used to access any external or internal glazed surface more than 9 m above the surface on which the ladder is sited.	D 4.8.3 ND 4.8.3

When cleaning a window from inside, a person should not have to sit or stand on a window sill or use other aids to reach the external face of the window.	D 4.8.3 ND 4.8.3

 It is ergonomically considered reasonable to apply a safe limit to downward reach of 610 mm and a safe limit to lateral and vertical reach as an arc with a radius of 850 mm measured from a point not more than 1.3 m above floor level.

Where the window is to be cleaned from a load-bearing surface such as a balcony or catwalk, there should be: • a means of safe access; and • a protective barrier not less than 1.1 m high to any edge of the surface or access to the surface which is likely to be dangerous.	D 4.8.3 ND 4.8.3

This method of cleaning is only appropriate where no part of the glazing is more than 4 m above the load-bearing surface.

Guarding of windows for cleaning

At heights of two storeys or more above ground level, where it is intended to clean the outside face of the glazing from inside the building, appropriate guarding should be provided to a height of at least 1.1 m above floor level.	D 4.8.4 ND 4.8.4

All guarding should: • be designed to conform to BS 6180: 1999; • be permanently fixed; • not be detachable to permit windows to open; • be designed so that it is not easily climbable by children.	D 4.8.4 ND 4.8.4

Access to manual controls

The location of any manual control device for openable windows and rooflights that is intended for operation by the occupants of a building should be: • installed in position that allows safe and convenient use; and	D 4.8.5 ND 4.8.6

- unless incorporating a restrictor or other protective   D 4.8.5
  device for safety reasons, controls should be operable   ND 4.8.6
  with one hand.

An openable window or rooflight that provides natural ventilation should have controls for opening, positioned at least 350 mm from any internal corner, projecting wall or similar obstruction and at a height of:

- not more than 1.7 m above floor level, where access to controls is unobstructed; or
- not more than 1.5 m above floor level, where access to controls is limited by a fixed obstruction of not more than 900 mm high which projects not more than 600 mm in front of the position of the controls, such as a kitchen base unit;
- not more than 1.2 m above floor level, in an unobstructed location, within an enhanced apartment or within accessible sanitary accommodation not provided with mechanical ventilation.

Energy

Every building must be designed and constructed in such a way that an insulation envelope is provided which reduces heat loss.

Maximum U-values

The design of a dwelling should involve extensive use of building-integrated or localized Low and Zero Carbon Technologies (LZCT) levels of thermal insulation.

The maximum U-values for building elements of the insulation   D 6.2
envelope should not be greater than those shown in Table 6.91.   LD 6.2

Table 6.91   Maximum U-values for building elements of the insulation envelope

Type of element	(a) Area-weighted average U-value (W/m^{2}K) for all elements of the same type	(b) Individual element U-value (W/m^{2}K)
Normal	2.2	3.3
Shell and fit-out buildings	1.8	3.3
High-usage entrance doors and display windows, and similar glazing	As required	As required
Conversion of heated buildings	1.8	
Constructing an extension	1.8	3.3

Alterations to the insulation envelope

Alterations that involve increasing the floor area and/or bringing parts of the existing building that were previously outwith the insulation envelope into the heated part of the dwelling are considered as extensions and/or conversions (regulation 4, schedule 2).

Where windows, doors and rooflights are being created or replaced, they should achieve the U-value recommended in Column (a) of Table 6.91 (an example of a compensatory approach is shown in Annex 6C to the Scottish Building Regulations).   D 6.2.11   ND 6.2.12

For secondary glazing, an existing window, after alteration should achieve a U-value of about $3.5\,W/m^2K$.   D 6.2.10

 Where the work relates only to one or two replacement windows or doors, to allow matching windows or doors be installed, the frame may be disregarded for assessment purposes, provided that the centre pane U-value for each glazed unit is $1.2\,W/m^2K$ or less.

Where additional windows, doors and rooflights are being created, the overall total area (including existing) should not exceed 25% of the total dwelling floor area and in the case of a heated communal room or other area (exclusively associated with the dwellings), it should not exceed 25% of the total floor area of these rooms/areas.   D 6.2.10

A building that was in a ruinous state should, after renovation, be able to achieve almost the level expected of new construction and in after an alteration of this nature to the insulation envelope, a roof should be able to achieve at least an average U-value of 0.35 and in the case of a wall or floor, $0.70\,W/m^2K$.   D 6.2.10

When alterations are carried out, attention should still be paid to limiting infiltration thermal bridging at junctions and around windows, doors and rooflights and limiting air infiltration.   D 6.2.11   ND 6.2.12

Conservatories

Although conservatories are attached to dwellings they are considered as stand-alone buildings.

For the glazing to conservatories of less than $50\,m^2$ floor area, a maximum U-value of 2.2 is recommended and for those $20\,m^2$ or less a maximum U-value of 3.3.   D 6.2.12

A conservatory should be thermally divided from the insulation envelope of the dwelling. — D 6.2.12

The dividing elements (wall, door, window and on the rare occasion floor) should have U-values equal to or better than the corresponding exposed elements in the rest of the dwelling. — D 6.2.12

Draught stripping for windows and doors which are part of the thermal division between the conservatory and the dwelling should be of a similar standard as the exposed windows and doors elsewhere in the dwelling. — D 6.2.12

Stand-alone buildings

Where the area of a communal room or other heated accommodation associated with a block of dwellings is less than $50\,m^2$, these rooms or accommodation may be treated as a stand-alone building.

Elements (including dividing elements) should have U-values equal to or better than those chosen for the rest of the building. — D 6.2.13

In these rooms or accommodation, the area of windows, doors, rooflights and roof windows should be limited to 25% of the total floor area of these parts — D 6.2.13

Controls for artificial and display lighting

Every artificial lighting system in a building that has a floor area of more than $50\,m^2$ should have controls which encourage the maximum use of daylight and minimize the use of artificial lighting during the times when rooms or spaces are unoccupied.

Lighting rows that are located adjacent to windows should ideally be controlled by photocells which monitor daylight and adjust the level of artificial lighting accordingly, either by switching or by dimming. — ND 6.2

For this reason, the Scottish Building Regulations does not offer any guidance on minimum or maximum area for windows, doors, rooflights and roof windows in dwellings and non-domestic buildings.

6.12 Doors

A door is a panel or barrier (usually hinged, sliding or electronic) that is used to cover an opening in a wall or partition going into a building or space. Doors allow passage between the inside and outside, and between internal rooms. The door is used to control the physical atmosphere within a space by enclosing it, excluding air drafts, so that interiors may be more effectively heated or cooled. Doors are significant in preventing the spread of fire.

6.12.1 Requirements

Fire

> *Every building must be designed and constructed in such a way*   2.4
> *that in the event of an outbreak of fire within the building:*
>
> - *the unseen spread of fire and smoke within concealed spaces*
>   *in its structure; and fabric is inhibited;*   2.9
> - *the occupants, once alerted to the outbreak of the fire, are*
>   *provided with the opportunity to escape from the building,*
>   *before being affected by fire or smoke.*
>
> *Every building must be:*   2.12
>
> - *accessible to fire appliances and fire service personnel;*   2.14
> - *designed and constructed in such a way that facilities are*
>   *provided to assist fire-fighting or rescue operations.*

Domestic buildings

> *Every domestic building must be designed and constructed in*   D 2.1
> *such a way that in the event of an outbreak of fire within the*
> *building, fire and smoke are inhibited from spreading beyond the*
> *compartment of origin until any occupants have had the time*
> *to leave that compartment and any mandatory fire containment*
> *measures have been initiated.*

Non-domestic buildings

> *Every non-domestic building, which is divided into more than*   ND 2.2
> *one area of different occupation, must be designed and*
> *constructed in such a way that in the event of an outbreak of fire*
> *within the building, fire and smoke are inhibited from spreading*
> *beyond the area of occupation where the fire originated.*

Environment

> *Every building must be designed and constructed in such a*   *3.14*
> *way that the air quality inside the building is not a threat to the*
> *health of the occupants or the capability of the building to resist*
> *moisture, decay or infestation.*

Safety

> *Every building must be designed and constructed in such a way*   *4.1*
> *that:*
>
> - *all occupants and visitors are provided with safe, convenient*
>   *and unassisted means of access to the building;*

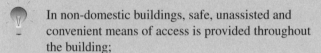

>   In non-domestic buildings, safe, unassisted and
> convenient means of access is provided throughout
> the building;

> - *in residential buildings, a proportion of the rooms intended*   *4.2*
>   *to be used as bedrooms must be accessible to a wheelchair*
>   *user;*
> - *in domestic buildings, safe and convenient means of access is*
>   *provided within common areas and to each dwelling;*
> - *in dwellings, safe and convenient means of access is*
>   *provided throughout the dwelling; and*
> - *in dwellings, unassisted means of access is provided to, and*
>   *throughout, at least one level.*

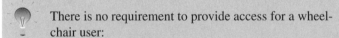

>   There is no requirement to provide access for a wheel-
> chair user:

> - in a non-domestic building not served by a lift, to a room,
>   intended to be used as a bedroom, that is not on an entrance
>   storey; or
> - in a domestic building not served by a lift, within common
>   areas and to each dwelling, other than on an entrance storey.

> - *people in and around the building are protected from injury*   *4.8*
>   *that could result from projections or moving*
>   *elements on the building;*
> - *fixed glazing in the building is not vulnerable to*
>   *breakage where there is the possibility of impact by*
>   *people in and around the building.*

Energy

*Every building must be designed and constructed in such a   6.2
way that an insulation envelope is provided which reduces
heat loss.*

6.12.2 Meeting the requirements

Fire

Every building must be designed and constructed in such
a way that in the event of an outbreak of fire within the
building:

- fire and smoke are inhibited from spreading beyond the   ND 2.1
  compartment of origin until any occupants have had
  the time to leave that compartment, and any mandatory
  fire containment measures have been initiated;

   This standard does not apply to domestic buildings.

- the occupants, once alerted to the outbreak of the fire,   D 2.9
  are provided with the opportunity to escape from the   ND 2.9
  building, before being affected by fire or smoke.

Every building must be: D 2.12

- accessible to fire appliances and fire service personnel;   ND 2.12
- designed and constructed in such a way that facilities   D 2.14
  are provided to assist fire-fighting or rescue operations.   ND 2.14

Environment

Every building must be designed and constructed in such a   D 3.23
way that a container for the storage of woody biomass fuel   ND 3.23
will inhibit fire from spreading to its contents from within,
or beyond the boundary.

This standard does not apply to portable containers.

Safety

Every building must be designed and constructed in such a way that:	D 4.2
• in non-domestic buildings, safe, unassisted and convenient means of access is provided throughout the building;	ND 4.2
• people in and around the building are protected from injury that could result from fixed glazing;	D 4.8 ND 4.8
• fixed glazing in the building is not vulnerable to breakage where there is the possibility of impact by people in and around the building.	
• all occupants and visitors are provided with safe, convenient and unassisted means of access to the building.	D 4.1 ND 4.1

There is no requirement to provide access for a wheelchair user to:

- a house, between either the point of access to or from any car parking within the curtilage of a building and an entrance to the house where it is not reasonably practicable to do so; or
- a common entrance of a domestic building not served by a lift, where there are no dwellings.

Energy

Every building must be designed and constructed in such a way that:	D 6.1 ND 6.1
• the energy performance is calculated in accordance with a methodology which is asset based, conforms with the European Directive on the Energy Performance of Buildings 2002/91/EC and uses UK climate data;	
• the energy performance of the building is capable of reducing carbon dioxide emissions;	
• an insulation envelope is provided which reduces heat loss.	D 6.2 ND 6.2

See Annex 6A to the Technical Handbooks for guidance on compensating *U*-values for doors.

Structure

Size and proportions of openings

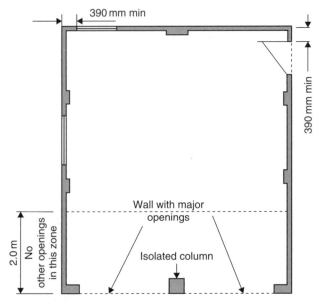

Figure 6.97 Size of openings in small single-storey, single-leaf buildings

No more than two major openings (maximum height 2.1 m, maximum width 5.0 m) are allowed in any one wall of the building or extension, either a single opening or the combined width of two openings.	D 1D40
No other openings shall be within 2.0 m of a wall containing a major opening.	D 1D40
The total size of openings in a wall not containing a major opening should not exceed 2.4 m².	D 1D40
There should not be more than one opening between piers.	D 1D40
The distance from a door to a corner should be at least 390 mm unless there is a corner pier.	D 1D40

Openings and service provisions

Openings through compartment walls and compartment floors can compromise their effectiveness and should be kept to a minimum and constructed to resist fire.

In domestic buildings, this can be achieved by following the guidance below.

A self-closing fire door with the same fire resistance duration as the separating wall should be installed.	D2.2.9
A self-closing fire door should not be fitted in a separating wall between two dwellings in different occupation.	D2.2.9
A lockable door to a cupboard or service duct with a floor area not more than $3\,m^2$ need not be self-closing.	D2.2.9
Self-closing fire doors can be fitted with hold-open devices as specified in BS 5839: Part 3: 1988 provided the door is not an emergency door, a protected door serving the only escape stair in the building (or the only escape stair serving part of the building) or a protected door serving a fire-fighting shaft.	D2.2.9
Hold-open devices shall deactivate on operation of the fire alarm.	D2.2.9
Electrically operated hold-open devices should deactivate: • on operation of an automatic fire alarm system; • on any loss of power to the hold-open device, apparatus or switch; • on operation of a manually operated switch fitted in a position at the door.	D2.2.9

In non-domestic buildings,

A self-closing fire door with the same fire resistance duration as the compartment wall should be installed in accordance with the recommendations in the Building Hardware Industry Federation, Code of Practice, 'Hardware for Timber Fire and Escape Doors' Issue 1, November 2000.	ND 2.1.14
For metal doorsets, reference should be made to the 'Code of Practice for fire resisting metal doorsets' published by the Door and Shutter Manufacturers' Association, 1999.	

Self-closing fire doors should **not** be installed: • in separating walls other than in buildings with common occupation; or • where the building is in the same occupation but in different use.	ND 2.2.6
Self-closing fire doors can be fitted with hold-open devices as specified in BS 5839: Part 3: 1988 **provided** the door is not an emergency door, a protected door serving the only escape stair in the building (or the only non-escape stair serving part of the building) or a protected door serving a fire-fighting shaft.	ND 2.1.14
Electrically operated hold-open devices should deactivate: • on operation of an automatic fire alarm system; • on any loss of power to the hold-open device, apparatus or switch; and • on operation of a manually operated switch fitted in a position at the door.	ND 2.1.14
A lockable door to a cupboard or service duct where the cupboard or the service duct has a floor area not more than $3\,\mathrm{m}^2$, need not be self-closing.	ND 2.1.14

 Note: A self-closing fire door in the enclosing structure of a fire-fighting shaft need only attain medium fire resistance duration.

Cavity barriers

In order to inhibit fire spread in a cavity, every cavity within a building should have cavity barriers with at least short fire resistance duration installed around the edges (i.e. around the head, jambs and sill) of an external door.	D 2.4.1 ND 2.4.1

Escape routes

Exits

In domestic buildings, there should be at least one escape route from: • the main entrance door of every flat or maisonette; and • the door of every communal room; and • every plant room.	D2.9.1

Doorway widths

The clear opening width at doorways (see Figure 6.98) depends on the number of people using the escape route.

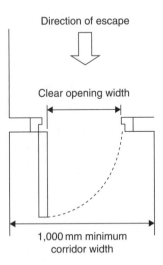

Figure 6.98 Doorway widths

Where the number of people using the escape route is:	ND 2.9.9

- not more than 225, the clear opening width of the doorway should be at least 850 mm.
- not more than 100, the clear opening width of the doorway should be at least 750 mm.

Direction of door openings

A side-hung door across an escape route may open against the direction of escape where the occupancy capacity in the building is sufficiently low.	ND 2.9.14
If the door is an emergency door or a door serving a place of special fire risk, the side-hung door should open in the direction of escape regardless of occupancy levels.	ND 2.9.14

Locks and hardware

Locks on exit doors or locks on doors across escape routes present difficulties when assessing the need for security against the need to allow safe egress from a building in the event of a fire.

If a door across an escape route has to be secured against entry when the building is occupied: D 2.9.14

- it should be fitted only with a lock or fastening which is readily operated, without a key, from the side approached by people making an escape;
- it should also have a notice, on the inside, explaining the operation of the opening device.

Security measures should not compromise the ability of the occupants to escape from a building in the case of fire.   ND 2.9.15

Where an exit door from a room, storey or a door across an escape route has to be secured against entry when the building or part of the building is occupied, it should only be fitted with a lock or fastening which is readily operated, without a key, from the side approached by people making their escape.   ND 2.9.15

Similarly, where a secure door is operated by a code, combination, swipe or proximity card, biometric data or similar means, it should also be capable of being overridden from the side approached by people making their escape (see also electrically operated locks).   ND 2.9.15

Electrically powered locks

Electrically powered locks should **not** be installed: ND 2.9.15

- on a protected door serving the only escape stair in the building;
- a protected door serving a fire-fighting shaft;
- on any door which provides the only route of escape from the building or part of the building.

Electrically powered locks on exit doors and doors across escape routes may be installed in buildings: ND 2.9.15

- which are inaccessible to the general public;
- when the building is accessible to the general public, the aggregate occupancy capacity of the rooms or storey served by the door does not exceed 60 persons.

  Staff in such areas will need to be trained both in the   ND 2.9.15
  emergency procedures and in the use of the specific
  emergency devices fitted (see clause 2.0.4).

| Electrically powered locks should return to the unlocked position: | ND 2.9.15 |

- on operation of the fire alarm system;
- on loss of power or system error;
- on activation of a manual door release unit.

| Where a locking mechanism is designed to remain locked in the event of a power failure or system error, the mechanism should not be considered appropriate for use on exit doors and doors across escape routes. | ND 2.9.15 |

Fire and smoke control in corridors

| For purposes of smoke control in a corridor, any door in the wall or screen should be a self-closing fire door. | ND 2.9.16 |

| If the corridor is a dead end more than 4.5 m long and provides access to a point from which more than one direction of escape is possible, it should be divided at that point or points, as shown in Figure 6.99. | ND 2.9.16 |

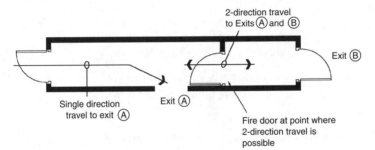

Figure 6.99 Fire and smoke control in corridors

Where the corridor provides at least two directions of escape and is more than 12 m in length between the exits it serves, it should be divided in the middle third of the corridor.	ND 2.9.16

Residential buildings

Any door in the wall of a residential building (other than a residential care building and/or hospital) room that is intended for use as sleeping accommodation (including any en-suite sanitary accommodation where provided) should be a suitable self-closing fire door with a short fire resistance duration.	ND 2.9.16

Protected lobbies

Any door in the wall dividing a protected lobby from the remainder of the protected zone should be a self-closing fire door with short fire resistance duration.	D 2.9.19 ND 2.9.21

Protected zones

Any door in the enclosing structure of a protected zone should be a self-closing fire door with at least short fire resistance duration.	D 2.9.20

For non-domestic buildings:

Any door in the enclosing structure of a protected zone should be a self-closing fire door with at least medium fire resistance duration.	ND 2.9.22
Any door in the wall separating the rooms/cupboard from the protected zone should be a self-closing fire door.	ND 2.9.23
A door to a cleaner's cupboard should have short fire resistance duration but need not be self-closing provided it is lockable.	ND 2.9.23
The doors separating the toilets or washrooms from the protected zone need not have fire resistance duration.	ND 2.9.23

Protected enclosures

Any door in the wall of a protected enclosure constructed within a dwelling should be a self-closing fire door with a short fire resistance duration.	D 2.9.21

External escape stairs

Every part of an external wall that includes a door 2 m from the escape stair should have short fire resistance duration unless it is a door opening from the top storey to the external escape stair.	D 2.9.22 ND 2.9.24

Obstacles

In domestic buildings, an escape route should not be by way of a sliding door.	D 2.9.7
In non-domestic buildings, an escape route should not be by way of a manual sliding door, other than one to which the public does not have access.	ND 2.9.7

Revolving doors and automatic doors can obstruct the passage of persons escaping and should not be placed across escape routes unless they are designed in accordance with BS 7036: 1996 and are either:

• arranged to fail safely to outward opening from any position of opening or; • provided with a monitored fail-safe system for opening the door from any position in the event of mains supply failure and also in the event of failure of the opening sensing device; and • open automatically from any position in the event of actuation of any fire alarm in the fire alarm zone within which the door is situated; and • permit easy manual opening from any position.	ND 2.9.7
Auditoriums with more than one exit should have at least: • one exit not less than two-thirds of the distance from any stage, screen or performing area to the back of the room; and • a gangway or exit door at each end of a row of more than 12 fixed seats.	ND 2.9.9

Fire safety

Access for fire service personnel

> Every elevation which is provided with vehicle or
> pedestrian access for fire service personnel should have
> a door giving access to the interior of the building.
>
> ND2.12.4

Fire-fighting shafts

> A self-closing fire door in the enclosing structure of a
> fire-fighting shaft need only attain medium fire resistance
> duration.
>
> ND 2.14.3

> The fire-fighting shaft should be provided with a ventilated
> fire-fighting lobby within the lobby shaft, with only one
> door to the room or storey it serves.
>
> ND 2.14.3

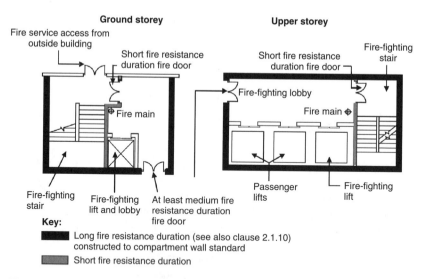

Figure 6.100 Examples of typical fire service facilities

Resistance to fire

The recommended fire resistance duration can be attained where the construction follows the guidance in the Columns 3, 4 and 5 of Table 6.92.

Table 6.92   Recommended fire resistance duration

Column 1	Column 2	Column 3	Column 4	Column 5	Column 6	Column 7
Construction	Fire resistance duration	British standards			European standards	Test exposure
		Load-bearing capacity (min)	Integrity (min)	Insulation (min)		
Fire door in a compartment wall, sub-compartment wall, separting wall or an internal wall or screen protecting routes of escape	Long Short Medium Long	120 None None None	None 30 60 120	None None None None	R 120 E 30 Sa E 60 Sa E 120 Sa	Each side separately when fitted in frame

Hospitals

A door from a hospital street to an adjoining compartment should: ND 2.B.3

- be located so that an alternative independent means of escape from each compartment is available; and
- **not** be located in the same sub-compartment as a door to a protected zone containing a stairway or lift.

Doors should be designed to accommodate bed-patient evacuation.   ND 2.B.3

Residential care buildings

Any door in the separating wall should be a self-closing fire door with medium fire resistance duration.   ND 2A1

Sanitary facilities

There should be a door separating a space containing a water closet (WC), or waterless closet, from a room or space used for the preparation or consumption of food, such as a kitchen or dining room.   D 3.12.1

In non-domestic buildings:

> Where unisex sanitary accommodation is located within   ND 3.12.1
> a separate space, for use by only one person at a time, it
> shall be provided with a door that can be secured from
> within for privacy.

Garages

The principal reason for ventilating garages is to protect the building users from the harmful effects of toxic emissions from vehicle exhausts.

> Where a garage is attached to a dwelling, communicating   D 3.14.11
> doors should have airtight seals.
>
>   Garages of less than 30 m² do **NOT** require the
> ventilation to be designed as a degree of
> fortuitous ventilation is created by the imperfect
> fit of 'up and over' doors or pass doors.

Bulk storage of woody biomass fuel

By its very nature woody biomass fuel is highly combustible and precautions need to be taken to reduce the possibility of the stored fuel igniting.

> Containers should have an outward-opening door   D 3.23.4
> incorporating containment to prevent the pellets escaping   ND 3.23.4
> when the door is opened.

Access to buildings

Accessible entrances

> An accessible entrance to a building should: D 4.1.7
>   ND 4.1.7
> • have a door leaf giving a clear opening width of at least
>   800 mm in accordance with Figure 6.101; and
> • if fitted with a door-closing device, be operable with
>   an opening force of not more than 30 N (for first 30°
>   of opening) and 22.5 N (for remainder of swing) when
>   measured at the leading edge of any door leaf; and
> • if not a powered door, have an unobstructed space to
>   the opening face of the door, next to the leading edge,
>   of at least 300 mm.

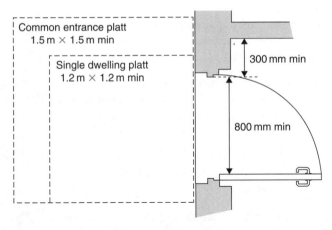

Figure 6.101 Accessible entrance door

 Note: The projection of ironmongery which extends across the width of a door leaf, such as an emergency push bar for escape or a horizontal grab rail, should be subtracted when calculating the clear opening width.

Common entrances

A common entrance to a domestic building should have a door entry system.	D 4.1.8

Accessible thresholds

To be accessible, a door should not present unnecessary barriers to use, such as a step or raised profile at a threshold that might present difficulties to a wheelchair user or be an entrapment or a trip hazard to an ambulant person, whether or not using a walking aid.

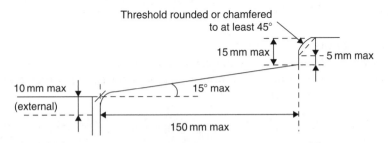

Figure 6.102 Generic threshold profile

Powered doors

Use of a powered door will improve accessibility at an entrance to a building. However, care should be taken to ensure that the form of such a door does not present any additional hazard or barrier to use.

| Powered doors should be controlled by either an automatic sensor, such as a motion detector, or a manual activation device, such as a push-pad. | ND 4.1.8 |

| Any manual control should be located at a height of between 750 mm and 1.0 m above ground level and at least 1.4 m from the plane of the door or, where the door opens towards the direction of approach, 1.4 m from the front edge of the open door leaf. | ND 4.1.8 |

| A manual control should contrast visually with the surface on which it is mounted. | ND 4.1.8 |

| In addition to the general recommendations for accessible entrances given in clause 4.1.7, a powered door should have: | ND 4.1.8 |

- signage to identify means of activation and warn of operation; and
- sensors to ensure doors open swiftly enough and remain open long enough to permit safe passage in normal use and to avoid the door striking a person passing through;
- if a swing door, identification of any opening vertical edge using visual contrast;
- if on an escape route, or forming a lobby arrangement where the inner door is also powered or lockable, doors that, on failure of supply will either fail 'open' or have a break-out facility permitting doors to be opened in direction of escape;
- guarding to prevent collision with, or entrapment by a door leaf, except where such guarding would prevent access to the door.

Internal doors

Doors within buildings should present as little restriction to passage as practicable and be constructed in a manner that does not present a hazard or a potential barrier to access.

| A door located within the common areas of a domestic building should: | D 4.2.4 ND 4.2.5 |

- if fitted with a threshold, have an accessible threshold;
- have a door leaf giving a clear opening width in accordance with Table 6.93;
- where across a circulation route or giving access to communal facilities, have a glazed vision panel in any opening leaf;

- have a door leaf that, if fitted with a door-closing device, be operable with an opening force of not more than 30 N (for first 30° of opening) and 22.5 N (for remainder of swing) when measured at the leading edge of the leaf;
- if not a powered door, have an unobstructed space to the opening face of the door, next to the leading edge, of at least 300 mm.

Table 6.93 Width of doors

Minimum corridor width at door (mm)	Minimum clear opening width (mm)
1,500	800
1,200	825
900	850

A door (other than a door to a cupboard or duct enclosure that is normally locked in a closed position) should not open onto a circulation route in a manner that creates an obstruction.	D 4.2.4 ND 4.2.5

In addition, for non-domestic buildings:

A clear glazed vision panel should be provided to any door across a corridor and - to a door between a circulation space and a room with an occupant capacity of more than 60; - to the outer door of a lobby leading solely to sanitary accommodation.	ND 4.2.5

Heavy door leafs and strong closing devices can make an otherwise accessible door impassable to many building users. The force needed to open and pass through a door, against a closing device, therefore should be limited.

A door should be capable of operating with an opening force of not more than 30 N (for first 30° of opening) and 22.5 N (for remainder of swing) when measured at the leading edge of the leaf.	ND 4.2.5

Accessibility within a storey of a dwelling

Each accessible level or storey within a dwelling should   D 4.2.6
have doors with a minimum clear opening width in
accordance with Table 6.94 (including any apartment,
kitchen or sanitary facility).

Table 6.94   Width of doors

Minimum corridor width at door (mm)	Minimum clear opening width (mm)
1,050	775
900	800

Stair landings

Clear space is needed to the head and foot of any stair flight to allow people
to move between a flight and an adjacent level surface safely.

On a private stair (other than on an intermediate landing,   D 4.3.6
common to two flights)

- a door to a cupboard or duct may open onto a top
  landing if, at any angle of swing, a clear space of at
  least 400 mm deep is maintained across the full width
  of the landing.
- a door may open on to a bottom landing, if, at any
  angle of swing, a clear space of at least 400 mm deep is
  maintained across the full width of the landing and the
  door swing does not encroach within space designated
  for future installation of a stair lift.

Stair landings serving outward-opening fully glazed doors

If a conservatory is not intended to be the accessible   D 4.3.8
entrance and has an outward-opening fully glazed door
and a landing (or has fully glazed doors leading from a
dwelling directly into a conservatory) then the landing
length should be in accordance with Figure 6.103.

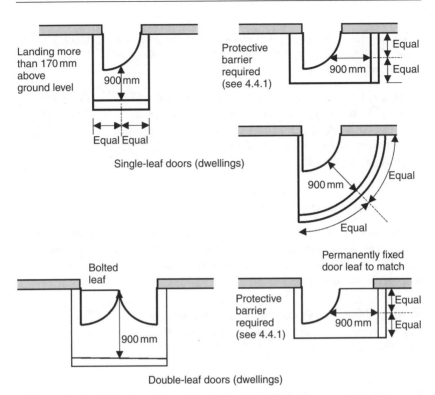

Figure 6.103 Landings serving outward-opening fully glazed doors

Safety and security

Door entry systems

Where a common entrance door, intended as a principal means of access to a building, is fitted with a locking device, a door entry system comprising a remote door release and intercom at the point of entry and a call unit within each dwelling served by that entrance should be installed.	D 4.5.3
Controls should contrast visually with surrounding surfaces and any numeric keypad should follow the 12-button telephone convention, with an embossed locater to the central '5' digit.	D 4.5.3

Collision with glazing

Glazing in certain locations is more vulnerable to human impact. Care should be taken in the selection of, particularly where glazed side panels may be mistaken for doors.

Glazing should be designed to resist human impact as set out in BS 6262: Part 4: 2005, where all, or part, of a pane is part of a door leaf.	D 4.8.2 ND 4.8.2

Unframed glazed doors (particularly powered doors) which operate on a pivot action should have any exposed vertical edges highlighted so as to contrast visually with surroundings, to assist in identifying the door edge when opening or in an open position.	D 4.8.2 ND 4.8.2

Building insulation envelope

Every building must be designed and constructed in such a way that an insulation envelope is provided which reduces heat loss and energy efficiency measures shall be provided that limit the heat loss through doors, etc., by suitable means of insulation.

The maximum U-values for building elements of the insulation envelope should not be greater than those shown in Table 6.95.	D 6.2.1 LD 6.2.1

Table 6.95   Maximum *U*-values for building elements of the insulation envelope

Type of element	(a) Area-weighted average U-value (W/m^2K) for all elements of the same type	(b) Individual element U-value (W/m^2K)
Normal	2.2	3.3
Shell and fit-out buildings	1.8	3.3

The opening areas in the building 'shell' should be designed in accordance with Table 6.96.	ND 6.2.3

Table 6.96   Maximum windows, doors and rooflight areas

Type of building	Windows and doors as a percentage of the area of the exposed wall
Residential (non-domestic)	30
Offices, shops and buildings for entertainment and assembly	40
Industrial and storage buildings	15
High-usage entrance doors and display windows, and similar glazing	As required

In a domestic building, the total area of windows, doors and rooflights, should not exceed 25% of the floor area of the dwelling created by the conversion of a heated building.

Alterations to the insulation envelope

For alterations it is more than likely that the existing construction will be from a different era, in building regulation terms and in certain cases (e.g. historic buildings), it may be necessary to adopt alternative energy efficiency measures which relate to the amount of alteration work being undertaken.

Alterations that involve increasing the floor area and/or bringing parts of the existing building that were previously outwith the insulation envelope into the heated part of the dwelling are considered as extensions and/or conversions (regulation 4, schedule 2).

Where additional windows, doors and rooflights are being created, the overall total area (including existing) should not exceed 25% of the total dwelling floor area and in the case of a heated communal room or other area (exclusively associated with the dwellings), it should not exceed 25% of the total floor area of these rooms/areas.	D 6.2.10

Conservatories

Although conservatories are attached to dwellings, they are normally considered as stand-alone buildings for the purpose of the Scottish Building Regulations.

The dividing elements (i.e. wall, door, window and on the rare occasion floor) should have U-values equal to or better than the corresponding exposed elements in the rest of the dwelling.	D 6.2.12
Draught stripping for doors which are part of the thermal division between the conservatory and the dwelling should be of a similar standard as the exposed windows and doors elsewhere in the dwelling.	D 6.2.12

Stand-alone buildings

Where the area of a communal room or other heated accommodation associated with a block of dwellings is less than $50\,m^2$, these rooms or accommodation may be treated as a stand-alone building and the area of windows, doors, rooflights and roof windows should be limited to 25% of the total floor area of these parts.	D 6.2.13

6.13 Stairs and ramps

Stairs, staircase, stairway and a flight of stairs are all names for a construction designed to bridge a large vertical distance by dividing it into smaller vertical distances, called steps. Stairways may be straight or round, or may consist of two or more straight pieces connected at angles. Special stairways include escalators and ladders. Alternatives to stairways are elevators, stairlifts and inclined moving sidewalks.

6.13.1 Requirements

Fire

> *Every building must be designed and constructed in such a way that in the event of an outbreak of fire within the building:*   **2.1**
>
> - *fire and smoke are inhibited from spreading beyond the compartment of origin until any occupants have had the time to leave that compartment, and any mandatory fire containment measures have been initiated;*
>
>   This standard does not apply to domestic buildings
>
> - *facilities are provided to assist fire-fighting or rescue operations;*   **2.14**
> - *the occupants, once alerted to the outbreak of the fire, are provided with the opportunity to escape from the building, before being affected by fire or smoke;*   **2.9**
> - *illumination is provided to assist in escape.*   **2.10**

Safety

> *Every building must be designed and constructed in such a way that:*   **4.2**
>
> - *in non-domestic buildings, safe, unassisted and convenient means of access is provided throughout the building;*
> - *in domestic buildings, safe and convenient means of access is provided within common areas and to each dwelling;*
> - *in dwellings, safe and convenient means of access is provided throughout the dwelling; and*
> - *in dwellings, unassisted means of access is provided to, and throughout, at least one level;*

- *every level can be reached safely by stairs or ramps;*   *4.3*
- *every sudden change of level that is accessible in, or*   *4.4*
  around, the building is guarded by the provision of
  pedestrian protective barriers;
- *people in and around the building are protected from*   *4.8*
  injury that could result from fixed glazing, projections
  or moving elements on the building;
- *(in non-domestic buildings) a safe and secure means of*   *ND 4.8*
  access is provided to a roof.

 There is no requirement to provide access for a wheelchair user:

- in a non-domestic building not served by a lift, to a
  room, intended to be used as a bedroom, that is not on
  an entrance storey; or
- in a domestic building not served by a lift, within
  common areas and to each dwelling, other than on an
  entrance storey.

Noise

Every building must be designed and constructed in such   *5.1*
a way that the transmission of noise from one dwelling
to another or another part of the building is limited to a
level that will not threaten the health of the occupants of
the dwelling or inconvenience them in the course of
normal domestic activities provided the source noise is
not in excess of that from normal domestic activities.

6.13.2 Meeting the requirements

Half of all accidents involving falls within and around buildings occur on stairways, with young children and elderly people being particularly at risk. This risk can be greatly reduced by ensuring that any change in level incorporates basic precautions to guard against accident and falls.

Every building must be designed and constructed in such a   D 4.3
way that every level can be reached safely by stairs or ramps.

Stairs and ramps should be constructed to be within limits   D 4.3.0
recognized as offering safe and convenient passage and
designed so that any person who is likely to use them can do so
comfortably and safely, with the minimum amount of difficulty.

Construction and measurement of stairs

The geometry of a stair flight can have a significant effect on the ability of people to use a stair safely and conveniently and limits should be placed on the rise and going of a stair, and steepness of pitch.

The pitch of a private stair flight may be steeper than that of a public flight (any other stair) in recognition that users, as occupants, will be more familiar with the stair through frequent use.

Rise, going, tread and pitch of stairs

To provide safe and convenient access, the rise, going, tread and pitch of a flight in a stair should be in accordance with Table 6.97.

Table 6.97   Rise, going, tread and pitch of stairs

Stair geometry – private stair				
Minimum rise	Maximum rise	Minimum going	Tread	Maximum pitch
100 mm	220 mm	225 mm	Not less than going	42°

Stair geometry – any other stair, including to a domestic building or within the common area of a building containing flats or maisonettes				
Minimum rise	Maximum rise	Minimum going	Tread	Maximum pitch
100 mm	170 mm	250 mm	Not less than going	34°

The following requirements also have to be observed:	D 4.3.2

D 4.3.2
ND 4.3.2

The following requirements also have to be observed:

- all rises in a flight should be of uniform height;
- in a straight flight, or in a part of a flight that is straight, measurement should be uniform along the centreline of the flight;
- where a flight consists partly of straight and partly of tapered treads, the going of the tapered treads should be uniform and should not be less than the going of the straight treads;
- the going measured at the narrow end of a tapered tread should be at least 50 mm (see Figure 6.104);
- the aggregate of the going and twice the rise should be at least 550 mm and not more than 700 mm;

 The maximum rise and minimum going on a private stair should not be used together as this will result in a pitch greater than the recommended maximum.

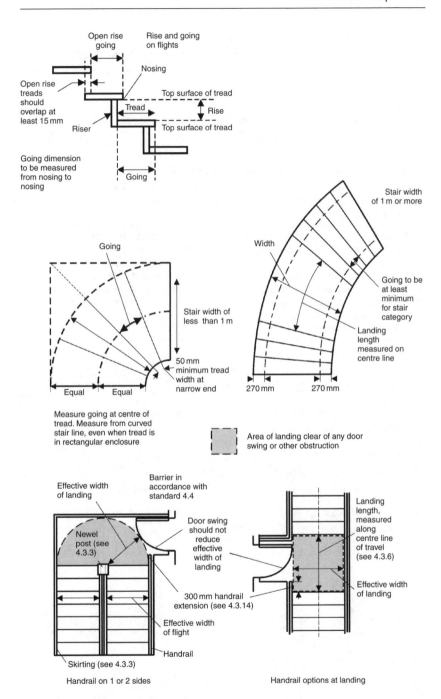

Figure 6.104 Measurement for stairs

Width of stair flights and landings

The clear, or effective, width of a stair should allow users to move up and down unhindered and, on stairs giving access to a dwelling or domestic building, permit people to pass on a flight.

The effective width of a stair should be in accordance with the recommendations of Table 6.98.	D 4.3.3 ND 4.3.3

Table 6.98   Effective widths of flights and landings

Private stair and stair wholly within shared residential accommodation	Any other stair
• 900 mm, such as from one storey to another or connecting levels within a storey; or • 600 mm where serving only sanitary accommodation and/or one room, other than accessible sanitary accommodation, a kitchen or an enhanced apartment.	• 1.0 m generally, such as to an external flight to a domestic building or a common access within a building containing flats or maisonettes; or • 900 mm to an external flight serving a single dwelling, to which the public have access.

The effective width of a private stair may be 800 mm where a continuous handrail is fitted to both sides of a flight.	D 4.3.3 ND 4.3.3
The projection of any stringer or newel post into this width should be not more than 30 mm.	D 4.3.3 ND 4.3.3
A stairlift may be fitted to a private stair and may project into the effective width of the stair but it must have at least one handrail and, when not in use, the installation should: • permit safe passage on the stair flight and any landing; and • not obstruct the normal use of any door, doorway or circulation space.	D 4.3.3

Number of rises in a flight

Generally a flight should have not more than 16 rises, and at least three rises.	D 4.3.4 ND 4.3.4
There may be fewer than three rises: • between an external door of a building and the ground or a balcony, conservatory, porch or private garage (unless it is other than at an accessible entrance); or • within an apartment other than where affecting provisions within an enhanced apartment; or	D 4.3.4 ND 4.3.4

- within sanitary accommodation (other than accessible sanitary accommodation); or
- between a landing and an adjoining level where the route of travel from the adjoining level to the next flight changes direction through 90° (i.e. on a quarter landing as the first step).

Risers and treads

All stairs providing access to and within buildings should be designed to be accessible by most persons with reduced mobility.

Open risers on a flight can be a hazard. When ascending a stair, people may be at risk of trapping the toes of shoes beneath projecting nosings, and of tripping as a result. In addition, many may feel a sense of insecurity when looking through spaces present between treads.	D 4.3.5 ND 4.3.5
A stair should have contrasting nosings to assist in identifying the position of treads and risers should be profiled to minimize tripping as shown in Figure 6.105.	D 4.3.5 ND 4.3.5

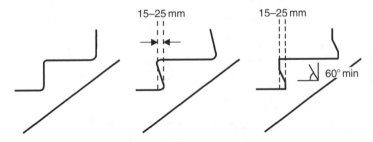

Figure 6.105 Step profile examples

Open risers should not be used unless a stair is intended for descent only, such as in a dedicated escape stair on an escape route.	D 4.3.5 ND 4.3.5
Small children can climb or fall through gaps in stair treads and the size of such gaps should be limited to prevent this. In a flight with open risers, the treads should overlap by at least 15 mm.	D 4.3.5 ND 4.3.5
Any opening between adjacent treads in a flight should be small enough to prevent the passage of a 100 mm sphere.	D 4.3.5 ND 4.3.5

Stair landings

Clear space is needed to the head and foot of any stair flight to allow people to move between a flight and an adjacent level surface safely. People may also wish to pause on stairs, particularly during ascent, and any intermediate landing should provide a temporary respite and be of a size to allow this whilst still permitting others to pass safely.

A stair landing should:	D 4.3.6
• be provided at the top and bottom of every flight (a single landing may, however, be common to two or more flights);	ND 4.3.6
• be level except, in external locations, for any minimal cross-fall necessary to prevent standing water;	
• have an effective width of not less than the effective width of the stair flight it serves;	
• be clear of any door swing or other obstruction, other than to a private stair, as noted below.	
The minimum length of a stair landing should be either 1.2 m or the effective width of the stair, whichever is less.	D 4.3.6 ND 4.3.6
On landings to external stair flights, where tactile paving is used, the minimum length of landing should be 1.2 m.	D 4.3.6 ND 4.3.6
Other than at an accessible entrance, a landing need not be provided to a flight of steps between the external door of:	D 4.3.6
• a dwelling and the ground, balcony, conservatory, porch or private garage (where the door slides or opens in a direction away from the flight and the total rise is not more than 600 mm);	
• a dwelling, or building ancillary to a dwelling, and the ground, balcony, conservatory or porch (where the change in level is not more than 170 mm, regardless of method of door operation).	
On a private stair (other than on an intermediate landing, common to two flights):	D 4.3.6
• a door to a cupboard or duct may open onto a top landing if, at any angle of swing, a clear space of at least 400 mm deep is maintained across the full width of the landing.	
• a door may open on to a bottom landing, if, at any angle of swing, a clear space of at least 400 mm deep is maintained across the full width of the landing and the door swing does not encroach within space designated for future installation of a stairlift.	

Additionally, for non-domestic buildings:

A landing need not be provided to a flight of steps between the external door landing of a building and the ground where:	ND 4.3.6

- the door is not an accessible entrance;
- the door slides or opens in a direction away from the flight;
- the rise of the flight is not more than 600 mm.

Warning surfaces to landings of external steps

A sudden and unguarded change of level on an access route can present a hazard to a person with a visual impairment. The use of 'corduroy' tactile paving identifies this hazard and advises users to 'proceed with caution'.

Corduroy tactile paving should:	D 4.3.7
	ND 4.3.7

- be provided on external routes serving more than one dwelling;
- be used to alert people to the presence of a flight of steps;
- be provided at the head and foot of any flight of external steps, forming a strip 800 mm deep, positioned 400 mm from the first step edge (as shown in Figure 6.106).
- (if on a landing mutual to a flight of steps and a ramp) lie outwith the landing area of any ramp flight, to prevent possible confusion which might lead to injury.

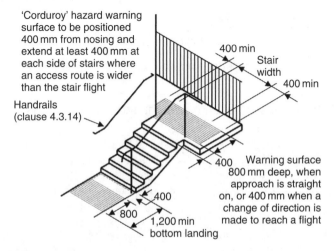

Figure 6.106 Use of corduroy tactile paving

Stair landings serving outward-opening fully glazed doors

> If a conservatory and/or similar is not intended to be the
> accessible entrance and has an outward-opening fully
> glazed door and a landing (or a fully glazed door leading
> from a dwelling directly into a conservatory) then the
> landing length should be in accordance with Figure 6.107.
>
> D 4.3.8

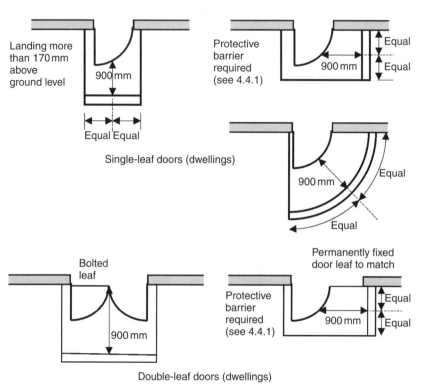

Figure 6.107 Landings serving outward-opening fully glazed doors

Stair flights consisting of both straight and tapered treads

> On that part of a flight consisting of tapered treads: D 4.3.9
> - the going of the tapered treads should be uniform and   ND 4.3.8
>   should not be less than the going of the straight treads:
> - at the inner end of the tread, the going should be at least
>   50 mm.
>
> In a flight less than 1 m wide the going should be measured   D 4.3.9
> at the centre line of the flight.   ND 4.3.8

In a flight 1 m wide or more the going should be measured at two points, 270 mm from each end of the tread; and the minimum going should be at least the going of the straight treads.	D 4.3.9 ND 4.3.8

Stair flights consisting wholly of tapered treads

Stairs formed from tapering treads, particularly where forming a spiral, can present greater difficulties in use for many people than straight flights. There should be an appropriate level of safety and amenity on such stairs, particularly where used as a primary means of access.

A flight consisting wholly of tapered treads, forming a helix or spiral, should be constructed in accordance with the guidance in BS 5395: Part 2: 1984 to give safe passage.	D 4.3.10 ND 4.3.9

Handrails to stairs and ramps

Handrails to a stair and ramp flights will provide support and assist safe passage.

A handrail should generally be provided to both sides of a stair or ramp flight.	D 4.3.14 ND 4.3.12
A handrail should generally be provided to both sides of any flight where there is a change of level of more than 600 mm, or where the flight on a ramp is longer than 2 m. However: <ul><li>handrails maybe omitted to the flight of a ramp, serving a single dwelling, where the change in level is less than 600 mm; and</li><li>a handrail need only be provided to one side on a flight of a private stair.</li></ul>	D 4.3.14 ND 4.3.12
A second handrail may be installed provided a clear width of 800 mm is maintained.	D 4.3.14 ND 4.3.12
A handrail should be fixed at a height of at least 840 mm and not more than 1.0 m, measured vertically above the pitch line of a flight on a stair or ramp and on a landing where a handrail is provided.	D 4.3.15 ND 4.3.14

Additionally, for domestic buildings:

Where a handrail is provided to only one side of a private stair flight, the side on which a handrail is not fixed should permit installation of a second handrail at a future date.	D 4.3.14

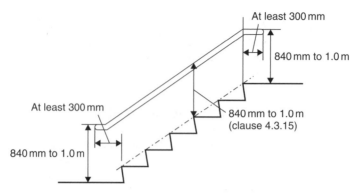

Figure 6.108 Handrails to stairs and ramps

A handrail on a stair or ramp flight should: D 4.3.14
- extend at least 300 mm beyond the top and bottom of the   ND 4.3.13
  flight as shown in Figure 6.108 (the 300 mm extension
  may be omitted where the handrail abuts a newel post);
  and
- have a profile and projection that will allow a firm grip;
  and
- end in a manner, such as a scrolled or wreathed end,
  that will not present a risk of entrapment to users; and
- contrast visually with any adjacent wall surface.

 It is only necessary to provide a handrail with a
profile and projection that will allow a firm grip on
a private stair or a ramp providing access within a
single dwelling, as users are likely to be familiar
with the layout and use of the flight.

A stair or ramp that is more than 2.3 m wide should be   D 4.3.14
divided by a handrail, or handrails, in such a way that each   ND 4.3.13
section is at least 1.1 m and not more than 1.8 m wide.

 This does not apply to a stair between an entrance
door to a building and ground level, where not
forming part of an escape route.

Headroom on stairs and ramps

A flight or landing on a stair or ramp should have clear   D 4.3.16
headroom (measured vertically from the pitch line of the   ND 4.3.15
flight or from the surface of the landing) of at least 2.0 m
extending over the whole of the effective width.

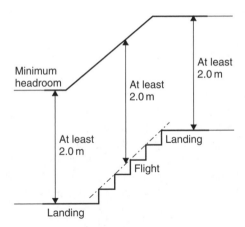

Figure 6.109 Headroom on stairs and ramps

Industrial stairs and fixed ladders

An industrial stair or a fixed ladder serving an area in any   D 4.3.17
building to which only limited access is provided (e.g.
a plant room) should be constructed so as to offer safe
passage.

Stairs and fixed ladders in agricultural buildings

A stair or fixed ladder in an agricultural building should   ND 4.3.17
offer safe passage.

Interruption of lateral support

Where a stair or other opening in a floor or roof is   D 1D37
alongside a supported wall and interrupts the continuity
of lateral support:

- the length of the opening should be not more than 3 m;
- if connection is via anchors, then these should be
  spaced less than 2 m on each side of the opening;
- if connection is not via anchors, then this should be
  provided throughout the length of each portion of the
  wall, and on each side of the opening.

 There should be **NO** other interruption of lateral support.

Openings for stairs

Where openings in floors are required for stairs: D 1F15

- the perimeter of the opening should be supported on all sides by load-bearing walls; or
- the floor joists should be strengthened by means of additional joists and trimmers as follows:
  - doubling up the trimming joists either side of the openings in floors parallel to the floor joists and connecting them together; or
  - installing trimmer joists perpendicular to floor joists.

The plan size of openings for stairs should be not more than 2.70 m parallel to the floor joists by 1.15 m perpendicular to the floor joists.

Collision with projections

Fixtures that project into, or open onto any place to which people have access can be a hazard and should be positioned, secured or guarded so that they do not present a risk to building users.

Where a building element (as beneath a stair flight) projects   D 4.8.1
into a circulation route or space, and any part of the   ND 4.8.1
obstruction is less than 2.0 m above the ground, guarding
should be provided to both highlight the hazard and prevent
collision with the building element.

Protective barriers

Protective barriers (such as a wall, partition or area of   D 4.4.1
fixed glazing) should be provided at the edge of every   ND 4.4.1
floor, stair, ramp, landing, raised floor or other raised area
to which people have access, where there is a difference in
level of 600 mm or more.

In and around domestic buildings, gaps in any protective barrier should not be large enough to permit a child to pass through. To ensure this:

- openings in a protective barrier should prevent the   D 4.4.2
  passage of a 100 mm diameter sphere; and

- a protective barrier (and any wall, partition or fixed glazing used instead of a barrier) should be secure, capable of resisting loads calculated in accordance with BS 6399: Part 1: 1996 and meet the height requirements shown in Table 6.99.

ND 4.4.2

Table 6.99 Height of pedestrian protective barriers

Location	Minimum height (mm)
At the edge of a floor in front of walls, partitions and fixed glazing incapable of withstanding the loads specified in BS 6399: Part 1: 1996	800
In front of an openable window	800
On a stair or ramp flight in a building or wholly within a dwelling	840
On a stair or ramp flight outwith a dwelling	900
To a gallery, landing or raised area within a dwelling	900
Directly in front of, or behind, fixed seating	800
All other locations	1,100

Airborne sound

Airborne sound resisting separating walls and separating floors should be provided between dwellings so that each dwelling is protected from noise emanating from the other one and from other parts, such as common stair enclosures and passages, solid waste disposal chutes, lift shafts, plant rooms, communal lounges and car parking garages.

D 5.1.1

No service pipes or ducts should pass between a dwelling and a common stairway, common passage or services enclosure.

D 5.1.3

Lighting

Any lighting point in a domestic building serving a stair should have controlling switches at, or in the immediate vicinity of, the stair landing on each storey.

D 4.5.1

In communal areas (particularly on stairs and ramps within a building) the possibility of slips, trips and falls and of collision with obstacles should be minimized. To achieve this:

Common areas should have artificial lighting capable of providing a uniform lighting level, at floor level, of not less than 100 lux on stair flights and landings and 50 lux elsewhere within circulation areas.	D 4.5.2
Lighting should not present sources of glare and should avoid creation of areas of strong shadow that may cause confusion or mis-step.	D 4.5.2
A means of automatic control should be provided to ensure that lighting is operable during the hours of darkness.	D 4.5.2

Escape stairs – domestic buildings

In a domestic building, the **normal** means of escape in the event of a fire will be by way of the internal stairs or other circulation areas.

Destination of escape routes

An escape route from a flat or maisonette should lead to a place of safety or an access deck directly or by way of an exit to an external escape stair.	D 2.9.5 ND 2.9.6

Flats entered from below the accommodation level

A flat at a storey height of more than 4.5 m which is entered from a storey below the level of the accommodation should be planned so that: • an alternative exit is provided; • all apartments are entered directly from a protected enclosure and the distance to be travelled from any door of an apartment to the head of the private stair is not more than 9 m; • the distance to be travelled from any point within the flat to the head of the private stair is not more than 9 m, and the direction of travel is away from cooking facilities.	D 2.9.10

Ducted warm air heating

Where a flat or maisonette has a storey at a height of more than 4.5 m, or a basement storey and is provided with a system of ducted warm air heating: — D 2.9.13

- transfer grilles should not be fitted between any room and the entrance hall or stair;
- joints between the duct and the surrounding construction should be sealed where a duct passes through any wall, floor or ceiling of an entrance hall or stair;
- the return air should be ducted back to the heater where warm air is ducted to an entrance hall or stair.

Mixed-use buildings

An escape stair serving flats or maisonettes should not communicate directly with a non-domestic building or communal facilities for a group of dwellings. — D 2.9.16

Where the escape stair is to be accessed from the common areas of the building, then that escape stair should be separated from the dwellings and where it serves the non-domestic accommodation at each level (including the topmost storey) by a protected lobby. — D 2.9.16

Where the building or part of the building has no storey at a height of more than 7.5 m and has only one escape route by way of an escape stair and there are no alternative escape routes from the building, protected lobbies should be provided at every level. — D 2.9.16

Where the building or part of the building has two or more escape routes, only one stair should communicate with both the domestic and non-domestic parts of the building and be provided with protected lobbies. — D 2.9.16

An escape stair which serves a flat or maisonette which is ancillary to a non-domestic building, may communicate with the non-domestic accommodation provided that: — D 2.9.16

- the escape stair is separated from the domestic and non-domestic accommodation by a protected lobby at every level;
- where the storey height of the flat or maisonette is more than 7.5 m an alternative escape route is available from the flat or maisonette;

- an alarm and detection system which is designed in accordance with BS 5839: Part 1: 2002 is installed in the common areas.

 Note: In this context '*ancillary*' includes caretakers', directors', supervisors' and similar flats or maisonettes.

Smoke control in corridors

Where a building has more than one escape stair and where a corridor, or part of a corridor, provides escape in only one direction, automatic opening ventilators should be provided in that part of the corridor which provides single direction escape.   D 2.9.17

Such ventilators should:

- provide for exhaust at or near ceiling level and for supply at or near floor level with a combined aggregate area of at least $1.5\,m^2$;
- be activated by automatic smoke detection fixed to the ceiling of the corridor and fitted with a manual override for fire service use.

Temporary waiting spaces

A protected zone enclosing an escape stair and an external escape stair should be provided with an unobstructed clear space capable of accommodating a wheelchair and measuring not less than $700 \times 1,200\,mm$ on every escape stair landing to which there is access from a storey.   D 2.9.18

 This space should **not** be used for any form of storage.

Protected lobbies

A protected lobby means a lobby within a protected zone but separated from the remainder of the protected zone so as to resist the movement of smoke from the adjoining accommodation to the remainder of the protected zone.

Where flats or maisonettes are served by only one escape   D 2.9.19
stair and there is no alternative means of escape from
the upper storeys, there should be a protected lobby with
automatic opening ventilators, at each storey within the
protected zone between the escape stair and the accommodation, including a parking garage and any other accommodation ancillary to the dwellings (see Figure 6.110).

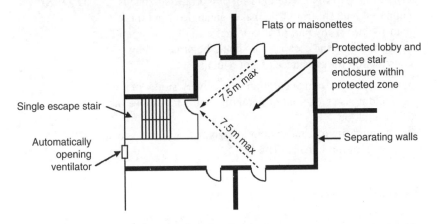

Figure 6.110 Single stair access to flats and maisonettes with any storey at a height of more than 7.5 m but not more than 18 m

A ventilated protected lobby need **not** be provided where   D 2.9.19d
there is more than one escape stair serving each dwelling.

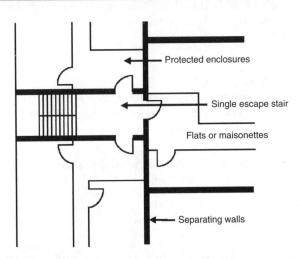

Figure 6.111 Single stair access to flats and maisonettes with every storey at a height of not more than 7.5 m

Protected zones

Each escape stair should be within a protected zone, except an external escape stair with a total rise of not more than 7.5 m, which leads directly to a place of a protected zone enclosing an escape stair should not enclose any room including a store room or any other ancillary rooms.	D 2.9.20
Where any part of an external wall of a protected zone is not more than 2 m from, and makes an angle of not more than 135° with any part of an external wall of another part of the building, the escape stair should be protected by construction for a distance of 2 m with a: • short fire resistance duration where every storey in the building is at a height of not more than 7.5 m above the ground; or • medium fire resistance duration where any storey is at a height of more than 7.5 m.	D 2.9.20

Protected enclosures

If a protected enclosure is constructed within a dwelling: • the walls should have a short fire resistance duration; and • any door in the wall should be a self-closing fire door with a short fire resistance duration.	D 2.9.21
In a house containing an apartment or kitchen in a storey at a height of more than 4.5 m, every stair should be in a protected enclosure.	D 2.9.31

External escape stairs

External escape stairs present additional hazards to people evacuating a building in the case of fire. This is because the escape stair will be exposed to the possible effects of inclement weather.

An external escape stair should not serve a building where the topmost storey height exceeds 7.5 m.	D2.9.22
An external escape stair should lead directly to a place of safety and be protected against fire from within the building.	D 2.9.22
Fire protection need not be provided to an external escape stair with a total rise not more than 1.6 m.	D 2.9.22

| Every part of an external wall (including a door, window or other opening) within 2 m from the escape stair, should have short fire resistance duration other than a door opening from the top storey to the external escape stair. | D 2.9.22 |

| Fire protection below the escape stair should be extended to the lowest ground level. | D 2.9.22 |

| Due to the likely smoke dissipation to atmosphere, service openings including ventilation ducts not more than 2 m from the escape stair may be protected by heat-activated sealing devices or systems. | D 2.9.22 |

Basements

| Where an escape stair also serves a basement storey, the protected zone enclosing the escape stair in the basement storey should be separated from the protected zone containing the escape stair serving the rest of the building, by a wall or screen (with or without a door) at the ground storey floor level. | D 2.9.28 |

Inner rooms

| Where a private stair serves a basement storey, the private stair should be in a protected enclosure. | D 2.9.29 |

Fire-fighting shafts

A fire-fighting shaft is an enclosure protected from fire in adjoining accommodation and contains an escape stair, and a fire-fighting lobby at every storey at which the fire-fighting shaft can be entered from the accommodation. The shaft may also contain a fire-fighting lift together with its machine room. These shafts are used in high buildings, deep buildings and in certain shops or storage buildings to assist fire service personnel to carry out fire-fighting and rescue operations.

| The enclosing structure of a fire-fighting shaft should have long fire resistance duration. | ND 2.1.9 |

Compartments

Compartment walls and compartment floors are intended to prevent fire passing from one compartment to another. Openings and service penetrations through these walls or floors can compromise their effectiveness and should be kept to a minimum.

Self-closing fire doors can be fitted with hold-open devices as specified in BS 5839: Part 3: 1988 **provided** the door is not an emergency door, a protected door serving the only escape stair in the building (or the only non-escape stair serving part of the building) or a protected door serving a fire-fighting shaft.	D2.2.9 ND 2.1.14
Electrically operated hold-open devices should deactivate: • on operation of an automatic fire alarm system; • on any loss of power to the hold-open device, apparatus or switch; and • on operation of a manually operated switch fitted in a position at the second door.	D2.2.9 ND 2.1.14

Fire shutters

Where an opening in a compartment floor contains a stair (but not an escape stair) and not more than two escalators: • a horizontal fire shutter may be installed which maintains the fire resistance duration of the compartment floor and is activated by a fusible link or other heat-sensitive device positioned to detect fire in the lower compartment; or • vertical fire shutters may be installed at each floor level (other than the topmost storey) which maintains the fire resistance duration of the compartment floor and are activated by smoke detection positioned to detect smoke in the lower compartment.	ND 2.1.14

Escape stairs – non-domestic buildings
Escape stair widths

 Note: Escape stair widths are calculated on the basis of the total number of available storey exits.

Individual widths

The effective width of an escape stair should be at least the width of any escape route giving access to it except that if the number of people using the escape route is not more than: • 225, it may be reduced to not less than 1.1 m; and • 100, it may be reduced to not less than 1 m.	ND 2.9.10

Note: The effective width of an escape stair is measured between handrails and clear of obstructions.

Aggregate width

Where the escape routes from a storey consist solely of escape stairs (and the storey has two or more escape stairs) the effective width of every escape stair in mm from that storey should be at least 5.3 times the appropriate capacity divided by the number of such stairs, less one.   ND 2.9.10

 An escape stair should NOT narrow in the direction of escape.

Combined escape routes

Where escape routes from a storey consist of a combination of escape stairs and other escape routes (see Figure 6.112) the effective width of any escape stair from that storey should be designed to take into account that proportion of the number of occupants on that storey who may escape by way of the other escape routes.   ND 2.9.10

Where the escape route from an escape stair is also the escape route from the ground storey and/or basement storey, the width of that escape route should be increased to take account of that proportion of the occupancy capacity from the ground storey and/or basement storey.   ND 2.9.10

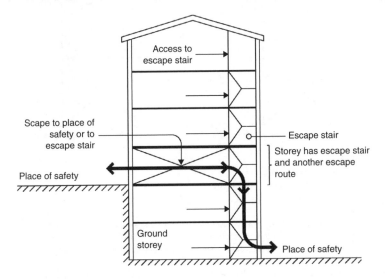

Figure 6.112 Combined escape routes

Appropriate capacity

The appropriate capacity of an escape stair in relation to
any storey above or below the place of safety is calculated
by one of the following methods and in accordance with
Figure 6.113:

ND 2.9.11

- the occupancy capacity of the part of the storey or
  room served by the escape stair;
- where the escape stair serves two or more storeys
  including any rooms, or part of a building, which is not
  divided by compartment floors, the occupancy capacity
  of the part of each storey, including any room, served
  at, and above or below as the case may be, the escape
  stair, less 20%.

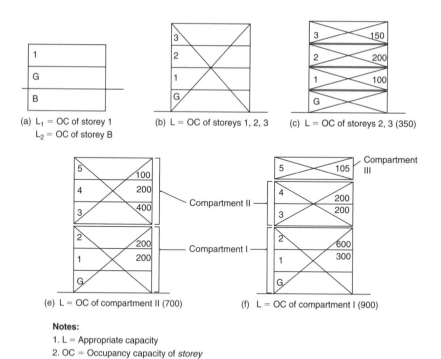

(a) L_1 = OC of storey 1
    L_2 = OC of storey B

(b) L = OC of storeys 1, 2, 3

(c) L = OC of storeys 2, 3 (350)

(e) L = OC of compartment II (700)

(f) L = OC of compartment I (900)

Notes:
1. L = Appropriate capacity
2. OC = Occupancy capacity of *storey*
3. G = *Ground storey*
4. B = *Basement storey*
5. Numbers 100, 500 etc. relate to occupancy capacity of each *storey*
6. Figure numbers refer to paragraph numbers in clause 2.9.11
7. For simplicity, the 20% reduction factor has been applied to above examples

Figure 6.113 Appropriate capacity

In a building, or part of a building, which is divided by one or more compartment floors, the total occupancy capacity, less 20%, of:

- each of the two adjacent upper storeys; or
- the compartment either above or below ground level served by the escape stair having the greatest occupancy capacity.

 Where a building or part of a building is designed on the basis of vertically phased evacuation, consideration should be given to the installation of an automatic fire detection and voice alarm system.

Independence of escape stairs

Where two protected zones enclosing escape stairs share a common wall, any access between them should be by way of a protected lobby.

ND 2.9.12

Where a room or storey needs two or more escape stairs, it should be possible to reach one escape stair without passing through the other.

ND 2.9.12

Escape routes in a central core

To reduce risk of smoke spreading to more than one escape stair, corridor or lobby, a building with more than one escape route contained in a central core should be planned so that:

ND 2.9.13

- the exits from the storey are remote from one another; and
- no two exits are approached from the same lift hall, common lobby or undivided corridor or linked by any of these other than through self-closing fire doors (Figure 6.113a).

Electrically powered locks

Electrically powered locks should not be installed on a protected door serving the only escape stair in the building.

ND 2.9.15

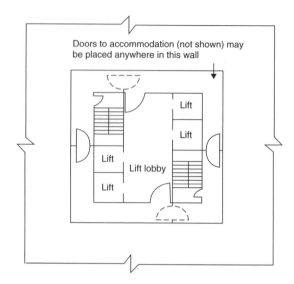

Figure 6.113a Escape routes in a central core

Protected lobbies

A protected lobby means a lobby within a protected zone but separated from the remainder of the protected zone so as to resist the movement of smoke from the adjoining accommodation to the remainder of the protected zone.

Where a building, or part of a building, has only one escape route by way of an escape stair, access to the escape stair should be by way of a protected lobby.	ND 2.9.21
In high-rise buildings where an escape stair serves a storey with a height greater than 18 m, access to the protected zone containing the escape stair should be by way of a protected lobby.	ND 2.9.21
The wall dividing a protected lobby from the remainder of the protected zone may have short fire resistance duration and any door in the wall should be a self-closing fire door with short fire resistance duration.	ND 2.9.21

Protected zones

A protected zone is that part of an escape route within a building (but not within a room) to which access is only by way of a protected door and there is an exit directly to a place of safety that is constructed as a compartment.

An escape stair should be within a protected zone, unless: • it connects two or more levels within a single storey where the difference in level between the highest and lowest level is not more than 1.8 m;	ND 2.9.22

- the escape stair is a fixed ladder; or
- an external escape stair with a total rise of not more than 1.6 m; or
- an external escape stair constructed in accordance with the regulations (see clause 2.9.24).

Galleries, catwalks or openwork floors

An escape stair from a gallery, catwalk (including lighting bridges) or openwork floor should have: ND 2.9.22

- an occupancy capacity of not more than 60; or
- an occupancy capacity of 61 to 100 and at least one route of escape is by way of a protected zone, an external escape stair or to another compartment.

If a gallery, catwalk or openwork floor serves more than one level within the room, an unenclosed escape stair may be provided between the floor of the room and the lowest gallery, catwalk or openwork floor only.   ND 2.9.22

Where any part of a protected zone enclosing an escape stair is not more than 2 m from, and makes an angle of not more than 135° with any part of an external wall of another part of the building, the escape stair should be protected for a distance of 2 m, by: ND 2.9.22

- short fire resistance duration if every storey in the building is less than 7.5 m above ground level;
- medium fire resistance duration if any storey is at a height of more than 7.5 m.

Rooms, toilets and washrooms in protected zones

Provided all parts of the building served by the escape stair have at least one other escape route: ND 2.9.23

- toilets, washrooms and a cleaner's cupboards not more than 3 m^2;
- reception rooms, an office and a general store room of not more than 10 m^2; and
- escalators

'may' be located within a protected zone.

External escape stairs

External escape stairs present additional hazards to people evacuating a building in the event of fire because the escape stair will be exposed to the possible effects of inclement weather and people who are unfamiliar with the escape routes can feel less confident using an unenclosed stair high above the ground. For these reasons:

An external escape stair should only serve a building where: • the topmost storey height is not more than 7.5 m; • the building or part of the building is not accessible to the general public; and • in the case of a residential care building or a hospital, the stair is intended to be used by staff only.	ND 2.9.24
An external escape stair should lead directly to a place of safety and be protected against fire from within the building. 💡 Fire protection need not be provided to an external escape stair with a total rise not more than 1.6 m.	ND 2.9.24
Every part of an external wall (including a door, window or other opening) not more than 2 m from the escape stair, should have short fire resistance duration.	ND 2.9.24
Fire protection below an escape stair should be extended to the lowest ground level.	ND 2.9.24
Service openings including ventilation ducts not more than 2 m from the escape stair may be protected by heat-activated sealing devices or systems.	ND 2.9.24

Temporary waiting spaces

The speed of evacuation of people with mobility problems can be much slower than able-bodied people and they should be provided with space to wait temporarily until it is safe to use the escape stair.

Temporary waiting spaces should **not** be used for any form of storage.	ND 2.9.25
A protected zone enclosing an escape stair and an external escape stair should be provided with an unobstructed clear space capable of accommodating a wheelchair and measuring not less than 700 mm × 1.2 m on every landing to which there is access from a storey.	ND 2.9.25

Fire evacuation in a hospital is based on progressive horizontal evacuation and therefore temporary waiting spaces in escape stairs need not be provided.

Escape stairs in basements

Where an escape stair also serves a basement storey, the protected zone enclosing the escape stair in the basement storey should be separated from the protected zone containing the escape stair serving the rest of the building, by a wall or screen, with or without a door, at the ground storey floor level.   ND 2.9.26

Note: The wall, screen and self-closing fire door where provided, should have medium fire resistance duration.

Combustibility

Every part of an escape stair (including landings) and the floor of a protected zone or protected lobby, should be constructed of non-combustible material, unless it is: ND 2.9.29

- an escape stair in shared residential accommodation;
- a handrail, balustrade or protective barrier on an escape stair;
- an escape stair which connects two or more levels within a single storey where the difference in height between the highest and lowest level is not more than 1.8 m;
- an escape stair from a gallery, catwalk or openwork floor;
- a floor finish (e.g. laminate flooring) applied to the escape stair (including landings) or to the floor of a protected zone or protected lobby.

Fire service facilities

Every building must be designed and constructed in such a way that every level can be reached safely by stairs or ramps.   D 4.3
ND 4.3

Facilities should be designed and installed within the building to assist the fire service in carrying out their fire-fighting or rescue operations as efficiently as possible.   ND 2.14.1

Fire service facilities – domestic
In domestic buildings

Flats and maisonettes with one escape stair shall provide the following fire-fighting facilities:	D 2A7

- a fire-fighting shaft;
- a fire-fighting lift; and
- a dry fire main with the outlet located in the fire-fighting lobby.

Flats or maisonettes which have two more escape stairs, shall provide the following fire-fighting facilities to at least two escape stairs positioned at least 20 m apart:	D 2A7

- fire-fighting shafts;
- fire-fighting lifts; and
- dry fire mains with the outlets located in the fire-fighting lobby.

Where flats and maisonettes have only one escape stair and any storey is at a height of more than 7.5 m, a protected lobby should be provided within the protected zone on every storey and should have an area of at least 5 m^2.	D 2.14.1
Within the protected lobby there should be an outlet from a dry fire main that will enable fire service personnel to attack the fire earlier.	D 2.14.1
Where flats or maisonettes have two or more escape stairs and any storey is at a height of more than 7.5 m, protected lobbies and dry fire mains should be provided to not less than two escape stairs positioned at least 20 m apart.	D 2.14.1
No point on any storey should be further from a dry riser outlet than one storey height and 60 m measured along an unobstructed route for the fire hose.	D 2.14.1
An escape stair within a protected zone should have either:	D 2.14.3 ND 2.14.5

- a ventilator of not less than 1 m^2 at the top of the stair; or
- an opening window at each storey with an openable area of 0.5 m^2.

Vertical circulation in common areas of domestic buildings

Stairs in common areas should be designed to be accessible to a person with reduced mobility.	D 4.2.5

There should be an accessible stair between each level of   D 4.2.5
a building.

Vertical circulation between storeys of a non-domestic building

Stairs within a building should be designed to be accessible to a person with reduced mobility.

There should be: ND 4.2.7
- an accessible stair between each level of a building;
- a means of unassisted access, other than a ramp, between storeys (except to specific areas where access by a lift need not be provided).

Access between storeys in a dwelling

Where a dwelling has accommodation on more than one   D 4.2.7
level, the levels containing accommodation should be
connected by a stair or ramp within the dwelling.

Unassisted access between storeys in a dwelling

Not everyone can use stairs unassisted and this may mean that the upper levels of a dwelling are not accessible to some occupants. To allow for future installation of a stairlift, any stair giving access to a principal living level or to accommodation greater than may be accessed via a 600 mm wide stair should:

- have an area of wall not less than 700 mm in length (or   D 4.2.8
  an equivalent space, adjacent to the bottom riser of a
  stair) that is clear of any obstruction, fitting or doorway, to allow for parking of a stairlift at rest position.

This space should not be less than 400 mm in depth,
and have a similar area of not less than 200 mm in
length, on the same side of the flight, at landing
level adjacent to the top nosing of the stair, to assist
in transfer at the upper level, allowing for projection
of a stairlift track.

Dwellings with limited entrance storey accommodation

Where there is no level or ramped access from the accessible entrance of a dwelling to the principal living level, the principal living level should be made accessible to as wide a range of occupants as possible and, accordingly:

- a stair, from an accessible entrance to the principal living level, should follow the guidance on rise, going and pitch for 'any other stair' (see clause 4.3.2);
- provision for installation of a stairlift should be made (see clause 4.2.8); and
- entrance level accommodation should contain an area of at least 800 mm wide by 1.1 m long that would permit storage of a wheelchair or pram and which is clear of any doorway, door swing, stair landing or space identified for a future stairlift installation.

D 4.2.10

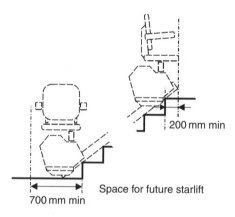

200 mm min

700 mm min

Space for future starlift

Figure 6.114 Future provision for unassisted access

High-rise buildings

A fire-fighting lift is a lift with additional protection and with controls to enable it to be used under the direct control of the fire service when fighting a fire.

The fire-fighting lift need not serve the top storey of a building where:

D 2A7

- the top storey is for service plant use only;
- access to the plant room is from an escape stair from the storey below;
- the foot of the escape stair is not more than 4.5 m from the fire-fighting lift; and
- a dry rising main is installed in the protected lobby of the escape stair.

> Automatic life safety fire suppression systems do not need   D 2A7
> to be installed in common spaces such as stairs, corridors,
> landings or communal facilities.

Fire service facilities – non-domestic

Facilities (such as fire mains, fire-fighting shafts and lifts, smoke clearance capability and safe bridgeheads) should be designed and installed within common escape stairs to assist the fire service in carrying out their fire-fighting or rescue operations as efficiently as possible.

Facilities on escape stairs

In complex buildings or high-rise buildings fire service personnel should be provided not only with good access and water supplies, but also with safe bridgeheads from which to work.

> Where a building has two or more escape stairs, fire-   ND 2.14.1
> fighting facilities should be provided to at least two escape
> stairs in accordance with Table 6.100 and positioned at
> least 20 m apart.

Table 6.100   Fire-fighting facilities on escape stairs

Storey height and depth of buildings	All buildings (other than those listed in column 3)	Shops, storage buildings and open-sided car parks where the area of any storey is more than 900 m² [3]
Basements at a depth more than 10 m	Fire-fighting shaft; Fire-fighting lift; Dry fire main (outlet located in fire-fighting lobby).	
Basements at a depth not more than 10 m	No provision	Fire-fighting shaft [1]; dry fire main (outlet located in fire-fighting lobby).
Topmost storey height not more than 7.5 m	No provision	Protected lobby; dry fire main (outlet located in protected lobby) [2].
Topmost storey height not more than 18 m Topmost storey height not more than 60 m	Protected lobby; Dry fire main (outlet located in protected lobby) [1]. Fire-fighting shaft; Dry fire main (outlet located in fire-fighting lobby).	

Notes:
1. The protected lobby should have an area of at least 5 m².
2. A fire-fighting lift need not serve the top storey of a building where:
   - the top storey is for service plant use only; and
   - access to the plant room is from an escape stair from the storey below; and
   - the foot of the escape stair is not more than 4.5 m from the fire-fighting lift; and
   - dry rising mains are installed in protected lobbies of escape stair.
3. For open-sided car parks, the dry fire main may be located in the protected zone enclosing the escape stair.

> Where there is only one escape stair in the building, then ND 2.14.1
> fire-fighting facilities need only be provided to that stair.

Fire-fighting shafts

A fire-fighting shaft contains an escape stair, a fire-fighting lobby at every storey (to allow the fire-fighting shaft to be entered) and, if provided, a fire-fighting lift together with its machine room. Figure 6.115 provides some examples of typical designs.

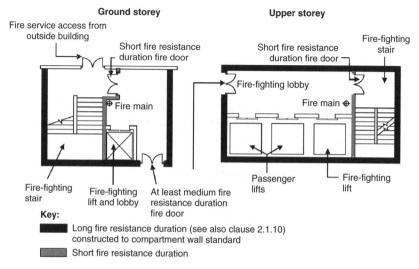

Figure 6.115 Examples of typical fire service facilities

> The lobby should be capable of providing a safe bridge- ND 2.14.3
> head for the fire-fighters to work as well as providing
> access from the escape stair to the accommodation and to
> any associated fire-fighting lift.

Ventilation of escape stairs

Ventilation of the escape stairs, protected lobbies and common access corridors is important to assist fire service personnel during fire-fighting operations and for smoke clearance purposes after the fire has been extinguished.

An escape stair within a protected zone should have either: ND 2.14.5

- a ventilator of not less than $1\,m^2$ at the top of the stair; or
- an opening window at each storey with an openable area of $0.5\,m^2$.

Hospitals and residential care buildings

The maximum travel distance from any point within a   ND 2.A.3
compartment should be not more than $64\,m$ to: ND 2.B.3

- each of two adjoining compartments; or
- an adjoining compartment and an escape stair or a final exit; or
- an adjoining compartment and a final exit; or
- an escape stair and a final exit.

Emergency lighting should be installed in a protected   ND 2.A.4
zone or unprotected zone in a single stair building of two
storeys or more and an occupancy capacity of 10 or more.

In a hospital where a storey is divided into three or more   ND 2.B.3
compartments, each compartment should have exits to:

- a compartment and a hospital street; or
- a compartment and an escape stair; or
- a compartment and a final exit.

Note: A hospital street is a protected zone in a hospital provided to assist in facilitating circulation and horizontal evacuation, and to provide a fire-fighting bridgehead.

At upper storey level there should be access to at least two   ND 2.B.3
escape stairs accessed from separate sub-compartments,
located such that:

- the distance between escape stairs is not more than $64\,m$; and
- the distance of single direction of travel within the hospital street is not more than $15\,m$; and
- the distance from a compartment exit to an escape stair is not more than $32\,m$.

A door from a hospital street to an adjoining compartment should: ND 2.B.3

- be located so that an alternative independent means of escape from each compartment is available; and
- not be located in the same sub-compartment as a door to a protected zone containing a stairway or lift.

Every escape stair opening into the hospital street should be located so that the travel distance from an escape stair exit to a door leading directly to a place of safety is not more than 64 m.   ND 2.B.3

An escape route from a hospital department to which patients have access should be to: ND 2.B.3

- a place of safety; or
- a protected zone; or
- an unprotected zone in another compartment or subcompartment.

In patient sleeping accommodation, an escape stair width should be not less than 1,300 mm and designed so as to facilitate mattress evacuation.   ND 2.B.3

The landing configuration should follow the guidance in Table 6.101.   ND 2.B.3

Table 6.101   Stair and landing configuration for mattress evacuation, in mm

Stair width	Minimum landing width	Minimum landing depth
1,300	2,800	1,850
1,400	3,000	1,750
1,500	3,200	1,550
1,600	3,400	1,600
1,700	3,600	1,700
1,800	3,800	1,800

In an area where a 15 s response time would be considered hazardous (e.g. a stairway), emergency lighting should be provided by battery back-up giving a response time of not more than 0.5 s.   ND 2.B.3

Where a hospital with a hospital street has two or more escape stairs, facilities should be provided in accordance with Table 6.102.   ND 2.B.6

Table 6.102   Facilities on escape stairs in hospitals with hospital streets

Storey height and depth of hospital	Facilities on escape stairs
Basements at a depth more than 10 m	Fire-fighting lift Fire-fighting shaft Dry fire main (outlet located at every departmental entrance)
Basements at a depth not more than 10 m	Dry fire main (outlet located at every departmental entrance)
Topmost storey height not more than 18 m	Dry fire main (outlet located at every departmental entrance)
Topmost storey height not more than 0 m	Fire-fighting shaft Fire-fighting lift Dry fire main (outlet located at every departmental entrance)

A fire-fighting lift need not serve the top storey of a building   ND 2.B.6
where:

- the top storey is for service plant use only;
- access to the plant room is from an escape stair from the storey below;
- the foot of the escape stair is not more than 4.5 m from the fire-fighting lift; and
- dry rising mains are installed in the protected lobbies of the escape stair.

Shopping centres

An automatic life safety fire suppression system should   ND 2.C.7
not be installed in a stairway enclosure in an enclosed
shopping centre.

6.14 Lifts and platforms

A lift is a device used for raising and lowering people or goods. A platform usually refers to some kind of standing surface used to support things, or give them stability or visibility.

6.14.1 Requirements

Fire

Every building, which is divided into more than one area of different occupation, must be designed and constructed in such a way that in the event of an outbreak of fire within the building, fire and smoke are inhibited from spreading beyond the area of occupation where the fire originated.	D 2.2
Every building must be designed and constructed in such a way that in the event of an outbreak of fire within the building:	2.9

- *the occupants, once alerted to the outbreak of the fire, are provided with the opportunity to escape from the building, before being affected by fire or smoke;*
- *facilities are provided to assist fire-fighting or rescue operations;*   2.14
- *fire and smoke are inhibited from spreading beyond the compartment of origin until any occupants have had the time to leave that compartment and any mandatory fire containment measures have been initiated.*   2.1

   This standard does not apply to domestic buildings.

Safety

Every building must be designed and constructed in such a way that:	
	D 4.3

- *every level can be reached safely by stairs or ramps;*
- *in non-domestic buildings, safe, unassisted and convenient means of access is provided throughout the building;*   D 4.2
- *in residential buildings, a proportion of the rooms intended to be used as bedrooms must be accessible to a wheelchair user;*
- *in domestic buildings, safe and convenient means of access is provided within common areas and to each dwelling;*
- *in dwellings, safe and convenient means of access is provided throughout the dwelling; and*
- *in dwellings, unassisted means of access is provided to, and throughout, at least one level.*

There is no requirement to provide access for a
wheelchair user:

- in a non-domestic building not served by a lift, to a
  room, intended to be used as a bedroom, that is not on
  an entrance storey; or

- in a domestic building not served by a lift, within
  common areas and to each dwelling, other than on an
  entrance storey.

Noise

Every building must be designed and constructed in such a   5.1
way that each wall and floor separating one dwelling from
another, or one dwelling from another part of the building,
or one dwelling from a building other than a dwelling,
will limit the transmission of noise to the dwelling to a
level that will not threaten the health of the occupants
of the dwelling or inconvenience them in the course of
normal domestic activities provided the source noise is
not in excess of that from normal domestic activities.

 This standard does not apply to:

- fully detached houses; or
- roofs or walkways with access solely for maintenance, or
  solely for the use, of the residents of the dwelling below.

6.14.2 Meeting the requirements

Escape routes

In all types of buildings, an escape route should **not** be by way of a lift.	D 2.9.7 ND 2.9.7

Structure

Lift wells and lift shafts

A lift well should be enclosed by compartment walls with medium fire resistance duration; and where the lift well is not the full height of the building, a compartment floor with medium fire resistance duration.	ND 2.1.11

If a localized capping system is constructed at the head of a lift shaft, then the system should be fire protected on both sides and be sufficiently robust to protect against premature collapse of the roof structure.	ND 2.1.15

Separating walls
In domestic buildings

Every lift well should be enclosed by separating walls with medium fire resistance duration.	D 2.2.6
Where the lift well does not extend the full height of the building, the lift well should form a junction with a separating floor with medium fire resistance duration.	D 2.2.6

Fire safety

Where a lift is installed, the landing controls and lift car controls should be of a type that do not operate on heat or pressure resulting from a fire.	D 2.2.6

Fire service facilities
Fire-fighting shafts and lifts should be designed and installed within common escape stairs to assist the fire service in carrying out their fire-fighting or rescue operations as efficiently as possible. Figure 6.116 provides some examples of typical designs.

A fire-fighting lift should be located within a fire-fighting shaft and have controls to enable it to be used under the direct command of the fire service.	ND-F-2.14.4
The lift should be entered only from:	ND-F-2.14.4
• a fire-fighting lobby having not more than one door to the room or storey it serves; or • an open access balcony or an access deck.	
To enable fire service personnel to attack the fire earlier, safe bridgeheads should ideally be linked to specially protected lifts and dry rising mains.	ND 2.14.1
The lobby should be capable of providing a safe bridgehead for the fire-fighters to work as well as providing access from the escape stair to the accommodation and to any associated fire-fighting lift.	ND 2.14.3

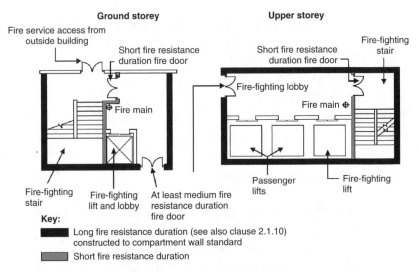

Figure 6.116 Examples of typical fire service facilities

Fire service facilities in high-rise domestic buildings and residential care buildings

In high-rise domestic buildings and residential care buildings, a fire-fighting lift is a lift with additional protection and controls to enable it to be used under the direct control of the fire service when fighting a fire.

The fire-fighting lift need not serve the top storey of a building where:	D 2A7

- the top storey is for service plant use only; and
- access to the plant room is from an escape stair from the storey below; and
- the foot of the escape stair is not more than 4.5 m from the fire-fighting lift; and
- a dry rising main is installed in the protected lobby of the escape stair.

A fire-fighting lift:	D C2A7

- should be constructed in accordance with BS 5588: Part 5: 2004;
- should only be entered from a fire-fighting lobby (having not more than one door to the room or storey it serves) or from an open access balcony or an access deck.

Flats and maisonettes with one escape stair shall provide the following fire-fighting facilities:	D 2A7

- a fire-fighting shaft;
- a fire-fighting lift; and
- a dry fire main with the outlet located in the fire-fighting lobby.

Flats or maisonettes which have two more escape stairs,   D 2A7
shall provide the following fire-fighting facilities to at least
two escape stairs positioned at least 20 m apart:

- fire-fighting shafts;
- fire-fighting lifts; and
- dry fire mains with the outlets located in the fire-fighting
  lobby.

Fire service facilities in hospitals

A fire-fighting lift need not serve the top storey of a building   ND 2.B.6
where:

- the top storey is for service plant use only; and
- access to the plant room is from an escape stair from
  the storey below;
- the foot of the escape stair is not more than 4.5 m from
  the fire-fighting lift; and
- dry rising mains are installed in the protected lobbies
  of the escape stair.

Fire-fighting facilities in enclosed shopping centres

Automatic fire detection and alarm systems in enclosed shopping centres can
increase significantly the level of safety of the occupants.

On the operation of the fire alarm, all escalators should   ND 2.C.5
come to a controlled halt and lifts should return to the
ground storey (or exit level).

Safety

Horizontal circulation in common areas of domestic buildings

There should be level or ramped access within the common   D 4.2.1
areas of a domestic building:

- from a common entrance to the entrance of any dwelling or
  communal facilities on the entrance storey and to any passenger lift; and
- where a passenger lift is installed, from the passenger lift
  to any dwelling and to any communal facilities on an upper
  storey.

A stairlift may be fitted to a private stair and may project into   D 4.3.3
the effective width of the stair but it must have at least one
handrail and, when not in use, the installation should:

- permit safe passage on the stair flight and any landing; and
- not obstruct the normal use of any door, doorway or
  circulation space.

Vertical circulation in common areas of domestic buildings

A building containing flats or maisonettes above four   D 4.2.5
storeys in height should have a passenger lift serving each
level of the building that contains a common entrance, an
entrance to a dwelling or communal facilities.

Passenger lifts should: D 4.2.5

- have a clear landing at least 1.5 m by 1.5 m in front of
  any lift entrance door;
- have automatic lift door(s), with a clear opening width
  of at least 800 mm, fitted with sensors that will prevent
  injury from contact with closing doors;
- be at least 1.1 m wide by 1.4 m deep and contain a
  horizontal handrail, of a size and section that is easily
  gripped, 900 mm above the floor on each wall not
  containing a door;
- have a mirror on the wall facing the doors, above handrail height, to assist a wheelchair user if reversing out
  (unless the lift car has through access);
- have tactile storey selector buttons and, in a lift serving
  more than two storeys, visual and voice indicators of
  the storey reached;
- have controls on each level served, between 900 mm
  and 1.1 m above the landing, and within the lift car on
  a side wall between 900 mm and 1.1 m above the car
  floor and at least 400 mm from any corner;
- on the landing of each level served, have tactile call
  buttons and visual and tactile indication of the storey
  level;
- have lift doors, handrails and controls that contrast
  visually with surrounding surfaces;
- have a signalling system which gives notification that
  the lift is answering a landing call;
- have a system which permits adjustment of the dwell
  time after which the lift doors close, once fully opened,
  to suit the level of use; and

- have a means of two-way communication, operable by a person with a hearing impairment, that allows contact with the lift if an alarm is activated, together with visual indicators that an alarm has been sounded and received.

Vertical circulation between storeys of a non-domestic building

Stairs within a building should be designed to be accessible to a person with reduced mobility.

In addition to a stair, a ramp or lifting device should be provided to every change of level within a storey, except to specific areas (such as an upper storey of a hotel not meant for accommodation) where access by a lift or ramp need not be provided (see clause ND 4.2.1).	ND 4.2.8
Generally, unassisted access between storeys should be provided by a passenger lift, with the installation meeting the recommendations of BS EN 81-70: 2003.	ND 4.2.7

Any lifting device should be designed and installed to include the following general provisions: ND 4.2.7

- a clear landing at least 1.5 m by 1.5 m in front of any lift entrance door;
- controls on each level served, between 900 mm and 1.1 m above the landing, and within the lift car on a side wall between 900 mm and 1.1 m above the car floor and at least 400 mm from any corner;
- on the landing of each level served, tactile call buttons and visual and tactile indication of the storey level;
- lift doors, handrails and controls that contrast visually with surrounding surfaces;
- a signalling system which gives notification that the lift is answering a call made from a landing; and
- a means of two-way communication, operable by a person with a hearing impairment, that allows contact with the lift if an alarm is activated, together with visual indicators that an alarm has been sounded and received.

Passenger lifts

In addition to general provisions for lifting devices, a passenger lift should be provided with:

ND 4.2.7

- automatic lift door(s), with a clear opening width of at least 800 mm, fitted with sensors that will prevent injury from contact with closing doors;
- a lift car at least 1.1 m wide by 1.4 m deep;
- within the overall dimensions of the lift car, a horizontal handrail, of a size and section that is easily gripped, located 900 mm above the floor on any wall not containing a door;
- within a lift car not offering through passage, a mirror on the wall facing the doors, above handrail height, to assist a wheelchair user in reversing out;
- within the lift car, tactile storey selector buttons and, in a lift serving more than two storeys, visual and voice indicators of the storey reached; and
- a system which permits adjustment of the dwell time after which the lift doors close, once fully opened, to suit the level of use.

Powered lifting platforms

In addition to general provisions for lifting devices, a powered lifting platform should:

ND 4.2.7

- if serving a storey to which the public have access, have a platform size of 1,100 mm wide by 1,400 mm deep and a clear opening width to any door of 850 mm; or
- if serving any other storey, have a platform size of at least 1,050 mm wide by 1,250 mm deep and a clear opening width to any door of 800 mm;
- be fully contained within a liftway enclosure;.
- have a operational speed of not more than 0.15 m/s;
- be operated by a continuous pressure type control, of a form operable by a person with limited manual dexterity;
- be provided with a horizontal handrail, of a size and section that is easily gripped, 900 mm above the floor fitted to at least one side of the platform; and
- be provided with permanent and clear operating instructions located adjacent to or within the platform.

Where a powered lifting platform is used, this may be   ND 4.2.8
without a liftway enclosure where vertical travel is not
more than 2.0 m.

Future installation of means of unassisted access between storeys in a dwelling

Not everyone can use stairs unassisted and this may mean that the upper levels of a dwelling are not accessible to some occupants. To meet this possibility, provision should be made for future installation of a means of unassisted access, both within a storey and between storeys.

To allow for future installation of a stairlift, any stair giving   D 4.2.8
access to a principal living level or to accommodation
greater than may be accessed via a 600 mm wide stair
should have an area of wall not less than 700 mm in length
(or an equivalent space, adjacent to the bottom riser of a
stair) that is clear of any obstruction, fitting or doorway, to
allow for parking of a stairlift at rest position.

This space should not be less than 400 mm in depth;
and have a similar area of not less than 200 mm in length,
on the same side of the flight, at landing level adjacent to
the top nosing of the stair, to assist in transfer at the upper
level, allowing for projection of a stairlift track.

Dwellings with limited entrance storey accommodation

Where there is no level or ramped access from the accessible entrance of a dwelling to the principal living level, the principal living level should be made accessible to as wide a range of occupants as possible and, accordingly:

- provision for installation of a stairlift should be made;   D 4.2.10
  and
- entrance level accommodation should contain an area of
  at least 800 mm wide by 1.1 m long that would permit
  storage of a wheelchair or pram and which is clear of
  any doorway, door swing, stair landing or space
  identified for a future stairlift installation.

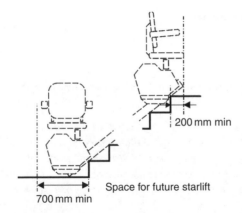

Figure 6.117 Future provision for unassisted access

Stair landings

On a private stair (other than on an intermediate landing, common to two flights) a door may open on to a bottom landing, if, at any angle of swing, a clear space of at least 400 mm deep is maintained across the full width of the landing and the door swing does not encroach within space designated for future installation of a stairlift.   D 4.3.6

Airborne sound
The following requirement should be met in all dwellings other than those that are totally detached.

Airborne sound resisting separating walls and separating floors should be provided between dwellings so that each dwelling is protected from noise emanating from lift shafts and plant rooms.   D 5.1.1

6.15 Corridors

A passageway is any covered walkway between rooms or buildings. On the other hand, a corridor is a path or guided way and usually refers to an interior passageway in modern buildings.

6.15.1 Requirements

Fire

> *Every building must be designed and constructed in such*   2.9
> *a way that:*
>
> - *in the event of an outbreak of fire within the building,*
>   *the occupants, once alerted to the outbreak of the fire,*   2.14
>   *are provided with the opportunity to escape from the*
>   *building, before being affected by fire or smoke;*
> - *facilities are provided to assist fire-fighting or rescue*
>   *operations.*

Safety

> *Every building must be designed and constructed in such*   4.2
> *a way that:*
>
> - *in non-domestic buildings, safe, unassisted and*
>   *convenient means of access is provided throughout*
>   *the building;*
> - *in residential buildings, a proportion of the rooms*
>   *intended to be used as bedrooms must be accessible to*
>   *a wheelchair user;*
> - *in domestic buildings, safe and convenient means of*
>   *access is provided within common areas and to each*
>   *dwelling;*
> - *in dwellings, safe and convenient means of access is*
>   *provided throughout the dwelling; and*
> - *in dwellings, unassisted means of access is provided to,*
>   *and throughout, at least one level.*
>
> 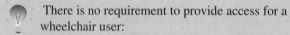 There is no requirement to provide access for a
>   wheelchair user:
>
> - in a non-domestic building not served by a lift, to a
>   room, intended to be used as a bedroom, that is not on
>   an entrance storey; or
> - in a domestic building not served by a lift, within
>   common areas and to each dwelling, other than on an
>   entrance storey.

6.15.2 Meeting the requirements

Fire and smoke control in corridors

For purposes of smoke control: ND 2.9.16

- a corridor should be subdivided with a wall or screen with a short fire resistance duration;
- any door in the wall or screen should be a self-closing fire door.

If the corridor is a dead end more than 4.5 m long and ND 2.9.16
provides access to a point from which more than one direction of escape is possible, it should be divided at that point or points, as shown in Figure 6.120.

Where the corridor provides at least two directions of escape ND 2.9.16
and is more than 12 m in length between the exits it serves, it should be divided in the middle third of the corridor.

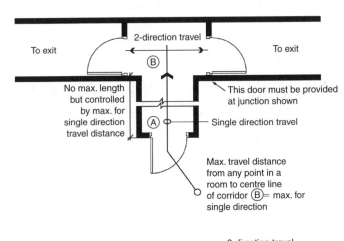

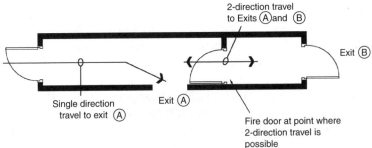

Figure 6.118 Fire and smoke control in corridors

Where a building has more than one escape stair and
where a corridor, or part of a corridor, provides escape in
only one direction, automatic opening ventilators should
be provided in that part of the corridor which provides
single direction escape.

D 2.9.17

Such ventilators should:

- provide for exhaust at or near ceiling level and for
  supply at or near floor level with a combined aggregate
  area of at least $1.5\,m^2$; and
- be activated by automatic smoke detection fixed to
  the ceiling of the corridor and fitted with a manual
  override for fire service use.

High-rise and residential care buildings

Automatic life safety fire suppression systems do not need
to be installed in common spaces such as stairs, corridors
and landings.

D 2A7

Enclosed shopping centres

Where a service corridor is used for means of escape
directly from a shop or shops, the unobstructed width
should be based on the total number of occupants of the
largest shop that evacuates into the corridor, plus an
additional width of 1 m to allow for goods in transit.

ND 2.C.3b

Where a service corridor is used as an escape route, it
should **not** be used for any form of storage.

ND 2.C.3b

Smoke clearance

Ventilation of common access corridors is important to assist fire service personnel during fire-fighting operations and for smoke clearance purposes after
the fire has been extinguished. In domestic buildings:

Every access corridor (or part of an access corridor) in
a building containing flats or maisonettes, should be
provided with openable ventilators.

D 2.14.3

The ventilators should provide exhaust at or near ceiling level and supply air at or near floor level with a combined aggregate opening area of at least 1.5 m².   D 2.14.3

Safety measures

Accessibility within a storey of a dwelling

To ensure facilities within a dwelling can be reached and used by occupants each accessible level or storey within a dwelling should have: D 4.2.6

- corridors with an unobstructed width of at least 900 mm;
- corridors that are large enough to accommodate an unobstructed area of 1.1 m by 800 mm which, where a door being used opens into the corridor, is oriented in the direction of entry and is clear of the door swing; and
- doors with a minimum clear opening width in accordance with Table 6.103 (including any apartment, kitchen or sanitary facility).

Table 6.103 Width of doors

Minimum corridor width at door (mm)	Minimum clear opening width (mm)
1,050	775
900	800

Escape routes in a central core

To reduce the risk of smoke spreading to more than one escape stair, corridor or lobby a building with more than one escape route contained in a central core should be planned so that: ND 2.9.13

- the exits from the storey are remote from one another; and
- no two exits are approached from the same lift hall, common lobby or undivided corridor or linked by any of these other than through self-closing fire doors (Figure 6.119).

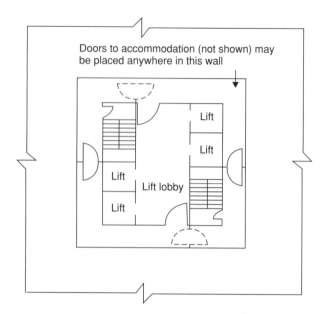

Figure 6.119 Escape routes in a central core

Horizontal circulation in common areas of domestic buildings

Circulation routes within common areas should allow safe and convenient passage and provide space for manoeuvring at junctions and when passing through doorways. It is essential, therefore, that **all** corridors should have a minimum width of at least 1.2 m.	D 4.2.1

Floor surfaces

Floor surfaces of corridors and circulation areas within a non-domestic building should be uniform, permit ease in manoeuvring, and be of a material and finish that, when clean and dry, will provide a level of traction that will minimize the possibility of slipping.	D 4.2.2 ND 4.2.3
Where there is a change in the characteristics of materials on a circulation route, such as from a tile to carpet finish, transition should be level and, where reasonably practicable, differing surfaces should contrast visually to identify the change in material and reduce the potential for trips.	D 4.2.2 ND 4.2.3

Table 6.104   Width of doors

Minimum corridor width at door (mm)	Minimum clear opening width (mm)
1,500	800
1,200	825
900	850

Internal doors

Doors within buildings should present as little restriction to passage as practicable and be constructed in a manner that does not present a hazard or a potential barrier to access.

A door located within a corridor in the common areas of a domestic building should have a door leaf giving a clear opening width in accordance with Table 6.104.	D 4.2.4 ND 4.2.5

In addition, for non-domestic buildings:

A clear glazed vision panel should be provided to any door across a corridor.	ND 4.2.5

Corridors (in non-domestic buildings)

Corridors: • should be wide enough to allow two-way traffic and manoeuvring at junctions or when passing through doorways (for example, a clear width of 1.8 m is the minimum that will allow two wheelchair users to pass safely); • should have an unobstructed width of at least 1.2 m. • should be have manoeuvring or passing spaces of not less than 1.8 m in length and width and be free of obstructions.	ND 4.2.2
An obstruction such as a radiator may project up to 100 mm, reducing corridor width to not less than 1.1 m, if it is less than 900 mm long, and guarding should be provided to any exposed edge of such an area.	ND 4.2.2

6.16 Lobbies

A lobby is a room in a building which is used for entry from the outside that is sometimes referred to as a foyer or an entrance hall and which is frequently used for social purposes and places of commerce. Research is currently underway to develop scales to measure lobby atmosphere in order to improve hotel lobby design.

6.16.1 Requirements

Fire

> *Every building must be designed and constructed in such a way that:*   2.9
>
> - *in the event of an outbreak of fire within the building, the occupants, once alerted to the outbreak of the fire, are provided with the opportunity to escape from the building, before being affected by fire or smoke;*   2.14
> - *facilities are provided to assist fire-fighting or rescue operations.*

Safety

> *Every building must be designed and constructed in such a way that in non-domestic buildings, safe, unassisted and convenient means of access is provided throughout the building.*   4.2

 There is no requirement to provide access for a wheelchair user:

- in a non-domestic building not served by a lift, to a room, intended to be used as a bedroom, that is not on an entrance storey; or
- in a domestic building not served by a lift, within common areas and to each dwelling, other than on an entrance storey.

Energy

> *Every building must be designed and constructed in such a way that the energy performance certificate is displayed in a prominent place within the building.*   D 6.9

6.16.2 Meeting the requirements

Dimensions

> Any lobby at the entrance to or within the common areas of a domestic building and/or the entrance to a non-domestic building should allow a person to pass through whilst remaining clear of the swing of doors.
>
> D 4.2.3
> ND 4.2.4
>
> **Note:** A rectangular area, outwith any door swing, of at least 1.6 m long by 750 mm wide will permit safe passage of, for example, a person in a wheelchair and a companion.

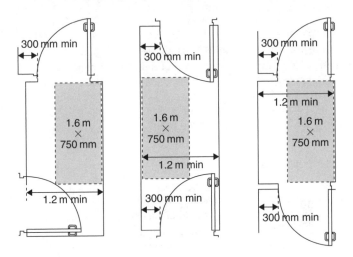

Figure 6.120 Accessible lobby dimensions

> Where either door can be secured by a locking device, a lobby should not be less than 1.5 m wide to enable a wheelchair or pram to be turned around should passage be denied.
>
> D 4.2.3
> ND 4.2.4

Protected lobbies

A protected lobby means a lobby within a protected zone but separated from the remainder of the protected zone so as to resist the movement of smoke from the adjoining accommodation to the remainder of the protected zone.

Non-domestic buildings

Where a building, or part of a building, has only one escape route by way of an escape stair, access to the escape stair should be by way of a protected lobby.	ND 2.9.21
In high-rise buildings where an escape stair serves a storey with a height greater than 18 m, access to the protected zone containing the escape stair should be by way of a protected lobby.	ND 2.9.21
The wall dividing a protected lobby from the remainder of the protected zone may have short fire resistance duration and any door in the wall should be a self-closing fire door with a short fire resistance duration.	ND 2.9.21
The floor of a protected lobby should be constructed of non-combustible material, unless it is a floor finish (e.g. laminate flooring) applied to the floor of a protected zone or protected lobby.	ND 2.9.29
A place of special fire risk should only be accessed from a protected zone by way of a protected lobby.	ND 2.9.20

Mixed-use buildings

Where the escape stair is to be accessed from the common areas of the building, then that escape stair should be separated from the dwellings and where it serves the non-domestic accommodation at each level (including the topmost storey) by a protected lobby.	D 2.9.16
Where the building or part of the building has no storey at a height of more than 7.5 m and has only one escape route by way of an escape stair and there are no alternative escape routes from the building, protected lobbies should be provided at every level.	D 2.9.16
Where the building or part of the building has two or more escape routes, only one stair should communicate with both the domestic and non-domestic parts of the building and be provided with protected lobbies.	D 2.9.16
An escape stair which serves a flat or maisonette which is ancillary to a non-domestic building, may communicate with the non-domestic accommodation provided that the escape stair is separated from the domestic and non-domestic accommodation by a protected lobby at every level.	D 2.9.16

Domestic buildings

A protected lobby should be constructed within a protected zone.	D 2.9.19
The wall dividing the protected lobby from the rest of the protected zone: • should have at least a short fire resistance duration for integrity only. Any door in the wall should be a self-closing fire door with a short fire resistance duration.	D 2.9.19
Where flats or maisonettes are served by only one escape stair and there is no alternative means of escape from the upper storeys, there should be a protected lobby with automatic opening ventilators, at each storey within the protected zone between the escape stair and the accommodation, including a parking garage and any other accommodation ancillary to the dwellings (see Figure 6.121).	D 2.9.19

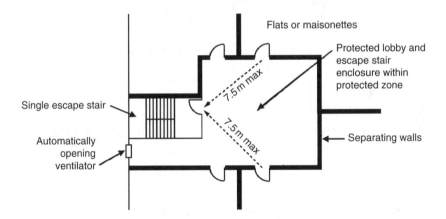

Figure 6.121 Single stair access to flats and maisonettes with any storey at a height of more than 7.5 m but not more than 18 m

The lobby protection should afford people who choose to evacuate the building, additional time to pass the fire floor in relative safety.	D 2.9.19
A ventilated protected lobby need **not** be provided: • for flats and maisonettes entered from an open access balcony or access deck which has an opening(s) to the external air extending over at least four-fifths of its length and at least one-third of its height;	D 2.9.19a

- where the protected zone provides access to not more   D 2.9.19b
  than 12 dwellings in total and no storey is at a height
  of more than 7.5 m and there are not more than four
  dwellings on each storey and each dwelling has within
  it a protected enclosure (see Figure 6.122);
- at the topmost storey;   D 2.9.19c
- where there is more than one escape stair serving each   D 2.9.19d
  dwelling.

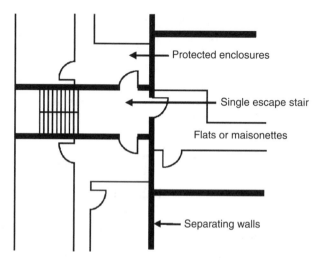

Figure 6.122 Single stair access to flats and maisonettes with every storey at a height of not more than 7.5 m

Where automatic opening ventilators are recommended,   D 2.9.19
they should:

- be activated by automatic smoke detection fixed to the
  ceiling of the protected lobby and fitted with a manual
  override for fire service use;
- be evenly spaced and at least 500 mm from any side of
  the lobby or corridor;
- have a detector located not more than 5 m from
  any change of direction in the lobby or corridor
  exceeding 45°.

Dry risers are normally located in protected lobbies within   D 2.14.2
protected zones.   ND 2.14.2

Note: Any part of a lobby or corridor divided from any other part by a beam or other obstruction projecting more than 600 mm below the soffit of the ceiling shall be deemed to be a separate lobby or corridor.

Escape stairs

Where flats and maisonettes have only one escape stair and any storey is at a height of more than 7.5 m, a protected lobby should be provided within the protected zone on every storey and should have an area of at least 5 m².	D 2.14.1 ND 2.14.1
Within the protected lobby there should be an outlet from a dry fire main that will enable fire service personnel to attack the fire earlier.	D 2.14.1 ND 2.14.1
Where flats or maisonettes have two or more escape stairs and any storey is at a height of more than 7.5 m, protected lobbies and dry fire mains should be provided to not less than two escape stairs positioned at least 20 m apart.	D 2.14.1 ND 2.14.1
In non-domestic buildings where two protected zones enclosing escape stairs share a common wall, any access between them should be by way of a protected lobby.	ND 2.9.12

Fire-fighting shafts

A fire-fighting shaft contains an escape stair and a fire-fighting lobby at every storey (to allow the fire-fighting shaft to be entered). Figure 6.123 provides some examples of typical designs.

The fire-fighting shaft should be provided with a ventilated fire-fighting lobby within the lobby shaft, with only one door to the room or storey it serves.	ND 2.14.3
The lobby should be capable of providing a safe bridgehead for the fire-fighters to work as well as providing access from the escape stair to the accommodation and to any associated fire-fighting lift.	ND 2.14.3

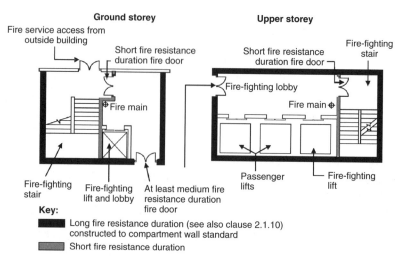

Figure 6.123 Examples of typical fire service facilities

Fire-fighting lifts

> The lift should be entered only from a fire-fighting lobby   ND-F-2.14.4
> having not more than one door to the room or storey it
> serves.

Smoke clearance

Ventilation of protected lobbies is important to assist fire service personnel during fire-fighting operations and for smoke clearance purposes after the fire has been extinguished.

Additional guidance for high-rise domestic and residential care buildings

The lobby is intended to provide a safe bridgehead for the fire-fighters to work and it provides access from the escape stair to the accommodation and to any associated fire-fighting lift.

> Flats and maisonettes with one escape stair shall provide a   D 2A7
> dry fire main with the outlet located in the fire-fighting lobby.
>
> Flats or maisonettes which have two more escape stairs,   D 2A7
> shall provide the following fire-fighting facilities to at least
> two escape stairs positioned at least 20 m apart:
> - fire-fighting shafts;
> - fire-fighting lifts;
> - dry fire mains with the outlets located in the fire-fighting lobby.

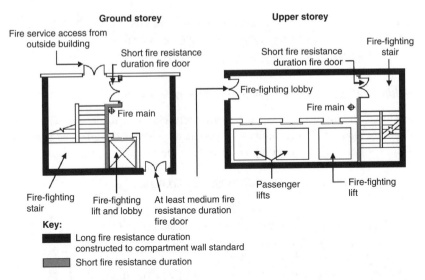

Figure 6.124 Fire-fighting shafts

The shaft should be provided with a ventilated fire-fighting lobby within the shaft, having not more than one door to the room or storey it serves.	D 2A7
The lobby should be constructed and ventilated in accordance with BS 5588: Part 5: 2004.	2A7
A dry rising main should be installed in the protected lobby of the escape stair.	D 2A7
A fire-fighting lift should **only** be entered from a fire-fighting lobby (having not more than one door to the room or storey it serves) or from an open access balcony or an access deck.	D 2A7

Additional guidance for hospitals

Entrances to an intensive therapy unit should be either from a hospital street or through a lobby, enclosed with the same fire resistance duration as that recommended for a sub-compartment.	ND 2.B.1
Where an escape stair in a protected zone serves an upper storey containing a department to which patients have access, access to the protected zone should be by way of a protected lobby, or via the hospital street if the height of the storey is not more than 18 m.	ND 2.B.3

> A fire-fighting lift need not serve the top storey of a   ND 2.B.6
> building where dry rising mains are installed in the pro-
> tected lobbies of the escape stair.

Additional guidance for enclosed shopping centres

Enclosed shopping centres can be extremely complex to design. There are large fire loads and large numbers of people all within a complicated series of spaces where most people only know one way in or out.

Energy certificates in non-domestic buildings

> Buildings with an area of over 1,000 m² occupied by public   ND 6.9.3
> authorities and by institutions providing public services to
> a large number of persons and therefore frequently visited
> by these persons, must have an energy certificate (no more
> than 10 years old) placed in a prominent place such as an
> area of wall which is clearly visible to the public in the
> main entrance lobby or reception.

6.17 Entrances and access

The Building (Scotland) Regulations is very specific about exits, passageways and gangways and local authorities are required to consult with the fire authority to ensure that proposed methods of ingress deemed satisfactory (depending on the type of building). The purpose for which the building is going to be used needs to be considered as each case can be different. In particular, is the building going to be:

- a theatre, hall or other public building that is used as a place of public resort;
- a restaurant, shop, store or warehouse to which members of the public are likely to be admitted;
- a club;
- a school;
- a church, chapel or other place of worship?

At all times, the means of ingress and egress and the passages and gangways, while persons are assembled in the building, are to be kept free and unobstructed.

You are required by the Building (Scotland) Regulations to ensure that all courts, yards and passageways giving access to a house, industrial or commercial building (not maintained at the public expense) are capable of allowing satisfactory drainage of its surface or subsoil to a proper outfall.

The local authority can require the owner of any of the buildings to complete such works as may be deemed necessary.

6.17.1 Requirements

Fire

Every building must be designed and constructed in such a way that:	*2.9*
• *in the event of an outbreak of fire within the building, the occupants, once alerted to the outbreak of the fire, are provided with the opportunity to escape from the building, before being affected by fire or smoke;*	
• *it is accessible to fire appliances and fire service personnel.*	*2.12*

Environment

Every private wastewater treatment plant or septic tank serving a building must be designed and constructed in such a way that it will ensure the safe temporary storage and treatment of wastewater prior to discharge.	*3.8*

Safety

Every building must be designed and constructed in such a way that all occupants and visitors are provided with safe, convenient and unassisted means of access to the building, and:	*4.1* *4.2*
• *in non-domestic buildings, safe, unassisted and convenient means of access is provided throughout the building;*	
• *in residential buildings, a proportion of the rooms intended to be used as bedrooms must be accessible to a wheelchair user;*	
• *in domestic buildings, safe and convenient means of access is provided within common areas and to each dwelling;*	
• *in dwellings, safe and convenient means of access is provided throughout the dwelling;*	
• *in dwellings, unassisted means of access is provided to, and throughout, at least one level.*	

There is no requirement to provide access for a wheelchair user: in a non-domestic building not served by a lift, to a room, intended to be used as a bedroom, that is not on an entrance storey; or in a domestic building not served by a lift, within common areas and to each dwelling, other than on an entrance storey.

Every building must be designed and constructed in such a way that: 4.3

- *every level can be reached safely by stairs or ramps;*
- *every sudden change of level that is accessible in, or*   4.4
  around, the building is guarded by the provision of
  pedestrian protective barriers;   4.8
- *people in and around the building are protected from*
  injury that could result from fixed glazing, projections
  or moving elements on the building;
- *fixed glazing in the building is not vulnerable to*
  breakage where there is the possibility of impact by
  people in and around the building;
- *both faces of a window and rooflight in a building are*
  capable of being cleaned such that there will not be
  a threat to the cleaner from a fall resulting in severe
  injury;

Every building accessible to vehicular traffic must be   4.12
designed and constructed in such a way that every change
in level is guarded.

Every non-domestic building must be designed and   4.7
constructed in such a way that:

- *a safe and secure means of access is provided to a*
  roof; and manual controls for ventilation and for   4.8
  electrical fixtures can be operated safely;
- *it is provided with aids to assist those with a hearing*
  impairment.

Energy

Every building must be designed and constructed in such a   6.9
way that an energy performance certificate for the building
is affixed to the building, indicating the approximate annual
carbon dioxide emissions and energy usage of the building
based on a standardized use of the building.

6.17.2 Meeting the requirements

Dwellings with limited entrance storey accommodation

Where there is no level or ramped access from the accessible entrance of a dwelling to the principal living level, the principal living level should be made accessible to as wide a range of occupants as possible and, accordingly:

- a stair, from an accessible entrance to the principal living level, should follow the guidance on rise, going and pitch for 'any other stair' (see clause 4.3.2); and
- provision for installation of a stairlift should be made (see clause 4.2.8);
- entrance level accommodation should contain an area of at least 800 mm wide by 1.1 m long that would permit storage of a wheelchair or pram and which is clear of any doorway, door swing, stair landing or space identified for a future stairlift installation.

D 4.2.10

Where the entrance level of such a dwelling contains two or more apartments:

- there should also be an accessible toilet on the entrance level in addition to accessible sanitary facilities on the principal living level;
- there should be level or ramped access from the accessible entrance of the dwelling to this accessible toilet and at least one of the apartments on the entrance storey.

Additional guidance for hospitals

Entrances to an intensive therapy unit should be either from a hospital street or through a lobby, enclosed with the same fire resistance duration as that recommended for a sub-compartment.

ND 2.B.1

Access for desludging

Wastewater treatment plants should be inspected monthly to check they are working correctly and all private wastewater treatment plants and/or septic tanks should be provided with an access for desludging.

The desludging tanker should be provided with access to a working area that:

- will provide a clear route for the suction hose from the tanker to the tank;

D 3.8.6
ND 3.8.6

- is not more than 25 m from the tank where it is not more than 4 m higher than the invert level of the tank;
- is sufficient to support a vehicle axle load of 14 tonnes.

Vehicle protective barriers

Where vehicles are introduced into a building, measures should be taken to protect people from any additional risks presented.

Every building accessible to vehicular traffic must be designed and constructed in such a way that every change in level is guarded.	D 4.12 ND 4.12
Where areas subject to vehicular traffic are at a level higher than adjacent areas, such as on ramps or platforms, precautions should be taken to ensure that vehicles cannot fall to a lower level.	D 4.12.0 ND 4.12.0
If vehicles have access to a floor, roof or ramp that forms part of a building, a vehicle protective barrier should be provided to the edge of any such area that is above the level of any adjoining floor, ground or any other route for vehicles.	D 4.12.1 ND 4.12.1
The designer should, wherever possible, avoid introducing projections on the vehicular face of the barrier and should also consider methods of redirecting vehicles in such a way as to cause minimum damage after impact.	D 4.12.1 ND 4.12.1
A vehicle protective barrier should be: - capable of resisting loads calculated in accordance with BS 6399: Part 1: 1996; - of a height at least that given in Table 6.105.	D 4.12.1 ND 4.12.1

Table 6.105 Height of vehicle protective barriers

Location	Minimum height (mm)
Floor or roof edge	400
Ramp edge	600

Setting-down points

For the convenience of a person arriving at a building in a vehicle driven by another, where a road is provided within the curtilage of a building, there should be a setting-down point close to a principal entrance of each building.

The setting-down point should be on a level surface, where the road gradient or camber is less than 1 in 50, with a dropped kerb between the road and an accessible route to the building.

ND 4.1.2

 Note: On a busy vehicular route, such as a public highway, a setting-down point should be positioned outwith the road carriageway.

Accessible routes

A person should be able to travel conveniently and without assistance to an entrance of a building.

D 4.1.3
ND 4.1.3

Regardless of how they arrive within the curtilage of a non-domestic building, a person should then be able to travel conveniently and without assistance to the entrance of a building.

ND 4.1.3

A route to an entrance should be provided that is accessible to everyone.

D 4.1.3
ND 4.1.3

An accessible route should contain no barriers, such as kerbs, steps or similar obstructions that may restrict access.

D 4.1.3
ND 4.1.3

Street furniture such as low-level bollards or chain-linked posts (which can present a hazard, particularly to a wheelchair user or a person with a visual impairment) should be located outwith the width of an accessible route.

D 4.1.3
ND 4.1.3

Use of low-level bollards or chain-linked posts should be avoided.

D 4.1.3
ND 4.1.3

There should be an accessible route to:

D 4.1.3
ND 4.1.3

- the accessible entrance of a single dwelling;
- the common entrance of a building containing flats or maisonettes;
- an accessible entrance of any dwelling not reached through a common entrance;
- the principal entrance to a non-domestic building; and
- to any other entrance that provides access for a particular group of people (for example, a staff or visitor entrance), from:
- a road; and
- any accessible car parking within the curtilage of the building.

There should also be an accessible route between accessible entrances of different buildings within the same curtilage.	ND 4.1.3
Level or gently sloping routes should be used wherever possible, in preference to ramps.	D 4.1.3 ND 4.1.3
An accessible route should be: • level (i.e. with a gradient of not more than 1 in 50); • gently sloping (i.e. with a gradient of more than 1 in 50 and not more than 1 in 20); • ramped (i.e. with a gradient of more than 1 in 20 and not more than 1 in 12.	D 4.1.3 ND 4.1.3
The cross-fall on any part of an accessible route should not exceed 1 in 40.	D 4.1.3 ND 4.1.3
Gently sloping gradients should be provided with level rest points of not less than 1.5 m in length, at intervals dependent on the gradient of the sloping surface.	D 4.1.3 ND 4.1.3
On a route serving more than one dwelling, any ramped access having a rise of more than 300 mm should be complemented by an alternative, stepped means of access.	D 4.1.3 ND 4.1.3
There may be stepped access to a route serving a single house where it is not reasonably practicable to construct an accessible route, such as on a steeply sloping site.	D 4.1.3
Where it is not reasonably practicable to construct an accessible route either from a road or from car parking within the curtilage of the dwelling, then a stepped access solution may be proposed.	D 4.1.3

Surfaces to accessible routes

The surface of an accessible route: • should be firm, uniform and of a material and finish that will permit ease in manoeuvring; • should provide a degree of traction that will minimize the possibility of slipping; • should take into account both anticipated use and environmental conditions.	D 4.1.4 ND 4.1.4
The surface of an accessible route should not offer a trip hazard or result in standing water.	

Surface elements such as drainage gratings and manhole covers, uneven surfaces (such as cobbles) or loose-laid materials (such as gravel) should not create a trip or entrapment hazard.	D 4.1.4 ND 4.1.4
Where a footpath is level with a road surface, such as at a dropped kerb, tactile paving should be used to warn a person with a visual impairment of the presence of a vehicular route.	D 4.1.4 ND 4.1.4

Length of accessible routes

The length of an accessible route to an accessible entrance of a building should be limited to 45 m.	D 4.1.5 ND 4.1.4

Width of accessible routes

The clear and unobstructed surface width of an accessible route should generally be at least 1.8 m, unless: • giving access to not more than 10 dwellings, where the minimum surface width may be not less than 1.2 m. • giving access to a single dwelling, where effective width may be not less than 900 mm.	D 4.1.6
Any gate across an accessible route should offer a clear opening width of at least 850 mm.	D 4.1.6 ND 4.1.6

Additionally, for non-domestic buildings:

Any part of an accessible route to a building from accessible parking spaces or a setting-down point should have a minimum surface width of 1.8 m.	ND 4.1.6
Elsewhere, the clear and unobstructed surface width of an accessible route should be not less than 1.2 m.	
To allow for passing, localized widening of any route narrower than 1.8 m wide to not less than 1.8 m should be made at any junction and change of direction and, where the whole length of the route is not visible, also at not more than 10 m intervals along the route.	ND 4.1.6
On an accessible route, a level footpath of not less than 1.0 m in width should be maintained to the rear of the slope to any dropped kerb.	ND 4.1.6

Accessible entrances

Each common entrance to a domestic building (and at least one entrance to a dwelling) should be accessible and designed to present as little restriction to passage as possible.   D 4.1.7

 Whilst an accessible entrance to a house is commonly the front or main entrance, an alternative entrance may be designated as the accessible entrance where this provides a more convenient or practical route into the dwelling.

An accessible entrance to a building should: D 4.1.7
ND 4.1.7

- have an unobstructed entrance platt of at least 1.2 m by 1.2 m, with a cross-fall of not more than 1 in 50, if required to prevent standing water;
- have a means of automatic illumination above or adjacent to the door;
- have an accessible threshold;
- have a door leaf giving a clear opening width of at least 800 mm in accordance with Figure 6.125;
- if fitted with a door-closing device, be operable with an opening force of not more than 30 N (for first 30° of opening) and 22.5 N (for remainder of swing) when measured at the leading edge of any door leaf;
- if not a powered door, have an unobstructed space to the opening face of the door, next to the leading edge, of at least 300 mm.

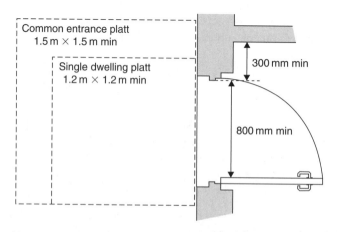

Figure 6.125 Accessible entrance door

Note: The projection of ironmongery which extends across the width of a door leaf, such as an emergency push bar for escape or a horizontal grab rail, should be subtracted when calculating the clear opening width.

In addition, for non-domestic buildings:

An entrance to a building that will be used as a principal means of access, including an entrance that provides access for a particular group of people, e.g. staff; or that offers a direct means of access between buildings, should be an accessible entrance, designed to present as little restriction to passage as possible.	ND 4.1.7
Where an intercom or entry control system is provided: <ul><li>it should be positioned between 900 mm and 1.2 m above floor level;</li><li>it should include an inductive coupler compatible with the 'T' setting on a personal hearing aid, together with a visual indicator that a call made has been received;</li><li>controls should contrast visually with surrounding surfaces and any numeric keypad should follow the 12-button telephone convention, with an embossed locater to the central '5' digit.</li></ul>	ND 4.1.7

Common entrances

A common entrance to a domestic building should have: <ul><li>an unobstructed entrance platt, measuring at least 1.5 m by 1.5 m, with a cross-fall of not more than 1 in 50 if required to prevent standing water;</li><li>a canopy, recessed entrance or similar means of protecting people entering the building from exposure to the elements;</li><li>a glazed vision panel, as described below;</li><li>a door entry system.</li></ul>	D 4.1.8
To assist in preventing collisions, a clear glazed vision panel or panels to a door should give a zone of visibility from a height of not more than 500 mm to at least 1.5 m above finished floor level. This may be interrupted by a solid element between 800 mm and 1.15 m above floor level. A vision panel is not needed to a powered door controlled by automatic sensors or where adjacent glazing offers an equivalent clear view to the other side of a door.	D 4.1.8

Door entry systems

Where a common entrance door, intended as a principal   D 4.5.3
means of access to a building, is fitted with a locking
device, a door entry system comprising a remote door
release and intercom at the point of entry and a call unit
within each dwelling served by that entrance should be
installed.

Controls should contrast visually with surrounding surfaces   D 4.5.3
and any numeric keypad should follow the 12-button
telephone convention, with an embossed locater to the
central '5' digit.

Powered doors

Use of a powered door will improve accessibility at an entrance to a building.
However, care should be taken to ensure that the form of such a door does not
present any additional hazard or barrier to use.

Powered doors should be controlled either by an automatic   ND 4.1.8
sensor, such as a motion detector, or by a manual activation
device, such as a push-pad.

Any manual control should be located at a height of   ND 4.1.8
between 750 mm and 1.0 m above ground level and at least
1.4 m from the plane of the door or, where the door opens
towards the direction of approach, 1.4 m from the front
edge of the open door leaf.

A manual control should contrast visually with the surface   ND 4.1.8
on which it is mounted.

In addition to the general recommendations for accessible   ND 4.1.8
entrances given in clause 4.1.7, a powered door should
have:

- means of activation and warning of operation;
- sensors to ensure doors open swiftly enough and remain
  open long enough to permit safe passage in normal use
  and to avoid the door striking a person passing through;
- if a swing door, identification of any opening vertical
  edge using visual contrast;
- if on an escape route, or forming a lobby arrangement
  where the inner door is also powered or lockable, doors
  that, on failure of supply will either fail 'open' or have
  a break-out facility permitting doors to be opened in
  direction of escape;

- guarding to prevent collision with, or entrapment by a door leaf, except where such guarding would prevent access to the door.

Signage to identify hearing enhancement systems

People with hearing loss should be able to access facilities in a building and to participate fully in activities such as conferences, meetings and entertainment. Three forms of hearing enhancement system are in common use, namely:

- audio frequency induction loop systems;
- infrared systems;
- radio systems.

A hearing enhancement system or similar device to assist a person with hearing loss should form part of a building installation and be provided to the principal reception desk, public counter or information point in any building to which the public have access. In larger buildings, with multiple entrances, there may be a number of these in different locations.
ND 4.7.1

The presence and type of hearing enhancement system installed should be indicated with clear signage at the entrance to any such room or at a service point.
ND 4.7.1

Accessible thresholds

To be accessible, a door should not present unnecessary barriers to use, such as a step or raised profile at a threshold that might present difficulties to a wheelchair user or be an entrapment or trip hazard to an ambulant person, whether or not using a walking aid.

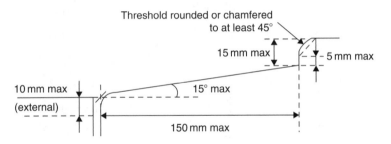

Figure 6.126 Generic threshold profile

An accessible threshold should ensure that: D 4.1.9
                                                                      ND 4.1.9
- it has been designed to prevent the ingress of rain;
- the surface of the platt is no more than 10 mm below the
  leading edge of any sill, with any exposed edge
  chamfered or rounded;
- an external sill or internal transition unit is at an angle
  no more than 15° from the horizontal and (if sloping) not
  more than 150 mm in length;
- the threshold is either level or of a height and form that
  will neither impede unassisted access by a wheelchair
  user nor create a trip hazard. (e.g. not more than 15 mm
  and with any vertical element that is more than 5 mm
  high being pencil-rounded or chamfered to an angle of
  not more than 45° from the horizontal).

 If the finished internal floor level is more than 15 mm
below the top of the threshold, an internal transition
unit, of not more than 15° to the horizontal, finishing
not more than 5 mm above the internal floor surface
may be used.

Lobbies

Any lobby at the entrance to or within the common areas of   D 4.2.3
a domestic building and/or the entrance to a non-domestic   ND 4.2.4
building should allow a person to pass through whilst
remaining clear of the swing of doors.

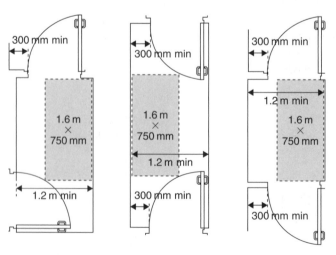

Figure 6.127 Accessible lobby dimensions

Note: A rectangular area, outwith any door swing, of at least 1.6 m long by 750 mm wide will permit safe passage of, for example, a person in a wheelchair and a companion.

Ducted warm air heating to the entrance hall

Where a flat or maisonette has a storey at a height of more than 4.5 m, or a basement storey, and is provided with a system of ducted warm air heating:

D 2.9.13

- transfer grilles should **not** be fitted between any room and the entrance hall or stair;
- supply and return grilles should be not more than 450 mm above floor level;
- if warm air is ducted to an entrance hall, the return air should be ducted back to the heater;
- if a duct passes through any wall, floor or ceiling of an entrance hall, all joints between the duct and the surrounding construction should be sealed.

Horizontal circulation in common areas of domestic buildings

There should be level or ramped access within the common areas of a domestic building:

D 4.2.1

- from a common entrance to the entrance of any dwelling or communal facilities on the entrance storey and to any passenger lift;
- where a passenger lift is installed, from the passenger lift to any dwelling and to any communal facilities on an upper storey.

Vertical circulation in common areas of domestic buildings

Stairs in common areas should be designed to be accessible to a person with reduced mobility.

D 4.2.5

There should be an accessible stair between each level of a building.

D 4.2.5

Level access, or access by a stair or ramp device should be provided to any storey, or part of a storey unless that storey:

D 4.2.5

- only contains fixed plant or machinery and the only normal visits to which are intermittent, for inspection or maintenance purposes;
- has restricted access for suitably trained persons for health and safety reasons (such as to walkways giving access only to machinery or to catwalks and working platforms, reached by industrial ladder).
- A building containing flats or maisonettes may be constructed without a passenger lift where not more than four storeys in height and where there is no dwelling with a principal living level at more than 10 m above either a common entrance level or the level of the lowest storey.
- A building containing flats or maisonettes above four storeys in height should have a passenger lift serving each level of the building that contains a common entrance, an entrance to a dwelling or communal facilities.

Vertical circulation between storeys of a non-domestic building

Stairs within a building should be designed to be accessible to a person with reduced mobility.

There should be:

ND 4.2.7

- an accessible stair between each level of a building;
- a means of unassisted access, other than a ramp, between storeys (except to specific areas where access by a lift need not be provided).

Generally, unassisted access between storeys should be provided by a passenger lift, with the installation meeting the recommendations of BS EN 81-70: 2003.

ND 4.2.7

Accessibility within a storey of a dwelling

To ensure facilities within a dwelling can be reached and used by occupants:

D 4.2.6

- each storey within a dwelling should be designed to be accessible;
- there should be safe and convenient access to and throughout each storey (unless that storey is used soley for storage be accessed via a 600 mm wide stair).

Each accessible level or storey within a dwelling should have:

- corridors with an unobstructed width of at least 900 mm;
- corridors that are large enough to accommodate an unobstructed area of 1.1 m by 800 mm which, where a door being used opens into the corridor, is oriented in the direction of entry and is clear of the door swing;
- doors with a minimum clear opening width in accordance with Table 6.106 (including any apartment, kitchen or sanitary facility).

Table 6.106   Width of doors

Minimum corridor width at door (mm)	Minimum clear opening width (mm)
1,050	775
900	800

There should be unassisted access to the basic accommodation needed in any dwelling.	D 4.2.6
The principal living level of a dwelling, normally also the entrance storey, should contain at least one enhanced apartment, a kitchen and accessible sanitary accommodation.	D 4.2.6

Access between storeys in a dwelling

Where a dwelling has accommodation on more than one level, the levels containing accommodation should be connected by a stair or ramp within the dwelling.	D 4.2.7

Unassisted access between storeys in a dwelling

Not everyone can use stairs unassisted and this may mean that the upper levels of a dwelling are not accessible to some occupants. To meet this possibility:

• provision should be made for future installation of a means of unassisted access, both within a storey and between storeys.	D 4.2.8

To allow for future installation of a stairlift, any stair giving access to a principal living level or to accommodation greater than may be accessed via a 600 mm wide stair should:

- have an area of wall not less than 700 mm in length (or an equivalent space, adjacent to the bottom riser of a stair) that is clear of any obstruction, fitting or doorway, to allow for parking of a stairlift at rest position. D 4.2.8

 This space should not be less than 400 mm in depth; and have a similar area of not less than 200 mm in length, on the same side of the flight, at landing level adjacent to the top nosing of the stair, to assist in transfer at the upper level, allowing for projection of a stairlift track.

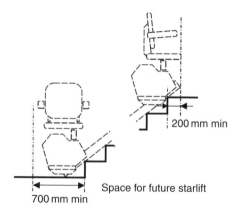

Figure 6.128 Future provision for unassisted access

Width of stair flights and landings

The clear, or effective, width of a stair should allow users to move up and down unhindered and, on stairs giving access to a dwelling or domestic building, permit people to pass on a flight.

The effective width of a stair should be in accordance with the recommendations of Table 6.107. D 4.3.3
ND 4.3.3

Risers and treads

All stairs providing access to and within buildings should be designed to be accessible by most persons with reduced mobility.

Table 6.107 Effective widths of flights and landings

Private stair and stair wholly within shared residential accommodation	Any other stair
<ul><li>900 mm, such as from one storey to another or connecting levels within a storey; or</li><li>600 mm where serving only sanitary accommodation and/or one room, other than accessible sanitary accommodation, a kitchen or an enhanced apartment.</li></ul>	<ul><li>1.0 m generally, such as to an external flight to a domestic building or a common access within a building containing flats or maisonettes; or</li><li>900 mm to an external flight serving a single dwelling, to which the public have access.</li></ul>

Open risers on a flight can be a hazard. When ascending a stair, people may be at risk of trapping the toes of shoes beneath projecting nosings, and of tripping as a result. In addition, many may feel a sense of insecurity when looking through spaces present between treads.	D 4.3.5 ND 4.3.5
A stair should have contrasting nosings to assist in identifying the position of treads and risers should be profiled to minimize tripping as shown in Figure 6.129.	D 4.3.5 ND 4.3.5

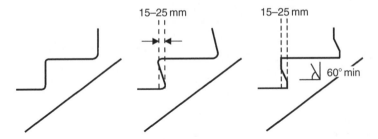

Figure 6.129 Step profile examples

Open risers should not be used unless a stair is intended for descent only, such as in a dedicated escape stair on an escape route.	D 4.3.5 ND 4.3.5
Small children can climb or fall through gaps in stair treads and the size of such gaps should be limited to prevent this.	D 4.3.5
In a flight with open rises, the treads should overlap by at least 15 mm.	ND 4.3.5
Any opening between adjacent treads in a flight should be small enough to prevent the passage of a 100 mm sphere.	D 4.3.5 ND 4.3.5

Warning surfaces to landings of external steps

A sudden and unguarded change of level on an access route can present a hazard to a person with a visual impairment. The use of 'corduroy' tactile paving identifies this hazard and advises users to 'proceed with caution'.

Corduroy tactile paving should be: • provided on external routes serving more than one dwelling; • used to alert people to the presence of a flight of steps; • provided at the head and foot of any flight of external steps, forming a strip 800 mm deep, positioned 400 mm from the first step edge (as shown in Figure 6.130); • if on a landing mutual to a flight of steps and a ramp tactile paving should lie outwith the landing area of any ramp flight, to prevent possible confusion which might lead to injury.	D 4.3.7 ND 4.3.7

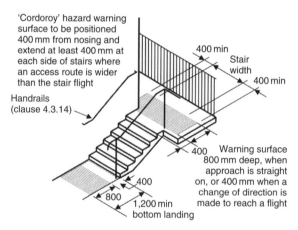

Figure 6.130 Use of corduroy tactile paving

Stair landings serving outward-opening fully glazed doors

If a conservatory and/or similar is not intended to be the accessible entrance and has an outward-opening fully glazed door and a landing (or a fully glazed door leading from a dwelling directly into a conservatory) then the landing length should be in accordance with Figure 6.131.	D 4.3.8

Stair flights consisting of both straight and tapered treads

On that part of a flight consisting of tapered treads: • the going of the tapered treads should be uniform and should not be less than the going of the straight treads; • at the inner end of the tread, the going should be at least 50 mm.	D 4.3.9 ND 4.3.8

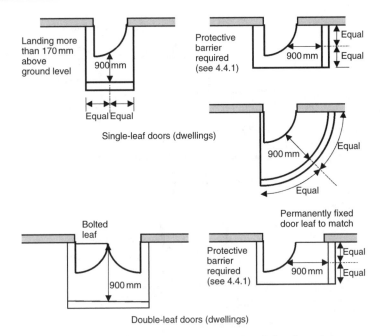

Figure 6.131 Landings serving outward-opening fully glazed doors

In a flight less than 1 m wide the going should be measured at the centre line of the flight.	D 4.3.9 ND 4.3.8
In a flight 1 m wide or more the going should be measured at two points, 270 mm from each end of the tread; and the minimum going should be at least the going of the straight treads.	D 4.3.9 ND 4.3.8

Single stair access to flats and maisonettes

Where flats or maisonettes are served by only one escape stair and there is no alternative means of escape from the upper storeys, there should be a protected lobby with automatic opening ventilators, at each storey within the protected zone between the escape stair and the accommodation, including a parking garage and any other accommodation ancillary to the dwellings (see Figure 6.132).

Access to basements

A basement storey which contains an apartment should be provided with either:	D 2.9.28

- an alternative exit from the basement storey, which provides access to the external air (below the adjoining ground) from which there is access to a place of safety at ground level; or
- an escape window in every basement apartment.

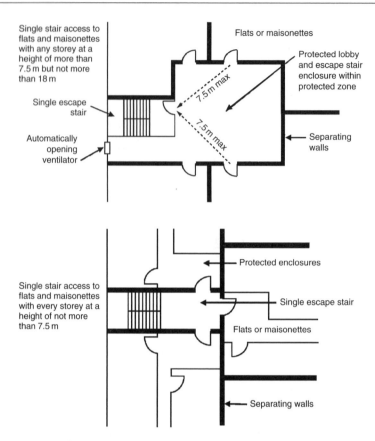

Figure 6.132 Single stair access to flats and maisonettes

Roof access

Raised tie roofs are designed for access and limited to maintenance/repair purposes. Collared roofs are designed for access for limited maintenance or repair purposes (Figure 6.133).

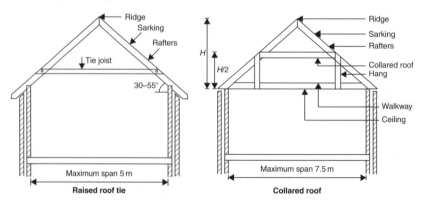

Figure 6.133 Raised and collared roofs

Working on roofs can be dangerous. Falls account for more deaths and serious injuries in the construction industry than any other cause.

Raised tie roofs designed for access (and limited to maintenance or repair purposes) should be capable of supporting a dead load of not more than 0.75 kN/m² and an imposed load not more than 1.5 kN/m² for truss centres of not more than 600 mm and a span not more than 5 m.	D 1F12
Collared roofs that have been designed for access for limited maintenance or repair purposes should support dead loads not more than 0.75 kN/m² and imposed load not more than 1.5 kN/m² for truss centres not more than 600 mm and a span not more than 7.5 m.	D 1F13
A means of safe and secure access should be provided to any roof of building unless it has eaves that, at any part, are at a height of less than 4.5 m above the adjacent ground.	ND 4.8.7
Where fixed ladders are provided and could be accessible to the public, the lowest section of the ladder (i.e. up to 4.5 m) should be demountable to prevent unauthorized access.	ND 4.8.7

Working on roofs

Where access to roofs is provided, precautions should be taken to limit the hazards presented by fragile roof surfaces.

There should be: • a clear visible warning identifying any part of a roof that is not capable of bearing a concentrated load of 0.9 kN on a 130 mm by 130 mm square; • the relevant hazard sign from BS 5499: Part 5: 2002	D 4.8

Split-level storey dwellings

Any change of level within a storey should not compromise access to facilities within the principal living level of a dwelling.	D 4.2.9
A storey may be split level provided a stepped change of level does not divide the accommodation forming the principal living level of a dwelling.	D 4.2.9

If a stepped change of level is proposed on an entrance storey containing the principal living level, the route from the accessible entrance of the dwelling to the accommodation forming the principal living level should be without a stepped change of level.

D 4.2.9

Fire precautions

Access for fire appliances

Every building must be accessible to fire appliances and fire service personnel.

D 2.12
ND 2.12

Vehicle access to the exterior of a building is required to allow high-reach appliances, such as turntable ladders and hydraulic platforms, to be used and to enable pumping appliances to supply water and equipment for fire-fighting and rescue activities.

Access from a public road should be provided for the fire service to assist fire-fighters in their rescue and fire-fighting operations.

D 2.12.1
ND 2.12.1

Vehicle access should be provided to at least one elevation of all buildings to assist in fire-fighting operations.

D 2.12.1
ND 2.12.1

In the case of flats or maisonettes with a common entrance, a vehicle access route for fire-fighting vehicles from a public road should be provided not more than 45 m from the common entrance.

D2.12.1

Fire service vehicles should not have to reverse more than 20 m from the end of an access road.

D 2.12.2
ND 2.12.2

 Note: In rural areas, access from a public road may not be possible to within 45 m of an entrance to the building.

Domestic buildings

Where dry fire mains are installed in a building, parking space should be provided for fire service vehicles a distance not more than 18 m from riser inlets.

D 2.12.1

Every house should be provided with a vehicle access route for fire-fighting vehicles from a public road to not more than 45 m from any door giving direct access to the interior of the dwelling.

D 2.12.1

Non-domestic buildings

The vehicle access route should be provided to the elevation or elevations where the principal entrance, or entrances, are located.	ND 2.12.1
Vehicle access is recommended to other elevations of a building where: • the building is a hospital; • the area of any compartment in a non-domestic building is more than 900 m²; • the building footprint has a perimeter more than 150 m.	ND 2.12.1

Access for fire service personnel

It is common practice for fire service personnel to enter a building through the normal entrance and fight the fire head on. This is termed 'offensive fire-fighting'.

In order to allow unobstructed access to a domestic building for fire service personnel, a paved (or equivalent) footpath at least 900 mm wide should be provided to the normal entrances of the building.	D 2.12.4 ND 2.12.4
In addition, where vehicle access is not possible to within 18 m of the dry riser inlets, a footpath should also be provided to the riser inlets.	D 2.12.4 ND 2.12.4
Every elevation which is provided with vehicle or pedestrian access for fire service personnel should have a door giving access to the interior of the building.	D 2.12.4 ND 2.12.4

Residential care buildings – fire alarm panel

A fire alarm control panel should be provided at the main entrance, or a suitably located entrance to the building agreed with the relevant authority.	ND 2.A.5

Hospital fire alarm panel

A main fire alarm control panel should be provided at the main entrance or at a suitably located secondary entrance to the building with repeater panels provided at all other fire service access points.	ND 2.B.5

Alteration and extension

> Where a dwelling is altered or extended: D 4.1.10
>
> - this work should not adversely affect an existing accessible entrance;
> - all works should ensure that any existing entrance remains accessible
>
> **Note:** If there is no existing accessible entrance, there is no need to provide one to the existing dwelling, or to the extension, as this will not result in the building failing to meet the standard to a greater degree.

Where the alteration of a building includes work to, or provision of, a new circulation area, guidance should be followed as far as is reasonably practicable.

> Where existing accommodation does not meet the   D 4.2.11
> provisions set out in guidance, it need not be altered to comply except for consequential work, required to ensure compliance with another standard (e.g. where an accessible entrance has been relocated and alterations are required to circulation space to maintain accessibility within the building.

Location of an Energy Performance Certificate

Non-domestic buildings

> Buildings with an area of over 1,000 m² occupied by   ND 6.9.3
> public authorities and by institutions providing public services to a large number of persons and therefore frequently visited by these persons, must have an energy certificate (no more than 10 years old) placed in a prominent place such as an area of wall which is clearly visible to the public in the main entrance lobby or reception.

6.18 Balconies

Where a building has obstructions such as balconies or other projections (that are usually supported by columns or console brackets and enclosed with a balustrade) the building line should be taken to be the outer edge of these balconies or other projections.

6.18.1 Requirements

Fire

Every building must be designed and constructed in such a way that:	2.9
• *in the event of an outbreak of fire within the building, the occupants, once alerted to the outbreak of the fire, are provided with the opportunity to escape from the building, before being affected by fire or smoke.*	
• *facilities are provided to assist fire-fighting or rescue operations.*	2.14

Environmental

Every building must be designed and constructed in such a way that the products of combustion are carried safely to the external air without harm to the health of any person through leakage, spillage, or exhaust nor permit the re-entry of dangerous gases from the combustion process of fuels into the building.	3.20

Safety

Every building must be designed and constructed in such a way that:	4.3
	4.4
• *every level can be reached safely by stairs or ramps;*	
• *every sudden change of level that is accessible in, or around, the building is guarded by the provision of pedestrian protective barriers;*	4.8
• *people in and around the building are protected from injury that could result from fixed glazing, projections or moving elements on the building;*	
• *fixed glazing in the building is not vulnerable to breakage where there is the possibility of impact by people in and around the building;*	

> • *both faces of a window and rooflight in a building are capable of being cleaned such that there will not be a threat to the cleaner from a fall resulting in severe injury;*
> • *a safe and secure means of access is provided to manual controls for ventilation and for electrical fixtures so they can be operated safely.*

6.18.2 Meeting the requirements

Means of escape

Travel distance

> In domestic buildings, the travel distance from a flat or maisonette, a communal room, or a plant room should meet the recommendations shown in Table 6.108.
>
> D 2.9.2

Table 6.108   Recommended travel distance in building with flats or maisonettes

Situation	Travel distance (m)	
	One direction of travel	More than one direction of travel
A storey at a height of not more than 7.5 m	7.5	32
A storey at a height of more than 7.5 m	7.5	32
A storey at any height with an access deck or open access balcony serving the dwellings	40	Unlimited

 Note: Travel distance is measured along the shortest route of escape from the main entrance door to the nearest protected door giving direct access to an escape stair or place of safety.

> In non-domestic buildings, the maximum travel distance from any point on a storey is related to the occupancy profile of the building users.
>
> ND 2.9.3

Open access balconies

> Where an open access balcony forms part of an escape route:
>
> D 2.9.6
>
> • it should have a medium fire resistance duration for the width of the escape route and for a further 3 m on either side of the escape route where appropriate;

- every wall not more than 2 m from either side of the escape route should have a short fire resistance duration up to a height of at least 1.1 m measured from the level of the escape route.

If an open access balcony is more than 2 m wide, any soffit above it should have a down-stand on the line of separation between each dwelling extending the full width of the balcony at 90° to the face of the building and extending at least 300 mm below any beam or down-stand parallel to the face of the building.	D 2.9.6
An access deck or open access balcony should have an opening or openings to the external air extending over at least four-fifths of its length and at least one-third of its height.	D 2.9.6
This space should not be used for any form of storage.	D 2.9.18

Fire precautions and facilities

Fire-fighting lifts

A fire-fighting lift should be located within a fire-fighting shaft and have controls to enable it to be used under the direct command of the fire service.	ND-F-2.14.4
The lift should be entered only from:	ND-F-2.14.4

- a fire-fighting lobby having not more than one door to the room or storey it serves; or
- an open access balcony or an access deck.

Smoke clearance

Where access to a flat or maisonette is from an open access balcony, openable ventilators need **not** be installed provided the balcony deck is open to the external air and the opening area extends over at least four-fifths of its length and at least one third of its height.

Flues discharging at low level where they may be within reach of people should be protected with a terminal guard.

Terminal discharges at low level

If the flue outlet is in a vulnerable position (such as where the flue discharges within reach of a balcony), it should be designed to prevent the entry of matter that could obstruct the flow of gases.	D 3.20.15 ND 3.20.15

Safety precautions

Cleaning of windows and rooflights

Falls account for most window cleaning accidents, and generally occur from loss of balance through over-extension of reach or due to breakage of part of the building fabric through improper use or access. For this reason:

Where a window is going to be cleaned from a load-bearing surface such as a balcony, there should be:	D 4.8.3 ND 4.8.3

- a means of safe access;
- a protective barrier not less than 1.1 m high to any edge of the surface or access to the surface which is likely to be dangerous.

 This method of cleaning is only appropriate where no part of the glazing is more than 4 m above the load-bearing surface.

Pedestrian protective barriers

Protective barriers are necessary to prevent an accidental fall at an unguarded change of level.

Any barrier should minimize the risk of persons falling or slipping through gaps in the barrier.	D 4.4.0 ND 4.4.0

 This is particularly important in all domestic buildings, where children will generally be present.

Number of rises in a flight

Generally a flight should have not more than 16 rises, and at least three rises.	D 4.3.4 ND 4.3.4
There may be fewer than three rises: between an external door of a building and a balcony.	D 4.3.4 ND 4.3.4

6.19 Water (and earth) closets, bathrooms and showers

All plans for buildings must include at least one water or earth closet unless the local authority is satisfied in the case of a particular building that one is not required (for example in a large garage separated from the house).

If you propose using an earth closet, the local authority cannot reject the plans unless they consider that there is insufficient water supply to that earth closet.

6.19.1 Requirements

Fire

Buildings must be designed and constructed in such a way that in the event of an outbreak of fire within the building, the occupants, once alerted to the outbreak of the fire, are provided with the opportunity to escape from the building, before being affected by fire or smoke.	*2.9*

Environment

Buildings must be designed and constructed with a surface water drainage system that will ensure the disposal of surface water without threatening the health and safety of the people in and around the building.	*3.6*
Every private wastewater treatment system serving a building must be designed and constructed in such a way that the disposal of the wastewater to ground is safe and is not a threat to the health of the people in or around the building.	*3.9*

Every building must be designed and constructed in such a way that:
- *sanitary facilities are provided for all occupants of, and visitors to, the building in a form that allows convenience of use and that there is no threat to the health and safety of occupants or visitors;*   *3.12*
- *the products of combustion are carried safely to the external air without harm to the health of any person through leakage, spillage, or exhaust nor permit the re-entry of dangerous gases from the combustion process of fuels into the building.*

Safety

Every building must be designed and constructed in such a way that:	
• *safe, unassisted and convenient means of access is provided throughout the building;*	*4.2*
• *the electrical installation does not:*	*4.5*
– *threaten the health and safety of the people in, and around, the building; and*	
– *become a source of fire;*	
• *a safe and secure means of access is provided;*	*4.8*
• *manual controls for ventilation and for electrical fixtures can be operated safely;*	
• *protection is provided for people in, and around, the building from the danger of severe burns or scalds from the discharge of steam or hot water.*	*4.9*

Energy

Every building must be designed and constructed in such a way that the heating and hot water service systems installed are energy efficient and are capable of being controlled to achieve optimum energy efficiency.	*6.3*

6.19.2 Meeting the requirements

Every building must be designed and constructed in such a way that sanitary facilities are provided for all occupants of (and visitors to) the building in a form that allows convenience of use and that there is no threat to the health and safety of occupants or visitors.	D 3.12
In a residential building (other than a residential care building and/or hospital) rooms intended for use as sleeping accommodation (including any en suite sanitary accommodation where provided) should be:	ND 2.9.16
• separated from an escape route by a wall providing short fire resistance duration; and	
• any door in the wall should be a suitable self-closing fire door with a short fire resistance duration.	

Sanitary facilities

Note: Although not recommending that sanitary facilities on the principal living level of a dwelling be designed to an optimum standard for wheelchair users, it should be possible for most people to use these facilities unassisted and in privacy.

Every building must be designed and constructed in such a way that sanitary facilities are provided for all occupants of, and visitors to, the building in a form that allows convenience of use and that there is no threat to the health and safety of occupants or visitors.	D 3.12 ND 3.12

Sanitary facilities – dwellings

Every dwelling should have sanitary facilities comprising at least one water closet (WC), or waterless closet, together with one wash hand basin per WC, or waterless closet, one bath or shower and one sink.	D 3.12.1
To allow for basic hygiene, a wash hand basin should **always** be close to a WC or waterless closet, either within a toilet, or located in an adjacent space providing the sole means of access to the toilet.	D 3.12.1
There should be a door separating a space containing a WC, or waterless closet, from a room or space used for the preparation or consumption of food, such as a kitchen or dining room.	D 3.12.1
If a waterless closet is installed it should be to a safe and hygienic design.	D 3.12.2

The principal living level should be made accessible to as wide a range of occupants as possible and, accordingly:

<ul><li>there should be an accessible toilet on the entrance level in addition to accessible sanitary facilities on the principal living level;</li><li>there should be level or ramped access from the accessible entrance of the dwelling to this accessible toilet and at least one of the apartments on the entrance storey.</li></ul>	D 4.2.6 D 4.2.10

Access to sanitary accommodation

To ensure that privacy can be maintained, the only accessible sanitary accommodation in a dwelling should **not** be en suite, reached through such an apartment.	D 3.12.1

Accessible sanitary accommodation

Bathrooms and toilets designed to minimum space standards can often create difficulties in use. As the ability of occupants can vary significantly, sanitary accommodation should be both immediately accessible and offer potential for simple alteration in the future.	D 3.12.3
A dwelling should have at least one accessible WC, or waterless closet, and wash hand basin and at least one accessible shower or bath.	D 3.12.3

Sanitary facilities should be: D 3.12.3

- located on the principal living level of a dwelling;
- of a size and form that allows unassisted use, in privacy, by almost any occupant;
- capable of being used by a person with mobility impairment or who uses a wheelchair.

An additional accessible toilet may be needed on the entrance level of a dwelling where this is not also the principal living level.	D 3.12.3

Accessible sanitary accommodation should have: D 3.12.3

- a manoeuvring space (at least 1.1 m long by 800 mm wide, oriented in the direction of entry, and clear of any door swing or other obstruction) that will allow a person to enter and close the door behind them; and
- unobstructed access at least 800 mm wide to each sanitary facility; and
- an activity space for each sanitary facility, as noted in Figure 6.134; and
- an unobstructed height above each activity space and above any bath or shower of at least 1.8 m above floor level; and
- walls adjacent to any sanitary facility that are of robust construction that will permit secure fixing of grab rails or other aids; and
- where incorporating a WC, space for at least one recognized form of unassisted transfer from a wheelchair to the WC.

An accessible shower room should be of a size that will accommodate both a level-access floor shower with a drained area of not less than 1.0 m by 1.0 m (or equivalent) or a 900 mm by 900 mm shower tray (or equivalent).   D 3.12.3

An accessible bathroom should be of a size that will accommodate a 1.7 m by 700 mm bath (or equivalent).   D 3.12.3

Within an accessible bathroom, it should be possible to replace the bath with an accessible shower without adversely affecting access to other sanitary facilities.   D 3.12.3

The activity space in front of a bath may be at any position along its length.   D 3.12.3

Where a dwelling has a bathroom or shower room on another level (which is not en suite to a bedroom) some occupants may not require the immediate provision for bathing on the principal living level. Where this is the case, the principal living level may instead have a separate, enclosed space of a size that, alone or by incorporation with the accessible toilet, will permit formation of an accessible shower room (as described above) at a future date.

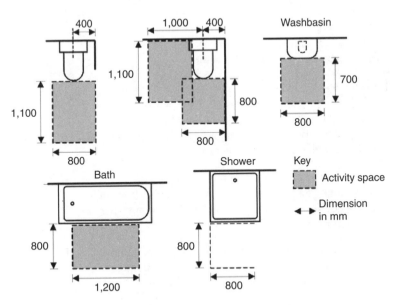

Figure 6.134 Activity spaces for accessible sanitary facilities

In this case:

● this space should have a drainage connection, positioned to allow installation of a floor shower or raised shower tray, sealed and terminated either immediately beneath floor level under a removable access panel or at floor level in a visible position;

- the structure and insulation of the floor in the area identified for a future floor shower should allow for the depth of an inset tray installation (all floors) and a 'laid to fall' installation (solid floors only); and
- if not adjacent to an accessible toilet and separated by an easily demountable partition, a duct to the external air should be provided to allow for later installation of mechanical ventilation.

Notes:

- Though commonly as shown, the activity space in front of a WC need not be parallel with the axis of the WC.
- Where allowing side transfer, a small wall-hung wash hand basin may project up to 300 mm into the activity space in front of the WC.
- The projecting rim of a wash hand basin may reduce the width of a route to another sanitary facility to not less than 700 mm.
- A hand-rinse basin should only be installed within a toilet and only **if** there is a full-size wash hand basin elsewhere in the dwelling.

Alteration and extensions

Additional sanitary facilities need not be provided as part of an extension to, or alteration of, a dwelling. However, an additional accessible toilet may be needed, if one does not exist on the entrance level of a dwelling.	D 3.12.5
If it is intended to install a new sanitary facility on the principal living level or entrance storey of a dwelling and there is not already an accessible sanitary facility of that type within the dwelling, the first new facility should be in accordance with the current standards.	D 3.12.5
If altering existing sanitary accommodation on the principal living level or entrance storey of a dwelling, any changes should at least maintain the level of compliance present before alterations.	D 3.12.5
Existing sanitary accommodation should only be removed or relocated where facilities at least equivalent to those removed will still be present within the dwelling.	D 3.12.5
A sanitary facility that is not an accessible facility may be altered or removed where the minimum provision for a dwelling is maintained.	D 3.12.5

Sanitary facilities – non-domestic buildings

To ensure safety, ease of use and hygiene, the following provisions should be made within **all** sanitary accommodation (Figure 6.135):

• sanitary facilities, fittings and surface finishes should be easily cleanable, to allow a hygienic environment to be maintained; and	ND 3.12.6
• to allow space for general use, where a door opens into a space containing a sanitary facility, there should be an unobstructed space of at least 450 mm in diameter between the sanitary facility and the door swing (Figure 6.135); and	
• a door fitted with a privacy lock should have an emergency release, operable from the outside and, if not sliding or opening outward, offer an alternative means of removal, to permit access in an emergency; and	
• a sanitary facility and any associated aid or fitting, such as a grab rail, should contrast visually with surrounding surfaces to assist in use by a person with a visual impairment.	

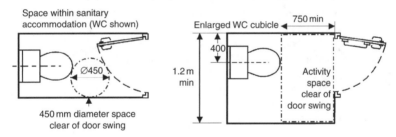

Figure 6.135 Space within sanitary accommodation and enlarged WC cubicle

At least one sanitary facility of each type provided within a building should be within accessible sanitary accommodation.	ND 3.12.7
Accessible sanitary accommodation should	ND 3.12.7
• be clearly identified by signage as accessible sanitary accommodation;	
• contain a manoeuvring space of at least 1.5 m by 1.5 m, clear of any obstruction, including a door swing, other than a wall-mounted wash hand basin which may project not more than 300 mm into this space;	

- be fitted with fixed and folding grab rails (see Figure 6.136);
- be fitted with an assistance alarm which can be operated or reset when using a sanitary facility and which is also operable from floor level;
- where more than one accessible sanitary facility of a type is provided within a building, offer both left- and right-hand transfer layouts.

The assistance alarm should have an audible tone, distin- ND 3.12.7
guishable from any fire alarm, together with a visual indicator, both within the sanitary accommodation and outside
in a location that will alert building occupants to the call.

Every toilet should: ND 3.12.6

- have a wash hand basin within either the toilet itself or in an adjacent space providing the sole means of access to the toilet;
- be separated by a door from any room or space used wholly or partly for the preparation or consumption of food;
- not open directly on to any room or space used wholly or partly for the preparation or consumption of food on a commercial basis:
- include at least one enlarged WC cubicle where sanitary accommodation contains four or more WC cubicles in a range.

Every bathroom or shower room should have a floor   ND 3.12.6
surface that minimizes the risk of slipping when wet.

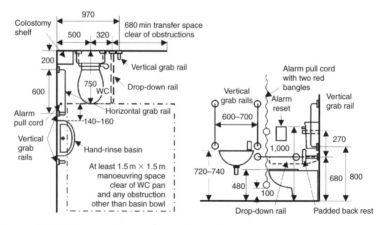

Figure 6.136 Provision within an accessible toilet

Internal doors
For non-domestic buildings:

A clear glazed vision panel should be provided to the outer door of a lobby leading solely to sanitary accommodation.	ND 4.2.5
Within residential buildings, such as hotels and halls of residence, sleeping accommodation which is accessible to a wheelchair user should be provided to at least one bedroom in 20 (or part thereof) and this accommodation should be provided with accessible sanitary accommodation.	ND 4.2.9

Accessible toilets

An accessible toilet should be provided in any building with toilet facilities.	ND 3.12.8
An accessible toilet should include a WC with: • a seat height of 480 mm, to assist in ease of transfer to and from a wheelchair; • a flush lever fitted to the transfer side of the cistern.	ND 3.12.8

Location of accessible toilets

Accessible toilets should be located where they can be reached easily and the horizontal distance from any part of a building to an accessible toilet should be not more than 45 m.	ND 3.12.9
Where there are no toilets on a storey, all occupied parts of that storey should be within 45 m of the nearest accessible toilet on an adjacent storey.	ND 3.12.9
Within the retail area of a large superstore or the concourse of a shopping mall, the distance from an accessible toilet may be increased to not more than 100 m, provided there are no barriers, such as pass doors or changes of level on the route and the location of the accessible toilet is well signposted.	ND 3.12.9

Accessible bathrooms and shower rooms

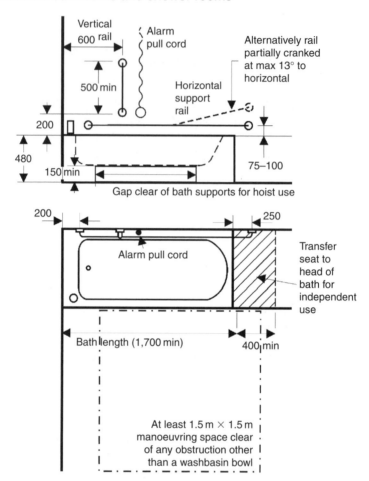

Figure 6.137 Provisions within an accessible bathroom

Within residential buildings or sports facilities (where bathing or showering forms an integral part of activities) a person should be able to use sanitary facilities in privacy, with or without assistance.	ND 3.12.10
In a building where baths or showers are provided, accessible sanitary accommodation should be provided at a ratio of 1 in 20 or part thereof, for each type of sanitary facility provided.	ND 3.12.10
An accessible bathroom should include a transfer space of at least 400 mm across the full width of the head of the bath.	ND 3.12.10

An accessible shower room should have: ND 3.12.10

- a dished floor of a gradient of not more than 1:50 discharging into a floor drain, or a proprietary level access shower with a drainage area of not less than 1.2 m by 1.2 m;
- a folding shower seat positioned 500 mm from a flanking wall and securely fixed, with a seat height that permits transfer to and from a wheelchair positioned outwith the showering area.

For most people, a level access shower is generally both easier and more convenient to use than a bath and, therefore, should always be included within a building where sanitary facilities for bathing are provided.

An accessible shower should be separate or screened from ND 3.12.10
other accommodation, to allow privacy when bathing.

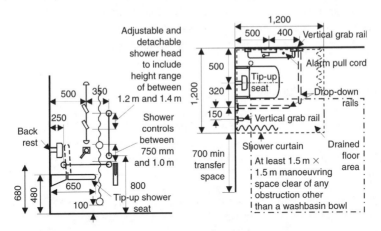

Figure 6.138 Provisions within an accessible shower room

A shower area without separating cubicles (such as ND 3.12.10
found within sporting facilities, etc.) which comprises
10 or more showers should include at least one communal
shower.

To avoid undue waiting times, where an accessible bath or ND 3.12.10
shower is combined with accessible toilet facilities, a
separate accessible toilet should also be provided.

Accessible changing facilities

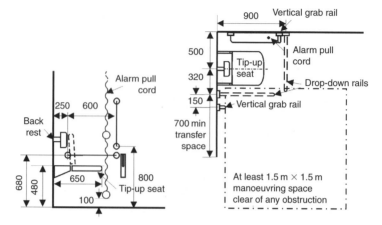

Figure 6.139 Provisions within an accessible changing facility

If an accessible changing facility is combined with an accessible showering facility, a second fold-down seat should be fitted out with the showering area and manoeuvring space to assist in drying and changing.	ND 3.12.11

Baby-changing facilities

A changing facility for babies should be provided in any building open to the public where such children will usually be present, such as in shops, assembly buildings and entertainment buildings.	ND 3.12.12
These facilities shall be accessible, both to accommodate a pram or buggy and to permit use by a person with mobility impairment.	ND 3.12.12
There should be baby-changing facilities in:	ND 3.12.12

- a shop or shopping mall with a total shop floor area of more than 1,000 m²;
- an assembly building accommodating more than 200 people;
- an entertainment building, including a restaurant, fast food outlet with seating or a licensed premises, accommodating more than 200 people.

Baby-changing facilities should be provided either as a
separate unisex facility or as a dedicated space within both
male and female sanitary accommodation and not within
an accessible toilet and should contain:

ND 3.12.12

- a manoeuvring space of at least 1.5 m by 1.5 m, clear
  of any obstruction, including a door swing, other than
  those noted below;
- a wash hand basin at height of between 720 mm and
  740 mm above floor level;
- a changing surface at a height of approximately
  750 mm, with a clear space of not less than 700 mm
  high beneath. This surface may overlap with a
  manoeuvring space by not more than 300 mm.

Rooms, toilets and washrooms in protected zones

Provided all parts of the building served by an escape
stair have at least one other escape route, toilets, washrooms and cleaners' cupboards not more than 3 m² may be
located within a protected zone.

ND 2.9.23

The walls/doors separating the toilets or washrooms from
the protected zone need not have a fire resistance duration.

ND 2.9.23

Number of sanitary facilities

The number of sanitary facilities provided within a building should be calculated from the maximum number of persons the building is likely to accommodate at any time, based upon the normal use of the building.

Separate male and female sanitary accommodation is usually
provided based upon the proportion of males and females
that will use a building, where this is known, or provide
accommodation for equal numbers of each sex otherwise.

ND 3.12.1

Unisex sanitary accommodation may be provided where each
sanitary facility, or a WC and wash hand basin, is located
within a separate space, for use by only one person at a time,
with a door that can be secured from within for privacy.

ND 3.12.1

In small premises, it is recognized that duplication of sanitary facilities
may not always be reasonably practicable and that they might be shared
between staff and customers. However, where practicable, it is good practice
for sanitary facilities for staff involved in the preparation or serving of food
or drink to be reserved for their sole use, with a separate provision made for

customers. Separate hand-washing facilities for such staff should always be provided.

> The numbers of sanitary facilities in schools should be provided in accordance with the tables in the School Premises (General Requirements and Standards) (Scotland) Regulations 1967, as amended.   ND 3.12.1

> Residential care buildings, day care centres and hospices, may be subject to additional standards set out in the relevant National Care Standards document for that service.   ND 3.12.1

> Accessible toilets should be provided within the overall number of sanitary facilities recommended in the various tables in the guidance to this standard, as either: ND 3.12.1
>
> - at least one unisex accessible toilet, accessed independently from any other sanitary accommodation; or
> - where provided within separate sanitary accommodation for males and females, at least one accessible toilet for each sex.
>
> The number required will be dependent upon travel distances within a building to an accessible toilet.

Provision in residential buildings

Public expectation of facilities in residential buildings has risen considerably over the years and en suite sanitary facilities now tend to be normal practice, though it is recognized that this may not be possible in all cases.

> Where sanitary accommodation is not en suite to bedrooms, it should be located directly off a circulation area, close to bedrooms and provided in accordance with Table 6.109.   ND 3.12.3

Table 6.109   Number of sanitary facilities in residential buildings

Sanitary facility	Number of sanitary facilities
WC	1 per 9 persons, or part thereof
WHB	1 per bedroom
Bath or shower	1 per 4 persons, or part thereof

For bedrooms, a wash hand basin should be en suite. An alternative ratio of one wash hand basin per four persons, or part thereof, may be used for dormitory sleeping accommodation.

Sanitary accommodation containing a bath or shower should **also** contain a WC and wash hand basin, in addition to the general provision for those sanitary facilities noted in the table.

ND 3.12.3

In a residential building, an accessible bedroom should be provided with accessible sanitary accommodation comprising a WC, wash hand basin and a bath or shower and should be en suite to each accessible bedroom, other than:

ND 3.12.3

- when altering or converting an existing building, where it is not reasonably practicable to provide en suite sanitary accommodation; or
- where sanitary facilities need to be kept separate for safety reasons, such as in a place of lawful detention.

Where accessible sanitary accommodation is not en suite, it should be located directly off a circulation area, close to any accessible bedroom, and should be clearly identified.

ND 3.12.3

Provision for staff

A building should be provided with sanitary facilities for staff in accordance with Table 6.110.

ND 3.12.2

Table 6.110   Number of sanitary facilities for staff

	Staff numbers	WC	WHB	Urinals
Male	1–15	1	1	1
	16–30	2	2	1
	31–45	2	2	2
	46–60	3	3	2
	61–75	3	3	3
	76–90	4	4	3
	91–100	4	4	4
	Over 100	1 additional WC, WHB and urinal for every additional 50 males, or part thereof		
Female (also male where no urinals are provided)	1–5	1		
	6–25	2	1	
	Over 25	1 additional WC, WHB and urinal for every additional 25 females (or males), or part thereof		

Provision for the public in shops and shopping malls

Sanitary accommodation for customers within shops and shopping malls should be: • clearly identified and located so that it may be easily reached; • provided on the entrance storey and, in larger buildings of more than two storeys, with a total sales floor area greater than 4,000 m², on every alternate storey; • in accordance with Table 6.111.	ND 3.12.4

One wash hand basin should be provided for each WC, plus one wash hand basin per five urinals, or part thereof.	ND 3.12.4
For shopping malls, the sum of the sales areas of all the shops in the mall should be calculated and used with this table. Sanitary facilities provided within a shop may be included in the overall calculation.	ND 3.12.4
If a shop has a restaurant or café, additional sanitary facilities to serve the restaurant should be provided.	ND 3.12.4

Provision for public in entertainment and assembly buildings

Provision for public in entertainment and assembly buildings should be in accordance with Table 6.112.	ND 3.12.5

In cinema-multiplexes and similar premises where the use of sanitary facilities will be spread through the opening hours, the level of sanitary facilities should normally be based upon 75% of total capacity. For single-screen cinemas, 100% occupancy is assumed.	ND 3.12.5
In the case of works and office canteens, the scale of provision may be reduced proportionally where there are readily accessible workplace sanitary facilities close to the canteen.	ND 3.12.5
In public houses, the number of customers should be calculated at the rate of 4 persons per 3 m² of effective drinking area (i.e. the total space of those parts of those rooms to which the public has access).	ND 3.12.5

Table 6.111   Number of sanitary facilities for people, other than staff, in shops

Building type		Sales area of shop	WCs	Urinals
Shops (Class 1) and shopping malls	Unisex	500 m²–1,000 m²	1	
		1,001 m²–2,000 m²	1	1
		2,001 m²–4,000 m²	1	2
		Over 4,000 m²	Plus 1 WC for each extra 2,000 m² of sales area, or part thereof	Plus 1 urinal for each extra 2,000 m² of sales area, or part thereof
	Female	1,000 m²–2,000 m²	2	
		2,001 m²–4,000 m²	5	
		Over 4,000 m²	Plus 2 WCs for each extra 2,000 m² of sales area, or part thereof	
	Male	1,000 m²–2,000 m²	1	1
Shops (Class 2) and shopping malls	Male	Over 4,000 m²	1 Plus 1 WC for each extra 3,000 m² of sales area, or part thereof	1 Plus 1 urinal for each extra 3,000 m² of sales area, or part thereof
	Female	1,000 m²–2,000 m²	3	
		2,001 m²–4,000 m²	1	
		Over 4,000 m²	Plus 1 WC for each extra 3,000 m² of sales area, or part thereof	

Notes:
For the purposes of this guidance, shop sales areas are classified as (other than those listed under Class 2):
- Class 1 supermarkets and department stores (all sales areas); shops for personal services such as hairdressing; shops for the delivery or uplift of goods for cleaning, repair or other treatment or for members of the public themselves carrying out such cleaning, repair or other treatment.
- Class 2 shop sales areas in shops trading predominantly in furniture, floor coverings, cycles, perambulators, large domestic appliances or other bulky goods or trading on a wholesale self-selection basis.

Public houses with restaurants and public music, singing and dancing licences, should be provided with sanitary facilities as for licensed bars.	ND 3.12.5
Sanitary facilities for spectators should be provided in accordance with buildings used for public entertainment.	ND 3.12.5

Table 6.112 Number of sanitary facilities for entertainment and assembly buildings

Building type		Number of people	WCs	Urinals
Buildings used for assembly or entertainment (e.g. places of worship, libraries, cinemas, theatres, concert halls and premises without licensed bars)	Male	1–100	1	2
		101–250	1	Plus 1 for each extra 50 males, or part thereof
		Over 250	Plus 1 for each extra 500 males, or part thereof	
	Female	1–40	3	
		40–70	4	
		71–100	5	
		Over 100	Plus 1 for each extra 500 males, or part thereof	
Restaurants, cafés, canteens and fast food outlets (where seating is provided)	Male	1–400	1 for every 100, or part thereof, plus 1 for each extra 250 males, or part thereof	1 for 50 males, or part thereof
		Over 400		
	Female	1–50	2	
		51–100	3	
		101–150	4	
		151–200	5	
		Over 200	6	
			Plus 1 for each extra 100 females, or part thereof	
Public houses and licensed bars	Male	1–75	1	2
		76–150	1	3
		Over 150	Plus 1 for each extra 150 males, or part thereof	Plus 1 for each extra 75 males, or part thereof
	Female	1–10	1	
		11–25	2	
		Over 25	Plus 1 for each extra 25 females, or part thereof	
Swimming pools (bathers only)	Male	1–100	2	1 per 20 males
		Over 100	Plus 1 for each extra 150 males, or part thereof	
	Female	1–25	2	
		Over 25	Plus 1 for each extra 25 females, or part thereof	

Sanitary pipework

Sanitary pipework should be constructed and installed in accordance with the recommendations in BS EN 12056-2: 2000 ND and according to the type of system used as shown in Table 6.113.	D 3.7.1 ND 3.7.1

Table 6.113   Sanitary discharge stack systems

System I	Single discharge stack system with partially filled branch discharge pipes
System II	Single discharge stack system with small bore discharge branch
System III	Single discharge stack system with full bore branch discharge pipes (the traditional system in use in the UK)
System IV	Separate discharge stack system (mainly found in various Continental European countries and unlikely to be appropriate for use in Scotland)

Sanitary appliances below flood level

The basements of approximately 500 buildings in Scotland are flooded each year when the sewers surcharge (the effluent runs back up the pipes because they are too small to take the required flow).

Wastewater from sanitary appliances and floor gullies below flood level should be drained by wastewater lifting plants (constructed in accordance with the requirements of BS EN 12056-4: 2000 or, where there is unlikely to be a risk to persons such as in a car park, via an anti-flooding device.	D 3.7.2 ND 3.7.2
Wastewater from sanitary appliances above flood level should not be drained through anti-flooding devices.	D 3.7.2 ND 3.7.2

Ventilation

Every building must be designed and constructed in such a way that the air quality inside the building is not a threat to the health of the occupants or the capability of the building to resist moisture, decay or infestation.

Ventilation should have the capability of:	D 3.14.1
• removing excess water vapour from areas where it is produced in significant quantities, such as bathrooms and shower rooms, to reduce the likelihood of creating conditions that support the germination and growth of mould, harmful bacteria, pathogens and allergens; • rapidly diluting pollutants and water vapour, where necessary, that are produced in sanitary accommodation.	ND 3.14.1
Mechanical extract ventilation should be provided for all toilets.	D 3.14.1 ND 3.14.1
Ventilation should be to the outside (i.e. external) air.	D 3.14.1 ND 3.14.1

Natural ventilation

Where a building is naturally ventilated, all moisture-producing areas such as bathrooms and shower rooms, should have the additional facility for removing such moisture before it can damage the building.	ND 3.14.2
Where rapid ventilation is provided (e.g. an opening window) some part of the opening should be at least 1.75 m above floor level.	ND 3.14.2

Trickle ventilators

Trickle ventilators should be provided in naturally ventilated areas to allow fine control of air movement.

A trickle ventilator serving a bathroom or shower room may open into an area that does not generate moisture, such as a bedroom or hallway, provided the area is fitted with a trickle ventilator.	D 3.14.5

Ventilation of dwellings

To reduce the effects of stratification of the air in a room, some part of the opening ventilator should be at least 1.75 m above floor level.	D 3.14.2
Most dwellings are naturally ventilated and ventilation should be provided in accordance with Table 6.114.	D 3.14.2

Table 6.114 Recommended ventilation of a dwelling

	Ventilation recommendations	Trickle ventilation $>10\,m^3/h/m^2$	Trickle ventilation $<10\,m^3/h/m^2$
Bathroom or shower room (with or without a WC)	Either: • mechanical extraction capable of at least 15 L/s (intermittent); or • a passive stack ventilation system.	$4,000\,mm^2$	$10,000\,mm^2$
Toilet	Either: • a ventilator with an opening area of at least 1/30th of the floor area it serves; or • mechanical extraction capable of at least 3 air changes per hour.	$4,000\,mm^2$	$10,000\,mm^2$

Ventilation of sanitary accommodation in non-domestic buildings

Any area containing sanitary facilities should be well ventilated, so that offensive odours do not linger.	ND 3.14.7
Measures should be taken to prevent odours entering other rooms (e.g. by providing a ventilated area between the sanitary accommodation and the other room or by using mechanical ventilation).	ND 3.14.7
No room containing sanitary facilities should communicate directly with a room for the preparation or consumption of food.	ND 3.14.7

Control of humidity

Control of moisture in bathrooms and shower rooms can be by active or passive means.	D 3.15.2

Conservatories and extensions built over existing windows and/or ventilators

If the conservatory or extension is constructed over an area that generates moisture, such as a bathroom or shower room, mechanical extract, via a duct or a passive stack ventilation system, should be provided direct to the outside air.	D 3.14.7 ND 3.14.4

Rainwater harvesting

Rain, as it falls on buildings, is soft, clear and largely free of contaminants. During collection and storage, however, there is potential for contamination. For this reason it is recommended that:

Recycled surface water is used **ONLY** for flushing water closets, car washing and garden taps.	D 3.6.7 ND 3.6.7

Greywater

Water reuse is becoming an accepted method of reducing demand on mains water and the use of greywater may be appropriate in some buildings for flushing of water closets. However, because greywater recycling systems require constant observation and maintenance they should only be used in buildings where a robust maintenance contract exists.

 The approval of Scottish Water is needed before any such scheme is installed.

Greywater discharge

Discharge of greywater may be via a water closet into a public sewer or public wastewater treatment plant provided under the Sewerage (Scotland) Act 1968, where it is reasonably practicable to do so.	ND 3.7.11

Greywater disposal

The disposal of greywater (from baths, showers, washbasins, sinks and washing machines) may be accomplished by an infiltration field.	D 3.9.3 ND 3.9.3

The area of an infiltration field can be calculated using the following formula:

$$A = P \times Vp \times 0.2$$

where:

A is the area of the sub-surface drainage trench, in m²;
P is the number of persons served; and
Vp is the percolation value obtained, in s/mm.

Hot water discharge from sanitary fittings

To prevent the development of *Legionellae* or similar pathogens, hot water within a storage vessel should be stored at a temperature of not less than 60°C and distributed at a temperature of not less than 55°C.	D 4.9.5 ND 4.9.5

To prevent scalding, the temperature of hot water, at point of delivery to a bath or bidet, should not exceed 48°C.	D 4.9.5 ND 4.9.5
A device or system limiting water temperature should allow flexibility in setting of a delivery temperature, up to a maximum of 48°C, in a form that is not easily altered by building users.	D 4.9.5 ND 4.9.5

Oil-firing appliances in bathrooms and bedrooms

There is an increased risk of carbon monoxide poisoning in bathrooms, shower rooms or rooms intended for use as sleeping accommodation, such as bed-sitters. Because of this:

open-flued oil-firing appliances should **not** be installed in these rooms or any cupboard or compartment connecting directly with these rooms;	D 3.20.5
where locating a combustion appliance in such rooms cannot be avoided, the installation of a room-sealed appliance would be appropriate.	ND 3.20.5

Gas-fired appliances in bathrooms and bedrooms

Regulation 30 of the Gas Safety (Installations and Use) Regulations 1998 has specific requirements for room-sealed appliances in these locations.

Electrical safety

Socket outlets in bathrooms and rooms containing a shower

In a bathroom or shower room, an electric shaver power outlet, complying with BS EN 60742: 1996 may be installed. Other than this, there should be **no** socket outlets and **no** means for connecting portable equipment.	D 4.5.4 ND 4.5.4
Where a shower cubicle is located in a room, such as a bedroom, any socket outlet should be installed at least 3 m from the shower cubicle.	D 4.5.4 ND 4.5.4

Lighting

A dwelling should have an electric lighting system providing at least one lighting point to every bathroom, toilet and other space having a floor area of 2 m² or more.	D 4.5.1

Access to manual controls

The location of any manual control device (i.e. for openable ventilators, windows and rooflights and for controls and outlets of electrical fixtures located on a wall or other vertical surface that is intended for operation by the occupants of a building) should be not more than 1.2 m above floor level within accessible sanitary accommodation not provided with mechanical ventilation.	D 4.8.5 ND 4.8.6

Fire detection and alarm systems in hospitals

In hospitals, a fire detection and alarm system need not be provided in sanitary accommodation.	ND 2.A.5 ND 2.B.5

Controls for heating circuits

Bathrooms or en suites which share a heating circuit with an adjacent bedroom: <ul><li>should provide heat only when the bedroom thermostat is activated;</li><li>should be fitted with an independent towel rail or radiator.</li></ul>	D 6.3.8

6.20 Combustion appliances

Combustion appliances and their associated flue-pipes, fireplaces and chimneys should be constructed and installed so as to reduce to a reasonable level the risk of people suffering burns or the building catching fire in consequence of their use.

6.20.1 Requirements

Fire

Every building, which is divided into more than one area of different occupation, must be designed and constructed in such a way that in the event of an outbreak of fire within the building, fire and smoke are inhibited from spreading beyond the area of occupation where the fire originated.	*2.2*

Every building must be designed and constructed in such   *2.6*
a way that in the event of an outbreak of fire within the
building, the spread of fire to neighbouring buildings is
inhibited.

Environment

Every building must be designed and constructed in such a   *3.14*
way that:
                                                               3.11
- *the air quality inside the building is not a threat to the*
  health of the occupants or the capability of the building
  to resist moisture, decay or infestation;
- *the size of any kitchen will ensure an accessible space*
  is provided to allow for the safe, convenient and
  sustainable drying of washing.

Standard 3.11 only applies to a dwelling.

Every building must be designed and constructed in such a   *3.17*
way that:
                                                               3.18
- *each fixed combustion appliance installation operates*
  safely;
                                                               3.19
- *any component part of each fixed combustion*
  appliance installation used for the removal of   *3.20*
  combustion gases will withstand heat generated as
  a result of its operation without any structural   *3.21*
  change that would impair the stability or
  performance of the installation;
- *any component part of each fixed combustion*
  appliance installation will not cause damage to
  the building in which it is installed by radiated,
  connected or conducted heat or from hot embers
  expelled from the appliance;
- *the products of combustion are carried safely to the*
  external air without harm to the health of any person
  through leakage, spillage, or exhaust nor permit the re-
  entry of dangerous gases from the combustion process
  of fuels into the building;
- *each fixed combustion appliance installation receives*
  air for combustion and operation of the chimney so that
  the health of persons within the building is not
  threatened by the build-up of dangerous gases as
  a result of incomplete combustion;

- *each fixed combustion appliance installation receives*   *3.22*
  air for cooling so that the fixed combustion appliance   *3.23*
  installation will operate safely without threatening the
  health and safety of persons within the building;
- *an oil storage installation, incorporating oil storage*
  tanks used solely to serve a fixed combustion
  appliance installation providing space heating or
  cooking facilities in a building, will inhibit fire from
  spreading to the tank and its contents from within, or
  beyond, the boundary;
- *a container for the storage of woody biomass fuel will*
  inhibit fire from spreading to its contents from within,
  or beyond the boundary.

Every building must be designed and constructed in such a   *3.24*
way that:

- *an oil storage installation, incorporating oil storage*
  tanks used solely to serve a fixed combustion appliance
  installation providing space heating or cooking
  facilities in a building will: reduce the risk of oil
  escaping from the installation; contain any oil spillage
  likely to contaminate any water supply, ground water,
  watercourse, drain or sewer; and permit any spill to be
  disposed of safely;
- *the volume of woody biomass fuel storage allows*
  the number of journeys by delivery vehicles to be
  minimized.

Liquefied petroleum gas storage

Every building must be designed and constructed in such a   *4.11*
way that each Liquefied Petroleum Gas (LPG) storage
installation, used solely to serve a combustion appliance
providing space heating, water heating, or cooking facilities,
will:

- *be protected from fire spreading to any liquefied*
  petroleum gas container; and
- *not permit the contents of any such container to form*
  explosive gas pockets in the vicinity of any container.

Fuel consumption meters

> *Every building must be designed and constructed in such*   *6.10*
> *a way that each part of a building designed for different*
> *occupation is fitted with fuel consumption meters.*

6.20.2 Meeting the requirements

Note: When reading this particular section of the book, you will probably notice that some of the requirements concerning chimneys have already been covered in previous sections (e.g. Chapter 6.10). This has been done in order to save the reader having to constantly turn back and re-read a previous page.

General requirements for combustion appliances

The correct installation of a heating appliance or design and installation of a flue can reduce the risk from combustion appliances and their flues from:

- endangering the health and safety of persons in or around a building;
- compromising the structural stability of a building;
- causing damage by fire.

> Every building must be designed and constructed in such a   D 3.17
> way that each fixed combustion appliance installation   ND 3.17
> operates safely.

The following guidance concerns **domestic** sized installations such as those comprising space and water heating or cooking facilities (with flues) and flueless appliances such as gas cookers and concerns:

- oil-firing appliances with an output rating not more than 45 kW;
- gas-fired appliances with a net input rating not more than 70 kW; and
- solid fuel appliances with an output rating not more than 50 kW.

A kitchen should be provided with space for a gas, electric or oil cooker or with a solid fuel cooker designed for continuous burning.

Fire safety

> Fire separation should form a complete barrier to the   D 2.2.0
> products of combustion: smoke, heat and toxic gases.

Fire resistance of external walls need not be provided for   2.6.1
an additional building to a dwelling (e.g. such as a carport,
covered area, greenhouse, summerhouse or swimming
pool enclosure) unless the building contains oil or liquefied
petroleum gas fuel storage.

Relationship to combustible materials

Combustion appliances and their component parts, particularly solid fuel
appliance installations, generate or dissipate considerable temperatures.
Certain precautions need to be taken to ensure that any high temperatures are
not sufficient to cause a risk to people and the building.

Every building must be designed and constructed in such   D 3.19
a way that any component part of each fixed combustion   ND 3.19
appliance installation will not cause damage to the building
in which it is installed by radiated, convected or conducted
heat or from hot embers expelled from the appliance.

Relationship of masonry chimneys to combustible material

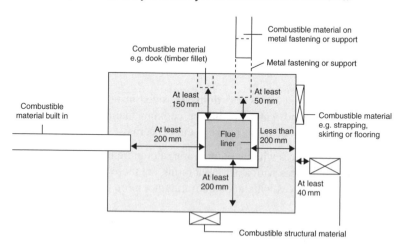

Figure 6.140 Plan view of a masonry chimney

Combustible material should not be located where the heat   D 3.19.1
dissipating through the walls of fireplaces or flues could   ND 3.19.1
ignite it.

All combustible materials should be located at least
200 mm from the surface surrounding a flue in a masonry
chimney.

D 3.19.1
ND 3.19.1

Exclusions are:

- a damp-proof course(s) firmly bedded in mortar;
- small combustible fixings may be located not less than
  150 mm from the surface of the flue;
- combustible structural material may be located not less
  than 40 mm from the outer face of a masonry chimney;
- flooring, strapping, sarking or similar combustible
  material may be located on the outer face of a masonry
  chimney.

Any metal fastening in contact with combustible material,
such as a joist hanger, should be at least 50 mm from the
surface surrounding a flue to avoid the possibility of the
combustible material catching fire due to conduction.

D 3.19.1
ND 3.19.1

Relationship of system chimneys to combustible material

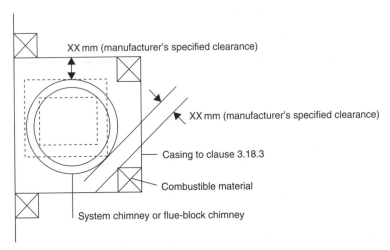

XX mm (manufacturer's specified clearance)

XX mm (manufacturer's specified clearance)

Casing to clause 3.18.3

Combustible material

System chimney or flue-block chimney

Figure 6.141 Plan of casing round a factory-made chimney

System chimneys do not necessarily require to be located
at such a distance from combustible material. It is the
responsibility of the chimney manufacturer to declare a
distance 'XX', as stipulated in BS EN 1856-1: 2003 and
BS EN 1858: 2003 as being a safe distance from the
chimney to combustible material.

D 3.19.2
ND 3.19.2

Relationship of metal chimneys to combustible material

There should be a separation distance where a metal chimney passes through combustible material. This is specified, as part of the designation string for a system chimney when used for oil or gas, as (Gxx), where xx is the distance in mm.

There is no need for a separation distance if the flue gases are not likely to exceed 100°C.

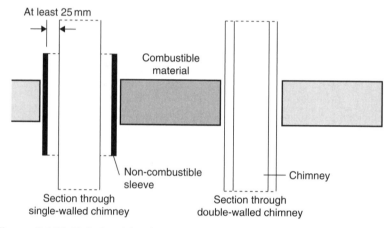

Figure 6.142 Relationship of metal chimneys to combustible material

Where no data is available, the separation distance for oil or gas applications with a flue gas temperature limit of T250 or less should be 25 mm from the outer surface of a single-walled chimney to combustible material measured from the surface of the inner wall of a double-walled chimney.	D 3.19.3 ND 3.19.3
There should be a separation distance of 25 mm from the outer surface of a single-walled chimney to combustible material (measured from the surface of the inner wall of a double-walled chimney) where the metal chimney runs in close proximity to combustible material.	D 3.19.3 ND 3.19.3

Relationship of flue-pipes to combustible material

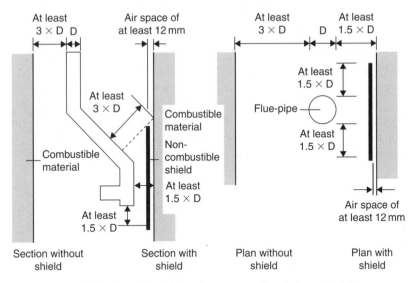

Figure 6.143 Relationship of flue-pipes to combustible material

To prevent the possibility of radiated heat starting a
fire, a flue-pipe should be separated from combustible
material by:

    D 3.19.4
    ND 3.19.4

- a distance according to the designation of the
  connecting flue-pipe in accordance with BS EN
  1856-2: 2005; or
- a distance equivalent to at least 3 times the diameter of
  the flue-pipe. However, this distance may be reduced:
- to 1.5 times the diameter of the flue-pipe, if there is a
  non-combustible shield provided in accordance with the
  following sketch; or
- to 0.75 times the diameter of the flue-pipe, if the
  flue-pipe is totally enclosed in non-combustible material
  at least 12 mm thick with a thermal conductivity of not
  more than 0.065 W/mK.

Relationship of hearths to combustible materials

Any part of a building that abuts or is adjacent to a hearth,
should be constructed in such a way as to minimize the
risk of ignition by direct radiation or conduction from a
solid fuel appliance located upon the hearth.

    D 3.19.8
    ND 3.19.8

A fireplace recess should be constructed of solid, non-combustible material and should incorporate a constructional hearth.

D 3.19.9
ND 3.19.9

The hearth should be located:

D 3.19.8
ND 3.19.8

- in a fireplace recess in accordance with BS 8303: Part 1: 1994; or
- (in any part of the dwelling, other than the floor) not more than 150 mm from the hearth, constructed of solid, non-combustible material in accordance with Figure 6.144 and Table 6.115.

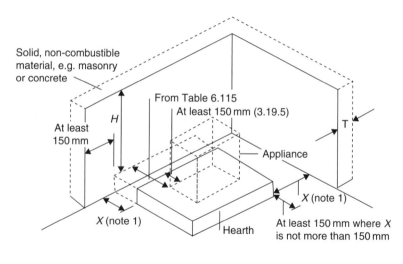

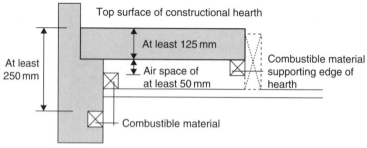

Figure 6.144 Location of hearths

Table 6.115   Hearth and appliance adjacent to any part of a building

Location of hearth or appliance	Thickness (T) of solid non-combustible material	Height (H) of solid non-combustible material
Where the hearth abuts a wall and the appliance is not more than 50 mm from the wall	200 mm	At least 300 mm above the appliance or 1.2 m above the hearth whichever is the greater
Where the hearth abuts a wall and the appliance is more than 50 mm but not more than 300 mm from the wall	75 mm	At least 300 mm above the appliance or 1.2 m above the hearth whichever is the greater
Where the hearth does not abut a wall and is not more than 150 mm from the wall	75 mm	At least 1.2 m above the hearth

Notes:

1. There is no requirement for protection of the wall where X is more than 150 mm.
2. All combustible material under a constructional hearth should be separated from the hearth by an air space of at least 50 mm. However, an air space is not necessary where:
   - the combustible material is separated from the top surface of the hearth by solid, non-combustible material of at least 250 mm;
   - the combustible material supports the front and side edges of the hearth.

Protection from combustion products

It is essential that flues continue to function effectively when in use **without** allowing the products of combustion to enter the building.

Every building must be designed and constructed in such a way that any component part of each fixed combustion appliance installation used for the removal of combustion gases will withstand heat generated as a result of its operation without any structural change that would impair the stability or performance of the installation.	D 3.18 ND 3.18
Combustion appliances, other than flueless appliances such as gas cookers, should incorporate, or be connected to, a flue-pipe and/or a chimney that will withstand the heat generated by the normal operation of the appliance.	D 3.18.1 ND 3.18.1
Chimneys and flue-pipes should be swept at least annually if smokeless solid fuel is burnt and more often if wood, peat and/or other high volatile solid fuel such as bituminous coal is burnt. Mechanical sweeping with a brush is the recommended method of cleaning.	D 3.18.1 ND 3.18.1

Every chimney should have such capacity, be of a height and location and with an outlet so located that the products of combustion are discharged freely and will not present a fire hazard.

D 3.18.1
ND 3.18.1

The surface of the flue should be essentially uniform, gas-tight and resistant to corrosion from combustion products.

D 3.18.1
ND 3.18.1

Chimneys should be constructed in accordance with:

- the recommendations of BS 6461: Part 1: 1984 for masonry chimneys; or
- the recommendations of BS 7566: Parts 1–4: 1992 for metal system chimneys; or
- BS 5410: Part 1: 1997 and OFTEC Technical Information Sheets TI/129, TI/132 and TI/135, where serving an oil-firing appliance; or
- BS 5440: Part 1: 2000, where serving a gas-fired appliance.

D 3.18.1
ND 3.18.1

Table 6.116 Recommended designation for chimneys and flue-pipes for use with oil-firing appliances with a flue gas temperature not more than 250°C

Appliance type	Fuel oil	Designation
Boiler including combination boiler – pressure jet burner	Class C2	T250 N2 D 1 Oxx
Cooker – pressure jet burner	Class C2	T250 N2 D 1 Oxx
Cooker and room heater – vaporizing burner	Class C2	T250 N2 D 1 Oxx
Cooker and room heater – vaporizing burner	Class D	T250 N2 D 2 Oxx
Condensing pressure jet burner appliances	Class C2	T160 N2 W 1 Oxx
Cooker – vaporizing burner appliances	Class D	T160 N2 W 2 Oxx

Table 6.117 Recommended designation for chimneys and flue-pipes for use with gas appliances

Appliance type	Fuel oil	Designation
Boiler – open-flued	Natural draught	T250 N2 D 1 Oxx
	Fanned draught	T250 P2 D 1 Oxx
	Condensing	T160 P2 W 1 Oxx
Boiler – room-sealed	Natural draught	T250 N2 D 1 Oxx
	Fanned draught	T250 P2 D 1 Oxx
Gas fire	Radiant/convector, ILFE or DFE	T250 N2 D 1 Oxx
Air heater	Natural draught	T250 N2 D 1 Oxx
	Fanned draught	T200 P2 D 1 Oxx
	SE duct	T450 N2 D 1 Oxx

Flues

A flue-pipe should be of a material that will safely discharge the products of combustion into the flue under all conditions that will be encountered.	D 3.18.5 ND 3.18.5
A flue-pipe serving a solid fuel appliance should be non-combustible and of a material and construction capable of withstanding the effects of a chimney fire without any structural change that would impair the stability and performance of the flue-pipe.	D 3.18.5 ND 3.18.5
A flue-pipe connecting a solid fuel appliance to a chimney should not pass through:	D 3.18.5 ND 3.18.5

- a roof space;
- an internal wall, although it is acceptable to discharge a flue-pipe into a flue in a chimney formed wholly or partly by a non-combustible wall;
- a ceiling or floor.

Design of flues

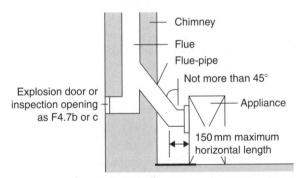

Section through appliance and flue-pipe

Figure 6.145 Design of flues

A combustion appliance should be connected to a chimney that discharges to the external air.	D 3.20.11 ND 3.20.11
Every solid fuel appliance should be connected to a separate flue.	D 3.20.11 ND 3.20.11
Every oil-firing appliance should be connected to a separate flue (unless the appliance has pressure jet burners and is connected into a shared flue).	D 3.20.11 ND 3.20.11

Every gas-fired appliance that requires a flue should connect into a separate flue.	D 3.20.11 ND 3.20.11
The flue of a natural draught appliance, such as a traditional solid fuel appliance, should offer the least resistance to the passage of combustion gases.	D 3.20.11 ND 3.20.11
The horizontal length of the back-entry flue-pipe at the point of discharge from the appliance should be not more than 150 mm.	D 3.20.11 ND 3.20.11
Where bends are essential, they should be angled at not more than 45° to the vertical.	D 3.20.11 ND 3.20.11
The flue should have no intermediate openings.	D 3.20.12 ND 3.20.12

Flue liners

A flue liner is the wall of the chimney that is in contact with the products of combustion. It can generally be of concrete, clay, metal or plastic depending on the designation of the application.

Stainless steel flexible flue liners meeting BS EN 1856-2: 2005 may be used for lining or re-lining flues for oil and gas appliances, and for lining flues for solid fuel applications provided that the designation is in accordance with the intended application.	D 3.18.6 ND 3.18.6
Single-skin, stainless steel flexible flue liners may be used for lining flues for gas and oil appliances.	D 3.18.6 ND 3.18.6
Double-skin, stainless steel flexible flue liners for multi-fuel use.	D 3.18.6 ND 3.18.6
Existing custom-built masonry chimneys may be lined or re-lined by one of the following flue liners: <ul><li>flexible, continuous length, single-skin stainless steel for lining or re-lining chimney flues for C2 oil and gas installations designated T250;</li><li>flexible, continuous length, double-skin stainless steel for lining or re-lining systems designated T400 for multi-fuel installations;</li><li>insulating concrete pumped in around an inflatable former;</li><li>spray-on or brush-on coating by specialist.</li></ul>	D 3.18.6 ND 3.18.6

Existing chimneys for solid fuel applications may also be re-lined using approved rigid metal liners or single-walled chimney products, an approved cast-*in situ* technique or an approved spray-on or brush-on coating. Approved products are listed in the HETAS Guide.

D 3.18.6
ND 3.18.6

Passive stack ventilation systems

The flue of an open-flued combustion appliance may serve as a passive stack ventilation system provided that either:

D 3.14.6

- the appliance is a solid fuel appliance and is the primary source of heating, cooking or hot water production; or
- the flue has an unobstructed area equivalent to a 125 mm diameter duct and the appliance's combustion air inlet and dilution air inlet are permanently open;
- the appliance is an oil-firing appliance which is a continually burning vaporizing appliance (only) such as a cooker or room heater and the room is fitted with a ventilator with a minimum free area of 10,000 mm^2.

A duct or casing forming a passive stack ventilation system serving a kitchen should be non-combustible.

D 3.14.6

Extract fans

Extract fans lower the pressure in a dwelling and may cause the spillage of combustion products from open-flued appliances.

In dwellings where it is intended to install open-flued combustion appliances and extract fans, the combustion appliances should be able to operate safely whether or not the fans are running.

D 3.17.8
ND 3.17.8

The installation of extract fans should be tested to show that combustion appliances operate safely whether or not fans are running.

D 3.17.8
ND 3.17.8

Where a kitchen contains an open-flued appliance, the extract rate of the fan for a gas-fired appliance should not exceed 20 L/s.

D 3.17.8
ND 3.17.8

Extract ventilation for solid fuel appliances should **not** generally be installed in the same room.

D 3.17.8
ND 3.17.8

Heating and cooking appliances

Heating and cooking appliances fuelled by oil, gas or solid fuel, all have the potential to cause carbon monoxide (CO) poisoning if they are poorly installed or commissioned, inadequately maintained or incorrectly used.

Every building must be designed and constructed in such a way that the products of combustion are carried safely to the external air without harm to the health of any person through leakage, spillage, or exhaust nor permit the re-entry of dangerous gases from the combustion process of fuels into the building.	D 3.20 ND 3.20

Combustion appliances – air for combustion

All combustion appliances need ventilation to supply them with oxygen for combustion.

Every building must be designed and constructed in such a way that each fixed combustion appliance installation receives air for combustion and operation of the chimney so that the health of persons within the building is not threatened by the build-up of dangerous gases as a result of incomplete combustion.	D 3.21 ND 3.21

Supply of air for combustion generally

- A room containing an open-flued appliance may need permanently open air vents.
- An open-flued appliance needs to receive a certain amount of air from outside dependent upon its type and rating.
- Infiltration through the building fabric may be sufficient but above certain appliance ratings permanent openings are necessary.

Ventilators for combustion should be located so that occupants are not provoked into sealing them against draughts and noise.	D 3.21.1 ND 3.21.1
Discomfort from draughts can be avoided by placing vents close to appliances, e.g. floor ventilators, by drawing air from intermediate spaces such as hallways or by ensuring good mixing of incoming air.	D 3.21.1 ND 3.21.1

Air vents should not be located within a fireplace recess except on the basis of specialist advice.	D 3.21.1 ND 3.21.1
Noise-attenuated ventilators may be needed in certain circumstances.	D 3.21.1 ND 3.21.1
Appliance compartments that enclose open-flued appliances should be provided with vents large enough to admit all the air required by the appliance for combustion and proper flue operation, whether the compartment draws air from the room or directly from outside.	D 3.21.1 ND 3.21.1

Combustion appliances – air for cooling

In some cases, combustion appliances may need air for cooling in addition to air for combustion. This air will keep control systems in the appliance at a safe temperature and/or ensure that casings remain safe to touch.

Every building must be designed and constructed in such a way that each fixed combustion appliance installation receives air for cooling so that the fixed combustion appliance installation will operate safely without threatening the health and safety of persons within the building.	D 3.22 ND 3.22

Fuel storage

The following guidance on oil relates only to its use solely where it serves a combustion appliance providing space heating or cooking facilities in a building. There is other legislation covering the storage of oils for other purposes.

Every building must be designed and constructed in such a way that: • an oil storage installation, incorporating oil storage tanks used solely to serve a fixed combustion appliance installation providing space heating or cooking facilities in a building, will inhibit fire from spreading to the tank and its contents from within, or beyond, the boundary; • a container for the storage of woody biomass fuel will inhibit fire from spreading to its contents from within, or beyond the boundary.	D 3.23 ND 3.23

 This standard does not apply to portable containers.

Note: Heating oils comprise Class C2 oil (kerosene) or Class D oil (gas oil) as specified in BS 2869: 2006.

Additional fire protection

The fuel feed system from the storage tank to the combustion appliance is also a potential hazard in the event of fire.

The fire valve on the fuel feed should be fitted in accordance with Clause 8.3 of BS 5410: Part 1: 1997.	D 3.23.2 ND 3.23.2
Oil pipelines located inside a building should be run in copper or steel pipe.	D 3.23.2 ND 3.23.2
Provision should therefore be made to prevent the tank becoming overgrown, such as a solid, non-combustible base in full contact with the ground.	D 3.23.2 ND 3.23.2
A base of concrete at least 100 mm thick or of paving slabs at least 42 mm thick that extends at least 300 mm beyond all sides of the tank would be appropriate.	D 3.23.2 ND 3.23.2

Note: Where the tank is within 1 m of the boundary and not more than 300 mm from a barrier or a wall of non-combustible construction type 7, short duration, the base need only extend as far as the barrier or wall.

Storage within a building

Where a storage tank is located inside a building, additional safety provisions should be made including:

• the place where the tank is installed should be treated as a place of special fire risk; • the space should be ventilated to the external air; • the space should have an outward-opening door that can be easily opened without a key from the side approached by people making their escape; • there should be sufficient space for access to the tank and its mountings and fittings; • a catchpit.	D 3.23.3 ND 3.23.3

Fuel storage – containment

Oil is a common and highly visible form of water pollution. Because of the way it spreads, even a small quantity can cause a lot of harm to the aquatic environment. Oil can pollute rivers, lochs, groundwater and coastal waters killing wildlife and removing vital oxygen from the water.

The UK government is required by this directive to prevent List I substances from entering groundwater and to prevent groundwater pollution by List II substances.

Every building must be designed and constructed in such a way that:

- an oil storage installation, incorporating oil storage   D 3.24
  tanks used solely to serve a fixed combustion appliance
  installation providing space heating or cooking
  facilities in a building will: reduce the risk of oil
  escaping from the installation; contain any oil spillage
  likely to contaminate any water supply, groundwater,
  watercourse, drain or sewer; and permit any spill to be
  disposed of safely;
- the volume of woody biomass fuel storage allows   ND 3.24
  the number of journeys by delivery vehicles to be
  minimized.

 This standard does not apply to portable containers.

Metering

To enable building operators to effectively manage fuel use, systems should be provided with fuel meters to enable the annual fuel consumption to be accurately measured.

Every building must be designed and constructed in such   6.10
a way that each part of a building designed for different
occupation is fitted with fuel consumption meters.

 This standard does not apply to:

- domestic buildings;
- communal areas of buildings in different occupation;
- district or block heating systems where each part of the
  building designed for different occupation is fitted with
  heat meters; or
- heating fired by solid fuel or biomass.

In non-domestic buildings

All buildings should be installed with fuel consumption metering in an accessible location, such as presently installed by the utility.	ND 6.10.1
Each area divided by separating walls and separating floors and designed for different occupation should be provided with a fuel meter to measure the fuel usage in each area.	ND 6.10.1
Where multiple buildings or fire separated units are served on a site by a communal heating appliance, the fuel metering shall be installed both at the communal heating appliance and heat meters at the individual buildings served.	ND 6.10.1
Metering shall be provided to measure the hours run, electricity generated, and the fuel supplied to a combined heat and power unit.	ND 6.10.1

Metering in existing buildings

Existing buildings which are a result of a conversion or in buildings where an extension or alteration is carried out should be installed with fuel consumption metering.	ND 6.10.2
Where conversions, extensions or alterations, result in the creation of two or more units, each unit should have a fuel meter installed in an accessible location.	ND 6.10.2
A fuel meter should be installed if a new fuel type or new boiler (where none existed previously) is installed.	ND 6.10.2

Oil-firing appliances

Oil-firing appliances should: • be constructed, installed, commissioned and serviced carefully to ensure that the entire installation operates safely;	D 3.17.5 ND 3.17.6
• be suitable for purpose and the class of oil used in the installation;	D 3.17.5 ND 3.17.6

- comply with the relevant OFTEC standard and should   D 3.17.5
  be installed in accordance with the recommendations in   ND 3.17.6
  BS 5410: Parts 1 and 2.

Fire valves should be fitted so as to cut off the supply of oil   D 3.17.5
remotely from the combustion appliance in the event of a   ND 3.17.6
fire starting in or around the appliance.

The valve should be located externally to the dwelling and   D 3.17.5
fitted in accordance with the recommendations in Section   ND 3.17.6
8.3 of BS 5410: Part 1: 1997 and OFTEC Technical
Information Sheet TI/138.

Relationship of oil-firing appliances to combustible material

A hearth is not required beneath an oil-firing appliance if   D 3.19.6
it incorporates a full-sized, rigid non-combustible base and   ND 3.19.6
does not raise the temperature of the floor beneath it to
more than 100°C under normal working conditions.

A floor-standing, oil-firing appliance should be positioned   D 3.19.6
on the hearth in accordance with Figure 6.146 and in such   ND 3.19.6
a way as to minimize the risk of ignition of any part of the
floor by direct radiation or conduction.

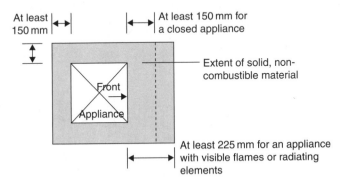

Plan of appliance on a hearth

Figure 6.146 Location of oil-fired appliance

 Note: The 150 mm does not apply where the appliance is located in a fire-place recess, nor does it apply where the back or sides of the hearth either abut or are carried into a solid, non-combustible wall.

An oil-firing appliance should be separated by: • a shield of non-combustible material at least 25 mm thick; or • an air space of at least 75 mm from any combustible material if the temperature of the back, sides or top of the appliance is more than 100°C under normal working conditions.	D 3.19.6 ND 3.19.6

Chimneys and flue-pipes serving oil-firing appliances

A chimney or flue-pipe serving an oil-firing appliance should be constructed to the recommendations of BS 5410: Part 1: 1997.	D 3.20.3 ND 3.20.3
Where the flue gas temperatures are more than 250°C, under normal working conditions, custom-built chimneys, system chimneys and flue-pipes should be designed and constructed for use with a solid fuel appliance.	D 3.20.3 ND 3.20.3
The cross-sectional area of a flue serving an oil-firing appliance should be in accordance with the recommendations in BS 5410: Part 1: 1997 and should be the same size as the appliance flue spigot.	D 3.20.9 ND 3.20.9
The outlet from a flue should be located externally at a safe distance from any opening, obstruction or combustible material.	D 3.20.18 ND 3.20.18
The outlets should be located in accordance with Figure 6.147 and Table 6.118.	D 3.20.18 ND 3.20.18

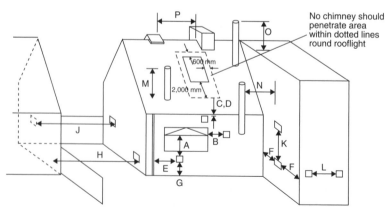

Figure 6.147 Oil-firing appliance flue outlets

Table 6.118   Oil-firing appliance flue outlets

Location		Minimum distance to terminal (mm)	
		Pressure jet	Vaporizing
A	Directly below an opening, air brick, opening window, etc.	600	Not allowed
B	Horizontally to an opening, air brick, opening window, etc.	600	Not allowed
C	Below a gutter, eaves or balcony with protection	75	Not allowed
D	Below a gutter, eaves or balcony without protection	600	Not allowed
E	From vertical sanitary pipework	300	Not allowed
F	From an internal or external corner	300	Not allowed
G	Above ground or balcony level	300	Not allowed
H	From a surface or boundary facing the terminal	600	Not allowed
J	From a terminal facing the terminal	1,200	Not allowed
K	Vertically from a terminal on the same wall	1,500	Not allowed
L	Horizontally from a terminal on the same wall	750	Not allowed
M	Above the highest point of an intersection with the roof	600	1,000
N	From a vertical structure to the side of the terminal	750	2,300
O	Above a vertical structure not more than 750 mm from the side of the terminal	600	1,000
P	From a ridge terminal to a vertical structure on the roof	1,500	Not allowed

Notes:
1. Appliances burning Class D oil should discharge the flue gases at least 2 m above ground level.
2. Terminating positions M, N and O for vertical balanced flues should be in accordance with manufacturer's instructions.
3. Vertical structure in N, O and P includes tank or lift rooms, parapets, dormers, etc.
4. Terminating positions A to L should only be used for appliances that have been approved for low-level flue discharge when tested in accordance with BS EN 303-1: 1999, OFS A100 or OFS A101.
5. Terminating positions should be at least 1,800 mm from an oil storage tank unless a wall with a non-combustible construction type 7, short duration and more than 300 mm higher and wider each side than the tank is provided between the tank and the terminating position.
6. Where a flue terminates not more than 600 mm below a projection and the projection is plastic or has a combustible finish, then a heat shield of at least 750 mm wide should be fitted.
7. The distance from an appliance terminal installed at right angles to a boundary may be reduced to 300 mm in accordance with Figure 6.148.
8. Where a terminal is used with a vaporizing burner, a horizontal distance of at least 2,300 mm should be provided between the terminal and the roof line.
9. Notwithstanding the dimensions above, a terminal should be at least 300 mm from combustible material.

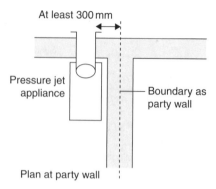

At least 300 mm

Pressure jet
appliance

Boundary as
party wall

Plan at party wall

Figure 6.148 Plan at party wall for an oil-firing appliance flue outlet

Supply of air for cooling to oil-firing appliances

An oil-firing appliance installed in an appliance compartment should have a supply of air for cooling by way of permanent ventilation, in addition to air for combustion, either direct to the open air or to an adjoining space.	D 3.22.2 ND 3.22.2

Fuel storage – oil tanks

Every fixed oil tank with a capacity of more than 90 L should be located at a distance from a building to reduce the risk of the fuel that is being stored from being ignited if there is a fire in the building.	D 3.23.1 ND 3.23.1
Some fire protection to, or for, the building is required if the oil tank is located close to the building. (See OFTEC Technical Information Sheet TI/136, Fire protection of oil storage tanks.)	D 3.23.1 ND 3.23.1
Precautions should be taken when an oil storage tank is located close to a boundary.	D 3.23.1 ND 3.23.1
The installation of a tank should not inhibit full development of a neighbouring plot.	D 3.23.1 ND 3.23.1
An oil tank with a capacity of more than 3,500 L should be located in accordance with the recommendations in BS 5410: Part 2: 1978.	D 3.23.1 ND 3.23.1
An oil tank with a capacity of not more than 3,500 L should be located in accordance with Table 6.119.	D 3.23.1 ND 3.23.1

Table 6.119   Location of oil storage tank not more than 3,500 litres capacity

Location of tank	Protection recommended	
	Buildings without openings	Buildings with openings
Not more than 1.8 m from any part of any building	Non-combustible base; and any part of the eaves not more than 1.8 m from the tank and extending 300 mm beyond each side of the tank must be non-combustible; and either: • any part of a building not more than 1.8 m from the tank should be of non-combustible construction short duration, or • a barrier [1]	Non-combustible base; and • any part of the eaves not more than 1.8 m from the tank and extending 300 mm beyond each side of the tank must be non-combustible; and • a barrier between the tank and any part of a building type 7, or not more than 1.8 m from the tank
More than 1.8 m from any building	Non-combustible base	
Not more than 760 mm from a boundary	Non-combustible base, and a barrier, or a wall with a non-combustible construction type 7, short duration	
More than 760 mm from a boundary	Non-combustible base	
Externally, wholly below ground	No protection required	

Note: 1. 'Barrier' means an imperforate, non-combustible wall or screen at least 300 mm higher and extending 300 mm beyond either end of the tank, constructed so as to prevent the passage of direct radiated heat to the tank.

Construction of oil storage tanks

Fixed oil storage tanks between 90 and 2,500 L and the fuel feed system connecting them to a combustion appliance should be strong enough to resist physical damage and corrosion so that the risk of oil spillage is minimized. Tanks should be constructed in accordance with Table 6.120.	D 3.24.1 ND 3.24.1

Table 6.120   Construction of oil storage tanks

Type of tank	Requirements
For a steel tank	BS 799: Part 5: 1987
For a steel tank, with or without integral bunding	The recommendations of OFTEC Technical Standard OFS T200
For a polyethylene tank with or without integral bunding	The recommendations of OFTEC Technical Standard OFS T100

Installation of oil storage tanks

Tanks of more than 2,500 L, and their associated pipework must be installed in accordance with the requirements of Regulation 6 of the Water Environment (Oil Storage) (Scotland) Regulations 2006.	D 3.24.2 ND 3.24.2
Note: Oil storage containers up to 2,500 L serving domestic buildings will be deemed to be authorized if they comply with the building regulations.	
Tanks with a capacity of more than 90 L but not more than 2,500 L and the fuel feed system connecting them to a combustion appliance should be installed in accordance with the recommendations of BS 5410: Part 1: 1997.	D 3.24.2 ND 3.24.2
Care should be taken to prevent leakage from pipework.	D 3.24.2 ND 3.24.2
Pipework should be run so as to provide the most direct route possible from the tank to the burner.	D 3.24.2 ND 3.24.2
Joints should be kept to a minimum and the use of plastic coated malleable copper pipe is recommended.	D 3.24.2 ND 3.24.2
Pipework should be installed in accordance with the recommendations in BS 5410: Parts 1: 1997 and Part 2: 1978.	D 3.24.2 ND 3.24.2

Secondary containment

If an oil tank is: • within 10 m of a watercourse; • located where spillage could run into an open drain or to a loose-fitting manhole cover; • within 50 m of a borehole or spring; • over ground where conditions are such that oil spillage could run off into a watercourse; • located in a position where the vent pipe outlet is not visible from the fill point, then these potential hazards should be borne in mind when considering the possible requirement for a catchpit (bund).	ND 3.24.30

Externally located, above-ground, oil tanks with a capacity
of not more than 2,500 L serving a domestic building should
be provided with a catchpit or be integrally bunded if the
tank is: D 3.24.3

- located within 10 m of the water environment
  (i.e. rivers, lochs, coastal waters);
- located where spillage could run into an open drain or
  to a loose-fitting manhole cover;
- within 50 m of a borehole or spring;
- over ground where conditions are such that oil spillage
  could run off into a watercourse;
- located in a position where the vent pipe outlet is not
  visible from the fill point.

Secondary containment should be provided where a tank is   D 3.24.3
within a building or wholly below ground.

Gas-fired appliances

Gas-fired appliance installations must: D 3.17.6
                                                                ND 3.17.7
- comply with the Gas Safety (Installation and Use)
  Regulations 1998; and
- be installed by a competent person.

Size of flues – gas-fired appliances

The area of a flue serving a gas-fired appliance should   D 3.20.10
have a size to ensure safe operation.   ND 3.20.10

A flue should be provided in accordance with Table 6.121.   D 3.20.10
                                                                ND 3.20.10

Table 6.121   Size of flues – oil-firing appliances

For a decorative fuel-effect gas appliance	Clause 9 of BS 5871: Part 3: 2005
For an inset live fuel-effect gas appliance	BS 5871: Part 2: 2005
For any other gas-fired appliance	BS 5440: Part 1: 2000

Relationship of gas-fired appliance to combustible material

A gas-fired appliance should be provided with a hearth as shown in Table 6.122.

Table 6.122 Gas-fired appliances – alternative hearth

Gas fire, convector heater and fire/back boiler	Clause 12 of BS 5871: Part 1: 2005
Inset live fuel-effect gas appliance	Clause 12 of BS 5871: Part 2: 2005
Decorative fuel-effect gas appliance	Clause 11 of BS 5871: Part 3: 2005
All other gas-fired appliances	The hearth should be a solid, heat-resistant, non-combustible, non-friable material at least 12 mm thick and at least the plan dimension shown in Figure 6.149

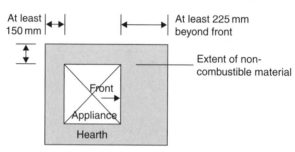

Plan of appliance on a hearth

Figure 6.149 Gas-fired appliances

A hearth need not be provided: • where every part of any flame or incandescent material in the appliance is at least 225 mm above the floor; or • where the appliance is designed not to stand on a hearth, such as a wall-mounted appliance or a gas cooker.	D 3.19.7 ND 3.19.7
A gas-fired appliance should be separated from any combustible material if the temperature of the back, sides or top of the appliance is more than 100°C under normal working conditions. Separation (see Figure 6.150) may be by: • a shield of non-combustible material at least 25 mm thick; or • an air space of at least 75 mm.	D 3.19.7 ND 3.19.7

 Note: A gas-fired appliance with a CE marking and installed in accordance with the manufacturer's written instructions may not require this separation .

Gas-fired appliances in bathrooms and bedrooms

 Regulation 30 of the Gas Safety (Installations and Use) Regulations 1998 has specific requirements for room-sealed appliances in these locations.

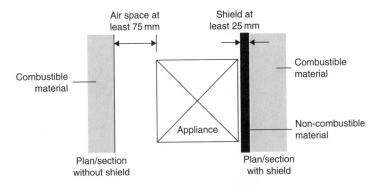

Figure 6.150 Gas-fired appliance separation from combustible material

Chimneys and flue-pipes serving gas-fired appliances

 A chimney or flue-pipe should be constructed and installed in accordance with BS 5440-1: 2000 and 'GE/UP/7: Edition 2', 'Gas installations in timber framed and light steel framed buildings'.

D 3.20.4
ND 3.20.4

> Some metal chimneys are specifically designed for use with gas-fired appliances and should **not** be used for solid fuel appliances because of the higher temperatures and greater corrosion risk.

The outlet from a flue should be located externally at a safe distance from any opening, obstruction or combustible material and in accordance with Figure 6.151 and Table 6.123.

D 3.20.19
ND 3.20.19

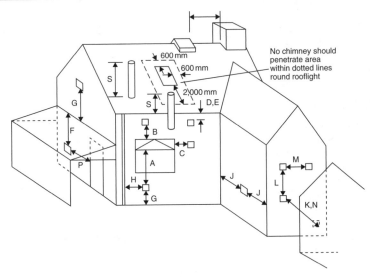

Figure 6.151 Gas-fired appliance flue outlets

Table 6.123   Gas-fired appliance flue outlets

Location		Minimum distance to terminal in millimetres			
		Balanced flue, room-sealed appliance		Open flue	
		Natural draught	Fanned draught	Natural draught	Fanned draught
A	Directly below an opening, air brick, opening window, etc.	(0–7 kW) 300 (>7–14 kW) 600 (>14–32 kW) 1,500 (>32–70 kW) 2,000	300	n/all	300
B	Above an opening, air brick, opening window, etc.	(0–32 kW) 300 (>32–70 kW) 600	300	n/all	300
C	Horizontally to an opening, air brick, opening window, etc.	(0–7 kW) 300 (>7–14 kW) 400 (>14–70 kW) 600	300	n/all	300
D	Below a gutter, or sanitary pipework	300 [2]	75 [1]	n/all	75 [1]
E	Below the eaves	300 [2]	200	n/all	200
F	Below a balcony or carport roof	600	200	n/all	200
G	Above ground, roof or balcony level	300	300	n/all	300
H	From vertical drain/ soil pipework	300	150 [3]	n/all	150
J	From an internal or external corner	600	300	n/all	200
K	From a surface or boundary facing the terminal [4]	600	600 [5]	n/app	600
L	Vertically from terminal on same wall	1,500	1,500	n/app	1,500
M	Horizontally from terminal on same wall	300	300	n/app	300
N	From a terminal facing the terminal	600	1,200 [6]	n/app	1,200
P	From an opening in a carport (e.g. door, window) into the building	1,200	1,200	n/app	1,200
R	From a vertical structure on the roof [7]	n/app	n/app	[Note 8]	n/app
S	Above an intersection with the roof	n/app	[Note 9]	[Note 10]	150

Table 6.123 (Continued)

Notes:
1. Notwithstanding the dimensions in the table, a terminal serving a natural draught and fanned draught appliance of more than 3 kW heat input, should be at least 300 mm and 150 mm respectively from combustible material.
2. Where a natural draught flue terminates not more than 1 m below a plastic projection or not more than 500 mm below a projection with a painted surface, then a heat shield at least 1 m long should be fitted.
3. This dimension may be reduced to 75 mm for appliances of up to 5 kW heat input.
4. The products of combustion should be directed away from discharging across a boundary.
5. The distance from a fanned draught appliance terminal installed at right angles to a boundary may be reduced to 300 mm in accordance with Figure 6.152.
6. The distance of a fanned flue terminal located directly opposite an opening in a dwelling should be at least 2 m.
7. Vertical structure includes a chimney-stack, dormer window, tank room, lift motor room or parapet.
8. 1,500 mm if measured to a roof terminal.
9. To manufacturer's instructions.
10. As Table 2 in BS 5440-1: 2000.
    n/all = not allowed; n/app = not applicable.

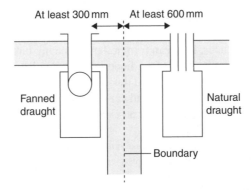

Figure 6.152 Plan at party wall for a gas-fired appliance flue outlet

Supply of air for combustion to gas-fired appliances

A gas-fired appliance installed in a room or space should have a supply of air for combustion and an air supply should be provided in accordance with Table 6.124.	D 3.21.4 ND 3.21.4

Supply of air for cooling to gas-fired appliances

A gas-fired appliance installed in an appliance compartment should have supply of air for cooling.	D 3.22.3 ND 3.22.3
Air for cooling should be provided in accordance with the recommendations in BS 5440: Part 2: 2000 for a gas-fired appliance located in an appliance compartment.	D 3.22.3 ND 3.22.3

Table 6.124   Supply of air for combustion to gas-fired appliances

For a decorative fuel-effect gas appliance	BS 5871: Part 3: 2005
For an inset live fuel-effect gas appliance	BS 5871: Part 2: 2005
For any other gas-fired appliance	BS 5440: Part 2: 2000

Fuel storage – LPG

Two forms of Liquefied Petroleum Gas (LPG) that are generally available in the UK are commercial butane and commercial propane and they have the following characteristics and potential hazards:

- LPG is stored as a liquid under pressure.
- It is colourless and its weight as a liquid is approximately half that of the equivalent volume of water.
- LPG vapour is denser than air, commercial butane being about twice as heavy as air. Therefore the vapour may flow along the ground and into drains, sinking to the lowest level of the surroundings and may therefore be ignited at a considerable distance from the source of the leakage.
- When mixed with air, LPG can form a flammable mixture.
- Leakage of small quantities of the liquefied gas can give rise to large volumes of vapour/air mixture and thus cause considerable hazard.
- Owing to its rapid vaporization and consequent lowering of temperature, LPG, particularly in liquid form, can cause severe frost burns if brought into contact with the skin.
- A container that has held LPG and is 'empty' may still contain LPG in vapour form and is thus potentially dangerous.

The type, size and location of an LPG storage installation will determine the factors that should be addressed in the construction of the facility, to comply with health and safety requirements.

The operation of properties where LPG is stored or is in use are subject to legislation enforced by both the Health and Safety Executive (HSE) and the local authority.

Every building must be designed and constructed in such a   4.11
way that each liquefied petroleum gas storage installation,
used solely to serve a combustion appliance providing
space heating, water heating, or cooking facilities, will:

- be protected from fire spreading to any liquefied
  petroleum gas container; and
- not permit the contents of any such container to form
  explosive gas pockets in the vicinity of any container.

 This standard does not apply to a liquefied
petroleum gas storage container, or containers, for
use with portable appliances.

LPG storage tanks

A liquefied petroleum gas storage tank, together with any associated pipe-work, should be designed, constructed and installed in accordance with the requirements set out in the Liquefied Petroleum Gas Association (LPGA) Code of Practice 1: 'Bulk LPG Storage at Fixed Installations'.

LPG storage tanks in excess of 4 tonnes (9,000 L) capacity are uncommon in domestic applications.

Every tank should be separated from a building, boundary, or fixed source of ignition, to:	
• (in the event of fire) reduce the risk of fire spreading to the tank; and • enable safe dispersal in the event of venting or leaks.	D 4.11.2 ND 4.11.2
Tanks should be situated outdoors, in a position that will not allow accumulation of vapour at ground level.	D 4.11.2 ND 4.11.2
Ground features such as open drains, manholes, gullies and cellar hatches, within the separation distances given in column A of Table 6.125, should be sealed or trapped to prevent the passage of LPG vapour.	D 4.11.2 ND 4.11.2
Tanks should be separated from buildings, boundaries or fixed sources of ignition in accordance with Table 6.125.	D 4.11.2 ND 4.11.2

Table 6.125   Separation distances for liquefied petroleum gas storage tanks

Maximum capacity (in tonnes)		Minimum separation distance for above-ground tanks (in metres)		
Of any single tank	Of any group of tanks	From a building, boundary or fixed source of ignition to the tank		
		A no fire wall	B with fire wall	Between tanks
0.25	0.8	2.5	0.3	1.0
1.1	3.5	3.0	1.5	1.0
4.0	12.5	7.5	4.0	1.0

Fire walls

For vessels up to 1.1 tonne capacity, the fire wall need be no higher than the top of the pressure relief valve and may form part of the site boundary.

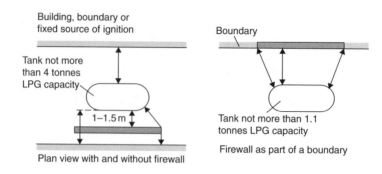

Building, boundary or fixed source of ignition

Boundary

Tank not more than 4 tonnes LPG capacity

1–1.5 m

Plan view with and without firewall

Tank not more than 1.1 tonnes LPG capacity

Firewall as part of a boundary

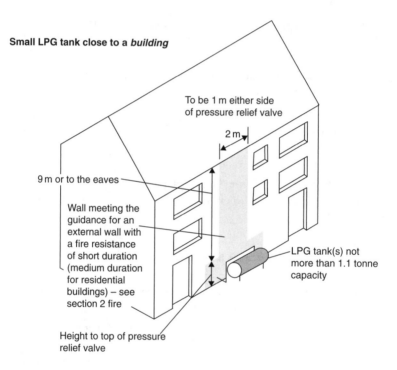

Small LPG tank close to a building

To be 1 m either side of pressure relief valve

2 m

9 m or to the eaves

Wall meeting the guidance for an external wall with a fire resistance of short duration (medium duration for residential buildings) – see section 2 fire

LPG tank(s) not more than 1.1 tonne capacity

Height to top of pressure relief valve

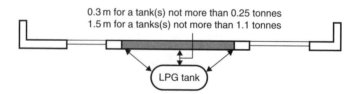

0.3 m for a tank(s) not more than 0.25 tonnes
1.5 m for a tanks(s) not more than 1.1 tonnes

LPG tank

Figure 6.153 Separation or shielding of an LPG tank from building, boundary or fixed source of ignition

For vessels up to 1.1 tonne capacity located closer to a building than the separation distance in column A of Table 6.125, the fire wall should form part of the wall of the building in accordance with Figure 6.153.

D 4.11.2
ND 4.11.2

Where a group of tanks are sited together, the number of tanks in a group should not exceed 6 and the total storage capacity of the group should not exceed that given for any group of tanks in Table 6.125.

D 4.11.2
ND 4.11.2

Motor vehicles under the control of a site occupier should be parked at least 6 m from LPG tanks or the separation distance in column A of Table 6.125, whichever is the smaller.

D 4.11.2
ND 4.11.2

Motor vehicles not under site control (e.g. those belonging to members of the public) should be parked no closer than the separation distance in column A of Table 6.125.

D 4.11.2
ND 4.11.2

LPG storage cylinders

Any installation should enable cylinders to stand upright, secured by straps or chains against a wall outside the building.

D 4.11.3
ND 4.11.3

Cylinders should be positioned on a firm, level base such as concrete at least 50 mm thick or paving slabs bedded on mortar, and located in a well-ventilated position at ground level, so that the cylinder valves will be:

D 4.11.3
ND 4.11.3

- at least 1 m horizontally and 300 mm vertically from openings in the buildings or from heat source such as flue terminals or tumble dryer vents;
- at least 2 m horizontally from untrapped drains, unsealed gullies or cellar hatches unless an intervening wall not less that 250 mm high is present.

Cylinders should be readily accessible, reasonably protected from physical damage and located where they do not obstruct exit routes from the building.

D 4.11.3
ND 4.11.3

Solid fuel appliances

Solid fuel appliance installations should be constructed and installed carefully to ensure that the entire installation operates safely.	D 3.17.4 ND 3.17.5
Appliances should be fit for purpose and for the type of fuel burnt.	D 3.17.4 ND 3.17.5

Relationship of solid fuel appliance to combustible material

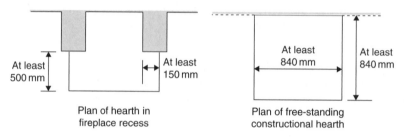

Plan of hearth in fireplace recess

Plan of free-standing constructional hearth

Figure 6.153a Relationship of solid fuel appliance to combustible material

A solid fuel appliance should: • sit on a hearth; • be located on the hearth such that protection will be offered from the risk of ignition of the floor by direct radiation, conduction or falling embers; in accordance with Figure 6.154.	D 3.19.5 ND 3.19.5

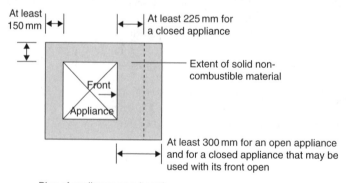

Plan of appliance on a hearth

Figure 6.154 Locating a solid fuel appliance

A solid fuel appliance should be provided with a solid, non-combustible hearth that will prevent the heat of the appliance from igniting combustible materials in accordance with Figure 6.155.	D 3.19.5 ND 3.19.5

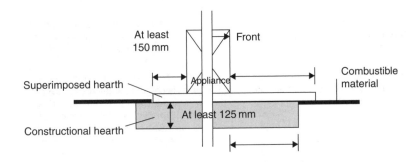

Section through superimposed hearth

Figure 6.155 Solid fuel appliance – superimposed hearth

Chimneys and flue-pipes serving solid fuel appliances

A flue in a chimney should be separated from every other flue and extend from the appliance to the top of the chimney.	D 3.20.1 ND 3.20.1
Every flue should be surrounded by non-combustible material that is capable of withstanding the effects of a chimney fire, without any structural change that would impair the stability or performance of the chimney.	D 3.20.1 ND 3.20.1
The size of a flue serving a solid fuel appliance should be at least the size shown in Table 6.126.	D 3.20.8 ND 3.20.8

Table 6.126   Minimum areas of flues

Appliance	Minimum flue size
Fireplace with an opening more than 500 mm × 550 mm, or a fireplace exposed on 2 or more sides	15% of the total face area of the fireplace opening(s)
Fireplace with an opening not more than 500 mm × 550 mm	200 mm diameter or rectangular/square flues having the same cross-sectional area and a minimum diameter not less than 175 mm

(Continued)

Table 6.126 (Continued)

Appliance	Minimum flue size
Closed appliance with rated output more than 30 kW but not more than 50 kW, burning any fuel	175 mm diameter or rectangular/square flues having the same cross-sectional area and a minimum diameter not less than 150 mm
Closed appliance with rated output not more than 30 kW, burning any fuel	150 mm diameter or rectangular/square flues having the same cross-sectional area and a minimum diameter not less than 125 mm
Closed appliance with rated output not more than 20 kW that burns smokeless or low volatiles fuel	125 mm diameter or rectangular/square flues having the same cross-sectional area and a minimum diameter not less than 100 mm for straight flues or 125 mm for flues with bends or offsets

Flue outlets for solid fuel appliances

The outlet from a flue should be located externally at a safe distance from any opening, obstruction or flammable or vulnerable materials and in accordance with Figure 6.156.	D 3.20.17 ND 3.20.17

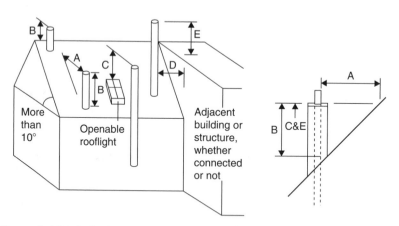

Figure 6.156 Solid fuel appliance flue outlets

Where:
A is 2.3 m horizontally clear of the weather skin;
B is 1.0 m provided A is satisfied or 600 mm where above the ridge;
C is 1.0 m above the top of any flat roof; and 1.0 m above any openable rooflight, dormer or ventilator, etc., within 2.3 m measured horizontally; and
E must be at least 600 mm where D is not more than 2.3 m.

Notes:
• Horizontal dimensions are to the surface surrounding the flue.
• Vertical dimensions are to the top of the chimney terminal.

Flue terminals in close proximity to roof coverings that are easily ignitable, such as thatch or shingles, should be located outside Zones A and B in Figure 6.157.	D 3.20.17 ND 3.20.17

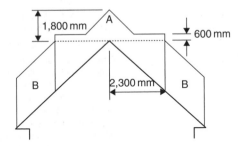

Figure 6.157 Location of flue terminals relative to easily ignitable roof coverings

Where:
- Zone A is at least 1.8 m vertically above the weather skin, and at least 600 mm above the ridge; and
- Zone B is at least 1.8 m vertically above the weather skin, and at least 2.3 m horizontally from the weather skin.

Supply of air for combustion to solid fuel appliances

A solid fuel appliance installed in a room or space should have a supply of air for combustion by way of permanent ventilation either direct to the open air or to an adjoining space (including a sub-floor space) that is itself permanently ventilated direct to the open air.	D 3.21.2 ND 3.21.2
An air supply should be provided in accordance with Table 6.127.	D 3.21.2 ND 3.21.2

Bulk storage of woody biomass fuel

By its very nature woody biomass fuel is highly combustible and precautions need to be taken to reduce the possibility of the stored fuel igniting.

Table 6.127   Supply of air for combustion

Type of appliance	Minimum ventilation opening sizes [2]
Open appliance without a throat [1]	A permanent air entry opening or openings with a total free area of 50% of the cross-sectional area of the flue
Open appliance with a throat [1]	A permanent air entry opening or openings with a total free area of 50% of the throat opening area
Any other solid fuel appliance	A permanent air entry opening or openings with a total free area of 550 mm² for each kW of combustion appliance rated output more than 5 kW. (A combustion appliance with an output rating of not more than 5 kW has no minimum requirement, unless stated by the appliance manufacturer)

Notes:
1. Where a draught stabilizer is fitted to a solid fuel appliance, or to a chimney or flue-pipe in the same room as a solid fuel appliance, additional ventilation opening should be provided with a free area of at least 300 mm²/kW of solid fuel appliance rated output.
2. Nominal fire size is related to the free opening width at the front of the fireplace opening.

To ensure maximum energy from the fuel, storage should be designed to be damp free and improve or maintain the moisture content of the fuel at time of delivery.	D 3.23.4 ND 3.23.4
To inhibit the spread of fire to the contents, bulk storage for wood fuels should be in containers in accordance with Table 6.128.	D 3.23.4 ND 3.23.4

Once a year any dust that has collected in the store should be removed.	D 3.23.4 ND 3.23.4
Storage containers for wood pellets, where they are to be pumped from a transporter to the container, should include a protective rubber mat over the wall to reduce damage to the pellets when they hit the wall.	D 3.23.4 ND 3.23.4
Containers should have an outward-opening door incorporating containment to prevent the pellets escaping when the door is opened.	D 3.23.4 ND 3.23.4

| To maintain fire-proof storage and prevent back-burning there should be an interruption to the fuel transport system normally by use of a star-feeder or chute for the fuel to fall into the boiler (see BS EN 303-5: 1999). | D 3.23.4 ND 3.23.4 |
| The woody biomass fuel should be stored separately from the boiler that the fuel feeds for fire safety reasons. | D 3.23.4 ND 3.23.4 |

Table 6.128 Bulk storage of woody biomass fuel

Location of store	Protection recommended
External and not more than 1.8 m from any part of any building	• Any part of the building eaves not more than 1.8 m from the container or storage space and extending 300 mm beyond each side of the container or storage space should be non-combustible; and • A barrier
External not more than 1 m from any boundary	The container or storage frame should be constructed to have short fire resistance duration to its boundary walls
Within a building	• separated from the building with internal wall constructions providing short fire resistance duration type 4; and • any door to be outward opening providing short fire resistance duration type 6; and • separated from the building with floor constructions providing short fire resistance duration type 2; and • external walls constructed that provide short fire resistance duration type 7 or type 8 as appropriate

Fuel storage containers – dwellings

In order to best exploit the advantages achieved through the use of woody biomass as low carbon technology it is recommended that wood fuel storage provision is of a size that will ensure deliveries need not be made at intervals of less that 3 months for bulk storage and 6 months for small installations.

Table 6.129 provides recommended size of storage for a variety of different dwelling types that will permit a large enough volume to be delivered whilst minimizing vehicle movements.

Table 6.129 Bulk woody biomass fuel storage: 100% heating (primary) and DHW

Dwelling size	Wood pellets	Wood chips	Logs – stacked
$<80\,m^2$	$1.5\,m^3$	$3.5\,m^3$	$3\,m^3$
$80–160\,m^2$	$2\,m^3$	$5\,m^3$	$4\,m^3$
$>160\,m^2$	$3\,m^3$	$6\,m^3$	$5\,m^3$

Table 6.130 provides recommended size of storage for secondary heating for a variety of dwelling types.

Table 6.130   Woody biomass fuel storage: secondary heating

Dwelling size	Wood pellets	Wood chips	Logs – stacked
<80 m²	0.3 m³ (9 bags)	1 m³	0.5 m³
80–160 m²	0.5 m³ (13 bags)	1.5 m³	1 m³
>160 m²	0.7 m³ (16 bags)	2 m³	1 m³

Fuel storage containers – non-domestic buildings

In order to best exploit the advantages achieved through the use of woody biomass as low carbon technology it is recommended that wood fuel storage provision is of a size that will ensure bulk deliveries need not be made at intervals of less than 1 month.

The following table provides recommendations for some building types of varying size on the storage recommendations that will permit a large enough volume to be delivered whilst minimizing vehicle movements (Table 6.131).

Table 6.131   Woody fuel storage recommendations for 100% heating

Building heat demand		Wood chips		Wood pellets	
Type (m)	Annual	Fuel required (m³/year)	Storage	Fuel required (m³/year)	Storage
Education (400–2,000)	90–450	110–565	10–60 m³	40–210	5–20 m³
Industrial (100–2,000)	20–360	25–450	5–55 m³	10–165	5–20 m³
Office (100–2,000)	20–420	25–525	5–55 m³	10–195	5–15 m³
Education (400–2,000)	90–450	110–565	10–60 m³	40–210	5–20 m³
Industrial (100–2,000)	20–360	25–450	5–55 m³	10–165	5–20 m³

6.21 Hot water storage systems

In domestic buildings it is normal that systems of up to 500 L storage capacity and a power input not exceeding 45 kW are used.

6.21.1 Requirements

Safety

Every building must be designed and constructed in such a way that protection is provided for people in, and around, the building from the danger of severe burns or scalds from the discharge of steam or hot water.   4.9

Energy

> Every building must be designed and constructed in such a way that:
>
> - the heating and hot water service systems installed are energy   6.3
>   efficient and are capable of being controlled to achieve
>   optimum energy efficiency;
> - temperature loss from heated pipes, ducts and vessels, and   6.4
>   temperature gain to cooled pipes and ducts, is resisted.

Information

> The occupiers of a building must be provided with written   6.8
> information by the owner on the operation and maintenance
> of the building services and energy supply systems.

6.21.2 Meeting the requirements

 It is not intended that the following guidance should be applied to storage systems with a capacity of less than 15 L, to systems used solely for space heating or to any system used for an industrial or a commercial process.

> Every building must be designed and constructed in such a   D 4.9.0
> way that protection is provided for people in, and around, the
> building from the danger of severe burns or scalds from the
> discharge of steam or hot water.
>
> Safety devices installed to protect from hazards such as   D 4.9.0
> scalding or the risk of explosion of unvented systems should
> be maintained to ensure correct operation

Solid fuel boilers

These should be thermostatically controlled to reduce the burning rate of the fuel, by varying the amount of combustion air to the fire.

> For safety reasons, a suitable heat bleed (slumber circuit) from   D 6.3.8
> the system should be formed (e.g. a gravity fed radiator
> without a TRV or a hot water cylinder that is connected
> independent of any controls).
>
> For hot water systems, unless the cylinder is forming the slumber   D 6.3.8
> circuit, a thermostatically controlled valve should be fitted.

Appliance efficiency

The following table shows recommended minimum thermal efficiencies for domestic hot water systems (Table 6.132).

Table 6.132   Domestic hot water systems

System type		Minimum thermal efficiencies (based on gross calorific value)
Direct-firing	Natural gas	73%
	LPG	74%
	Oil	75%
Indirect-firing	Natural gas	80%
	LPG	81%
	Oil	82%

There is no minimum thermal efficiency specified for electric domestic hot water heaters.

Domestic hot water heating controls

Note: Although this guidance refers only to non-domestic buildings, hot water systems, to confuse matters, are generally referred to as 'Domestic Hot water' (DHW) systems!

A DHW system should have controls that will switch off the heat when the water temperature required by the occupants has been achieved and during periods when there is no demand for hot water.	ND 6.3.7
The controls shown in Tables 6.133 and 6.134 should be observed for all DHW systems.	ND 6.3.7

Gas/oil-firing systems

Table 6.133   Gas/oil-firing systems

Systems	Controls
Direct	• Automatic thermostat control to shut off the burner/primary heat supply when the desired temperature of the hot water has been reached
Indirect	• Automatic thermostat control to shut off the burner/primary heat supply when the desired temperature of the hot water has been reached • High-limit thermostat to shut off primary flow if system temperature too high • Time control

Electric DHW systems

Table 6.134   Electric DHW systems

Control system	Point of use	Locally	Central	Instantaneous
Automatic thermostat control to interrupt the electrical supply when the desired storage temperature has been reached	Yes	Yes	Yes	No
High-limit thermostat (thermal cut-out) to interrupt the energy supply if the system temperature gets too high	Yes	Yes	Yes	No
Manual reset in the event of an over temperature trip	Yes	Yes	Yes	No
A 7-day time-clock or Building Management System (BMS) interface should be provided to ensure bulk heating of water using off-peak electricity	No	Yes	Yes	No
High-limit thermostat (thermal cut-out) to interrupt the energy supply if the outlet temperature gets too high	No	No	No	Yes
Flow sensor that only allows electrical input should sufficient flow through the unit be achieved	No	No	No	Yes

A DHW system (other than a system with a solid fuel boiler) should have controls that will switch off the heat when the water temperature required by the occupants has been achieved and during periods when there is no demand for hot water.	ND 6.3.7
Independent time and temperature control of hot water circuits should be provided along with a boiler interlock (refer to Table 6.135) to ensure that the boiler and pump only operate when there is a demand for heat.	D 6.3.8
Zone controls are not considered necessary for single apartment dwellings.	D 6.3.8
For hot water systems in large dwellings, more than one hot water circuit each with independent time and temperature control should be provided.	D 6.3.8

A hot water system (other than for combi boilers with storage capacity 15 L or less) should have controls that will switch off the heat when the water temperature required by the occupants has been achieved and during periods when there is no demand for hot water.

D 6.3.8

Controls for dry space heating and hot water systems

Zone controls are not considered necessary for single apartment dwellings.

For large dwellings with a floor area over 150 m², independent time and temperature control of multiple space heating zones is recommended.

D 6.3.9

Each zone (not exceeding 150 m²) should have a room thermostat, and a single multi-channel programmer or multiple heating zone programmers.

D 6.3.9

Controls for combined warm air and hot water systems

The following controls should be provided:

D 6.3.10

- independent time control of both the heating and hot water circuits (achieved by means of a cylinder thermostat and a timing device, wired such that when there is no demand for hot water both the pump and circulator are switched off);
- pumped primary circulation to the hot water cylinder;
- a hot water circulator interlock (achieved by means of a cylinder thermostat and a timing device, wired such that when there is no demand from the hot water both the pump and circulator are switched off); and
- time control by the use of either:
  - a full programmer with separate timing to each circuit;
  - two or more separate timers providing timing control to each circuit;
- a programmable room thermostat(s) to the heating circuit(s);or
- a time switch/programmer (two channel) and room thermostat.

Insulation

Insulation of pipes and ducts

Hot water pipes serving a space heating system should be thermally insulated against heat loss unless the use of such pipes or ducts always contributes to the heating demands of the room or space.	D 6.4.1 ND 6.4.1
Pipes that are used to supply hot water to appliances within a building should be insulated against heat loss.	D 6.4.1 ND 6.4.1

Insulation of vessels

A hot water storage vessel should be insulated against heat loss.	D 6.4.2 ND 6.4.2
Where an unvented hot water system is installed, additional insulation should be considered to reduce the heat loss that can occur from the safety fittings and pipework.	D 6.4.2 ND 6.4.2

Unvented hot water systems

Small unvented hot water storage systems – specification

An unvented hot water storage system should be: • designed and installed to prevent the temperature of the stored water at any time exceeding 100°C and to provide protection from malfunctions of the system; • in accordance with the recommendations of BS 7206: 1990.	D 4.9.2 ND 4.9.2
A unit or package should have fitted (or supplied for fitting by the installer): • a check valve to prevent backflow; • a pressure control valve to suit the operating pressure of the system; • an expansion valve to relieve excess pressure; • an external expansion vessel or other means of accommodating expanded heated water.	D 4.9.2 ND 4.9.2

Additional to any thermostatic control that is fitted to maintain the temperature of the stored water at around 60°C, a
unit or package should have a minimum of 2 temperature-
activated devices operating in sequence comprising:
- a non-self-resetting thermal cut-out;
- a temperature relief valve.

D 4.9.2
ND 4.9.2

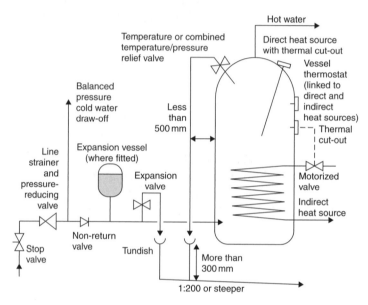

Figure 6.158 Unvented hot water storage system – indirect example

A temperature-operated, non-self-resetting, energy cut-out
complying with BS 3955: 1986 should be fitted to the vessel.

D 4.9.2
ND 4.9.2

In the event of thermostat failure, heating to the water in the vessel should
stop before the temperature rises to the critical level required for operation of
the temperature relief valve.

In indirectly heated vessels, the non-self-resetting thermal
cut-out should operate a motorized valve, or other similar
device, to shut off the flow from the heat source.

D 4.9.2
ND 4.9.2

On directly heated vessels or where an indirectly heated
vessel has an alternative direct method of water heating
fitted, a non-self-resetting thermal cut-out device should
be provided for each direct source.

D 4.9.2
ND 4.9.2

The temperature relief valve should be located directly on the storage vessel.	D 4.9.2 ND 4.9.2
The relief valve should have a discharge capacity rating at least equal to the rate of energy (power in kilowatts) input to the heat source.	D 4.9.2 ND 4.9.2

Large unvented hot water storage systems – specification

An unvented hot water storage system should be designed and installed to prevent the temperature of the stored water at any time exceeding 100°C and to provide protection from malfunctions of the system.

Where the system has a power input of less than 45 kW, safety devices (such as thermal cut-outs and discharge pipework) should be provided.	ND 4.9.3
Where the system has a power input greater than 45 kW, safety devices should also include temperature or combined temperature/pressure relief valves capable of a combined discharge rating at least equal to the power input of the system.	ND 4.9.3

Installation of unvented hot water storage systems

The installation of an unvented hot water storage system should be carried out by a person with appropriate training and practical experience including current membership of a registration scheme operated by a recognized professional body such as the Scottish and Northern Ireland Plumbing Employers Federation (SNIPEF) and the Construction Industry Training Board (CITB).

Concerning the installation of an unvented hot water storage system:

• the installer should be a competent person and, on completion, the labelling of the installation should identify the installer;	D 4.9.1 ND 4.9.1
• the installed system should be meet the recommendations of BS 7206: 1990 or be the subject of an approval by a notified body;	
• certification of the unit or package should be recorded by permanent marking and a warning label which should be visible after installation. A comprehensive installation/user manual should be supplied;	

- the tundish and discharge pipework should be correctly located and fitted by the installer and the final discharge point should be visible and safely positioned where there is no risk from hot water discharge;
- the operation of the system under discharge conditions should be tested to ensure provision is adequate.

Discharge from unvented hot water storage systems

The removal of discharges of water from the system can be considered in three parts:

Relief valve to tundish

Each relief valve should discharge into a metal pipe not less than the nominal outlet size of the valve.	D 4.9.3 ND 4.9.4
The discharge pipe should have an air-break, such as a tundish, not more than 500 mm from the vessel relief valve and located in an easily visible location within the same enclosure.	D 4.9.3 ND 4.9.4
Bulb discharge pipes from more than one relief valve may be taken through the same tundish.	D 4.9.3 ND 4.9.4
Pipework should be installed so that any discharge will be directed away from electrical components should the discharge outlet become blocked.	D 4.9.3 ND 4.9.4

Tundish to final discharge point

The discharge pipe from the tundish to final discharge point should be of a material, usually copper, capable of withstanding water temperatures of up to 95°C and be at least one pipe size larger than the outlet pipe to the relief valve.	D 4.9.3 ND 4.9.4
A vertical section of pipe, at least 300 mm long, should be provided beneath the tundish before any bends to the discharge pipe; thereafter the pipe should be appropriately supported to maintain a continuous fall of at least 1 in 200 to the discharge point.	D 4.9.3 ND 4.9.4

The pipework should have a resistance to the flow of water no greater than that of a straight pipe 9 m long unless the pipe bore is increased accordingly. Guidance on sizing of pipework from the tundish to the final discharge point is shown in Table 6.135.

D 4.9.3
ND 4.9.4

Table 6.135   Size of discharge pipework

Valve outlet size	Minimum size of discharge pipe to tundish	Minimum size of discharge pipe from tundish	Maximum resistance allowed, expressed as a length of straight pipe, i.e. no elbows or bends	Equivalent resistance created by the addition of each elbow or bend
G1/2	15 mm	22 mm	Up to 9 m	0.8 m
		28 mm	Up to 18 m	1.0 m
		35 mm	Up to 27 m	1.4 m
G3/4	22 mm	28 mm	Up to 9 m	1.0 m
		35 mm	Up to 18 m	1.4 m
		42 mm	Up to 27 m	1.7 m
G1	28 mm	35 mm	Up to 9 m	1.4 m
		42 mm	Up to 18 m	1.7 m
		54 mm	Up to 27 m	2.3 m

Discharge pipe termination

The pipe termination should be in a visible location and installed so that discharge will not endanger anyone inside or outside the building.

D 4.9.3
ND 4.9.4

Ideally, the final discharge point should be above the water seal to an external gully and below a fixed grating. Other methods for terminating the final discharge point would include:

D 4.9.3
ND 4.9.4

• up to 100 mm above external surfaces such as car parks, grassed areas or hard standings; a wire cage or similar guard should be provided to both prevent contact with discharge and protect the outlet from damage, whilst maintaining visibility;

- at high level into a hopper and downpipe of a material, such as cast iron, appropriate for a hot water discharge with the end of the discharge pipe clearly visible;
- onto a flat roof or pitched roof clad in a material capable of withstanding high-temperature discharges of water, such as slate/clay/concrete tiles or metal sheet, with the discharge point a minimum of 3 m from any plastic guttering system that would collect such discharges.

Note: Discharge at high level may be possible if the discharge outlet is terminated in such a way as to direct the flow of water against the external face of a wall. However, evidence of the minimum height of the outlet above any surface to which people have access and the distance needed to reduce the discharge to a non-scalding level should be established by test or otherwise.

Discharge of steam or hot water

In a domestic building, any vent or overflow pipe of a hot water system should be positioned so that any discharge will not endanger anyone inside or outside the building.	D 4.9.4
To prevent the development of *Legionella* or similar pathogens, hot water within a storage vessel should be stored at a temperature of not less than 60°C and distributed at a temperature of not less than 55°C.	D 4.9.5 ND 4.9.5

Hot water discharge from sanitary fittings

To prevent scalding, the temperature of hot water, at point of delivery to a bath or bidet, should not exceed 48°C.	D 4.9.5 ND 4.9.5
A device or system limiting water temperature should allow flexibility in setting of a delivery temperature, up to a maximum of 48°C, in a form that is not easily altered by building users.	D 4.9.5 ND 4.9.5

Solar water heating

Solar water heating has low or zero carbon dioxide emissions, little or no associated running costs and is inherently energy efficient.

Solar roof panels should be regarded as forming part of the roof covering and as such should be able to resist ignition from an external source.   2.8

Location and orientation for optimum energy efficiency and to avoid overshading should be considered.   D 6.3.6

All pipes of a solar water heating primary system should be insulated.   D 6.4.1 / ND 6.4.1

Controls for solar water heating

Controls should be provided to: D 6.3.11

- optimize the useful energy gain from the solar collectors into the system's dedicated storage vessel(s);
- minimize the accidental loss of stored energy by the solar hot water system;
- ensure that hot water produced by auxiliary heat sources is not used when adequate grade solar pre-heated water is available;
- guard against the adverse affects of excessive primary temperatures and pressures;
- limit the inlet temperature of any separate domestic hot water heating appliance (such as a combi boiler).

Energy efficiency

Every building must be designed and constructed in such a way that the hot water systems installed are energy efficient and are capable of being controlled to achieve optimum energy efficiency.   D 6.3

Dry central heating systems

Where a gas-fired circulator is incorporated in the warm-air unit to provide domestic hot water, it should be of a type that is able to deliver full and part load efficiency.   D 6.3.5

Heat pumps – hot water systems

Heat pump unit controls should include: D 6.3.8

- control of water temperature for the distribution system;
- control of water pumps (integral or otherwise);

- defrost control of external airside heat exchanger (for air to water units);
- control of outdoor fan operation (for air to water units);
- protection for water flow failure;
- protection for high water temperature;
- protection for high refrigerant pressure; and
- protection for external air flow failure (on air to water units).

Controls which are not integral to the unit should include:

- room thermostat to regulate the space temperature and interlocked with the heat pump unit operation; and
- timer to optimize operation of the heat pump.

Work on existing buildings

Where alterations are being made to an existing heating/ hot water system or a new or replacement heating/hot water system is being installed in an existing dwelling (or building consisting of dwellings), such alterations should not allow the heating system as a whole to be downgraded in terms of energy efficiency or compromised from a safety point of view.	D 6.3.12
Where a new or replacement boiler or hot water storage vessel is installed, or where existing systems are extended, new or existing pipes that are accessible or exposed as part of the work should be insulated as for new systems.	D 6.4.3 ND 6.4.3

Written information

For a domestic building, written information concerning the operation and maintenance of the hot water system should be made available for the use of the occupier.	D 6.8.1
For non-domestic buildings, a logbook containing information of energy system operation and maintenance (such as building services plant and controls) to ensure that the building user can optimize the use of fuel, shall be provided.	ND 6.8.1

6.22 Heating systems

In the design of all buildings, the energy efficiency of the heating plant is an important part of the package of measures which contributes to the overall building carbon dioxide emissions.

Ideally the system should have sufficient zone, time and temperature controls to ensure that the heating system only provides the desired temperature when the building is occupied.

6.22.1 Requirements

Environment

Every building must be designed and constructed in such a way that:

- *it can be heated.*   *3.13*
- *the air quality inside the building is not a threat to the*   *3.14*
  health of the occupants or the capability of the building
  to resist moisture, decay or infestation.
- *the products of combustion are carried safely to the*   *3.20*
  external air without harm to the health of any person
  through leakage, spillage, or exhaust nor permit the
  re-entry of dangerous gases from the combustion
  process of fuels into the building.
- *an oil storage installation, incorporating oil storage*   *3.23*
  tanks used solely to serve a fixed combustion appliance
  installation providing space heating in a building, will:
- *inhibit fire from spreading to the tank and its contents*
  from within, or beyond, the boundary.
- *it will:*   *3.24*
  – reduce the risk of oil escaping from the installation;
  – contain any oil spillage likely to contaminate any
    water supply, ground water, watercourse, drain or
    sewer; and
  – permit any spill to be disposed of safely;
- *the energy performance of the building is capable of*   *6.1*
  reducing carbon dioxide emissions;

Energy

Every building must be designed and constructed in such a   *6.2*
way that:

- *an insulation envelope is provided which reduces heat*
  loss;

- *the heating and hot water service systems installed are*   6.3
  *energy efficient and are capable of being controlled to
  achieve optimum energy efficiency.*
- *temperature loss from heated pipes, ducts and vessels,*   6.4
  *and temperature gain to cooled pipes and ducts, is
  resisted.*
- *energy supply systems and building services which use*   6.7
  *fuel or power for heating the internal environment are
  commissioned to achieve optimum energy efficiency.*
- *an energy performance certificate for the building*   6.9.
  *is affixed to the building, indicating the approximate
  annual carbon dioxide emissions and energy usage
  of the building based on a standardized use of the
  building;*
- *each part of a building designed for different*   6.10
  occupation is fitted with fuel consumption meters.

The occupiers of a building must be provided with written   6.8
*information by the owner in the operation and maintenance
of the building services and energy supply systems.*

6.22.2 Meeting the requirements

This standard applies only to a dwelling.

Heating, ventilation and thermal insulation should be considered as part of a total design that takes into account all heat gains and losses. Failure to do so can lead to inadequate internal conditions (e.g. condensation and mould and the inefficient use of energy due to overheating).

Every building must be designed and constructed in such a way that it can be heated.	D 3.13

Every dwelling should have some form of fixed heating system, or alternative that is capable of maintaining a temperature of 21°C in at least one apartment and 18°C elsewhere, when the outside temperature is −1°C.	D 3.13.1

There is no need to maintain these temperatures in storage rooms with a floor area of not more than 4 m².

Alternative heating systems may involve a holistic design approach to the dwelling and can include the use of natural sources of available energy such as the sun, wind and the geothermal capacity of the earth.

Where there are elderly or infirm occupants in a dwelling the   D 3.13.2
capability of the heating system to maintain an apartment
at a temperature higher than 21°C is a sensible precaution.

Heating appliances

Safe operation

The correct installation of heating appliances (and the design and installation
of a flue) can reduce the risk from combustion appliances and their flues from:

- endangering the health and safety of persons in or around a building;
- compromising the structural stability of a building;
- causing damage by fire.

Domestic space and water heating facilities should be designed in accordance
with Table 6.136.

**Table 6.136 Recommended designation for chimneys and flue-pipes for use
with oil-firing appliances with a flue gas temperature not more than 250°C**

Appliance type	Fuel oil	Designation
Boiler including combination boiler – pressure jet burner	Class C2	T250 N2 D 1 Oxx
Cooker – pressure jet burner	Class C2	T250 N2 D 1 Oxx
Cooker and room heater – vaporizing burner	Class C2	T250 N2 D 1 Oxx
Cooker and room heater – vaporizing burner	Class D	T250 N2 D 2 Oxx
Condensing pressure jet burner appliances	Class C2	T160 N2 W 1 Oxx
Cooker – vaporizing burner appliances	Class D	T160 N2 W 2 Oxx

Removal of products of combustion

Heating appliances fuelled by solid fuel, oil or gas all have the potential to
cause carbon monoxide (CO) poisoning if they are poorly installed or commissioned, inadequately maintained or incorrectly used.

Insulation of pipes, ducts and vessels

Thermal insulation to heating pipes and ducts will improve energy efficiency
by preventing:

- uncontrolled heat loss from such equipment;
- an uncontrolled rise in the temperature of the parts of the building where
  such equipment is situated.

Every building must be designed and constructed in such   6.4
a way that temperature loss from heated pipes, ducts and
vessels, and temperature gain to cooled pipes and ducts, is
resisted.

Work on existing buildings

Where a new or replacement boiler storage vessel is installed, or where existing systems are extended, new or existing pipes that are accessible or exposed as part of the work should be insulated as for new systems.	D 6.4.3 ND 6.4.3
Where alterations are being made to an existing heating system or a new or replacement heating system is being installed in an existing dwelling (or building consisting of dwellings), such alterations should not allow the heating system as a whole to be downgraded in terms of energy efficiency or compromised from a safety point of view.	D 6.3.12

Conservatories

As a conservatory which is heated will be inefficient in energy terms, the general guidance to occupiers is that they should be heated as little as possible.

A conservatory with heating installed should have controls (e.g. such as a TRV to the radiator) to regulate it from the rest of the dwelling.	D 6.3.13

Heating systems

In the design of all buildings, the energy efficiency of the heating plant is an important part of the package of measures which contributes to the overall building carbon dioxide emissions.

Ideally the system should have sufficient zone, time and temperature controls to ensure that the heating system only provides the desired temperature when the building is occupied.

Every building must be designed and constructed in such a way that the heating and hot water service systems installed are energy efficient and are capable of being controlled to achieve optimum energy efficiency.	6.3
In non-domestic buildings:	ND 6.3.0
• a heating system boiler should be correctly sized to ensure energy efficiency; • where future heating capacity is required, consideration should be given to providing additional space for extra plant: and • pipework or ductwork should be configured to allow for the future loading.	

Wet central heating efficiency
Gas and oil

Boilers and appliances installed in a dwelling or building consisting of dwellings should have minimum appliance efficiencies as shown in Table 6.137.

D 6.3.1

Table 6.137   Gas and oil wet central heating efficiency

Heating system	Efficiency
Gas and oil central heating boilers (natural gas or LPG)	SEDBUK 86% i.e. condensing boiler
Gas or oil (twin burner) range cooker central heating boilers (www.rangeefficiency.org.uk)	SEDBUK 75%
Gas-fired fixed independent space heating appliances used as primary space heating	58% gross
Oil-fired fixed independent space heating appliances used as primary space heating	60% gross

Each appliance should be capable of providing independent temperature control in areas with different heating needs. This could be independent or in conjunction with room thermostats or other appropriate temperature sensing devices.

D 6.3.8

Solid fuel

The appliance efficiency should be at least that required for its category as designated by the Heating Equipment Testing Approval Scheme (HETAS) as given in Table 6.138.

D 6.3.2

Table 6.138   Solid fuel wet central heating efficiency

Category	Appliance type	Efficiency (gross calorific value)
D	Open fires with high-output boilers	63% (trapezium) 65% (rectangle)
F	Room heaters and stoves with boilers	67%
G	Cookers with boilers	50% (not more than 3.5 kW) 60% (3.5–7.5 kW)
J	Independent boilers (including pellet and log boilers)	65% (batch fed) 70–75% (automatic anthracite)

Electric

For the most efficient use of electrical supplies it is recommended that an electric flow boiler is used to provide space heating alone, with the bulk of the hot water demand of the dwelling being supplied by a directly heated water heater utilizing 'off-peak' electricity tariffs.

Electric flow boilers should be constructed to meet the requirements of the Low Voltage Directive and Electromagnetic Compatibility Directive, preferably shown by a third party electrical approval, e.g. British Electrotechnical Approvals Board (BEAB) or similar.	D 6.3.3
Vented copper hot water storage vessels associated with the system should meet BS 1566: 2002 or BS 3198: 1981.	D 6.3.3

Controls for wet space heating systems

Independent time and temperature control of heating circuits should be provided along with a boiler interlock (refer to Table 6.139) to ensure that the boiler and pump only operate when there is a demand for heat.	D 6.3.8
Zone controls are not considered necessary for single apartment dwellings.	D 6.3.8
For large dwellings with a floor area over 150 m independent time and temperature control of multiple space heating zones is recommended.	D 6.3.8
Each zone (not exceeding 150 m) should have a room thermostat, and a single multi-channel programmer or multiple heating zone programmers.	D 6.3.8

Table 6.139 Controls for combis, CPSU boilers and electric boilers

Type of control	Means to achieve
Boiler control	• Boiler interlock • Automatic bypass valve
Time control	• Time switch (7 day for space heating) • Full programmer for electric
Room temperature control	• TRVs (all radiators except in rooms with room thermostats or where 'heat bleed' required) • Room thermostat(s)

Table 6.140 Controls for other boilers

Type of control	Means to achieve
Boiler control	• Boiler interlock (for solid fuel as advised by manufacturer) • Automatic bypass valve
Time control	• Full programmer (7 day for space and hot water)
Room temperature control	• As above table
Cylinder control	• Cylinder thermostat plus 2 port valves or a 3 port valve • Separately controlled circuits to cylinder and radiators with pumped circulation
Pump control	• Pump overrun timing device as required by manufacturer

An electric flow boiler should be fitted with a flow temperature control and be capable of modulating the power input to the primary water depending on space and heating conditions.

D 6.3.8.

Dry central heating systems

For a new gas-fired warm air system, the appliance should meet the recommendations of BS EN 778: 1998 or BS EN 1319: 1999, depending on the design of the appliance.

D 6.3.5

Where a gas-fired circulator is incorporated in the warm air unit to provide domestic hot water, it should be of a type that is able to deliver full- and part-load efficiency.

D 6.3.5

For heat pump warm air systems, minimum clearances adjacent to all airflow paths should be maintained.

D 6.3.5

For ground to air and water to air systems, constant water flow should be maintained through the heat pump.

D 6.3.5

Controls for dry space heating systems

Zone controls are not considered necessary for single apartment dwellings.

For large dwellings with a floor area over $150\,m^2$, independent time and temperature control of multiple space heating zones is recommended.

D 6.3.9

Each zone (not exceeding 150 m^2) should have a room thermostat, and a single multi-channel programmer or multiple heating zone programmers.

D 6.3.9

Warm air systems

Ducted warm air heating

Where a flat or maisonette has a storey at a height of more than 4.5 m, or a basement storey, and is provided with a system of ducted warm air heating:

D 2.9.13

- transfer grilles should not be fitted between any room and the entrance hall or stair;
- supply and return grilles should be not more than 450 mm above floor level;
- where warm air is ducted to an entrance hall or stair, the return air should be ducted back to the heater;
- where a duct passes through any wall, floor or ceiling of an entrance hall or stair, all joints between the duct and the surrounding construction should be sealed;
- there should be a room thermostat in the living room, at a height more than 1,370 mm and not more than 1,830 mm, with an automatic control which will turn off the heater, and actuate any circulation fan should the ambient temperature rise to more than 35°C;
- where the system recirculates air, smoke detectors should be provided in every extract duct to cause the recirculation of air to stop and direct all extract air to the outside of the building in the event of fire.

Warm air ducts should be thermally insulated against heat loss unless the use of such pipes or ducts always contributes to the heating demands of the room or space.

D 6.4.1
ND 6.4.1

Electric warm air systems

Time and temperature control should be provided either integral to the heater or external, using either:

D 6.3.9

- a time switch/programmer and room thermostat;
- a programmable room thermostat.

Gas-fired warm air systems (without water heating)

Time and temperature control should be provided using: D 6.3.9

- controls outwith the heater: time switch/programmer and room thermostat, or programmable room thermostat;
- controls integrated with heater: time-switch/programmer and room temperature sensor linked to heater firing and fan speed control.

Heat pumps

Warm air system controls should include: D 6.3.9

- control of room air temperature (integral or otherwise); and
- control for secondary heating (if fitted) (on air to air systems).

System efficiency (warm and hot water)

All heat pumps are at their most efficient when the source temperature is as high as possible, the heat distribution temperature is as low as possible and pressure losses are kept to a minimum.

If radiators are used they should be high-efficiency radiators with high water volume.   D 6.3.4

Supply water temperatures should be in the range 40°C to 55°C to radiators, 30°C to 40°C to an underfloor heating system and 35°C to 45°C to fan coil units.   D 6.3.4

Electrically driven heat pumps should have a coefficient of performance of not less than 2.0 when operating at the heating system design condition.   D 6.3.4

Heat pump controls

In new buildings, where space heating is provided by heating only (heat pumps or reverse cycle heat pumps) the controls shown in Table 6.141 should be observed.   ND 6.3.5

Table 6.141   Heat pump controls

Source	System	Minimum controls package
All types	All technologies	On/off zone control. If the unit serves a single zone, and for buildings with a floor area of 150 m or less the minimum requirement is achieved by default time control.
Air to air	Single package	Controls package for all types above plus heat pump unit controls to include: • control of room air temperature (if not provided externally); • control of outdoor fan operation; • defrost control of external airside heat exchanger; • control for secondary heating (if fitted).
Air to air	Split system Multi-split system Variable refrigerant flow system	Controls package for all types above plus heat pump unit controls to include: • control of room air temperature (if not provided externally); • control of outdoor fan operation; • defrost control of external airside heat exchanger; • control for secondary heating (if fitted).
Water or ground to air	Single package energy transfer systems (matching heating/cooling demand in buildings)	Controls package for all types above plus heat pump unit controls to include: • control of room air temperature (if not provided externally); • control of outdoor fan operation for cooling tower or dry cooler (energy transfer systems); • control for secondary heating (if fitted) on air to air systems; • control of external water pump operation.
Air to water Water or ground to air	Single package Split package	Controls package for all types above plus heat pump unit controls to include: • control of water pump operation (if not provided externally); • control of outdoor fan operation for cooling tower or dry cooler (energy transfer systems); • control for secondary heating (if fitted); • control of external water pump operation.
Gas engine driven heat pumps	Multi-split Variable refrigerant flow	Controls package for all types above plus heat pump unit controls to include: • control of room air temperature (if not provided externally); • control of outdoor fan operation; • defrost control of external airside heat exchanger; • control for secondary heating (if fitted).

For all systems in Table 6.141, additional controls should   ND 6.3.5
include room thermostats (if not integral heat pump) to
regulate the space temperature and interlocked with the
heat pump operation.

Controls for combined warm air and hot water systems

The following controls should be provided: D 6.3.10

- independent time control of both the heating and
  hot water circuits (achieved by means of a cylinder
  thermostat and a timing device, wired such that when
  there is no demand for hot water both the pump and
  circulator are switched off);
- pumped primary circulation to the hot water cylinder;
- a hot water circulator interlock (achieved by means of
  a cylinder thermostat and a timing device, wired such
  that when there is no demand from the hot water both
  the pump and circulator are switched off); and
  - time control by the use of either:
  - a full programmer with separate timing to each circuit;
  - two or more separate timers providing timing control
    to each circuit;
- a programmable room thermostat(s) to the heating
  circuit(s);or
- a time switch/programmer (two channel) and room
  thermostat.

Electric storage heaters

Electric storage heater controls should include: D 6.3.9

- automatic charge control (able to detect the internal
  or external temperature and adjust the charging of the
  heater accordingly); and
- temperature control: heaters (with manual controls for
  adjusting the rate of heat release from the appliance).

Underfloor heating
Hot water underfloor heating

The following controls should be fitted to ensure safe   D 6.3.8
system operating temperatures:

- a separate flow temperature high-limit thermostat
  for warm water systems connected to any high water
  temperature heat supply; and

- a separate means of reducing the water temperature to the underfloor heating system.

The minimum recommendations for room temperature, time and boiler controls are shown in Table 6.142.   D 6.3.8

Table 6.142   Controls for underfloor heating

Type of control	Means to achieve
Room temperature control	• Thermostats for each room (adjacent rooms with similar functions may share • Weather compensating controller
Time control	• Automatic setback of room temperature during unoccupied periods/at night
Boiler control	• Boiler interlock

Bathrooms or en-suites which share a heating circuit with an adjacent bedroom: D 6.3.8

- should provide heat only when the bedroom thermostat is activated;
- should be fitted with an independent towel rail or radiator.

Electric underfloor heating

For electric storage, direct acting systems and under-tile systems programmable room timer/thermostats with manual override feature room controls are recommended for all heating zones, with air and floor (or floor void) temperature sensing capabilities to be used individually or combined.

A storage system should have: D 6.3.9

- anticipatory controllers for controlling low-tariff input charge with external temperature sensing and floor temperature sensing;
- a manual override facility should be available for better user control.

Controls for storage systems with room timer/thermostats should take advantage of low-tariff electricity.   D 6.3.9

Fuel storage

The guidance on oil relates only to its use solely where it serves a combustion appliance providing space heating (or cooking) facilities in a building. There is other legislation covering the storage of oils for other purposes. Heating oils comprise Class C2 oil (kerosene) or Class D oil (gas oil) as specified in BS 2869: 2006.

Every building must be designed and constructed in such a way that an oil storage installation, incorporating oil storage tanks used solely to serve a fixed combustion appliance installation providing space heating (or cooking facilities) in a building, will inhibit fire from spreading to the tank and its contents from within, or beyond, the boundary.	D 3.23 ND 3.23

Oil storage

Every building must be designed and constructed in such a way that an oil storage installation, incorporating oil storage tanks used solely to serve a fixed combustion appliance installation providing space heating facilities in a building will: • reduce the risk of oil escaping from the installation; • contain any oil spillage likely to contaminate any water supply, groundwater, watercourse, drain or sewer; and • permit any spill to be disposed of safely.	D 3.24 ND 3.24

Liquefied petroleum gas storage

The operation of properties where LPG is stored or is in use are subject to legislation enforced by both the Health and Safety Executive (HSE) and the local authority.

Every building must be designed and constructed in such a way that each liquefied petroleum gas storage installation, used solely to serve a combustion appliance providing space heating, water heating (or cooking facilities) will: • be protected from fire spreading to any liquefied petroleum gas container; and • not permit the contents of any such container to form explosive gas pockets in the vicinity of any container.	D 4.11 ND 4.11

Woody biomass storage

By its very nature woody biomass fuel is highly combustible and precautions need to be taken to reduce the possibility of the stored fuel igniting.

To maintain fire-proof storage and prevent back-burning there should be an interruption to the fuel transport system normally by use of a star-feeder or chute for the fuel to fall into the boiler (see BS EN 303-5: 1999).	D 3.23.4 ND 3.23.4
The woody biomass fuel should be stored separately from the boiler that the fuel feeds for fire safety reasons.	D 3.23.4 ND 3.23.4

Boilers

Oil storage tanks used solely to serve a fixed combustion appliance installation providing space heating facilities in a building will:	D 3.23 ND 3.23
• inhibit fire from spreading to the tank and its contents from within, or beyond, the boundary; • reduce the risk of oil escaping from the installation; • contain any oil spillage likely to contaminate any water supply, groundwater, watercourse, drain or sewer; and • permit any spill to be disposed of safely.	D 3.24 ND 3.24
Every building must be designed and constructed in such a way that the products of combustion are carried safely to the external air without harm to the health of any person through leakage, spillage, or exhaust nor permit the re-entry of dangerous gases from the combustion process of fuels into the building.	D 3.20 ND 3.20

Terminal discharge from condensing boilers

 The condensate plume from a condensing boiler can cause damage to external surfaces of a building if the terminal location is not carefully considered. The manufacturer's instructions should be followed.

Solid fuel boilers

These should be thermostatically controlled to reduce the burning rate of the fuel, by varying the amount of combustion air to the fire.	D 6.3.8
For safety reasons, a suitable heat bleed (slumber circuit) from the system should be formed (e.g. a gravity-fed radiator without a TRV or a hot water cylinder that is connected independent of any controls).	D 6.3.8
For hot water systems, unless the cylinder is forming the slumber circuit, a thermostatically controlled valve should be fitted.	D 6.3.8

Boiler plant controls

When installing boiler plant in new buildings the minimum control package shown in Table 6.143 should be observed. ND 6.3.4

Table 6.143 Minimum controls for new boilers or multiple-boiler systems (depending on boiler plant output or combined boiler plant output)

Boiler plant output and controls package	Minimum controls
Less than 100 kW (Package A)	Timing and temperature demand control which should be zone-specific where the building floor area is greater than 150 m². Weather compensation except where a constant temperature supply is required.
100–500 kW (Package B)	Controls package A above plus: Optimal start/stop control is required with night set-back or frost protection outside occupied periods. Boiler with two-stage high/low firing facility or multiple boilers should be installed to provide efficient part-load performance. For multiple boilers, sequence control should be provided and boilers, by design or application, should have limited heat loss from non-firing modules, for example by using isolation valves or dampers. Individual boilers, by design or application, should have limited heat loss from non-firing modules, for example by using isolation valve or dampers.
Greater than 500 kW (Package C)	Controls package A and B above plus the burner controls should be fully modulating for gas-fired boilers or multi-stage for oil-fired boilers.

Electric boiler plant controls

Table 6.144 Electric boiler controls

System	Controls
Boiler temperature control	The boiler should be fitted with a flow temperature control and be capable of modulating the power input to the primary water depending on space heating conditions. Buildings with a total usable floor area up to 150 m² should be divided into at least two zones with independent temperature control. For buildings with a total usable floor area greater than 150 m², sub-zoning of at least two space heating zones must be provided, each having separate timing and temperature controls, by either; • multiple heating zone programmers; or • a single multi-channel programmer.

(Continued)

Table 6.144   (Continued)

System	Controls
Zone temperature control	Separate temperature control of zones within the building, using either: • room thermostats or programmable room thermostats in all zones; • a room thermostat or programmable room thermostat in the main zone and individual radiator; • controls such as Thermostatic Radiator Valves (TRVs) on all radiators in the other zones; or • a combination of the above.
Time control of space and water heating	Time control of space and water heating should be provided by either: • a full programmer with separate timing to each circuit; • two or more separate timers providing timing control to each circuit; or • programmable room thermostat(s) to the heating circuit(s), with separate timing of each circuit.

Additional guidance for hospitals

Boiler houses should: ND 2.B.1

- never be directly below, nor directly adjoin, the operating theatres, intensive therapy units or special care baby units; and
- be provided with a fire suppression system if they are directly below, or directly adjoin, any other hospital department to which patients have access.

Additional enclosed shopping centres

On the operation of the fire alarm (subject to the 4 min   ND 2.C.5
grace period where appropriate) all air moving systems, mains and pilot gas outlets, combustion air blowers and gas, electrical and other heating appliances in the reservoir are shut down.

Efficiency and credits

Appliances installed in a building should be energy efficient.   ND 6.3.1

Designers of non-domestic buildings may wish to consider using heating efficiency credits when designing systems incorporating boilers, warm air heaters, radiant heaters and heat pumps to exceed the minimum efficiency specified.

Examples of how this is achieved are given in Annex 6F to the Scottish Building Regulations.

Appliance efficiency

The following tables recommend efficiencies for:

- minimum boiler seasonal efficiency for heating plant;
- minimum thermal efficiency for gas- and oil-fired warm air systems and radiant heaters;
- coefficient of performance (COP) for heat pumps; and
- maximum permissible specific fan power for air distribution systems.

Table 6.145   Boiler seasonal efficiency in new buildings

Fuel type	Boiler system	Minimum boiler seasonal efficiency (based on gross calorific value)
Gas (natural)	Single	84%
Gas (LPG)	Multiple	80% for any individual boiler and 84% for the overall multi-boiler system
Oil		

Table 6.146   Minimum boiler seasonal efficiency (based on gross calorific value)

Fuel type	Minimum effective heat generating seasonal efficiency (based on gross calorific value)	Effective heat generating seasonal efficiencies and boiler seasonal efficiency in existing buildings
Gas (natural)	84%	80%
Gas (LPG)	85%	81%
Oil	86%	82%

Table 6.147   Gas- and oil-firing warm air systems – minimum thermal efficiency

System	Minimum thermal efficiency (based on gross calorific value)
Gas-firing forced convection heater without a fan complying with EN 621	80%
Fan-assisted gas-firing forced convection complying with EN 1020	80%
Direct gas-firing forced convection heater complying with EN 525	90%
Oil-firing forced convection	80%

Table 6.148   Radiant heaters – minimum thermal efficiency

System	Minimum thermal efficiency (based on gross calorific value)
Luminous (flueless)	85.5%
Non-luminous (flueless)	85.5%
Non-luminous (flued)	73.8%
Multi-burner radiant heaters	80%

Table 6.149   Heat pump coefficient of performance (COP)

System	Minimum heating COP (at design condition)
All types except absorption heat pumps and gas engine heat pumps	2.0
Absorption heat pumps	0.5
Gas engine driven heat pumps	1.0

CHPQA Quality Index (CHP(QI))

CHPQA is a registration and certification scheme which serves as an indicator of the energy efficiency and environmental performance of a CHP (Combined Heat and Power) scheme, relative to the generation of the same amounts of heat and power by separate, alternative means.

The required minimum combined heat and power quality index for all types of CHP should be 105.   ND 6.3.3

There is no minimum combined heat and power quality index specified for electric (primary) heating.

The CHP unit should operate as the lead heat generator and be sized to supply no less than 45% of the annual heating demand.   ND 6.3.3

CHP may be used as the main or supplementary heat source in community heating or district heating schemes.   ND 6.3.3

Space heating controls (general)

If space heating is to be intermittent (and does not make use of off-peak electricity) the system should only operate when the building is normally occupied or is about to be occupied.   ND 6.3.8

Commissioning building services

Commissioning [i.e. in terms of achieving the levels of energy efficiency that the component manufacturers expect from their product(s)] should also be carried out with a view to ensuring the safe operation of the system.

Every building must be designed and constructed in such a way that energy supply systems and building services which use fuel or power for heating the internal environment are commissioned to achieve optimum energy efficiency.   D 6.7   ND 7.7

Table 6.150   Primary and secondary electric heating system controls (other than electric boilers)

System	Controls
Electric warm air system	Time and temperature control, either integral to the heater system or external: • time switch/programmer and room thermostat; or • programmable room thermostat. For buildings with a total usable floor area greater than 150 m more than one space heating circuit should be provided, each having separate timing and temperature control via: • multiple heating zone programmers; or • a single multi-channel programmer.
Electric radiant heater	Zone or occupancy control. Connection to a passive infrared detector (electric radiant heaters can provide zone heating or be used for a scheme). Common electric radiant heaters include the quartz or ceramic type.
Panel/skirting heater	Local time and temperature control heater: Time control provided by: • a programmable time switch integrated into the appliance; or • a separate time switch; or • individual temperature control provided by integral thermostats or by separate room thermostat.
Storage heaters	Charge control (automatic control of input charge with the ability to detect the internal temperature and adjust the charging of the heater accordingly). Temperature control (manual controls for adjusting the rate of heat release from the appliance such as adjustable damper or some other thermostatically controlled means).
Fan/fan convector heaters	Local fan control (a switch integrated into the appliance or a separate remote heaters switch.

Written information

For a domestic building, written information concerning the operation and maintenance of the heating system (together with any decentralized power generation equipment) should be made available for the use of the occupier.   D 6.8.1

> For non-domestic buildings, a logbook containing   ND 6.8.1
> information of energy system operation and maintenance
> (such as building services plant and controls) to ensure
> that the building user can optimize the use of fuel, shall
> be provided.

 CIBSE Technical Memorandum 31 (TM31)
provides guidance on the presentation of a logbook,
and the logbook information should be presented in
this or a similar manner.

Calculating the carbon dioxide emissions for a certificate

> Where a building contains multiple dwellings a rating   D 6.9.1
> is required for each individual dwelling, taking into   ND 6.9.1
> consideration the percentage of low-energy lighting
> and the type of heating that has been installed.

Metering

To enable building operators to effectively manage fuel use, systems should be provided with fuel meters to enable the annual fuel consumption to be accurately measured.

> Where multiple buildings or fire separated units are served   ND 6.10.1
> on a site by a communal heating appliance, the fuel metering
> shall be installed both at the communal heating appliance
> and heat meters at the individual buildings served.
>
> A fuel meter should be installed if a new fuel type or new   ND 6.10.2
> boiler (where none existed previously) is installed.

 See Annex 6E (Guidance concerning efficiency of wet central heating systems) and methods for determining seasonal boiler efficiency of commercial boilers for use in wet central heating systems as follows:

- natural gas boilers;
- liquid petroleum gas (LPG) boilers; and
- oil-firing boilers.

6.23 Electrical safety

According to the aims of the Building (Scotland) Regulations 'the design of all buildings should ensure access and usability and reduce the risk of accident'. To guarantee electrical safety, buildings should be designed to ensure that electrical installations are safe in terms of the hazards likely to arise from defective installations, namely fire and loss of life or injury from electric shock or burns.

In time, this electrical section will probably be updated similarly to the current English/Welsh Building Regulations.

6.23.1 Requirements

Fire

> *Every building must be designed and constructed in such a way that in the event of an outbreak of fire within the building, illumination is provided to assist in escape.*   *2.10*
>
> *Every building except a building which:*   *2.11*
>
> - *is a dwelling;*
> - *is a residential building; or*
> - *is an enclosed shopping centre,*
>
> *must be designed and constructed in such a way that in the event of an outbreak of fire within the building, the occupants are alerted to the outbreak of fire.*

Environment

> *Every building must be designed and constructed in such a way that natural lighting is provided to ensure that the health of the occupants is not threatened.*   *D 3.16*
>
> Standard 3.12 only applies to a dwelling.
>
> *Every building must be designed and constructed in such a way that sanitary facilities are provided that cause no threat to the health and safety of occupants or visitors.*   *3.12*

Safety

> *Every building must be designed and constructed in such a way that:*   *4.1*
>
>   *4.2*
>
> - *all occupants and visitors are provided with safe, convenient and unassisted means of access to and throughout the building.*

- *the electrical installation does not:*   4.5
  - *threaten the health and safety of the people in, and around, the building;*
  - *become a source of fire.*

This standard does not apply to an electrical installation:

- serving a building or any part of a building to which the Mines and Quarries Act 1954 or the Factories Act 1961 applies; or
- forming part of the works of an undertaker to which regulations for the supply and distribution of electricity made under the Electricity Act 1989.

Every building must be designed and constructed in such a way that: 4.6

- *electric lighting points and socket outlets are provided to ensure the health, safety and convenience of occupants and visitors;*
- *it is provided with aids to assist those with a hearing impairment;*   4.7
- *electrical fixtures can be operated safely;*   4.8
- *protection is provided for people in, and around, the building from the danger of severe burns or scalds from the discharge of steam or hot water.*   4.9

Energy

Every building must be designed and constructed in such a way that: 6.2

- *an insulation envelope is provided which reduces heat loss;*   6.3
- *heating and hot water service systems installed are energy efficient and are capable of being controlled to achieve optimum energy efficiency;*   6.5
- *any artificial or display lighting installed is energy efficient and is capable of being controlled to achieve optimum energy efficiency.*

This standard does not apply to:

- process and emergency lighting components in a building;
- communal areas of domestic buildings;
- alterations in dwellings.

- *energy supply systems and building services which use*   6.7
  power for heating, lighting, ventilating and cooling
  the internal environment and heating the water, are
  commissioned to achieve optimum energy efficiency.

  This standard does not apply to:

- major power plants serving the National Grid;
- the process and emergency lighting components of a
  building;
- heating provided solely for the purpose of frost
  protection;
- energy supply systems used solely for industrial and
  commercial processes, leisure use and emergency use
  within a building.

6.23.2 Meeting the requirements

Electrical safety

The overall requirement of this standard (which applies to fixed installations in buildings) is to ensure that electrical installations are safe in terms of the hazards likely to arise from defective installations, namely fire, electric shock and burns or other personal injury.

An installation consists of the electrical wiring and associated components and fittings, including all permanently secured equipment, but excluding portable equipment and appliances.

Every building must be designed and constructed in such a   D 4.5
way the electrical installation does not:

- threaten the health and safety of the people in, and around,
  the building;
- become a source of fire.

  This standard does not apply to an electrical installation:

- serving a building or any part of a building to which the
  Mines and Quarries Act 1954 or the Factories Act 1961
  applies;
- forming part of the works of an undertaker to which
  regulations for the supply and distribution of electricity
  made under the Electricity Act 1989.

Installations should: D 4.5.0
                         ND 4.5.0

- safely accommodate any likely maximum demand; and
- incorporate appropriate automatic devices for protection
  against overcurrent or leakage;

- provide means of isolating parts of the installation or equipment connected to it, as are necessary for safe working and maintenance.

Every building must be designed and constructed in such a way that:	D 4.6

- electric lighting points and socket outlets are provided to ensure the health, safety and convenience of occupants and visitors;
- it is provided with aids to assist those with a hearing impairment; — ND 4.7
- electrical fixtures can be operated safely; — D 4.8
- the artificial or display lighting installed is energy efficient and is capable of being controlled to achieve optimum energy efficiency; — D 6.5
- energy supply systems and building services which use power for heating, lighting, ventilating and cooling the internal environment and heating the water, are commissioned to achieve optimum energy efficiency. — D 6.7

Emergency lighting

Emergency lighting is lighting designed to come into, or remain in, operation automatically in the event of a local and general power failure.

Emergency lighting should be installed in buildings or parts of a building considered to be at higher risk such as:	D 2.10.3 ND 2.10.3

- high-rise buildings, buildings with basements or in rooms where the number of people is likely to exceed 60;
- in a protected zone and an unprotected zone in a building with any storey at a height of more than 18 m;
- in a room with an occupancy capacity of more than 60;
- in an underground car park;
- in a protected zone or unprotected zone serving a basement storey;
- in a place of special fire risk (other than one requiring access only for the purposes of maintenance) and any protected zone or unprotected zone serving it;
- in any part of an air-supported structure;
- in a protected zone or unprotected zone serving a storey which has at least two storey exits in the following buildings:
- entertainment, assembly, factory, shop, multi-storey storage (Class 1), single-storey storage (Class 1) with a floor area more than 500 m²;

- a protected zone or unprotected zone serving a storey in a multi-storey non-residential school;
- a protected zone or unprotected zone serving any storey in an open-sided car park.

Emergency lighting should be installed in accordance with BS 5266: Part 1: 1999.	D 2.10.3 ND 2.10.3
In the case of a building with a smoke and heat exhaust ventilation system, the emergency lighting should be sited below the smoke curtains or installed so that it is not rendered ineffective by smoke filled reservoirs.	D 2.10.3 ND 2.10.3
Basements should (ideally) be installed with a mechanical smoke and heat extraction system incorporating a powered smoke and heat exhaust ventilator which has a capacity of at least 10 air changes per hour.	ND 2.14.7

Escape route lighting

Escape routes should be illuminated to aid the safe evacuation of a building in an emergency.	D 2.10.1 ND 2.10.1

Whilst in non-domestic buildings, it is mandatory that escape routes should be illuminated to aid the safe evacuation of a building in an emergency, specifically dedicated escape lighting is not necessary within dwellings as it is assumed the occupants will have a degree of familiarity with the layout, and escape routes only begin at the door to the dwelling.

However, in buildings containing flats and maisonettes, the common escape routes should be illuminated to assist the occupants of the building to make their way to a place of safety.

Every part of an escape route should have artificial lighting supplied by a fire protected circuit that provides a level of illumination not less than that recommended for emergency lighting.	D 2.10.1 ND 2.10.1
Where artificial lighting serves a protected zone, it should be via a protected circuit separate from that supplying any other part of the escape route.	D 2.10.1 ND 2.10.1
Artificial lighting supplied by a protected circuit need not be provided if a system of emergency lighting is installed.	D 2.10.1 ND 2.10.1

Note: A protected circuit is a circuit originating at the main incoming switch or distribution board, the conductors of which are protected against fire.

Escape routes should be capable of being illuminated when the building is in use.	D 2.10.2 ND 2.10.2

Escape lighting – enclosed shopping centres

An enclosed shopping centre should be provided with emergency lighting in all mall areas and all protected zones and unprotected zones.	ND 2.C.4
Emergency lighting should be installed so that it is not rendered ineffective by smoke-filled reservoirs.	ND 2.C.4

Escape lighting – residential care buildings

Every protected zone or unprotected zone should be provided with emergency lighting.	2A6
Emergency lighting should be installed in: • a room with an occupancy capacity of more than 10 (and any protected zone or unprotected zone serving such a room); • a protected zone or unprotected zone serving a storey which has two exits, other than a storey in a building not more than two storeys high with a combined floor area of not more than 300 m² and an occupancy capacity of not more than 10; • a protected zone or unprotected zone in a single stair building of two storeys or more and an occupancy capacity of 10 or more.	ND 2.A.4

Escape lighting – hospitals

Essential lighting circuits should be installed throughout a hospital and designed to provide not less than 30% of the normal lighting level.	ND 2.B.4
In an area where a 15 s response time would be considered hazardous (e.g. a stairway), emergency lighting should be provided by battery back-up giving a response time of not more than 0.5 s.	ND 2.B.3

The distribution boards for essential and non-essential
circuits may be in the same location but should be in
separate cabinets.   ND 2.B.4

Natural lighting in domestic buildings

In the case of domestic buildings, every building must be designed and constructed so that an adequate standard of daylighting should be available in
all habitable rooms in dwellings to allow domestic activities to be carried out
conveniently and safely.

Every domestic building must be designed and constructed   D 3.16
in such a way that natural lighting is provided to ensure
that the health of the occupants is not threatened.

Every apartment should have a translucent glazed opening,   D 3.16.1
or openings, of an aggregate glazed area equal to at least
1/15th of the floor area of the apartment and located in an
external wall or roof or in a wall between the apartment
and a conservatory.

Emergency exits

Locks on emergency exits should be capable of being released by people
who may be unfamiliar with the building and have received no training in the
emergency procedures or the types of exit devices used in the building (e.g.
panic exit devices operated by a horizontal bar to BS EN: 1125: 1997).

Electrically powered locks

Electrically powered locks for emergency exits should **not**   ND 2.9.15
be installed:

- on a protected door serving the only escape stair in the
  building;
- a protected door serving a fire-fighting shaft;
- on any door which provides the only route of escape
  from the building or part of the building.

Electrically powered locks on exit doors and doors across   ND 2.9.15
escape routes **may** be installed in buildings:

- which are inaccessible to the general public;
- when the building is accessible to the general public,
  the aggregate occupancy capacity of the rooms or
  storeys served by the door does not exceed 60 persons.

Staff in such areas will need to be trained both in the emergency procedures and in the use of the specific emergency devices fitted (see clause 2.0.4).

Electrically powered locks should return to the unlocked position: ND 2.9.15

- on operation of the fire alarm system;
- on loss of power or system error;
- on activation of a manual door release unit.

Where a locking mechanism is designed to remain locked in the event of a power failure or system error, the mechanism should not be considered appropriate for use on exit doors and doors across escape routes.   ND 2.9.15

Where a secure door is operated by a code, combination, swipe or proximity card, biometric data or similar means, it should also be capable of being overridden from the side approached by people making their escape.   ND 2.9.15

Self-closing fire doors can be fitted with hold-open devices as specified in BS 5839: Part 3: 1988 **provided** the door is **not** an emergency door, a protected door serving the only escape stair in the building (or the only non-escape stair serving part of the building) or a protected door serving a fire-fighting shaft.

Obstacles

Revolving doors and automatic doors (as they can obstruct the passage of persons escaping) should not be placed across escape routes unless they are designed in accordance with BS 7036: 1996 and open automatically from any position in the event of actuation of any fire alarm in the fire alarm zone within which the door is situated.   ND 2.9.7

Smoke alarms
Smoke alarms – dwellings

If a fire does begin in a dwelling then early detection and warning to the occupants can play a vital role in increasing their chances of escape. This is particularly important as the occupants may well be asleep.

A dwelling where no storey is more than $200\,m^2$ should be provided with: D 2.11.1

- one or more smoke alarms located on each storey; plus
- a standby supply to BS 5446: Part 1: 2000.

A dwelling with any storey area more than 200 m² should be provided with a fire detection and alarm system designed and installed in accordance with BS 5839: Part 6: 2004 for a Grade C Type LD2 installation.   D 2.11.3

The standby power supply for the smoke alarm should take the form of a primary battery, a secondary battery or a capacitor.   D 2.11.2

The capacity of the standby supply should be sufficient to power the smoke alarm when the mains power supply is off for at least 72 h while giving a visual warning of mains power supply being off.   D2.11.2

There should remain sufficient capacity to provide a warning of smoke for a further 4 min.   D 2.11.2

An audible warning should be given at least once every minute where the capacity of the standby power supply falls below the recommended standby duration when the mains power supply is on and persist for at least 30 days when the mains power supply is off.   D 2.11.2

A smoke alarm should be ceiling mounted and located: D 2.11.2

- in a circulation area which will be used as a route along which to escape;
- not more than 7 m from the door to a living room or kitchen; and
- not more than 3 m from the door to a room intended to be used as sleeping accommodation;
- where the circulation area is more than 15 m long;
- not more than 7.5 m from another smoke alarm on the same storey;
- at least 300 mm away from any wall or light fitting, heater or air-conditioning outlet;
- on a surface which is normally at the ambient temperature of the rest of the room or circulation area in which the smoke alarm is situated.

Where more than one smoke alarm is installed in a dwelling they should be interconnected so that detection of a fire by any one of them operates the alarm signal in all of them.   D 2.11.2

A smoke alarm should be permanently wired to a circuit.   D 2.11.2

The mains supply to the smoke alarm should take the form of either:

- an independent circuit at the dwelling's main distribution board, in which case no other electrical equipment should be connected to this circuit (other than a dedicated monitoring device installed to indicate failure of the mains supply to the smoke alarms); or
- a separately electrically protected, regularly used local lighting circuit.

D 2.11.2

Smoke alarms may be interconnected by 'hard wiring' on a single final circuit.

D 2.11.2

Any smoke alarm in a dwelling which forms part of residential accommodation with a warden or supervisor should have a connection to a central monitoring unit so that in the event of fire the warden or supervisor can identify the dwelling concerned.

D 2.11.2

Assistance alarm – sanitary accommodation

Accessible sanitary accommodation should be fitted with an assistance alarm which:

- can be operated or reset when using a sanitary facility and which is also operable from floor level;
- should have an audible tone, distinguishable from any fire alarm, together with a visual indicator, both within the sanitary accommodation and outside in a location that will alert building occupants to the call.

ND 3.12.7

Automatic fire detection and alarm systems – non-dwellings

Owing to the special fire precautions that are required within residential care buildings, hospitals and enclosed shopping centres, it is recommended that automatic fire detection is installed.

Automatic fire shutters should not be activated by a fire alarm signal from a remote location or outwith the fire alarm zone.

ND 2.1.14

Fire detection and alarm system – buildings in different occupation

For buildings in different occupation: ND 2.2.2

- a separating wall or separating floor should be provided between parts of a building where they are in different occupation;
- if each unit is under the control of an individual tenant, employer or self-employed person, then separating walls and separating floors should be provided between the areas intended for different occupation;
- if this is not possible, then the building should have a common fire alarm system/evacuation strategy and the same occupancy profile.

Residential buildings (other than residential care buildings and hospitals)

Residential buildings present particular problems in the event of an outbreak of fire because the occupants may be asleep and will not be aware that their lives may be at risk. A higher level of automatic fire detection coverage is recommended in residential care buildings and hospitals to give occupants and staff the earliest possible warning of an outbreak of fire and allow time for assisting occupants to evacuate the building in an emergency.

For this reason,

Residential buildings (other than residential care buildings and hospitals) should be provided with an automatic fire detection and alarm; ND 2.11.1

- the fire alarm should be activated upon the operation of:
  - manual call points;
  - automatic detection; or
  - the operation of any automatic fire suppression system installed;
- the audibility level of the fire alarm sounders should be as specified in BS 5839: Part 1: 2002.

In shared residential accommodation that is designed to provide sleeping accommodation for not more than six persons and has no sleeping accommodation below ground level or above first floor level, a domestic system comprising smoke alarms with a standby supply may be installed.

Residential care buildings

An automatic fire detection and alarm system (designed ND 2.A.5
and installed in accordance with the guidance in BS 5839:
Part 1: 2002 Category L1) should be installed in every
residential care building.

Detection need not be provided in the following locations: ND 2.A.5

- sanitary accommodation;
- a lockable cupboard with a plan area not more than
  1 m^2;
- a void and roof space which contain only mineral insulated wiring, or wiring laid on metal trays, or in metal
  conduits, or metal/plastic pipes used for water supply,
  drainage or ventilating ducting.

The fire alarm should be activated on the operation of ND 2.A.5
manual call points, automatic detection or the operation of
the automatic life safety fire suppression system.

The building should be divided into detection zones not ND 2.A.5
extending beyond a single compartment.

The audibility level of the fire alarm sounders should be ND 2.A.5
as specified in BS 5839: Part 1: 2002 except in a place of
lawful detention including prisons, the alarm need not be
sounded throughout the entire building.

A fire alarm control panel should be provided at the main ND 2.A.5
entrance, or a suitably located entrance to the building
agreed with the relevant authority.

On the actuation of the fire alarm, a signal should be ND 2.A.5
transmitted automatically to the fire service, either directly
or by way of a remote centre.

Mixed-use buildings

An escape stair which serves a flat or maisonette which is D 2.9.16
ancillary to a non-domestic building, may communicate
with the non-domestic accommodation provided that
an alarm and detection system which is designed in
accordance with BS 5839: Part 1: 2002 is installed in
the common areas.

Hospitals

An automatic fire detection and alarm system should be installed in every hospital to ensure that staff and patients are given the earliest possible warning of the outbreak of fire anywhere in the building.	ND 2.B.5
Detection need not be provided in the following locations:	ND 2.B.5

- sanitary accommodation;
- a lockable cupboard with a plan area not more than $1\,m^2$;
- a void and roof space which contain only mineral insulated wiring, or wiring laid on metal trays, or in metal conduits, or metal/plastic pipes used for water supply, drainage or ventilating ducting.

Manual fire alarm call points should be located and installed in accordance with BS 5839: Part 1: 2002.	ND 2.B.5
The fire alarm should be activated on the operation of manual call points, automatic detection or the operation of any automatic fire suppression system installed.	ND 2.B.5
The building should be divided into detection zones not extending beyond a single sub-compartment.	ND 2.B.5
The audibility level of the fire alarm sounders should follow the guidance in BS 5839: Part 1: 2002. However, in a hospital department to which patients have access, the audibility need only be $55\,dB(A)$ or $5\,dB(A)$ above the level of background noise, whichever is greater.	ND 2.B.5

- A main fire alarm control panel is provided at the main   ND 2.B.5
  entrance, or a suitably located secondary entrance to
  the building; and
- repeater panels should be provided at all other fire
  service access points.

On the actuation of the fire alarm, a signal should be transmitted automatically to the fire service, either directly or by way of a remote centre.	ND 2.B.5
In the case of a hospital designed to accommodate not more than 10 residents, one or more smoke alarms should be installed.	ND 2.B.5

Enclosed shopping centres

Automatic fire detection and alarm systems in enclosed shopping centres can increase significantly the level of safety of the occupants.

An enclosed shopping centre should be provided with an automatic fire detection and alarm system, designed and installed in accordance with the following recommendations: ND 2.11.4
ND 2.C.5

- the fire alarm should be activated upon the operation of the sprinklers; or
  - manual call points; or
  - on the activation of the alarm in a shop; or
  - on the activation of the alarm anywhere other than in an individual shop; or
  - on activation of sprinklers anywhere within the shopping centre;
- all areas of the shopping centre, including shops, should be alerted using a voice alarm system which follows the guidance in BS 5839: Part 8: 1998, however, individual shops may use conventional sounders within the shop itself.

The fire alarm system should be interfaced with other fire safety systems, to operate automatically in the correct zones.   ND 2.C.5

On the operation of the fire alarm: ND 2.C.5

- all escalators should come to a controlled halt and lifts should return to the ground storey (or exit level);
- all systems within the mall or shops which play amplified music are silenced;
- any smoke dampers installed to prevent the siphoning of smoke are activated; and
- (subject to the 4 min grace period where appropriate) all air moving systems, mains and pilot gas outlets, combustion air blowers and gas, electrical and other heating appliances in the reservoir are shut down.

The main fire alarm system control panel shall be installed within the control room and indicator (or repeater) panels shall be provided at each of the fire-fighting access points.   ND 2.C.5

Smoke and Heat Exhaust Ventilation Systems – shopping centres

A Smoke and Heat Exhaust Ventilation System (SHEVS) should be installed in the mall of an enclosed shopping centre and in shops with a storey area more than 1,300 m².

SHEVS should be designed in accordance with the
following where appropriate:

- the underside of the mall roof should be divided into
  smoke reservoirs, each of which should be not more
  than 2,000 m^2 in area and at least 1.5 m deep measured
  to the underside of the roof or to the underside of any
  high-level plant or ducts within the smoke reservoir or
  the underside of an imperforate suspended ceiling;
- smoke reservoirs should be formed by fixed or
  automatically descending smoke curtains which are
  no greater than 60 m apart, measured along the
  direction of the mall;
- smoke should not be allowed to descend to a height
  of less than 3 m above any floor level;
- each smoke reservoir should be provided with the
  necessary number of smoke ventilators or extract fans
  to extract the calculated volume of smoke produced,
  spaced evenly throughout the reservoir;
- where mechanical extraction is used, there should
  be spare fan capacity equivalent to the largest single
  fan in the reservoir which will operate automatically
  on the failure of any one of the fans, or which runs
  concurrently with the fans;
- any fans, ducts and reservoir screens provided should
  be designed to operate at the calculated maximum
  temperature of the smoke within the reservoir in which
  they are located, but rated to a minimum of 300°C for
  30 min;
- structures supporting any fans, ducts or reservoir
  screens should have the same performance level as the
  component to be supported;
- the fans or ventilators within the affected smoke
  reservoirs should operate:
  - on the actuation of any automatic fire suppression
    system; or
  - on the actuation of the smoke detection system
    within the reservoir; or
  - on the operation of more than one smoke detector
    anywhere in the shopping centre; or
  - following a delay not exceeding 4 min from
    initiation of the first fire alarm signal anywhere
    in the shopping centre;

- replacement air should be provided automatically on the operation of the ventilation or exhaust system at a level at least 0.5 m below the calculated level of the base of the smoke layer;
- any power source provided to any elements of the smoke and heat exhaust ventilation system should be connected by mineral insulated cables or by cables which are code A category specified in BS 6387: 1994 or by cables protected from damage to the same level;
- an automatically switched standby power supply provided by a generator should be connected to any fans provided as part of the smoke and heat ventilation system capable of simultaneously operating the fans in the reservoir affected and any of the two adjacent reservoirs;
- simple manual overriding controls for all smoke exhaust, ventilation and air input systems should be provided at all fire service access points and any fire control room provided;
- where outlets are provided with weather protection, they should open on the activation of the fan(s) or ventilators;
- smoke from areas adjoining the smoke reservoirs should only be able to enter one reservoir;
- where there is an openwork ceiling, the free area of the ceiling should not be less than 25% of the area of the smoke reservoir, or, for natural ventilation, 1.4 times the free area of the roof mounted fire ventilator above (3 times where the height from floor to roof ventilator is more than 12 m), whichever free area is the greater, and be evenly distributed to prevent an unbalanced air flow into the reservoir; and
- when a natural ventilation system is used and the smoke reservoir includes a suspended ceiling, other than an openwork ceiling, the free area of the ventilator opening in the suspended ceiling, or any ventilator grille in the ceiling, should not be:
  - less than 1.4 times (3 times where the height from floor to roof ventilator is more than 12 m) that of the roof mounted fire ventilator above in the case of a ventilator opening; or
  - 2 times (3.5 times where the height from floor to roof ventilator is more than 12 m) for any ventilator grille.

Voice alarms

Where a building or part of a building is designed on the basis of vertically phased evacuation, consideration should be given to the installation of an automatic fire detection and voice alarm system.   ND 2.9.11

Electrical installations and fixtures

Electricity, when properly used, is a safe and convenient source of energy for heat, light and power within buildings. However, misuse may lead to significant harm to individuals and buildings alike.

Risk of fire from an electrical installation should be minimized and should not create the risk of fire, burns, shock or other injury to people.

An electrical installation should be designed, constructed, installed and tested such that it is in accordance with the recommendations of BS 7671: 2001, as amended.   D 4.5.1
ND 4.5.1

Outlets and controls of electrical fixtures

The location of any controls and outlets of electrical fixtures located on a wall or other vertical surface that is intended for operation by the occupants of a building should be: D 4.8.5
ND 4.8.6

• installed in position that allows safe and convenient use; and
• unless incorporating a restrictor or other protective device for safety reasons, controls should be operable with one hand.

Outlets and controls of electrical fixtures and systems such as sockets, switches, fire alarm call points and timer controls or programmers, etc., should be positioned at least 350 mm from any internal corner, projecting wall or similar obstruction and (unless the need for a higher location can be demonstrated) not more than 1.2 m above floor level.   D 4.8.5
ND 4.8.6

Light switches should be positioned at a height of between 900 mm and 1.1 m above floor level.   D 4.8.5
ND 4.8.6

Standard switched or unswitched socket outlets and outlets for other services such as telephone or television should be positioned at least 400 mm above floor level.   D 4.8.5
ND 4.8.6

Above an obstruction, such as a worktop, fixtures should be at least 150 mm above the projecting surface.   D 4.8.5
ND 4.8.6

Where socket outlets are concealed, such as to the rear of white goods in a kitchen, separate switching should be provided in an accessible position, to allow appliances to be isolated.	D 4.8.5 ND 4.8.6

Extra-low-voltage installations

Any circuit which is designed to operate at or below extra-low voltage should be protected against both direct and indirect contact with any other circuit operating at higher than extra-low voltage.	D 4.5.2 ND 4.5.2
Extra-low voltage is defined as not more than 50 V alternating current or 120 V direct current, measured between conductors or to earth. This might include installations for alarm or detection purposes, or for transmission of sound, vision, data or power.	D 4.5.2 ND 4.5.2

Installations operating above low voltage

Any circuit which is designed to operate at a voltage higher than low voltage should be provided with a cut-off switch for use in emergency in accordance with the recommendations of BS 7671: 2001, as amended.	D 4.5.3 ND 4.5.3
Such installations are not usual in domestic buildings.	
A fireman's switch, in a conspicuous position, should be provided to any circuit supplying exterior electrical installations or internal discharge lighting installations (including luminous tube signage) operating at a voltage exceeding low voltage.	D 4.5.3 ND 4.5.3

 Note: Low voltage is defined as not more than 1,000 V alternating current or 1,500 V direct current, measured between conductors or not more than 600 V alternating current or 900 V direct current between conductors and earth.

Artificial and display lighting

Every building must be designed and constructed in such a way that the artificial or display lighting installed is energy efficient and is capable of being controlled to achieve optimum energy efficiency.

This standard does not apply to:

- process and emergency lighting components in a building;
- communal areas of domestic buildings; or
- alterations in dwellings.

When designing a lighting system consideration should be given to the advances in lighting technology, particularly with light emitting diodes (LED) technology; and the system design should accommodate future upgrading with minimal disruption to the building fabric and services. | ND 6.5.0

A building with a floor area of more than 50 m^2 and installed with artificial lighting (other than emergency lighting or specialist process lighting which is intended to illuminate specialist tasks within a space, rather than the space itself) should have general purpose artificial lighting systems which are designed to be energy efficient. | ND 6.5.1

Display lighting systems such as artificial lighting that: | ND 6.5.2

- highlights a merchandising display (e.g. in retail premises);
- highlights an exhibit (e.g. in a museum or art gallery); or
- is used in spaces intended for public entertainment (e.g. dance halls, auditoria and cinemas), but
- does not include any specialist process lighting within the space;
- and which is installed in a building with a floor area of more than 50 m,

should be designed to be energy efficient.

Controls for artificial and display lighting

Every artificial lighting system in a building that has a floor area of more than 50 m^2 should have controls which encourage the maximum use of daylight and minimize the use of artificial lighting during the times when rooms or spaces are unoccupied. | ND 6.2

There are a number of alternative ways (or a combination of ways) that will achieve the objectives of the standard for general artificial lighting, such as ensuring that:

- the radial distance on plan from any local switch to any luminaire it controls is not more than 6 m, or twice the height of the luminaire above the floor if this is greater;
- switches are readily accessible (e.g. located on circulation routes);

- switches can be operated by ultrasonic, infrared or other remote control handsets;
- if lighting rows are located adjacent to windows, they can be controlled by photocells which monitor daylight and adjust the level of artificial lighting accordingly, by either switching or dimming;
- automatic switching turns the lighting on or off when it senses the presence or absence of occupants.

Buildings used for industrial, retail, assembly, entertainment or other similar uses and in areas where continuous lighting is required by the occupants of the building during hours of operation, the control can be by way of time switching or daylight-linked photo-electric switching.	ND 6.2
For managed spaces in these buildings (such as cinema/theatre, sports hall, restaurant, passenger terminal, museum/gallery, foyer, large kitchen and shop) centralized manual switching may be used.	
Dedicated circuits should be provided so that display lighting can be switched off (e.g. using timers, etc.) at times when people will not be inspecting exhibits or merchandise or occupying the spaces used for public entertainment.	ND 6.2

Air-supported structures

An air-supported structure is a structure which has a space-enclosing, single-skin membrane anchored to the ground and kept in tension by internal air pressure so that it can support applied loading.

An air-supported structure should be designed and constructed so that inflation equipment includes a standby power system which will start up automatically on any failure of the main power supply, is independent of the main power supply, and includes weather protected, non-return dampers in the ducts, outside the structure.	ND 2.9.28

Unvented hot water storage systems

An unvented hot water storage system should be designed and installed to prevent the temperature of the stored water at any time exceeding 100°C and to provide protection from malfunctions of the system.

Where the system has a power input of less than 45 kW, safety devices (such as thermal cut-outs and discharge pipework) should be provided.	ND 4.9.3

Where the system has a power input greater than 45 kW, ND 4.9.3
safety devices should also include temperature or
combined temperature/pressure relief valves capable of
a combined discharge rating at least equal to the power
input of the system.

Pipework should be installed so that any discharge will be D 4.9.3
directed away from electrical components should the ND 4.9.4
discharge outlet become blocked.

Powered doors

Use of a powered door will improve accessibility at an entrance to a building.
However, care should be taken to ensure that the form of such a door does not
present any additional hazard or barrier to use.

Powered doors should be controlled by either an automatic ND 4.1.8
sensor, such as a motion detector, or a manual activation
device, such as a push-pad.

Any manual control should be located at a height of ND 4.1.8
between 750 mm and 1.0 m above ground level and at least
1.4 m from the plane of the door or, where the door opens
towards the direction of approach, 1.4 m from the front
edge of the open door leaf.

A manual control should contrast visually with the surface ND 4.1.8
on which it is mounted.

A powered door should have: ND 4.1.8

- signage to identify means of activation and warn of
  operation;
- sensors to ensure doors open swiftly enough and
  remain open long enough to permit safe passage in
  normal use and to avoid the door striking a person
  passing through;
- if a swing door, identification of any opening vertical
  edge using visual contrast;
- if on an escape route, or forming a lobby arrangement
  where the inner door is also powered or lockable, doors
  that, on failure of supply will either fail 'open' or have
  a break-out facility permitting doors to be opened in
  direction of escape;
- guarding to prevent collision with, or entrapment by a
  door leaf, except where such guarding would prevent
  access to the door.

Additional guarding may be needed to prevent collision with, or entrapment by, a powered door leaf.	D 4.8.1 ND 4.8.1

 A vision panel is not needed to a powered door controlled by automatic sensors or where adjacent glazing offers an equivalent clear view to the other side of a door.

Electrically operated hold-open devices

Self-closing fire doors can be fitted with hold-open devices as specified in BS 5839: Part 3: 1988 **provided** the door is **not** an emergency door, a protected door serving the only escape stair in the building (or the only non-escape stair serving part of the building) or a protected door serving a fire-fighting shaft.

Electrically operated hold-open devices should deactivate: • on operation of an automatic fire alarm system; • on any loss of power to the hold-open device, apparatus or switch; and • on operation of a manually operated switch fitted in a position at the door.	D2.2.9 ND 2.1.14

Lifting devices between storeys of a non-domestic building

Any lifting device should be designed and installed to include the following general provisions: • a clear landing at least 1.5 m by 1.5 m in front of any lift entrance door; • controls on each level served, between 900 mm and 1.1 m above the landing, and within the lift car on a side wall between 900 mm and 1.1 m above the car floor and at least 400 mm from any corner; • on the landing of each level served, tactile call buttons and visual and tactile indication of the storey level; and • lift doors, handrails and controls that contrast visually with surrounding surfaces; • a signalling system which gives notification that the lift is answering a call made from a landing; • a means of two-way communication, operable by a person with a hearing impairment, that allows contact with the lift if an alarm is activated, together with visual indicators that an alarm has been sounded and received.	ND 4.2.7

In addition to general provisions for lifting devices, a powered lifting device should:

ND 4.2.7

- if serving a storey to which the public have access, have a platform size of 1,100 mm wide by 1,400 mm deep and a clear opening width to any door of 850 mm;
- if serving any other storey, have a platform size of at least 1,050 mm wide by 1,250 mm deep and a clear opening width to any door of 800 mm;
- be fully contained within a liftway enclosure;
- have a operational speed of not more than 0.15 m/s;
- be operated by a continuous pressure type control, of a form operable by a person with limited manual dexterity;
- be provided with a horizontal handrail, of a size and section that is easily gripped, 900 mm above the floor fitted to at least one side of the platform; and
- be provided with permanent and clear operating instructions located adjacent to or within the platform.

Where a powered lifting platform is used, this may be without a liftway enclosure where vertical travel is not more than 2.0 m.

Intercom and entry control systems

Where an intercom or entry control system is provided:

ND 4.1.7

- it should be positioned between 900 mm and 1.2 m above floor level;
- it should include an inductive coupler compatible with the 'T' setting on a personal hearing aid, together with a visual indicator that a call made has been received;
- controls should contrast visually with surrounding surfaces and any numeric keypad should follow the 12-button telephone convention, with an embossed locater to the central '5' digit.

Assistance alarms in residential buildings

Within residential buildings, such as hotels and halls of residence, sleeping accommodation which is accessible to a wheelchair user should be provided to at least one bedroom in 20 (or part thereof) and this accommodation should include an assistance alarm that can be operated or reset from a bedspace, and which is also operable from floor level.

ND 4.2.9

The alarm should have an audible tone distinguishable from a fire alarm and a visual indicator provided both within and outside the bedroom and should also give an alert at a location where staff will be on duty.

Hearing enhancement systems – non-domestic buildings

A variety of hearing enhancement systems are commonly used within buildings including induction loops, infrared and radio transmission systems. The type of system and performance sought should be considered at an early stage in the design process.

Every building must be designed and constructed in such a way that it is provided with aids to assist those with a hearing impairment.	ND 4.7

Socket outlets in bathrooms and rooms containing a shower

In a bathroom or shower room, an electric shaver power outlet, complying with BS EN 60742: 1996, may be installed. Other than this, there should be no socket outlets and no means for connecting portable equipment.	D 4.5.4 ND 4.5.4
Where a shower cubicle is located in a room, such as a bedroom, any socket outlet should be installed at least 3 m from the shower cubicle.	D 4.5.4 ND 4.5.4

Electrical fixtures in domestic buildings

During daylight, lighting levels within a building are generally much less than those outdoors and it is important to avoid hazardous situations that may be created by the nature of the lighting itself including insufficient light sources, glare, gloom and shadows.

Today, with ever more electrical appliances being used in homes, an adequate provision of power points reduces the possibility of both overloading of individual sockets, risking fire and the creation of trip hazards from use of extension cabling.

Every building must be designed and constructed in such a way that electric lighting points and socket outlets are provided to ensure the health, safety and convenience of occupants and visitors.	D 4.6

Lighting

A dwelling should have an electric lighting system providing at least one lighting point to every circulation space, kitchen, bathroom, toilet and other space having a floor area of $2\,m^2$ or more.	D 4.5.1
Any lighting point serving a stair should have controlling switches at, or in the immediate vicinity of, the stair landing on each storey.	D 4.5.1

Lighting in common areas of domestic buildings

In communal areas (particularly on stairs and ramps within a building) the possibility of slips, trips and falls and of collision with obstacles should be minimized. To achieve this:

Common areas should have artificial lighting capable of providing a uniform lighting level, at floor level, of not less than 100 lux on stair flights and landings and 50 lux elsewhere within circulation areas.	D 4.5.2
Lighting should not present sources of glare and should avoid creation of areas of strong shadow that may cause confusion or mis-step.	D 4.5.2
A means of automatic control should be provided to ensure that lighting is operable during the hours of darkness.	D 4.5.2

Door entry systems

Where a common entrance door, intended as a principal means of access to a building, is fitted with a locking device, a door entry system comprising a remote door release and intercom at the point of entry and a call unit within each dwelling served by that entrance should be installed.	D 4.5.3
Controls should contrast visually with surrounding surfaces and any numeric keypad should follow the 12-button telephone convention, with an embossed locater to the central '5' digit.	D 4.5.3

Passenger lifts – common areas of domestic buildings

Passenger lifts should: D 4.2.5

- have a clear landing at least 1.5 m by 1.5 m in front of any lift entrance door;
- have automatic lift door(s), with a clear opening width of at least 800 mm, fitted with sensors that will prevent injury from contact with closing doors;
- be at least 1.1 m wide by 1.4 m deep and contain a horizontal handrail, of a size and section that is easily gripped, 900 mm above the floor on each wall not containing a door;
- have a mirror on the wall facing the doors, above handrail height, to assist a wheelchair user if reversing out (unless the lift car has through access);
- have tactile storey selector buttons and, in a lift serving more than two storeys, visual and voice indicators of the storey reached;
- have controls on each level served, between 900 mm and 1.1 m above the landing, and within the lift car on a side wall between 900 and 1.1 m above the car floor and at least 400 mm from any corner;
- on the landing of each level served, have tactile call buttons and visual and tactile indication of the storey level;
- have lift doors, handrails and controls that contrast visually with surrounding surfaces;
- have a signalling system which gives notification that the lift is answering a landing call;
- have a system which permits adjustment of the dwell time after which the lift doors close, once fully opened, to suit the level of use;
- have a means of two-way communication, operable by a person with a hearing impairment, that allows contact with the lift if an alarm is activated, together with visual indicators that an alarm has been sounded and received.

Socket outlets

To reduce the possibility of overheating and the risk of fire, a dwelling should be provided with at least the following number of 13A socket outlets:

- four within each apartment;

- six within the kitchen (at least three of which should be situated above worktop level in addition to any outlets provided for floor-standing white goods or built-in appliances);
- an additional four anywhere in the dwelling (including at least one within each circulation area on a level or storey).

D 4.5.4

Sockets may be installed as single or double outlets, to give the recommended number of outlets in each space.

D 4.5.4

Heat systems

All heat pumps are at their most efficient when the source temperature is as high as possible, the heat distribution temperature is as low as possible and pressure losses are kept to a minimum.

Electrically driven heat pumps should have a coefficient of performance of not less than 2.0 when operating at the heating system design condition.

D 6.3.4

Time and temperature controls should be provided using:

D 6.3.9

- a programmable time switch and thermostat integral to the appliance; or
- a separate time switch and separate room thermostats.

Electric warm air systems

Time and temperature control should be provided either integral to the heater or external, using either:
- a time switch/programmer and room thermostat; or
- a programmable room thermostat.

D 6.3.9

Electric underfloor heating

For electric storage, direct acting systems and under-tile systems programmable room timer/thermostats with manual override feature room controls are recommended for all heating zones, with air and floor (or floor void) temperature sensing capabilities to be used individually or combined.

A storage system should have:

D 6.3.9

- anticipatory controllers for controlling low-tariff input charge with external temperature sensing and floor temperature sensing;

> • a manual override facility should be available for better   D 6.3.9
> user control.
>
> Controls for storage systems with room timer/thermostats
> should take advantage of low-tariff electricity.

Controls for combined warm air and hot water systems

> The following controls should be provided: D 6.3.10
>
> • independent time control of both the heating and
> hot water circuits (achieved by means of a cylinder
> thermostat and a timing device, wired such that when
> there is no demand for hot water both the pump and
> circulator are switched off);
> • pumped primary circulation to the hot water cylinder;
> • a hot water circulator interlock (achieved by means of
> a cylinder thermostat and a timing device, wired such
> that when there is no demand from the hot water both
> the pump and circulator are switched off); and
> • time control by the use of either:
> • a full programmer with separate timing to each circuit;
> • two or more separate timers providing timing control to
> each circuit;
> • a programmable room thermostat(s) to the heating
> circuit(s); or
> • a time switch/programmer (two channel) and room
> thermostat.

Electric boiler controls
The boiler should be fitted with a flow temperature control and be capable of
modulating the power input to the primary water depending on space heating
conditions.

Artificial and display lighting
Artificial lighting accounts for a high proportion of the electricity used in
most buildings but careful lighting design (such as using natural daylight) can
reduce carbon dioxide emissions, internal heat gains and running costs.

> Every building must be designed and constructed in such a   D 6.5
> way that the artificial or display lighting installed is energy
> efficient and is capable of being controlled to achieve
> optimum energy efficiency.

Standard D 6.5 does not apply to:

- process and emergency lighting components in a building;
- communal areas of domestic buildings; or
- alterations in dwellings.

A minimum of 50% of the fixed light fittings and lamps installed in a dwelling should be low-energy type and fitting may be either:	D 6.5.1
	D 6.5.1

- dedicated fittings with a separate control gear and which will only take fluorescent lamps; or
- lamps with integrated control gear (e.g. bayonet or Edison screw base lamps).

6.24 Kitchens and utility rooms

Although the main function of a kitchen is cooking, it can be the centre of other activities as well, especially within homes, depending on its size, furnishing and equipment. If a washing machine is present, washing and drying laundry is also done in the kitchen. The kitchen may also be the place where the family eats, provided it is large enough. Sometimes, it is the most comforting room in a house, where family and visitors tend to congregate.

6.24.1 Requirements

Fire

Every building must be designed and constructed in such a way that in the event of an outbreak of fire within the building, the occupants, once alerted to the outbreak of the fire, are provided with the opportunity to escape from the building, before being affected by fire or smoke.	2.9
	2.11

Environment

Every building must be designed and constructed in such a way that:	*D 3.11*

- *the size of any apartment or kitchen will ensure the welfare and convenience of all occupants and visitors; and*
- *an accessible space is provided to allow for the safe, convenient and sustainable drying of washing;*

> - *sanitary facilities are provided for all occupants of,*   3.12
>   *and visitors to, the building in a form that allows*
>   *convenience of use and that there is no threat to the*
>   *health and safety of occupants or visitors;*
> - *the air quality inside the building is not a threat to*   3.14
>   *the health of the occupants or the capability of the*
>   *building to resist moisture, decay or infestation;*
> - *there will not be a threat to the building or the health*   3.15
>   *of the occupants as a result of moisture caused by*
>   *surface or interstitial condensation;*
> - *each fixed combustion appliance installation operates*   3.17
>   *safely.*

Noise

> *Safe, unassisted and convenient means of access should be*   4.2.
> *provided throughout the building.*
>
> *Every building must be designed and constructed in such*   4.5
> *a way the electrical installation does not:*
>
> - *threaten the health and safety of the people in, and*
>   *around, the building; and*
> - *become a source of fire.*

6.24.2 Meeting the requirements

Access

Accessibility within a storey of a dwelling

> To ensure facilities (such as a kitchen or sanitary facility)   D 4.2.6
> within a dwelling can be reached and used by occupants,
> each accessible level or storey within a dwelling should have:
>
> - corridors with an unobstructed width of at least
>   900 mm; and
> - doors with a minimum clear opening width in
>   accordance with Table 6.151.

Table 6.151   Width of doors

Minimum corridor width at door (mm)	Minimum clear opening width (mm)
1,050	775
900	800

The principal living level of a dwelling, normally also the entrance storey, should contain at least one kitchen and accessible sanitary accommodation.	D 4.2.6

Flats entered on the accommodation level

If a flat has a story height of more than 4.5 m, the sleeping accommodation (and that part of the circulation area which serves the sleeping accommodation and the exit to the flat) should be separated from any kitchen by a construction with the fire resistance equivalent to that of a protected enclosure.	D 2.9.9
If a flat has a storey height of more than 7.5 m and the distance to be travelled from any point within the flat to the exit is more than 15 m, an alternative exit from the living accommodation should be provided.	D 2.9.9

Protected enclosures – houses

In a house containing a kitchen in a storey that is greater than 4.5 m:	D 2.9.31
• every stair should be in a protected enclosure;	
• every storey higher than 7.5 m should have an alternative exit.	D 2.9.32

Additional guidance for hospitals

Kitchens should:	ND 2.B.1
• never be directly below, nor directly adjoin, the operating theatres, intensive therapy units or special care baby units; and	
• be provided with a fire suppression system if they are directly below, or directly adjoin, any other hospital department to which patients have access.	

Ventilation

Every building must be designed and constructed in such a way that the air quality inside the building is not a threat to the health of the occupants or the capability of the building to resist moisture, decay or infestation.

Ventilation should have the capability of:	D 3.14.1 ND 3.14.1

- providing outside air to maintain indoor air quality sufficient for human respiration;
- removing excess water vapour from areas where it is produced in significant quantities, such as kitchens, utility rooms, bathrooms and shower rooms to reduce the likelihood of creating conditions that support the germination and growth of mould, harmful bacteria, pathogens and allergens;
- removing pollutants that are a hazard to health from areas where they are produced in significant quantities, such as non-flued combustion appliances;
- rapidly diluting pollutants and water vapour, where necessary, that are produced in apartments and sanitary accommodation.

Ventilation should be to the outside (i.e. external) air.	D 3.14.1 ND 3.14.1

Ventilation of dwellings

To reduce the effects of stratification of the air in a room, some part of the opening ventilator should be at least 1.75 m above floor level.	D 3.14.2
Most dwellings are naturally ventilated and ventilation should be provided in accordance with Table 6.152.	D 3.14.2

Table 6.152 Recommended ventilation of a dwelling

Ventilation recommendations		Trickle ventilation $>10\,m^3/h/m^2$	Trickle ventilation $<10\,m^3/h/m^2$
Kitchen	Either: • mechanical extraction capable of at least 30 L/s (intermittent) above a hob; or • mechanical extraction capable of at least 60 L/s (intermittent) if elsewhere; or • a passive stack ventilation system.	4,000 mm²	10,000 mm²
Utility room	Either: • mechanical extraction capable of at least 30 L/s (intermittent) above a hob; or • a passive stack ventilation.	4,000 mm²	10,000 mm²

Passive stack ventilation systems

A passive stack ventilation system uses a duct running from a kitchen ceiling to a terminal on the roof to remove any moisture-laden air.

A duct or casing forming a passive stack ventilation system serving a kitchen should be non-combustible.	D 3.14.6

Extract fans

Combustion appliances should be capable of operating safely regardless of whether the extract fans are running.	D 3.17.8 ND 3.17.8
For solid fuel appliances, extract ventilation should generally not be installed in the same room.	D 3.17.8 ND 3.17.8
For a gas-fired appliance, where a kitchen contains an open-flued appliance, the extract rate of the fan should not exceed 20 L/s.	D 3.17.8 ND 3.17.8

Electrical safety

Smoke alarms

Smoke alarms should be ceiling mounted and located: • in a circulation area which will be used as a route along which to escape; • not more than 7 m from the door to a kitchen; and • not more than 3 m from the door to a room intended to be used as sleeping accommodation; • on a surface which is normally at the ambient temperature of the rest of the room or circulation area in which the smoke alarm is situated.	D 2.11.2

 Where more than one smoke alarm is installed in a dwelling they should be interconnected so that detection of a fire by any one of them operates the alarm signal in all of them.

Lighting

A dwelling should have an electric lighting system providing at least one lighting point to every circulation space, kitchen, bathroom, toilet and other space having a floor area of 2 m^2 or more.	D 4.5.1

Light switches should be positioned at a height of between 900 mm and 1.1 m above floor level.	D 4.8.5
Standard switched or unswitched socket outlets and outlets for other services such as telephone or television should be positioned at least 400 mm above floor level.	ND 4.8.6

Socket outlets

To reduce the possibility of overheating and the risk of fire, a dwelling should be provided with at least six 13A socket outlets within the kitchen, at least three of which should be situated above worktop level in addition to any outlets provided for floor-standing white goods or built-in appliances.	D 4.5.4
Sockets may be installed as single or double outlets, to give the recommended number of outlets in each space.	D 4.5.4
Where socket outlets are concealed, such as to the rear of white goods in a kitchen, separate switching should be provided in an accessible position, to allow appliances to be isolated.	D 4.8.5 ND 4.8.6

Access to manual controls

The location of any manual control device (i.e. for openable ventilators, windows and rooflights and for controls and outlets of electrical fixtures located on a wall or other vertical surface that is intended for operation by the occupants of a building) should be: • installed in position that allows safe and convenient use; and • unless incorporating a restrictor or other protective device for safety reasons, controls should be operable with one hand.	D 4.8.5 ND 4.8.6
An openable window or rooflight that provides natural ventilation should have controls for opening, positioned at least 350 mm from any internal corner, projecting wall or similar obstruction and at a height of: • not more than 1.7 m above floor level, where access to controls is unobstructed; or	D 4.8.5 ND 4.8.6

- not more than 1.5 m above floor level, where access to controls is limited by a fixed obstruction of not more than 900 mm high which projects not more than 600 mm in front of the position of the controls, such as a kitchen base unit;
- not more than 1.2 m above floor level, in an unobstructed location, within an enhanced apartment or within accessible sanitary accommodation not provided with mechanical ventilation.

Outlets and controls of electrical fixtures and systems (such as such as sockets, switches, fire alarm call points and timer controls or programmers, etc.) should be positioned at least 350 mm from any internal corner, projecting wall or similar obstruction and (unless the need for a higher location can be demonstrated) not more than 1.2 m above floor level.	D 4.8.5 ND 4.8.6
Above an obstruction, such as a worktop, fixtures should be at least 150 mm above the projecting surface.	D 4.8.5 ND 4.8.6

Facilities in dwellings

Activity spaces and provision of services

A dwelling should have a kitchen and (to be accessible) this should be on the principal living level.	D 3.11.3
Space should be provided within the kitchen to: • assist use by a person with mobility impairment to make use of its facilities; and • offer flexibility for future alteration.	D 3.11.3
The layout should include an unobstructed manoeuvring space of at least a 1.5 m by 1.5 m square or an ellipse of 1.4 m by 1.8 m.	D 3.11.3
A door may open across this manoeuvring space but a clear space of at least 1.1 m long by 800 mm wide, oriented in the direction of entry into the room, should remain unobstructed, to allow an occupant to enter and close the door.	D 3.11.3
A wall-supported worktop or similar obstruction, the underside of which is at least 750 mm above floor level, may overlap the manoeuvring space by not more than 300 mm.	D 3.11.3

Where a kitchen is within the same room as an apartment, the manoeuvring space for the kitchen may project into the apartment but should **not** overlap with the separate manoeuvring space of an enhanced apartment.   D 3.11.3

A kitchen should be provided with space for a gas, electric or oil cooker or with a solid fuel cooker designed for continuous burning.   D 3.11.3

The space should accommodate such piping, cables or other apparatus as will allow the appliance to operate.   D 3.11.3

A cooker should have an activity space to allow access to, and safe use of, an oven, as shown in Figure 6.159.   D 3.11.3

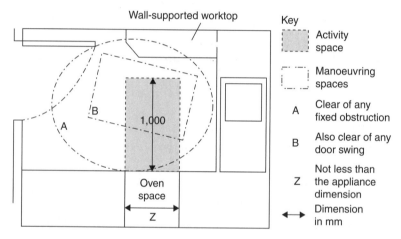

Figure 6.159 Space provision within a kitchen

An activity space need not be provided in front of a hob or microwave oven.   D 3.11.3

Activity spaces within a kitchen should have an unobstructed height of at least 1.8 m.   D 3.11.4

Kitchen storage of at least 1 m³ should be provided either within or adjacent to the kitchen.   D 3.11.3

Additional storage may be required depending on the local authority's recycling policy.   D 3.11.3

Every building must be designed and constructed in such a   D 3.11
way that:

- the size of any kitchen will ensure the welfare and convenience of all occupants and visitors; and
- an accessible space is provided to allow for the safe, convenient and sustainable drying of washing.

Drying of washing

Drying washing indoors can produce large amounts of water vapour that needs to be removed before it can damage the building fabric or generate mould growth that can be a risk to the health of occupants.

Where it is reasonably practicable, an accessible space for   D 3.11.6
the drying of washing should be provided for every house
on ground immediately adjacent to, and in the same
occupation as, the house. The area provided should allow
space for at least 1.7 m of clothes line per apartment.

Since weather is unreliable in Scotland, a designated space   D 3.11.6
for the drying of washing should be provided in every
dwelling, in addition to the external space.

The designated space may be either: D 3.11.6

- capable of allowing a wall-mounted appliance which may, for example be fixed over a bath; or
- capable of allowing a ceiling-mounted pulley arrangement; or
- a floor space in the dwelling on which to set out a clothes horse.

The designated space should have a volume of at least 1 m³   D 3.11.6
and should have no dimension less than 700 mm.

The designated space should allow space for at least 1.7 m   D 3.11.6
of clothes line per apartment.

The location of the designated space should not restrict   D 3.11.6
access to any other area or appliance within the dwelling
nor obstruct the swing of any door.

Control of humidity

If the average relative humidity within a room stays at or above 70% for a long period of time, the localized relative humidity at external wall surfaces will be higher and is likely to support the germination and growth of moulds.

Control of moisture in the kitchen and utility can be by   D 3.15.2
active or passive means.

Refuse facilities

Where communal solid waste storage is located within a building, such as where a refuse chute is utilized, the storage area should have provision for washing down and draining the floor into a wastewater drainage system.	D 3.25.4

Sanitary facilities

Non-domestic buildings

In non-domestic buildings (particularly small premises) it is recognized that duplication of sanitary facilities may not always be reasonably practicable and that they might be shared between staff and customers. However, where practicable, it is good practice for sanitary facilities for staff involved in the preparation or serving of food or drink to be reserved for their sole use, with a separate provision made for customers. Separate hand-washing facilities for such staff should always be provided.

No room containing sanitary facilities should communicate directly with a room for the preparation or consumption of food.	ND 3.14.7
Every toilet should:	ND 3.12.6

- have a wash hand basin within either the toilet itself or in an adjacent space providing the sole means of access to the toilet;
- be separated by a door from any room or space used wholly or partly for the preparation or consumption of food;
- not open directly on to any room or space used wholly or partly for the preparation or consumption of food;
- on a commercial basis include at least one enlarged WC cubicle where sanitary accommodation contains four or more WC cubicles in a range.

Domestic buildings

There should be a door separating a space containing a WC, or waterless closet, from a room or space used for the preparation or consumption of food, such as a kitchen or dining room.	D 3.12.1

6.25 Extensions

Mathematically speaking, 'an extension of some structure, is another structure which contains the original structure'. The same could be said about a building extension. The only difference is that it must fully meet the requirements of the current Scottish Building Regulations.

6.25.1 Requirements

See Technical Handbooks Annex 6A (Compensating *U*-values for windows, doors and rooflights) and Annex 6B (Compensatory approach – heat loss example) for further guidance on extensions.

Construction

Constructions and/or conversions shall be carried out so that the work complies with the applicable requirements and mandatory standards shown in the six sections of the Technical Handbooks (namely structure, fire safety, environment, safety, noise and energy).	*0.9.2* *0.12.2*

Environment

Buildings must be designed and constructed in such a way that there will not be a threat to the health of people in or around the building due to the emission and containment of radon gas.	*3.2*
Every building (and hard surface within the curtilage of a building) must be designed and constructed with a surface water drainage system that will:	*3.6*
• ensure the disposal of surface water without threatening the building and the health and safety of the people in and around the building; *• have facilities for the separation and removal of silt, grit and pollutants.*	
Wastewater drainage systems serving a building must be designed and constructed so as to ensure the removal of wastewater from the building without threatening the health and safety of the people in and around the building, and:	*3.7*
• that facilities for the separation and removal of oil, fat, grease and volatile substances from the system are provided;	

- *that discharge is to a public sewer or public wastewater treatment plant, where it is reasonably practicable to do so;*
- *where discharge to a public sewer or public wastewater treatment plant is not reasonably practicable that discharge is to a private wastewater treatment plant or septic tank.*

The last point of Standard 3.7 does not apply to a dwelling.

Every building must be designed and constructed in such a way that: 3.10

- *there will not be a threat to the building or the health of the occupants as a result of moisture from precipitation penetrating to the inner face of the building;*
- *sanitary facilities are provided for all occupants of, and visitors to, the building in a form that allows convenience of use and that there is no threat to the health and safety of occupants or visitors;* 3.12
- *the air quality inside the building is not a threat to the health of the occupants or the capability of the building to resist moisture, decay or infestation.* 3.14
- *the size of any apartment or kitchen will ensure the welfare and convenience of all occupants and visitors;* 3.11
- *an accessible space is provided to allow for the safe, convenient and sustainable drying of washing;*
- *natural lighting is provided to ensure that the health of the occupants is not threatened.* 3.16

Safety

Every building must be designed and constructed in such a way that: 4.1

- *all occupants and visitors are provided with safe, convenient and unassisted means of access to the building.*
- *each part of a building designed for different occupation is fitted with fuel consumption meters.* 4.2
- *in non-domestic buildings, safe, unassisted and convenient means of access is provided throughout the building;*
- *in residential buildings, a proportion of the rooms intended to be used as bedrooms is accessible to a wheelchair user;*

- *in domestic buildings, safe and convenient means of access is provided within common areas and to each dwelling;*
- *in dwellings, safe and convenient means of access is provided throughout the dwelling;*

  There is no requirement to provide access for a wheelchair user:

- in a non-domestic building not served by a lift, to a room, intended to be used as a bedroom, that is not on an entrance storey; or
- in a domestic building not served by a lift, within common areas and to each dwelling, other than on an entrance storey.

Energy

Every building must be designed and constructed so that an insulation envelope is provided which reduces heat loss.	6.2
The occupiers of a building must be provided with written information by the owner:	6.8
• *on the operation and maintenance of the building services and energy supply systems; and* • *where any air-conditioning system in the building is subject to a time-based interval for inspection of the system.*	

6.25.2 Meeting the requirements

Conversion of heated buildings

Where conversion of a heated building is to be carried out, the insulation envelope should be examined and upgraded (if necessary) in accordance with Table 6.153.	D 6.2.7 ND 6.2.8

In a domestic building, the total area of windows, doors and rooflights, should not exceed 25% of the floor area of the dwelling created by the conversion of a heated building.	D 6.2.7

Extensions to the insulation envelope

When constructing an extension, the majority of the work will be new and seldom will there be the need to construct to a lesser specification as is sometimes the case for alteration work.

Table 6.153 Maximum *U*-values for building elements of the insulation envelope

Type of element	Area-weighted average *U*-value (W/m²K) for all elements of the same type
Wall	0.70
Floor	0.70
Roof	0.35
New and replacement windows, doors, rooflights	1.8

Where the insulation envelope of a non-domestic building is extended, the new building fabric should be designed in accordance with Table 6.153.

ND 6.2.10

Where the insulation envelope of a domestic building is extended, the area of windows, doors, rooflights and roof windows should be limited to 25% of the floor area of the extension.

D 6.2.9

Where the insulation envelope of a dwelling or a building consisting of dwellings is extended the new building fabric should be designed in accordance with Table 6.153.

D 6.2.9

 For ease of understanding, the *U*-values (area-weighted average *U*-values) in Table 6.153 are summarized in Figure 6.65.

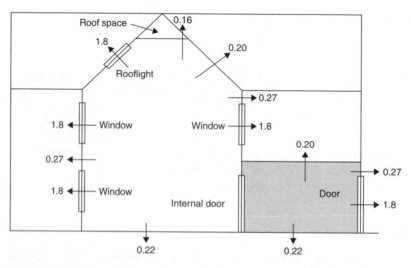

Figure 6.160 Maximum *U*-values for building elements of the insulation envelope

Note: The extension is the shaded portion; the existing dwelling is in elevation behind.

The U-values for the elements involved in the work may be varied provided that the area-weighted overall U-value of all the elements in the extension is no greater than that of a 'notional' extension. (See Annex 6B of the Scottish Building Regulations for an example of this approach.)

D 6.2.9

For non-domestic buildings, where the insulation envelope is extended, the new opening areas should be designed in accordance with Table 6.154.

ND 6.2.10

Table 6.154 Maximum U-values for building elements of the insulation envelope

Type of element	(a) Area-weighted average U-value (W/m²K) for all elements of the same type	(b) Individual element U-value (W/m²K)
Wall	0.27	0.70
Floor	0.22	0.70
Pitched roof (insulation between ceiling ties or collars)	0.16	0.35
Flat roof or pitched roof (insulation between rafters or roof with integral insulation)	0.20	0.35
Windows, doors, rooflights	1.8	3.3

Alterations to the insulation envelope

Alterations that involve increasing the floor area and/or bringing parts of the existing building that were previously outwith the insulation envelope into the heated part of the dwelling are considered as extensions.

The infill of an existing opening of approximately 4 m² or less in the building fabric should have a U-value which matches at least that of the remainder of the surrounding element and in the case of a wall or floor it should not be worse than 0.70 W/m²K (for a roof, not worse than 0.35 W/m²K).

D 6.2.11
ND 6.2.12

The infill of an existing opening of greater area (than approximately 4 m²) in the building fabric should have a U-value which achieves those in Column (a) of Table 6.154.

D 6.2.11
ND 6.2.12

Where the alteration causes an existing internal part or other element of a building to form the insulation envelope, that part of the building (including any infill construction) should have U-values which achieve those in Column (a) of Table 6.154.	D 6.2.11 ND 6.2.12
Where windows, doors and rooflights are being created or replaced, they should achieve the U-value recommended (an example of a compensatory approach is shown in Annex 6C to the Scottish Building Regulations).	D 6.2.11 ND 6.2.12
For secondary glazing, an existing window, after alteration should achieve a U-value of about $3.5\,\mathrm{W/m^2 K}$.	D 6.2.10
Where the work relates only to one or two replacement windows or doors, to allow matching windows or doors be installed, the frame may be disregarded for assessment purposes, provided that the centre pane U-value for each glazed unit is $1.2\,\mathrm{W/m^2 K}$ or less.	
Where additional windows, doors and rooflights are being created, the overall total area (including existing) should not exceed 25% of the total dwelling floor area and in the case of a heated communal room or other area (exclusively associated with the dwellings), it should not exceed 25% of the total floor area of these rooms/areas.	D 6.2.10
A building that was in a ruinous state should, after renovation, be able to achieve almost the level expected of new construction and after an alteration of this nature to the insulation envelope, a roof should be able to achieve at least an average U-value of 0.35 and in the case of a wall or floor, $0.70\,\mathrm{W/m^2 K}$.	D 6.2.10
When alterations are carried out, attention should still be paid to limiting infiltration thermal bridging at junctions and around windows, doors and rooflights and limiting air infiltration.	D 6.2.11 ND 6.2.12

Structure

Extensions to existing buildings

Where the wall and its foundation is to be constructed against an existing wall then the foundation should be treated as if it were an extension to an existing building (Figure 6.161).

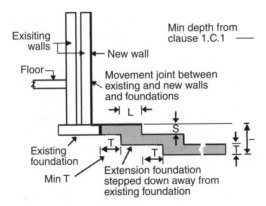

Figure 6.161 Foundations for extensions

Where the depth of the existing foundation is less than

- to the selected rock or soil bearing stratum;
- 450 mm to the underside of foundations;
- 600 mm to the underside of foundations where clay soils are present; (depending which is the greater) then for extensions no greater than of two storeys connected to existing buildings

• the depth of the extension foundation should match the depth of the existing foundation at the interface and step down as shown in Figure 6.161;	1C4
• the initial step down in the underside of the new foundation should not commence until the horizontal distance from the vertical face of the existing foundation is at least the foundation thickness, T.	1C5

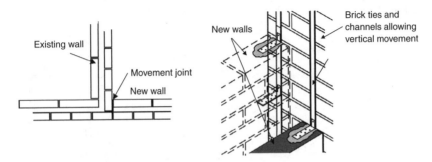

Figure 6.162 Movement joints and brick ties

To minimize the risk of differential settlement occurring between the extension and the existing structure, the following should be considered:

- movement joints should be placed between the existing and new foundations and walls to accommodate any differential settlement;   1C5
- for soil types I–III the strip foundation widths should be as per Table 6.155;
- for soil types IV, V and VI the strip foundation widths shown in Table 6.155 should be increased by 25%;

 Additional information is provided in BRE GBG 53 'Foundations for low-rise building extensions'.

- where a new roof connects to an existing roof, the loads from the new roof should be carried down to the new foundation and no new loads should be carried by the existing structure.   1C5

Size of extensions to domestic buildings

For single-storey, single-leaf extensions to domestic buildings (including garages and outbuildings):

the height H should be not more than the limits shown in Figure 6.163.   1D2b

Note: H is measured from the top of the foundation or from the underside of the floor slab where this provides effective lateral restraint.

Single-leaf external walls

The single leaf of an external wall to a single-storey, single-leaf domestic building and/or extension should be at least 90 mm thick.

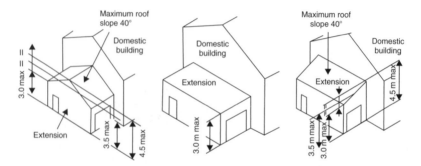

Figure 6.163 Dimensions of extensions to domestic buildings

Table 6.155 Minimum width of strip footings

Type of ground (including engineering fill)	Condition of ground	Applicable field test	Total load of load-bearing walling not more than (kN/m)					
			20	30	40	50	60	70
			Minimum width of strip foundation WF (mm)					
I Rock	Not inferior to sandstone, limestone or firm chalk	Requires at least a pneumatic or other mechanical operated pick for excavation	At least equal to the width of the wall					
II Gravel or sand	Medium dense	Requires a pick for excavation. Wooden peg 50 mm² in cross-section, hard to drive beyond 150 mm	250	300	400	500	600	650
III Clay or sandy clay	Stiff	Can be indented slightly by thumb	250	300	400	500	600	650
IV Clay or sandy clay	Firm	Thumb makes impression easily	300	350	450	600	750	850
V Sand, silty sand or clayey sand	Loose	Can be excavated by a spade. Wooden peg 50 mm² in cross-section can be easily driven in	400	600	×	×	×	×
VI Silt, clay, sandy clay or silty clay	Soft	Finger pushed in up to 10 mm	450	600	×	×	×	X
VII Silt, clay, sandy clay or silty clay	Very soft	Finger easily pushed in up to 25 mm	×	×	×	×	×	×

Size and proportions of openings

Other than windows and a single-leaf door that meet the requirements shown in Figure 6.164:

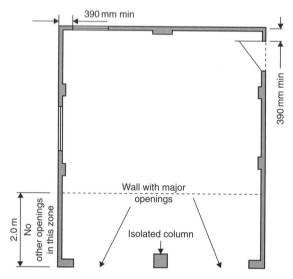

Figure 6.164 Size of openings in small single-storey, single-leaf buildings

No more than two major openings (maximum height 2.1 m, maximum width 5.0 m) are allowed in any one wall of the extension, either a single opening or the combined width of two openings.	1D40
No other openings shall be within 2.0 m of a wall containing a major opening.	1D40
The total size of openings in a wall not containing a major opening should not exceed 2.4 m².	1D40
There should not be more than one opening between piers.	1D40
The distance from a window or a door to a corner should be at least 390 mm unless there is a corner pier.	1D40

Wall thicknesses and piers

• The walls should be at least 90 mm thick; • pier sizes ($A_p \times B_p$) should be at least 390 mm by 190 mm or 327 mm by 215 mm depending on the size of the masonry units;	1D41

- isolated columns should be at least 325 mm by 325 mm ($C_C \times C_C$).

Walls which do not contain a major opening but are more than 2.5 m long or wide should be bonded or tied to piers for their full height at not more than 3 m centres as shown in Figure 6.165.

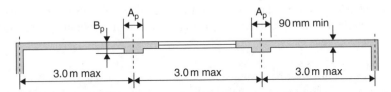

Figure 6.165 Wall thicknesses and piers

Differential movement

The allowances shown in Figure 6.166 should be made for differential movement between timber and masonry construction. 1E50

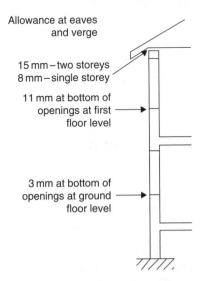

Figure 6.166 Allowances for differential movement

The roof of any timber-frame extension that is connected to
an existing traditional masonry wall should be supported on a
timber bearer connected to the existing wall.

1E50

Site preparation

Protection from radon gas

All new extensions and alterations to dwellings that are built in areas where
there might be radon concentration, may need to incorporate precautions
against radon gas.

Every building must be designed and constructed in such a
way that there will not be a threat to the health of people in
or around the building due to the emission and containment
of radon gas.

D 3.2
ND 3.2

Soakaways serving small extensions

Soakaways have been the traditional method of disposal of surface water from
buildings and paved areas where no mains drainage exists.

A soakaway serving an extension should be tested in accordance
with the percolation test method.

D 3.6.5
ND 3.6.5

Precipitation

The outer leaf of a previously external wall will become an internal wall and
any moisture that enters the cavity could collect and cause serious damage to
the building.

Where the building is located in an exposed location or
where the existing construction might allow the passage
of rain through either facing brick or a poorly rendered
masonry wall, the use of a cavity tray along the line of
the roof of the extension may be appropriate.

D 3.10.4
ND 3.10.4

In sheltered situations or where the detailing can prevent
damage to the building as a result of rain penetration
a raggled flashing (chased into the wall) may be
sufficient.

D 3.10.4
ND 3.10.4

Extensions to domestic buildings

Wastewater drainage

A careful check should be made before breaking into an existing   D 3.7.6
drain to ensure it is the correct one (e.g. whether wastewater to
surface water or vice versa) and a further test carried out after
connection, such as a dye test, to confirm correct connection.

Provision of new enhanced apartment

Where an extension of a building includes work to, or provision of a new,
enhanced apartment on the principal living level of the dwelling, and there is
not already an enhanced apartment on that level:

enhanced apartments should: D 3.11.5
    D 3.11.2

- have a floor area of at least $12\,m^2$ and a length and width
  at least 3.0 m;
- this area should exclude any space less than 1.8 m in
  height and any portion of the room designated as a
  kitchen;
- contain a unobstructed manoeuvring space of at
  least a 1.5 m by 1.5 m square or an ellipse of at least
  1.4 m by 1.8 m, which may overlap with activity spaces.

Sanitary facilities

Every building must be designed and constructed in such a way that sanitary
facilities are provided for all occupants of, and visitors to, the building in a
form that allows convenience of use and that there is no threat to the health
and safety of occupants or visitors.

Additional sanitary facilities need not be provided as part   D 3.12.5
of an extension to, or alteration of, a dwelling. However, an
additional accessible toilet may be needed, if one does not
exist on the entrance level of a dwelling.

If it is intended to install a new sanitary facility on the   D 3.12.5
principal living level or entrance storey of a dwelling and
there is not already an accessible sanitary facility of that
type within the dwelling, the first new facility should be in
accordance with the current standards.

If altering existing sanitary accommodation on the principal   D 3.12.5
living level or entrance storey of a dwelling, any changes
should at least maintain the level of compliance present
before alterations.

Existing sanitary accommodation should only be removed or relocated where facilities at least equivalent to those removed will still be present within the dwelling.	D 3.12.5
A sanitary facility that is not an accessible facility may be altered or removed where the minimum provision for a dwelling is maintained.	D 3.12.5

Extensions built over existing windows

Constructing an extension over an existing window, or ventilator, will effectively result in an internal room, restrict air movement and could significantly reduce natural ventilation to that room.

If an extension is built over a ventilator, then a new ventilator should be provided to the room.	D 3.14.7
If the extension is constructed over an area that generates moisture, such as a kitchen, bathroom, shower room or utility room, mechanical extract, via a duct if necessary or a passive stack ventilation system should be provided direct to the outside air.	D 3.14.7

Extensions built over a glazed opening to a room

An extension constructed over a glazed opening to a room, because of its greater solidity, can seriously restrict daylight from entering the dwelling and the existing room and extension should be treated a single room.

The area of the translucent glazed opening to the extension should be at least 1/15th of the combined floor area of the existing room and the extension.	D 3.16.3
The area of the translucent glazed opening to the extension should be at least 1/15th of the combined floor area of the existing room and the extension.	D 3.16.3
A new translucent glazed opening should be provided to the existing room but, where this is not practicable, the wall separating the two rooms should be opened up to provide a single space.	D 3.16.3
To ensure sufficient 'borrowed light' is provided, the opening area between the existing room and the extension should be not less than 1/10th of the total combined area of the existing room and the extension.	D 3.16.3

Means of access

Where a dwelling is altered or extended: D 4.1.10

- this work should not adversely affect an existing accessible entrance;
- all works should ensure that any existing entrance remains accessible;

 Note: If there is **no** existing accessible entrance, there is no need to provide one to the existing dwelling, or to the extension, as this will not result in the building failing to meet the standard to a greater degree.

Access within buildings

Where the alteration of a building includes work to, or provision of, a new circulation area, the following guidance should be followed as far as is reasonably practicable.

Where existing accommodation does not meet the provisions D 4.2.11 set out in guidance, it need not be altered to comply except for consequential work, required to ensure compliance with another standard (e.g. where an accessible entrance has been relocated and alterations are required to circulation space to maintain accessibility within the building).

Extensions to non-domestic buildings

Extensions built over existing windows

Constructing an extension over an existing window, or ventilator, will effectively result in an internal room, will restrict air movement and could significantly reduce natural ventilation to that room.

A new ventilator **and** trickle ventilator should be provided ND 3.14.4 to the existing room.

 BUT if this is not reasonably practicable (e.g. if virtually the entire external wall of the room is covered by the extension) the new extension should be treated as part of the existing room as opposed to being a separate internal room.

Because an extension will be relatively airtight, the ND 3.14.4 opening area between the two parts of the room should be not less than 1/15th of the total combined area of the existing room plus the extension.

If the extension is constructed over an area that generates moisture, such as a kitchen, bathroom, shower room or utility room, mechanical extract, via a duct if necessary, should be provided direct to the outside air.	ND 3.14.4

Access within buildings

A ramp which forms part of an escape route should have an effective width of not less than 1.2 m.	ND 4.3.12

Completion of work – building logbook

On completion of the extension or alteration to the building services system, the commissioning information should be updated in the logbooks.	ND 6.8.2

Metering

Every building must be designed and constructed in such a way that each part of a building designed for different occupation is fitted with fuel consumption meters.

Buildings where an extension or alteration is carried out should be installed with fuel consumption metering.	ND 6.10.2
Where extensions result in the creation of two or more units, each unit should have a fuel meter installed in an accessible location.	ND 6.10.2
A fuel meter should be installed if a new fuel type or new boiler (where none existed previously) is installed.	ND 6.10.2

6.26 Conservatories

Conservatories, similarly, are another form of extension to an existing building and must fully meet the requirements of the current Scottish Building Regulations.

6.26.1 Requirements

Fire

> *Every building must be designed and constructed in such a*   2.6
> *way that in the event of an outbreak of fire within the building,*
> *the spread of fire to neighbouring buildings is inhibited.*

Environment

> *Every building must be designed and constructed in such a*   3.10
> *way that:*
>
> - *there will not be a threat to the building or the health of*
>   *the occupants as a result of moisture from precipitation*
>   *penetrating to the inner face of the building;*
> - *the air quality inside the building is not a threat to the*   3.14
>   *health of the occupants or the capability of the building to*
>   *resist moisture, decay or infestation;*
> - *natural lighting is provided to ensure that the health of the*   3.16
>   *occupants is not threatened.*

Safety

> *Every building must be designed and constructed in such a*   D 4.3
> *way that every level can be reached safely by stairs or ramps.*

Energy

> *Every building must be designed and constructed in such*   6.3
> *a way that the heating and hot water service systems*
> *installed are energy efficient and are capable of being*
> *controlled to achieve optimum energy efficiency.*

6.26.2 Meeting the requirements

Location of conservatories

> A conservatory may be constructed over a translucent glazed   D 3.16.2
> opening to a room in a dwelling provided that the area of
> the glazed opening of the internal room so formed is at least
> 1/15th of the floor area of the internal room.

Conservatories and extensions built over existing windows

Constructing a conservatory over an existing window, or ventilator, will effectively result in an internal room, restrict air movement and could significantly reduce natural ventilation to that room.

A conservatory may be constructed over a ventilator serving a room in a dwelling provided that the ventilation of the conservatory is to the outside air and has an opening area of at least 1/30th of the total combined floor area of the internal room so formed and the conservatory.	D 3.14.7
The ventilator to the internal room should have an opening area of at least 1/30th of the floor area of the room.	D 3.14.7
Trickle ventilators should be provided relevant to the overall areas created.	D 3.14.7
If a conservatory is built over a ventilator, then a new ventilator should be provided to the room.	D 3.14.7
If the conservatory is constructed over an area that generates moisture, such as a kitchen, bathroom, shower room or utility room, mechanical extract, via a duct if necessary or a passive stack ventilation system should be provided direct to the outside air.	D 3.14.7

External walls

The outer leaf of a previously external wall will become an internal wall and any moisture that enters the cavity could collect and cause serious damage to the building.

Where the existing building is located in an exposed location or where the existing construction might allow the passage of rain either through facing brick or a poorly rendered masonry wall, the use of a cavity tray along the line of the roof of the conservatory may be appropriate.	D 3.10.4
In sheltered situations or where the detailing can prevent damage to the building as a result of rain penetration a raggled flashing (chased into the wall) may be sufficient.	D 3.10.4

Fire resistance of external walls

The installation of an automatic fire suppression system greatly reduces the amount of radiant heat flux from a fire through an unprotected opening.

External walls of a conservatory attached to a dwelling should have short fire resistance duration, if more than 1 m from the boundary.   D 2.6.1

Ventilation of conservatories

With large areas of glazing, conservatories attract large amounts of the sun's radiation that can create unacceptable heat build-up. Efficient ventilation therefore is very important to ensure a comfortable environment.

A conservatory should have a ventilator or ventilators with an opening area of at least 1/5th of the floor area it serves.   D 3.14.3

To minimize the effects of heat build-up, it is recommended that at least 30% of the ventilator area provision is located on the roof or at as high a level as possible.   D 3.14.3

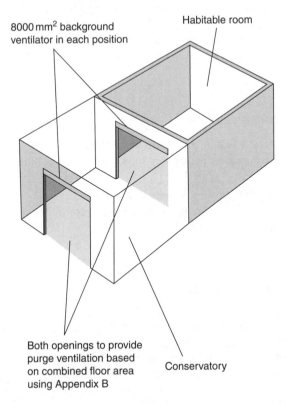

8000 mm² background ventilator in each position

Habitable room

Both openings to provide purge ventilation based on combined floor area using Appendix B

Conservatory

Figure 6.167 A habitable room ventilated through a conservatory

Escape windows

'Escape windows' such as a window or door (i.e. french window) situated in an external wall or roof are openable windows that are large enough for occupants to escape through and should be provided from every apartment from which the occupants can make their escape by lowering themselves from the window. They should have an unobstructed openable area that is at least 0.33 m²; and be at least 450 mm high and 450 mm wide.

If a conservatory is located below an escape window consideration should be given to the design of the conservatory roof to withstand the loads exerted from occupants lowering themselves onto the roof in the event of a fire.	D 2.6.4

Stair landings serving outward-opening fully glazed doors

If a conservatory and/or similar is not intended to be the accessible entrance and has an outward-opening fully glazed door and a landing (or a fully glazed door leading from a dwelling directly into a conservatory) then the landing length should be in accordance with Figure 6.168.	D 4.3.8

Heating of conservatories

As a conservatory which is heated will be inefficient in energy terms, the general guidance to occupiers is that they should be heated as little as possible.

A conservatory with heating installed should have controls (e.g. such as a TRV to the radiator) to regulate it from the rest of the dwelling.	D 6.3.13

Natural lighting for apartments

Every apartment should have a translucent glazed opening, or openings, of an aggregate glazed area equal to at least 1/15th of the floor area of the apartment and located in an external wall or roof or in a wall between the apartment and a conservatory.	D 3.16.1

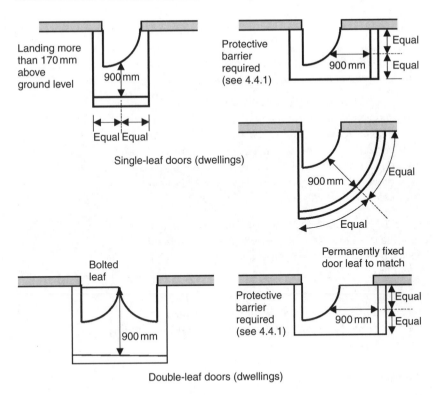

Single-leaf doors (dwellings)

Double-leaf doors (dwellings)

Figure 6.168 Landings serving outward-opening fully glazed doors

6.27 Garages and outbuildings

Although heating of garages and outbuildings is not an absolute necessity, the ventilation of these sorts of buildings is extremely important.

6.27.1 Requirements

General

Construction shall be carried out so that the work complies   0.9.2
with the applicable requirements and mandatory standards
shown in the six sections of the Technical Handbooks (namely
structure, fire safety, environment, safety, noise and energy).

Fire

Every building must be designed and constructed in such a way that in the event of an outbreak of fire within the building:	*2.1*
• *fire and smoke are inhibited from spreading beyond the compartment of origin until any occupants have had the time to leave that compartment and any mandatory fire containment measures have been initiated;*	
• *the occupants, once alerted to the outbreak of the fire, are provided with the opportunity to escape from the building, before being affected by fire or smoke.*	*2.9*
Every building, which is divided into more than one area of different occupation, must be designed and constructed in such a way that in the event of an outbreak of fire within the building:	*2.2*
• *fire and smoke are inhibited from spreading beyond the area of occupation where the fire originated.*	
• *the spread of fire to neighbouring buildings is inhibited.*	*2.6*

Environment

Every building, and hard surface within the curtilage of a building, must be designed and constructed with a surface water drainage system that will:	*3.6*
• *ensure the disposal of surface water without threatening the building and the health and safety of the people in and around the building;*	
• *have facilities for the separation and removal of silt, grit and pollutants.*	
Every building must be designed and constructed in such a way that the air quality inside the building is not a threat to the health of the occupants or the capability of the building to resist moisture, decay or infestation.	*3.14*

Safety

Every building must be designed and constructed in such a way that every level can be reached safely by stairs or ramps.	*D 4.3*

Noise

Every building must be designed and constructed in such a   5.1
way that each wall and floor separating one dwelling from
another, or one dwelling from another part of the building, or
one dwelling from a building other than a dwelling, will limit
the transmission of noise to the dwelling to a level that will
not threaten the health of the occupants of the dwelling or
inconvenience them in the course of normal domestic activities
provided the source noise is not in excess of that from normal
domestic activities.

 This standard does not apply to:

* fully detached houses;
* roofs or walkways with access solely for maintenance, or
  solely for the use of the residents of the dwelling below.

Energy

Every building must be designed and constructed in such a way   6.2
that an insulation envelope is provided which reduces heat loss.

 This standard does not apply to buildings which are
ancillary to dwellings, other than conservatories, which
are either unheated or provided with heating which is
solely for the purpose of frost protection.

6.27.2 Meeting the requirements

Structure

Size of single-storey, single-leaf extensions to domestic buildings
For single-storey, single-leaf extensions to domestic buildings (including
garages and outbuildings) (Figure 6.169):

* The height *H* should be not more than the limits shown   D 1D2b
  in Figure 6.169.

 Note: *H* is measured from the top of the foundation
or from the underside of the floor slab where this
provides effective lateral restraint.

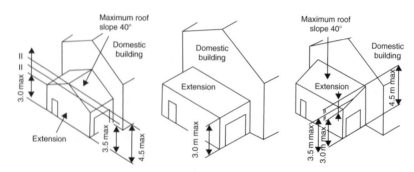

Figure 6.169 Dimensions of extensions to domestic buildings

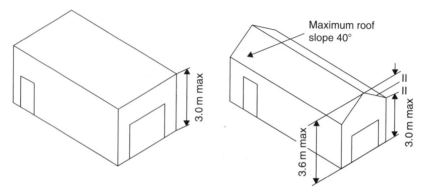

Figure 6.170 Dimensions of single-storey, single-leaf buildings

Size of single-storey, single-leaf garages

For single-storey, single-leaf buildings forming a garage or outbuilding within the curtilage of a dwelling:

- The height *H* should be not more than 3 m;   D 1D2b
- The length should be not more than 9 m.

 Note: *H* is measured from the top of the foundation or from the underside of the floor slab where this provides effective lateral restraint

Size and proportions of openings

The following guidance is equally applicable to:

- single-storey, single-leaf extensions to domestic buildings (including garages and outbuildings);
- single-storey, single-leaf buildings forming a garage or outbuilding within the curtilage of a dwelling.

Other than windows and a single-leaf door that meet the requirements shown in Figure 6.171:

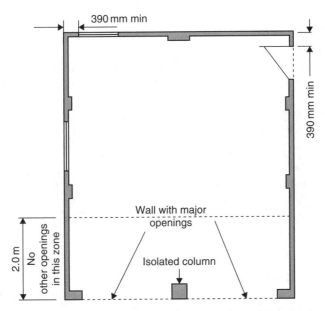

Figure 6.171 Size of openings in small single-storey, single-leaf buildings

No more than two major openings (maximum height 2.1 m, maximum width 5.0 m) are allowed in any one wall of the building or extension, either a single opening or the combined width of two openings.	D 1D40
No other openings shall be within 2.0 m of a wall containing a major opening.	D 1D40
The total size of openings in a wall not containing a major opening should not exceed 2.4 m².	D 1D40
There should not be more than one opening between piers.	D 1D40
The distance from a window or a door to a corner should be at least 390 mm unless there is a corner pier.	D 1D40

Fire resistance

Fire separation, forming a complete barrier to the products of combustion: smoke, heat and toxic gases, should be provided between dwellings and between dwellings and any common spaces.

A separating wall or separating floor with short fire resistance   D 2.2.6
duration should be provided between an integral or attached
garage and a dwelling in the same occupation.

Fire resistance of external walls

The installation of an automatic fire suppression system greatly reduces the
amount of radiant heat flux from a fire through an unprotected opening.

External walls of a garage wall should have short fire   D 2.6.1
resistance duration, if more than 1 m from the boundary.

Fire resistance duration need not be provided for an   D 2.6.1
additional building to a dwelling (such as a carport, covered
area, greenhouse, summerhouse or swimming pool enclosure)
unless the building contains oil or liquefied petroleum gas
fuel storage.

Fire-resisting ceilings

Where a fire-resisting ceiling, including a suspended ceiling, contributes to
the fire resistance duration of a compartment:

- the ceiling should not be easily demountable;   ND 2.1.15
- openings and service penetrations in the ceiling should   ND 2.1.15
  be protected;
- the ceiling should form a junction with the compartment   ND 2.1.15
  wall (or sub-compartment wall)

Environmental discharges into a drainage system

Where a discharge into a traditional drainage system   D 3.6.9
contains silt or grit, for example from a hard standing   ND 3.6.9
with car wash facilities:

- there should be facilities for the separation of such
  substances;
- removable grit interceptors should be incorporated into
  the surface water gully pots to trap the silt or grit.

| Where a discharge into a drainage system contains oil, grease or volatile substances (e.g. from a vehicle repair garage) there should be facilities for the separation and removal of such substances. | ND 3.6.9 |

 The use of emulsifiers to break up any oil or grease in the drain is **not** recommended as they can cause problems further down the system.

Ventilation of garages

The principal reason for ventilating garages is to protect the building users from the harmful effects of toxic emissions from vehicle exhausts.

| Where a garage is attached to a dwelling: | D 3.14.11 |

- the separating construction should be as airtight as possible;
- communicating doors should have airtight seals (in certain circumstances a lobby arrangement may be appropriate).

| Garages of less than 30 m² do not require the ventilation to be designed as a degree of fortuitous ventilation is created by the imperfect fit of 'up and over' doors or pass doors. | D 3.14.11 |

| A garage with a floor area of at least 30 m² but not more than 60 m² used for the parking of motor vehicles should have provision for natural or mechanical ventilation. | D 3.14.11 |

| Ventilation should be provided in accordance with the following guidance: | D 3.14.11 |

- where the garage is naturally ventilated, by providing at least two permanent ventilators, each with an open area of at least 1/3,000th of the floor area they serve, positioned to encourage through ventilation with one of the permanent ventilators being not more than 600 mm above floor level;
- where the garage is mechanically ventilated, by providing a system:
  - capable of continuous operation, designed to provide at least two air changes per hour;
  - independent of any other ventilation system;
  - constructed so that two-thirds of the exhaust air is extracted from outlets not more than 600 mm above floor level.

Ventilation of small garages

A garage with a floor area of at least $30\,m^2$ but not more than $60\,m^2$ used for the parking of motor vehicles should have provision for natural or mechanical ventilation as follows:

ND 3.14.8

- where the garage is naturally ventilated, by providing at least two permanent ventilators, each with an open area of at least 1/3,000th of the floor area they serve, positioned to encourage through ventilation with one of the permanent ventilators being not more than $600\,mm$ above floor level;
- where the garage is mechanically ventilated, by providing a system:
  - capable of using any other ventilation system;
  - constructed so that two-thirds of the exhaust air is extracted from outlets not more than $600\,mm$ above floor level.

Ventilation of large garages

A garage with a floor area more than $60\,m^2$ for the parking of motor vehicles should have provision for natural or mechanical ventilation on every storey.

Ventilation should:

ND 3.14.9

- give carbon monoxide concentrations of not more than 30 parts per million averaged over an 8 h period;
- restrict peak concentrations of carbon monoxide at areas of traffic concentrations such as ramps and exits to not more than 90 ppm for periods not exceeding 15 min.

Ventilation may be achieved:

ND 3.14.9

- by providing openings in the walls on every storey of at least 1/20th of the floor area of that storey with at least half of such area in opposite walls to promote extract ventilation, if the garage is naturally ventilated;
- by providing mechanical ventilation system capable of at least six air changes per hour and at least 10 air changes per hour where traffic concentrations occur;
- where it is a combined natural/mechanical ventilation system, by providing:
  - openings in the wall on every storey of at least 1/40th of the floor area of the storey with at least half of such area in opposite walls;
  - a mechanical system capable of at least three air changes per hour.

Protected lobbies

A protected lobby means a lobby within a protected zone but separated from the remainder of the protected zone so as to resist the movement of smoke from the adjoining accommodation to the remainder of the protected zone.

> Where flats or maisonettes are served by only one escape stair and there is no alternative means of escape from the upper storeys, there should be a protected lobby with automatic opening ventilators, at each storey within the protected zone between the escape stair and the accommodation, including a parking garage and any other accommodation ancillary to the dwellings (see Figure 6.172).
>
> D 2.9.19

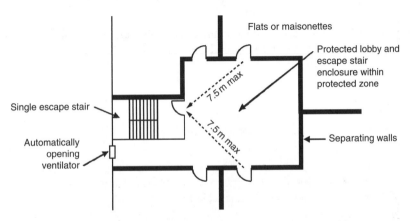

Figure 6.172 Single stair access to flats and maisonettes with any storey at a height of more than 7.5 m but not more than 18 m

> A ventilated protected lobby need **not** be provided:
> - for flats and maisonettes entered from an open access balcony or access deck which has an opening(s) to the external air extending over at least four-fifths of its length and at least one-third of its height;
>
>   D 2.9.19a
>
> - where the protected zone provides access to not more than 12 dwellings in total and no storey is at a height of more than 7.5 m and there are not more than four dwellings on each storey and each dwelling has within it a protected enclosure (see Figure 6.173);
>
>   D 2.9.19b
>
> - where there is more than one escape stair serving each dwelling.
>
>   D 2.9.19d

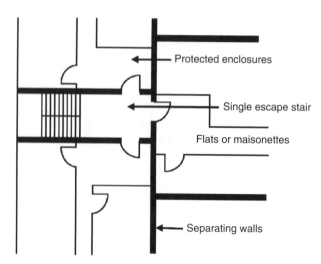

Figure 6.173 Single stair access to flats and maisonettes with every storey at a height of not more than 7.5 m

Stairs

Generally a flight should have not more than 16 rises.

There may be fewer than three rises between an external door of a building and the ground or a balcony, conservatory, porch or private garage (unless it is other than at an accessible entrance).	D 4.3.4 ND 4.3.4

Stair landings

Other than at an accessible entrance, a landing need not be provided to a flight of steps between the external door of: • a dwelling and the ground, balcony, conservatory, porch or private garage (where the door slides or opens in a direction away from the flight and the total rise is not more than 600 mm); or • a dwelling, or building ancillary to a dwelling, and the ground, balcony, conservatory, or porch (where the change in level is not more than 170 mm, regardless of method of door operation).	D 4.3.6

Noise

Airborne sound

Airborne sound resisting separating walls and separating floors   D 5.1.1
should be provided between dwellings so that each dwelling is
protected from noise emanating from the other one and from
other parts, such as common stair enclosures and passages,
solid waste disposal chutes, lift shafts, plant rooms, communal
lounges and car parking garages.

Impact sound resisting construction should be provided   D 5.1.1
between a dwelling and a roof that acts as a floor or a walkway
directly above the dwelling so that the dwelling below is
protected from sound emanating from the roof or walkway
above (e.g. roofs that act as floors are access decks, car parking,
escape routes and roof gardens).

Except where:

- where two houses are linked only by an imperforate
  separating wall between their ancillary garages, it is not
  necessary for the wall to be airborne sound resisting;
- where the wall between a dwelling and another part of
  the building is substantially open to the external air, it is
  not necessary for the wall to resist airborne sound
  transmission (e.g. a wall between a dwelling and an
  access deck);
- where the wall between a dwelling and another part of the
  building incorporates a fire door, it is not necessary for the
  door to be airborne sound resisting;
- when a roof or walkway is providing access solely for
  the purpose of maintenance or is solely for the use of the
  residents of the dwelling directly below, it is not necessary
  to provide impact sound resisting construction;
- in the case of a separating wall or separating floor between
  a dwelling and a private garage or a private waste storage
  area which is ancillary to the same dwelling, it is not
  necessary for the wall or floor to be airborne or impact
  sound resisting.

Flanking transmission

At the time of publication, air-tightness testing need only be carried out when better than routine air-tightness levels are declared at the Building Warrant application stage.

Conversions

Converting an unheated building (such as a barn, the roof space of a dwelling or a garage) for future domestic use can be quite a challenge in terms of achieving energy efficiency and because of this, the most demanding of measures are recommended when conversion occurs. Table 6.156 shows some examples of the work involved.

Table 6.156   Conversion of unheated buildings

Location	Work involved
Conversion of a roof space in a dwelling	This will usually involve extending the insulation envelope to include the gables, the collars, a part of the rafters and the oxters, as well as any new or existing dormer construction.
	The opportunity should also be taken at this time to upgrade any remaining poorly performing parts of the roof which are immediately adjacent to the conversion, for example, insulation to parts of the ceiling ties at the eaves.
Conversion of an unheated garage	This will usually involve extending the insulation envelope to include the existing floor, perimeter walls and the roof/ceiling to the new habitable part.
Conversion of a deep solum space into an apartment	This will usually involve extending the insulation envelope to include the solum/existing floor and perimeter walls to the new habitable part.

Where conversion of an unheated building (e.g. a barn) or part of a dwelling is to be carried out, the building should be treated as if it were an extension to the insulation envelope of a domestic or non-domestic building.	D 6.2.6 ND 6.2.7

6.28 Residential care buildings and hospitals

As hospitals and residential care buildings are designed for the provision of a care home service, school accommodation service and/or treatment of persons suffering from an illnesses, mental or physical disability or handicap, fire safety is of paramount importance.

6.28.1 Requirements

Fire

Every building must be designed and constructed in such a way that in the event of an outbreak of fire within the building:	*2.1*
• *fire and smoke are inhibited from spreading beyond the compartment of origin until any occupants have had the time to leave that compartment and any mandatory fire containment measures have been initiated;*	
• *the unseen spread of fire and smoke within concealed spaces in its structure and fabric is inhibited;*	*2.4*
• *the development of fire and smoke from the surfaces of walls and ceilings within the area of origin is inhibited;*	*2.5*
• *the spread of fire to neighbouring buildings is inhibited;*	*2.6*
• *the occupants, once alerted to the outbreak of the fire, are provided with the opportunity to escape from the building, before being affected by fire or smoke;*	*2.9*
• *fire and smoke will be inhibited from spreading through the building by the operation of an automatic life safety fire suppression system.*	*2.15*
Every building must be:	*2.12*
• *accessible to fire appliances and fire service personnel;*	
• *provided with a water supply for use by the fire service.*	*2.13*

6.28.2 Meeting the requirements

Fire precautions

Fire detection and alarm systems

An automatic fire detection and alarm system should be installed in every hospital to ensure that staff and patients are given the earliest possible warning of the outbreak of fire anywhere in the building.	ND 2.A.5 ND 2.B.5

Detection need not be provided in the following locations: • sanitary accommodation; • a lockable cupboard with a plan area not more than 1 m²; • a void and roof space which contain only mineral insulated wiring, or wiring laid on metal trays, or in metal conduits, or metal/plastic pipes used for water supply, drainage or ventilating ducting.	ND 2.A.5 ND 2.B.5
Manual fire alarm call points should be located and installed in accordance with BS 5839: Part 1: 2002.	ND 2.A.5 ND 2.B.5
The fire alarm should be activated on the operation of manual call points, automatic detection or the operation of the automatic life safety fire suppression system installed.	ND 2.A.5 ND 2.B.5
The building should be divided into detection zones not extending beyond a single compartment.	ND 2.A.5 ND 2.B.5
On the actuation of the fire alarm, a signal should be transmitted automatically to the fire service, either directly or by way of a remote centre.	ND 2.A.5 ND 2.B.5

Compartmentation

Residential care buildings and hospitals, with a total storey area more than the limits given in Tables 6.157 and 6.158, should be sub-divided by compartment walls and, where appropriate, compartment floors.

Internal linings

Rapid fire spread on internal linings in escape routes is particularly important because this could prevent the occupants from escaping.

Every room, fire-fighting shaft, protected zone or unprotected zone, should have wall and ceiling surfaces with a reaction to fire which follow the guidance in the Table 6.159.	ND 2.5.1

Fire resistance of external walls

Although there is a risk of fire spread on the external walls of a building, for most buildings it is only necessary to consider this if the external wall is in close proximity to the boundary. However, residential care buildings and hospitals present a greater risk because the mobility, awareness and understanding of the occupants could be impaired.

Table 6.157 Single-storey buildings and compartmentation between single-storey and multi-storey buildings where appropriate

Building use	Maximum total area of any compartment (m²)	Minimum fire resistance duration for compartmentation (if any)
Assembly building	6,000	Long
Entertainment building	2,000	Medium
Office	4,000	Medium
Open-sided car park	Unlimited	Not relevant
Residential care building, hospital	1,500	Medium
Shop	2,000	Long
Storage building (Class 1)	1,000	Long
Storage building (Class 2)	14,000	Long

In hospitals and residential care buildings,

> External walls should have at least the fire resistance duration   ND 2.6.1
> as shown in Table 6.160.

Roof space cavities above undivided spaces

Where there is sleeping accommodation, the material exposed in the cavity and the size of a cavity, should be controlled due to the nature of the risk. In such cases, the limits set in Table 6.161 should not be exceeded (ND 2.4.3).

Cavities above ceilings

> In hospitals and residential care buildings where a roof   ND 2.4.4
> space cavity or a ceiling void cavity extends over a room
> intended for sleeping (or over such a room and any other
> part of the building) cavity barriers should be installed on
> the same plane as the wall.

Compartmentation

> Every upper storey and every basement storey in a   ND 2.A.1
> residential care building and/or hospital should form   NB 2.B.1
> a separate compartment.

Table 6.158   Multi-storey buildings

Building use	Maximum total area of any compartment (m²)	Minimum fire resistance duration for compartmentation (if any)			
		Basements	The topmost storey of a building is at a height of not more than 7.5m above ground	The topmost storey of a building is at a height of not more than 18m above ground	The topmost storey of a building is at a height of more than 18m above ground
Assembly building	1,500 3,000 6,000	Medium Medium Long	Short Medium Long	Medium Medium Long	Long Long Long
Office	2,000 4,000 8,000	Medium Medium Long	Short Medium Long	Medium Medium Long	Long Long Long
Open sided car park	Unlimited	Medium	Short	Short	Medium
Residential care building and hospitals	1,500	Medium	Medium	Medium	Long
Shop	500 1,000 2,000	Medium Medium Long	Short Medium Long	Medium Medium Long	Long Long Long
Storage building (Class 1)	200 1,000	Medium Long	Medium Long	Medium Long	Long Long
Storage building (Class 2)	500 5,000	Medium Long	Medium Long	Medium Long	Medium Long

Table 6.159   Reaction to fire of wall and ceiling surfaces

Building	Residential care buildings and hospitals	Shops	All other buildings
Room not more than 30 m²	Medium risk	High risk	High risk
Room more than 30 m²	Low risk	Medium risk	Medium risk
Unprotected zone	Low risk	Low risk	Medium risk
Protected zone and fire-fighting shaft	Low risk	Low risk	Low risk

Table 6.160   Recommended fire resistance duration of external walls

Use of building	Not more than 1 m from the boundary		More than 1 m from the boundary	
	No fire suppression system	Fire suppression system	No fire suppression system	Fire suppression system
Assembly building	Medium	Medium	Medium	None
Entertainment building	Medium	Medium	Medium	Medium
Storage building	Medium	Medium	Medium	Medium
Residential care building and hospital	Medium	Medium	Medium	None
Shop	Medium	Medium	Medium	Medium
Office	Medium	Medium	Medium	None
Open-sided car park	Short	Short	None	None

Table 6.161   Recommended distance between cavity barriers in roof spaces above undivided spaces (m)

	Where surfaces are non-combustible or low-risk materials (m)	Where surfaces are medium-, high- or very high-risk materials (m)
Intended for sleeping	20	15
Not intended for sleeping	20	20

Means of escape

Number of exits

The number and distribution of fire exits and escape routes   ND 2.9.1
throughout a building should provide all building users
with the opportunity to escape safely from the building in
the event of an outbreak of fire.

Travel distance

The maximum travel distance from any point within a   ND 2.A.3
compartment should be not more than 64 m to: ND 2.B.3

- each of two adjoining compartments;
- an adjoining compartment and an escape stair or a final
  exit;
- an adjoining compartment and a final exit;
- an escape stair and a final exit.

In non-domestic buildings (such as hospitals and   ND 2.9.3
residential care buildings) the maximum travel distance
from any point on a storey is related to the occupancy
profile of the building users. The available directions of
travel given in Table 6.162 and the recommendations on
travel distance reflect this philosophy.

External escape stairs

An external escape stair should lead directly to a place of safety and be protected against fire from within the building.

An external escape stair should only be used in a residential   ND 2.9.24
care building or a hospital by the staff of those buildings.

Fire precautions

Residential care buildings and hospitals present particular problems in the
event of an outbreak of fire because the occupants may be asleep and will
not be aware that their lives may be at risk. Special fire precautions and a
higher level of automatic fire detection coverage are required for residential
care buildings and hospitals, so as to give occupants and staff the earliest possible warning of an outbreak of fire and allow time for assisting occupants to
evacuate the building in an emergency.

Table 6.162   Recommended travel distance related to available directions of travel

Occupancy profile	Building use	Recommended travel distance (m)	
		One direction of travel	More than one direction of travel
Very slow evacuation	Residential care buildings and hospitals	9	18
	Buildings primarily for disabled people, or people with learning difficulties		
	Swimming pools in air-supported structures		
Slow evacuation	Residential buildings (other than residential care buildings and hospitals)	15	32
	Entertainment buildings, assembly buildings, rooms or auditoriums with provision for fixed seating		
	Shops		
Medium evacuation	Offices, storage buildings and factories	18 [4]	45
	Open-sided car parks		

Fire service access

Access from a public road should be provided for the fire service to assist fire-fighters in their rescue and fire-fighting operations.

The vehicle access route should be provided to the elevation or elevations where the principal entrance, or entrances, is located.	ND 2.12.1
Vehicle access is recommended to other elevations if the building is a hospital.	ND 2.12.1

Residential care buildings

All residential buildings pose special problems because the occupants may be asleep when a fire starts. In residential care buildings, the problems are greater as the mobility, awareness and understanding of the occupants may also be impaired.

Additional guidance concerning residential care buildings is contained in Annex 2A to the Technical Handbook for non-domestic buildings.

A residential care building should have an automatic life safety fire suppression system designed and installed in accordance with guidance set out in Annex ND 2.A.	ND 2.15.2

Compartmentation

Compartments in a residential care building should be limited to a maximum area of 1,500 m².	ND 2.A.1
Sub-compartmentation and the enclosure of fire hazard rooms is to provide physical barriers to a fire, thus affording the staff and occupants additional time to evacuate the building safely.	ND 2.A.1
Every compartment in a residential care building should be divided into at least two sub-compartments by a sub-compartment wall with short fire resistance duration, so that each sub-compartment is not greater than 750 m².	ND 2.A.1
A sub-compartment wall can be constructed with combustible material (i.e. material that is low, medium, high or very high risk provided the wall has short fire resistance duration).	ND 2.A.1
Where a lower roof abuts an external wall, the roof should provide medium fire resistance duration for a distance of at least 3 m from the wall.	ND 2.A.1

Means of escape

At least two exits should be provided from any storey in a residential care building.	ND 2.A.3
Each sub-compartment should be provided with at least two exits by way of protected zones and/or unprotected zones to adjoining, but separate, compartments or sub-compartments.	ND 2.A.3
The security measures proposed should take account of the potential hazards imposed on residential care occupants and extra emphasis may need to be placed on management control and/or any automated life safety systems to ensure the safe evacuation of the building.	ND 2.A.3

Each compartment should be capable of holding the occupancy capacity of that compartment and the occupancy capacity of the largest adjoining compartment.	ND 2.A.3
Where the travel distance is measured to a protected door in a compartment wall or sub-compartment wall in a residential care building, the escape route should not pass through any of the fire hazard rooms.	ND 2.A.3
No room intended for sleeping should be used by more than four people.	ND 2.A.3

Escape lighting

Emergency lighting should be installed in: • a room with an occupancy capacity of more than 10 (and any protected zone or unprotected zone serving such a room); • a protected zone or unprotected zone serving a storey which has two exits, other than a storey in a building not more than two storeys high with a combined floor area of not more than $300\,m^2$ and an occupancy capacity of not more than 10; • a protected zone or unprotected zone in a single stair building of two storeys or more and an occupancy capacity of 10 or more.	ND 2.A.4

Fire detection and alarm systems

The audibility level of the fire alarm sounders should be as specified in BS 5839: Part 1: 2002 except in a place of lawful detention including prisons, the alarm need not be sounded throughout the entire building.	ND 2.A.5
A fire alarm control panel should be provided at the main entrance, or a suitably located entrance to the building agreed with the relevant authority.	ND 2.A.5

Automatic life safety fire suppression systems

Residential care buildings must be designed and constructed in such a way that, in the event of an outbreak of fire within the building, fire and smoke will be inhibited from spreading through the building by the operation of an automatic life safety fire suppression system.	ND 2.15

Note: This requirement was introduced in the United Kingdom following the Ronan Point disaster in 1968.

Residential care buildings should have an automatic life safety fire suppression system.	ND 2.A.6
Sprinkler heads should be 'quick response type' with a response time index (RTI) of not more than $50 \ (m/s)^{1/2}$ and a conductivity factor (c) of not more than $1 \ (m/s)^2$.	ND 2.A.6
Concealed or recessed pattern sprinkler heads should only be used with the approval of the Verifier.	ND 2.A.6

Additional guidance for hospitals

Additional guidance concerning hospitals is contained in Annex 2B to the Technical Handbook for non-domestic buildings.

Compartmentation

Compartment walls and compartment floors in hospitals should be constructed from materials which are non-combustible.	ND 2.1.12
Compartments in a hospital should be limited to a maximum area of $1,500 \ m^2$.	ND 2.B.1
To assist in the safe horizontal evacuation of the occupants in a hospital, every compartment should be divided into at least two sub-compartments by a sub-compartment wall with short fire resistance duration, so that no sub-compartment is more than $750 \ m^2$.	ND 2.B.1
Every storey at a height of more than 7.5 m containing departments to which patients have access, should either: • comprise at least four compartments, each of which should have an area of at least $500 \ m^2$; • have a hospital street and at least three other compartments.	ND 2.B.1
In a hospital, every storey at a height of more than 18 m containing departments to which patients have access, should either: • comprise at least three compartments, each of which should have an area of at least $500 \ m^2$; • have a hospital street and at least three other compartments, each of which should have an area of at least $500 \ m^2$.	ND 2.B.1

A compartment wall with medium fire resistance duration should be provided between different hospital departments and between a hospital department and a protected zone.

ND 2.B.1

Departments such as boiler houses, kitchens and laundries, etc., should:

ND 2.B.1

- never be directly below, nor directly adjoin, the operating theatres, intensive therapy units or special care baby units;
- be provided with a fire suppression system if they are directly below, or directly adjoin, any other hospital department to which patients have access.

Departments such as pharmacies, intensive therapy units, central sterile supplies and health records, etc., should be provided with an automatic fire suppression system where they are directly below, or directly adjoin, operating theatres, intensive therapy units, or special care baby units.

ND 2.B.1

- Intensive therapy units should be divided into at least two sub-compartments by sub-compartment walls with short fire resistance duration;
- all entrances to an intensive therapy unit should be either from a hospital street or through a lobby, enclosed with the same fire resistance duration as that recommended for a sub-compartment.

ND 2.B.1

Rooms such as chemical stores, laboratories, laundry rooms and X-ray film and record stores, etc., are considered to be hazardous and should be enclosed by walls providing a short fire resistance duration.

ND 2.B.1

Compartment walls or compartment floors in a hospital should be constructed of non-combustible material.

ND 2.B.1

Where a compartment wall or sub-compartment wall meets an external wall, there should be a 1 m wide strip of the external wall which has the same level of fire resistance duration as the compartment wall or sub-compartment wall, to prevent lateral fire spread.

ND 2.B.1

Where a lower roof abuts an external wall, the roof should provide a medium fire resistance duration for a distance of at least 3 m from the wall.

ND 2.B.1

Cavity barriers

Cavity barriers need not be provided to divide a cavity above an operating theatre and its ancillary rooms.	ND 2.B.2
Where cavity barriers are installed between a roof and a ceiling above an undivided space, the maximum limit of 20 m should be applied.	ND 2.B.2

Means of escape

The number of exits in a hospital should be calculated in accordance with Table 6.163 (increasing in proportion to the number of patient beds).	ND 2.B.3
In a hospital where a storey is divided into three or more compartments, each compartment should have exits to: • a compartment and a hospital street; • a compartment and an escape stair; • a compartment and a final exit.	ND 2.B.3
Each sub-compartment should be provided with at least two exits by way of protected zones and unprotected zones to adjoining, but separate, compartments or sub-compartments.	ND 2.B.3
Travel distance in a hospital should not exceed 15 m in one direction of travel and 32 m in more than one direction.	ND 2.B.3
The escape route should not pass through a fire hazard room (see 2.B.1).	ND 2.B.3

Table 6.163 Stair and landing configuration for mattress evacuation, in mm

Stair width	Minimum landing width	Minimum landing depth
1,300	2,800	1,850
1,400	3,000	1,750
1,500	3,200	1,550
1,600	3,400	1,600
1,700	3,600	1,700
1,800	3,800	1,800

Hospital streets

A hospital street is a protected zone in a hospital provided to assist in facilitating circulation and horizontal evacuation, and to provide a fire-fighting bridgehead.

A hospital street should have an unobstructed width of at least 3 m.	ND 2.B.3
It should be divided into at least three sub-compartments and not contain a shop or other commercial enterprise.	
At ground storey level, a hospital street should have at least two final exits.	ND 2.B.3
At upper storey level there should be access to at least two escape stairs accessed from separate sub-compartments, located such that: <ul><li>the distance between escape stairs is not more than 64 m;</li><li>the distance of single direction of travel within the hospital street is not more than 15 m;</li><li>the distance from a compartment exit to an escape stair is not more than 32 m.</li></ul>	ND 2.B.3
A door from a hospital street to an adjoining compartment should: <ul><li>be located so that an alternative independent means of escape from each compartment is available;</li><li>not be located in the same sub-compartment as a door to a protected zone containing a stairway or lift.</li></ul>	ND 2.B.3
Every escape stair opening into the hospital street should be located so that the travel distance from an escape stair exit to a door leading directly to a place of safety is not more than 64 m.	ND 2.B.3
An escape route from a hospital department to which patients have access should be to: <ul><li>a place of safety;</li><li>a protected zone;</li><li>an unprotected zone in another compartment or sub-compartment.</li></ul>	ND 2.B.3
Doors should be designed to accommodate bed-patient evacuation.	ND 2.B.3
The unobstructed width of every escape route intended for bed-patient evacuation should be at least 1,500 mm.	
In patient sleeping accommodation, an escape stair width should be not less than 1,300 mm and designed so as to facilitate mattress evacuation.	ND 2.B.3

The landing configuration should follow the guidance in
Table 6.163.   ND 2.B.3

Locks

Hospitals can present difficulties when assessing the risks associated with
security against the need to evacuate the building safely in the case of fire.

Some parts of hospitals could have patients who might put themselves at risk
and the security measures proposed should take account of the potential hazards
and extra emphasis may need to be placed on management control and/or any
automated life safety systems to ensure the safe evacuation of the building.

Protected lobbies

Where an escape stair in a protected zone serves an upper   ND 2.B.3
storey containing a department to which patients have
access, access to the protected zone should be by way of
a protected lobby, or via the hospital street if the height of
the storey is not more than 18 m.

Escape lighting

Essential lighting circuits should be installed throughout a   ND 2.B.4
hospital and designed to provide not less than 30% of the
normal lighting level.

In an area where a 15 s response time would be considered   ND 2.B.4
hazardous (e.g. a stairway), emergency lighting should be
provided by battery back-up giving a response time of not
more than 0.5 s.

The distribution boards for essential and non-essential   ND 2.B.4
circuits may be in the same location but should be in
separate cabinets.

Fire detection and alarm systems

The audibility level of the fire alarm sounders should   ND 2.B.5
follow the guidance in BS 5839: Part 1: 2002. However,
in a hospital department to which patients have access, the
audibility need only be 55 dB(A) or 5 dB(A) above the level
of background noise, whichever is greater.

- A main fire alarm control panel is provided at the main entrance, or a suitably located secondary entrance to the building;
- repeater panels should be provided at all other fire service access points.

ND 2.B.5

In the case of a hospital designed to accommodate not more than 10 residents, one or more smoke alarms should be installed

ND 2.B.5

Fire service facilities

Where a hospital with a hospital street has two or more escape stairs, facilities should be provided in accordance with Table 6.164.

ND 2.B.6

For fire hose, no point on any storey should be further from a fire-fighting outlet than one storey height, and 60 m measured along an unobstructed route.

ND 2.B.6

Different fire-fighting facilities should not be provided throughout the varying storey heights of a building.

ND 2.B.6

A fire-fighting lift need not serve the top storey of a building where:

ND 2.B.6

- the top storey is for service plant use only;
- access to the plant room is from an escape stair from the storey below;

Table 6.164 Facilities on escape stairs in hospitals with hospital streets

Storey height and depth of hospital	Facilities on escape stairs
Basements at a depth more than 10 m	• Fire-fighting lift • Fire-fighting shaft • Dry fire main (outlet located at every departmental entrance)
Basements at a depth not more than 10 m	• Dry fire main (outlet located at every departmental entrance)
Topmost storey height not more than 18 m	• Dry fire main (outlet located at every departmental entrance)
Topmost storey height not more than 0 m	• Fire-fighting shaft • Fire-fighting lift • Dry fire main (outlet located at every departmental entrance)

- the foot of the escape stair is not more than 4.5 m from the fire-fighting lift;
- dry rising mains are installed in the protected lobbies of the escape stair.

Every single-storey hospital with a hospital street should be   ND 2.B.6
provided with a dry fire main and the outlet should be located
in the hospital street at every hospital departmental entrance.

 A dry fire main need not be provided where no point within the storey (not being a protected zone) is more than 60 m measured along an unobstructed route for the fire hose, from the access point or points.

6.29 Shopping centres

A shopping mall or shopping centre is a building or set of buildings that contain a variety of retail units, with interconnecting walkways enabling visitors to walk easily from unit to unit. Most shopping centres are a hive of activity and fire safety precautions have to be fully observed.

6.29.1 Requirements

Fire

Every building in an enclosed shopping centre must be   2.11
designed and constructed in such a way that in the event of
an outbreak of fire within the building: 2.15

- *the occupants are alerted to the outbreak of fire;*
- *fire and smoke will be inhibited from spreading through the building by the operation of an automatic life safety fire suppression system.*

6.29.2 Meeting the requirements

 For additional guidance on enclosed shopping centres, see Annex 2.C to the Non-Domestic Technical Handbook.

Occupancy capacity

The occupancy capacity shall be obtained by dividing the area in square metres by the relevant occupancy load factor.

Occupancy capacity of the entire shopping centre

The overall occupancy capacity of the entire shopping centre is the sum of: ND 2.C.3a

- all mall areas up to a maximum width of 6 m (using an occupancy load factor of 0.7);
- all remaining areas beyond the 6 m (using an occupancy load factor of 2.0);
- food courts forming part of the mall (using an occupancy load factor of 1.0);
- all shops and all other use tenancies (using an occupancy load factor of 4.0).

Occupancy capacity of shops

When calculating the occupancy capacity of the individual shops (or other use tenancies) the occupancy load factors shown in Table 6.165 should be used as appropriate.

The aggregate unobstructed width (in mm) of all escape routes from a mall should be at least 2.65 multiplied by the occupancy capacity of the entire shopping centre as described above.   ND 2.C.3b

A shop of more than 1,300 m² in area at mall level, where the means of escape has been designed independently of the mall, need not be included in this calculation. Similarly, a kiosk not exceeding 25 m² which is not accessible to the public, or where the depth the kiosk from the mall is not more than 5 m, need not be included in the calculation.   ND 2.C.3b

Table 6.165 Occupancy capacity for shops

Description of room or space	Occupancy load factor
Shops trading predominantly in furniture, floor coverings, cycles, perambulators, large domestic appliances or other bulky goods or trading on a wholesale self-selection basis	7.0
Shop sales areas (other than those listed above) including supermarkets and department stores (all sales areas), *shops* for personal services such as hairdressing and *shops* for the delivery or uplift of goods for cleaning, repair or other treatment or for members of the general public themselves carrying out such cleaning, repair or other treatment	2.0

The unobstructed width of each individual exit from the mall should be at least 1.8 m.   ND 2.C.3b

The aggregate unobstructed exit width, in mm, from each shop and the unobstructed exit width of a route, or routes, which do not enter the mall, should be at least 5.3 times the occupancy capacity of the shop.   ND 2.C.3b

Crèches

A crèche that is provided within an enclosed shopping centre should be designed so that it is: ND 2.C.3b

- at ground level or exit level;
- not sited on a storey higher than those where parents or guardians may be located, unless escape is via the upper storey itself;
- located adjacent to an external wall and has at least two exits, one of which should be directly to a place of safety.

Fire precautions

Enclosed shopping centres can be extremely complex to design. There are large fire loads and large numbers of people all within a complicated series of spaces where most people only know one way in or out. Special fire precautions are, therefore, required within an enclosed shopping centre and it is recommended that automatic fire detection be installed throughout.

Every building in an enclosed shopping centre must be designed and constructed in such a way that in the event of an outbreak of fire within the building, the occupants are alerted to the outbreak of fire.   ND 2.11

In enclosed shopping centres: ND 2.11.4

- alternative life safety systems should be installed;
- automatic fire detection systems should be compatible with, and interact with, other mechanical and electrical equipment.

Automatic fire detection and alarm system

Automatic fire detection and alarm systems in enclosed shopping centres can increase significantly the level of safety of the occupants.

Every building in an enclosed shopping centre must be designed and constructed in such a way that, in the event of an outbreak of fire within the building, fire and smoke will be inhibited from spreading through the building by the operation of an automatic life safety fire suppression system.	ND 2.15

An enclosed shopping centre should be provided with an automatic fire detection and alarm system, designed and installed in accordance with the following recommendations:

ND 2.15.3
ND 2.C.5

- the fire alarm should be activated upon the operation of the sprinklers; or
  - manual call points; or
  - on the activation of the alarm in a shop; or
  - on the activation of the alarm anywhere other than in an individual shop; or
  - on activation of sprinklers anywhere within the shopping centre; or
- all areas of the shopping centre, including shops, should be alerted using a voice alarm system which follows the guidance in BS 5839: Part 8: 1998; however, individual shops may use conventional sounders within the shop itself.

The fire alarm system should be interfaced with other fire safety systems, to operate automatically in the correct zones.

ND 2.C.5

On the operation of the fire alarm:

ND 2.C.5

- all escalators should come to a controlled halt and lifts should return to the ground storey (or exit level);
- all systems within the mall or shops which play amplified music are silenced;
- any smoke dampers installed to prevent the siphoning of smoke are activated;
- (subject to the 4 min grace period where appropriate) all air moving systems, mains and pilot gas outlets, combustion air blowers and gas, electrical and other heating appliances in the reservoir are shut down.

The main fire alarm system control panel shall be installed within the control room and indicator (or repeater) panels shall be provided at each of the fire-fighting access points.

ND 2.C.5

Smoke and Heat Exhaust Ventilation Systems

A Smoke and Heat Exhaust Ventilation System (SHEVS) should be installed in:

ND 2.1.3

ND 2.C.1

- the mall of an enclosed shopping centre;
- shops with a storey area more than $1,300\,m^2$;
- large shops (other than in enclosed shopping centres), with a compartment area more than $5,600\,m^2$.

SHEVS should be designed in accordance with the following where appropriate:

ND 2.C.1

- the underside of the mall roof should be divided into smoke reservoirs, each of which should be not more than $2,000\,m^2$ in area and at least 1.5 m deep measured to the underside of the roof or to the underside of any high-level plant or ducts within the smoke reservoir or the underside of an imperforate suspended ceiling;
- smoke reservoirs should be formed by fixed or automatically descending smoke curtains which are no greater than 60 m apart, measured along the direction of the mall;
- smoke should not be allowed to descend to a height of less than 3 m above any floor level;
- each smoke reservoir should be provided with the necessary number of smoke ventilators or extract fans to extract the calculated volume of smoke produced, spaced evenly throughout the reservoir;
- where mechanical extraction is used, there should be spare fan capacity equivalent to the largest single fan in the reservoir which will operate automatically on the failure of any one of the fans, or which runs concurrently with the fans;
- any fans, ducts and reservoir screens provided should be designed to operate at the calculated maximum temperature of the smoke within the reservoir in which they are located, but rated to a minimum of 300°C for 30 min;
- structures supporting any fans, ducts or reservoir screens should have the same performance level as the component to be supported;

- the fans or ventilators within the affected smoke
  reservoirs should operate:
  - on the actuation of any automatic fire suppression
    system;
  - on the actuation of the smoke detection system
    within the reservoir;
  - on the operation of more than one smoke detector
    anywhere in the shopping centre;
  - following a delay not exceeding 4 min from
    initiation of the first fire alarm signal anywhere in
    the shopping centre;
- replacement air should be provided automatically on
  the operation of the ventilation or exhaust system at
  a level at least 0.5 m below the calculated level of the
  base of the smoke layer;
- any power source provided to any elements of the
  smoke and heat exhaust ventilation system should be
  connected by mineral insulated cables or by cables
  which are code A category specified in BS 6387: 1994
  or by cables protected from damage to the same level;
- an automatically switched standby power supply
  provided by a generator should be connected to any
  fans provided as part of the smoke and heat ventilation
  system capable of simultaneously operating the fans
  in the reservoir affected and any of the two adjacent
  reservoirs;
- simple manual overriding controls for all smoke
  exhaust, ventilation and air input systems should be
  provided at all fire service access points and any fire
  control room provided;
- where outlets are provided with weather protection,
  they should open on the activation of the fan(s) or
  ventilators;
- smoke from areas adjoining the smoke reservoirs
  should only be able to enter one reservoir;
- where there is an openwork ceiling, the free area of the
  ceiling should not be less than 25% of the area of the
  smoke reservoir, or, for natural ventilation, 1.4 times
  the free area of the roof mounted fire ventilator above
  (3 times where the height from floor to roof ventilator
  is more than 12 m), whichever free area is the greater,
  and be evenly distributed to prevent an unbalanced air
  flow into the reservoir;

- when a natural ventilation system is used and the   ND 2.C.1
  smoke reservoir includes a suspended ceiling, other
  than an openwork ceiling, the free area of the ventilator
  opening in the suspended ceiling, or any ventilator
  grille in the ceiling, should not be:
    - less than 1.4 times (3 times where the height from
      floor to roof ventilator is more than 12 m) that of the
      roof mounted fire ventilator above in the case of a
      ventilator opening;
    - 2 times (3.5 times where the height from floor to roof
      ventilator is more than 12 m) for any ventilator grille.

Automatic life safety fire suppression systems

An automatic life safety fire suppression system should   ND 2.C.7
be installed in an enclosed shopping centre and cover the
entire area, other than:

- a mall or part of a mall with a ceiling height more than
  10 m;
- a stairway enclosure;
- a car park;
- every area where sprinklers would prove to be a hazard
  (e.g. main electrical switchgear).

The level of fire suppression should be appropriate to   ND 2.C.7
the occupancies within the enclosed shopping centre and
should be determined on the basis of a risk assessment.

The type of sprinkler head should be a 'quick response'   ND 2.C.7
type with a response time index (RTI) of not more than 50.

Separating walls

A separating wall is not necessary between a shop and a   ND 2.C.2
mall except to shops having mall-level storey areas more
than 2,000 m² that are located opposite each other.

The mall width should at no point be less than 6 m.   ND 2.C.2

Means of escape

An enclosed shopping centre should be so designed that:

- from every part of the mall and from every mall-level shop there should be at least two directions of travel leading to a place of safety without passing through any space in single occupation;   ND 2.C.3
- each shop having a frontage to the mall should be provided with an alternative escape route that is not through the mall;
- the travel distances in the mall should be limited to 9 m in one direction of travel and 45 m in more than one direction of travel;
- in a shop, the travel distance is 15 m in one direction and 32 m in more than one direction.

Where a service corridor is used for means of escape directly from a shop or shops, the unobstructed width should be based on the total number of occupants of the largest shop that evacuates into the corridor, plus an additional width of 1 m to allow for goods in transit.   ND 2.C.3b

Where a service corridor is used as an escape route, it should not be used for any form of storage.   ND 2.C.3b

Escape lighting

An enclosed shopping centre should be provided with emergency lighting in all mall areas and all protected zones and unprotected zones.   ND 2.C.4

Emergency lighting should be installed so that it is not rendered ineffective by smoke-filled reservoirs.   ND 2.C.4

Fire service facilities

An enclosed shopping centre should have a fire control room: ND 2.C.6

- constructed as a separate compartment;
- have access points away from the discharge points for the general public;   ND 2.C
- be provided with an alternative means of escape;
- be located adjacent to a fire service access point and accessible from the open air.

> Every single-storey enclosed shopping centre should be   6
> provided with a dry fire main with an outlet located not
> more than 5 m from a fire service access point or points.

6.30 Building documentation and information to be provided

The Building (Scotland) Regulations require that all buildings shall be energy efficient and that occupiers of all buildings shall be provided with information regarding the maintenance and efficient operation of building services and energy supply systems.

6.30.1 Requirements

> *The occupiers of a building must be provided with written*   6.8
> *information by the owner:*
>
> - *on the operation and maintenance of the building serv-*
>   *ices and energy supply systems; and*
> - *where any air-conditioning system in the building is*
>   *subject to a time-based interval for inspection of the*
>   *system.*
>
>   This standard does not apply to:
>
> - major power plants serving the National Grid;
> - buildings which do not use fuel or power for heating,
>   lighting, ventilating and cooling the internal environment
>   and heating the water supply services;
> - the process and emergency lighting components of a
>   building;
> - heating provided solely for the purpose of frost
>   protection;
> - lighting, ventilation and cooling systems in a domestic
>   building;
> - energy supply systems used solely for industrial and
>   commercial processes, leisure use and emergency use
>   within a building.

Every building must be designed and constructed in such   6.9
a way that:

(a) *an energy performance certificate for the building*
    is affixed to the building, indicating the approximate
    annual carbon dioxide emissions and energy usage
    of the building based on a standardized use of the
    building;
(b) *the energy performance for the certificate is*
    calculated in accordance with a methodology
    which is asset-based, conforms with the European
    Directive 2002/91/EC and uses UK climate data;
(c) *the energy performance certificate is displayed in a*
    prominent place within the building.

 Note: We are reliably informed that it is intended
that Scottish Ministers will direct local authorities
to apply Standard 6.9 to **all** existing buildings (i.e.
being sold or rented out) using Section 25 (2) of
the Building (Scotland) Act 2003.

 Standard 6.9(c) only applies to buildings with a floor
area of more than $1,000\,m^2$, which are occupied by
public authorities and institutions providing public
services, which can be visited by the public.

 This standard does not apply to:

- buildings which do not use fuel or power for
  controlling the temperature of the internal environment;
- non-domestic buildings and buildings that are
  ancillary to a dwelling that are stand-alone having an
  area less than $50\,m^2$;
- conversions, alterations and extensions to buildings
  other than alterations and extensions to stand-alone
  buildings less than $50\,m^2$ that would increase the area
  to $50\,m^2$ or more;
- buildings involving the fit-out of the building shell
  which is the subject of a continuing requirement;
- limited-life buildings which have an intended life of
  less than 2 years.

Every building must be designed and constructed in such   6.10
a way that each part of a building designed for different
occupation is fitted with fuel consumption meters.

This standard does not apply to:

- domestic buildings;
- communal areas of buildings in different occupation;
- district or block heating systems where each part of the building designed for different occupation is fitted with heat meters;
- heating fired by solid fuel or biomass.

6.30.2 Meeting the requirements

Written information

Correct use and maintenance of building services equipment is essential if the benefits of enhanced energy efficiency are to be realized from such equipment. To achieve this it is essential that user and maintenance instructions together with all other relevant documentation are available to the occupier of the building.

Domestic buildings

For a domestic building, written information concerning the operation and maintenance of the heating and hot water service system (together with any decentralized power generation equipment) should be made available for the use of the occupier.   D 6.8.1

Non-domestic buildings

For non-domestic buildings, a logbook containing information of energy system operation and maintenance (such as building services plant and controls) to ensure that the building user can optimize the use of fuel, shall be provided.   ND 6.8.1

 CIBSE Technical Memorandum 31 (TM31) provides guidance on the presentation of a logbook, and the logbook information should be presented in this or a similar manner.

Work on existing buildings

Where alterations are carried out to building services on a piecemeal basis, the alterations might not result in optimum energy efficiency being attained for the whole system. Where this occurs:

• a list of recommendations which would improve the overall energy efficiency of the system should be provided;	D 6.8.2
• the person responsible for the commissioning of that part of the system should make available to the owner and occupier, a list of recommendations that will improve the overall energy efficiency of the system;	ND 6.8.2
• on completion of the extension or alteration to the building services system, the commissioning information should be updated in the logbooks.	ND 6.8.2

Energy Performance Certificates

EU Directive (2002/91/EC) requires Energy Performance Certificates (EPCs) to be made available to prospective owners and tenants when dwellings are constructed.

Calculating the carbon dioxide emissions for a certificate

The EU Directive allows energy performance to be evaluated in a number of different ways. Within Scotland, the method chosen is carbon dioxide using the Directive compliant methodology.

Note: This is the reason why the Scottish Building Regulations does not offer any guidance on minimum or maximum area for windows, doors, rooflights and roof windows in dwellings and non-domestic buildings.

For the purpose of establishing a rating for the Energy Performance Certificate for a new dwelling the values and specifications used to obtain a Building Warrant should be adopted.	D 6.9.1 ND 6.9.1
Note: In most cases SAP 2005 will be used for the new dwelling and the SBEM calculation tool for the new building.	
Where a building contains multiple dwellings a rating is required for each individual dwelling, taking into consideration the percentage of low-energy lighting and the type of heating that has been installed.	D 6.9.1 ND 6.9.1

Note: Accommodation up to $50\,m^2$ used by an occupant of a dwelling in their professional or business capacity should be considered as a part of the dwelling.

If a non-domestic building incorporates within it a dwelling (e.g. a caretaker's flat) a separate certificate should always be provided for the dwelling and reference should be made to the Domestic Technical Handbook.

ND 6.9.1

Information to be provided for dwellings

The Energy Performance Certificate should display the following information:

D 6.9.2

- the postal address of the building for which the certificate is issued;
- building type;
- the name of the SBSA protocol organization issuing the certificate (if applicable) and may include the member's membership number;
- the date of the certificate;
- the conditioned floor area of the dwelling;
- the main type of heating and fuel;
- the calculation tool used for certification;
- a specific indication of current CO_2 emissions and an indication of potential emissions;
- a seven-band scale in different colours representing the following bands of carbon dioxide emissions A, B, C, D, E, F and G, where A = excellent and G = very poor;
- the approximate energy use expressed in kWh per m^2 of floor area per annum;
- a list of cost-effective improvements (lower cost measures).

The Energy Performance Certificate should also include a statement to the effect:

D 6.9.2

'THIS CERTIFICATE MUST BE AFFIXED TO THE BUILDING AND NOT BE REMOVED UNLESS IT IS REPLACED WITH AN UPDATED VERSION'.

A model form for an Energy Performance Certificate for a dwelling is given on the SBSA website (www.sbsa.gov.uk).

Measures presented on the certificate must meet Scottish Building Regulations, and should be technically feasible and specific to the individual dwelling.

D 6.9.2

Certificates may give additional advice on projected energy costs and improvements that are cost-effective only when additional work is being carried out, e.g. providing insulation when replacing flat roof coverings.

D 6.9.2

Information to be provided for non-domestic buildings

The Energy Performance Certificate should display the following information:

ND 6.9.2

- the postal address of the building for which the certificate is issued;
- building type;
- the name of the SBSA protocol organization issuing the certificate (if applicable) and may include the member's membership number;
- the date of the certificate;
- the conditioned floor area of the dwelling;
- the main type of heating and fuel;
- the type of electricity generation;
- whether or not there is any form of building integrated renewable energy generation;
- the calculation tool used for certification;
- the type of ventilation system;
- a specific indication of current CO_2 emissions and an indication of potential emissions;
- a seven-band scale in different colours representing the following bands of carbon dioxide emissions: A, B, C, D, E, F and G, where A = excellent and G = very poor;
- the approximate energy use expressed in kWh per m^2 of floor area per annum;
- a list of cost-effective improvements (lower cost measures).

The Energy Performance Certificate should also include:

ND 6.9.2

- a statement to the effect:
  'THIS CERTIFICATE MUST BE AFFIXED TO THE BUILDING AND NOT BE REMOVED UNLESS IT IS REPLACED WITH AN UPDATED VERSION';

- if the building is a public building and over $1,000\,m^2$ in area, a statement to the effect:
  'THIS CERTIFICATE SHALL BE DISPLAYED IN A PROMINENT PLACE'

A model form for an Energy Performance Certificate for a dwelling is given on the SBSA website (www.sbsa.gov.uk).

Measures presented on the certificate must meet Scottish Building Regulations, and should be technically feasible and specific to the individual dwelling.	D 6.9.2
Certificates may give additional advice on projected energy costs and improvements that are cost-effective only when additional work is being carried out, e.g. providing insulation when replacing flat roof coverings.	D 6.9.2

Location of an Energy Performance Certificate
Domestic buildings

The Energy Performance Certificate should be indelibly marked and located in a position that is readily accessible (such as a cupboard containing the gas or electricity meter or the water supply stopcock) and it should be protected from weather and not easily obscured.	D 6.9.3
For conservatories and for other ancillary stand-alone buildings of less than $50\,m^2$ floor area, an Energy Performance Certificate need not be provided.	D 6.9.4
For those buildings of a floor area of $50\,m^2$ or more, the guidance in the Non-Domestic Technical Handbook should be followed and an additional certificate supplementing the one for the dwelling should be provided.	D 6.9.4

Non-domestic buildings

Buildings (see below) with an area of over $1,000\,m^2$ occupied by public authorities and by institutions providing public services to a large number of persons and therefore frequently visited by these persons, must have an energy certificate (no more than 10 years old) placed in a prominent place such as an area of wall which is clearly visible to the public in the main entrance lobby or reception.	ND 6.9.3.

Note: Examples of such buildings are:

- colleges (further education, higher education), universities;
- community centres;
- concert halls, theatres;
- crematoria;
- day centres;
- education centres, schools (nursery, primary, secondary, special);
- exhibition halls (multi-function centres);
- headquarters' buildings (of local authorities such as district councils, health and social services trusts and boards, education and library boards, etc.) where the public have an unqualified right of access (e.g. to attend council meetings, parliamentary meetings or other events to which the public have access);
- health centres, hospitals;
- hostels, halls of residence;
- law courts;
- leisure centres, swimming pools, sports pavilions;
- libraries, museums, art galleries;
- offices (passport office, motor tax office, benefits office, etc.) having a public counter and providing services directly to the public;
- outdoor centres;
- passenger terminals (rail, bus, sea and air);
- police stations (with a public counter);
- residential care buildings;
- visitor centres; and
- youth centres.

For all other buildings, the Energy Performance Certificate should be indelibly marked and located in a position that is readily accessible, protected from weather and not easily obscured, such as in a cupboard containing the gas or electricity meter or the water supply stopcock.	ND 6.9.3
For stand-alone ancillary buildings (such as a kiosk for a petrol filling station) of less than $50\,m^2$ floor area, an Energy Performance Certificate need not be provided.	ND 6.9.4
For stand-alone buildings of a floor area of $50\,m^2$ or more that are heated or cooled which are ancillary or subsidiary to the main building, a certificate should be provided, in addition to the one for the main building.	ND 6.9.4

Metering – non-domestic buildings

To enable building operators to effectively manage fuel use, systems should be provided with fuel meters to enable the annual fuel consumption to be accurately measured.

New buildings

All buildings should be installed with fuel consumption metering in an accessible location, such as presently installed by the utility.	ND 6.10.1
Each area divided by separating walls and separating floors and designed for different occupation should be provided with a fuel meter to measure the fuel usage in each area.	ND 6.10.1
Where multiple buildings or fire separated units are served on a site by a communal heating appliance, the fuel metering shall be installed both at the communal heating appliance and heat meters at the individual buildings served.	ND 6.10.1
Metering shall be provided to measure the hours run, electricity generated, and the fuel supplied to a combined heat and power unit.	ND 6.10.1

Existing buildings

Existing buildings which are a result of a conversion or in buildings where an extension or alteration is carried out should be installed with fuel consumption metering.	ND 6.10.2
Where conversions, extensions or alterations result in the creation of two or more units, each unit should have a fuel meter installed in an accessible location.	ND 6.10.2
A fuel meter should be installed if a new fuel type or new boiler (where none existed previously) is installed.	ND 6.10.2

Bibliography

Standards referred to

Compliance with a British, European or International Standard does not of itself confer immunity from legal obligations. These Standards can, however, provide a useful source of information which could be used to supplement or provide an alternative to the guidance given in the Technical Handbooks that form part of the Scottish Building Regulations.

When a Technical Handbook makes reference to a named standard, the relevant version of the standard is the one listed at the end of the publication. However, if this version of the standard has been revised or updated by the issuing standards body, the new version may be used as a source of guidance provided it continues to address the relevant requirements of the Regulations.

Drafts for Development (DDs) are not British Standards. They are issued in the DD series of publications and are of a provisional nature. They are intended to be applied on a provisional basis so that information and experience of their practical application may be obtained and the document developed. Where the recommendations of a DD are adopted then care should be taken to ensure that the requirements of the Regulations are adequately met.

Any observations that a user may have in relation to any aspect of a DD should be passed on to the BSI.

1.1 Standards

Title	Standard
Accommodation of building services in ducts. Code of practice.	BS 8313: 1989
Acoustics – measurement of sound insulation in buildings and of building elements:	BS EN ISO 140
Part 3: 1995 Laboratory measurement of airborne sound insulation of building elements.	
Part 4: 1998 Field measurements of airborne sound insulation between rooms.	
Part 6: 1998 Laboratory measurements of impact sound insulation of floors.	
Part 7: 1998 Field measurements of impact sound insulation of floors.	
Part 8: 1998 Laboratory measurements of the reduction of transmitted impact noise by floor coverings on a heavyweight standard floor.	

(*Continued*)

Title	Standard
Acoustics – rating of sound insulation in buildings and of building elements. Part 1: 1997 Airborne sound insulation. Part 2: 1997 Impact sound insulation.	BS EN ISO 717
Acoustics – sound absorbers for use in buildings.	BS EN ISO 11654: 1997
Acoustics – materials used under floating floors in dwellings.	BS EN 29052-1: 1992
Aggregates from natural sources for concrete.	BS 882: 1983
Air admittance valves for drainage systems. Requirements, test methods and evaluation of conformity.	BS EN 12380: 2002
Ancillary components for masonry. Part 1: 2001 Ties, tension straps, hangers and brackets. Part 2: 2001 Lintels. Part 3: 2001 Bed joint reinforcement of steel meshwork.	BS EN 845
Application of fire safety engineering principles to the design of buildings. Part 0: 2002 Guide to design framework and fire safety engineering procedures. Part 1: 2003 Initiation and development of fire within the enclosure of origin (Sub-system 1). Part 2: 2002 Spread of smoke and toxic gases within and beyond the enclosure of origin (Sub-system 2). Part 3: 2003 Structural response and fire spread beyond the enclosure of origin (Sub-system 3). Part 4: 2003 Detection of fire and activation of fire protection systems (Sub-system 4). Part 5: 2002 Fire service intervention (Sub-system 5). Part 6: 2004 Human factors. Life safety strategies. Occupation evaluation, behaviour and condition (Sub-system 6). Part 7: 2003 Probabilistic risk assessment.	BS 7974
Asbestos-cement pipes, joints and fittings for sewerage and drainage.	BS 3656: 1981 (1990)
Assessing exposure of walls to wind-driven rain. Code of practice.	BS 8104: 1992
Audio-Frequency Induction-Loop Systems (AFILS)	BS 7594: 1993
Building components and building elements. Thermal resistance and thermal transmittance – calculation method.	BS EN ISO 6946: 1997
Building drainage. Code of practice.	BS 8301: 1990
Building hardware – emergency exit devices operated by a lever handle or push pad. Requirements and test methods.	BS EN 179: 1998
Building hardware – panic exit devices operated by a horizontal bar – requirements and test methods.	BS EN 1125: 1997
Building materials and products – hygrothermal properties – tabulated design values.	BS EN 12524: 2000

(*Continued*)

Title	Standard
Building valves. Combined temperature and pressure relief valves, tests and requirements.	BS EN 1490: 2000
Buildings and Structures for Agriculture: Part 22: 2003 Code of practice for design, construction and loading. Part 50: 1993 Code of practice for design, construction and use of storage tanks and reception pits for livestock slurry. Part 80: 1990 Code of practice for design and construction of workshops, maintenance and inspection facilities.	BS 5502
Calcium silicate (sandlime and flintlime) bricks.	BS 187: 1978
Calibration and testing laboratory accreditation systems – general requirements for operation and recognition.	BS EN ISO/IEC 17011: 2004
Cast iron spigot and socket drain pipes and fittings.	BS 437: 1978
Cast iron spigot and socket flue or smoke pipes and fittings.	BS 41: 1973 (1981)
Cement Part 1: 2000 Composition, specifications and conformity criteria for common elements. Part 2: 2000 Conformity evaluation.	BS EN 197
Chimneys – chimney components – concrete flue blocks.	BS EN 1858: 2003
Chimneys – chimney components – concrete flue liners.	BS EN 1857: 2003
Chimneys – clay/ceramic flue liners – requirements and test methods.	BS EN 1457: 1999
Chimneys – components, concrete outer wall elements.	BS EN 12446: 2003
Chimneys – general requirements.	BS EN 1443: 2003
Chimneys – performance requirements for metal chimneys – system chimney products.	BS EN 1856-1: 2003
Chimneys (factory-made insulated). Part 1: 1990 (1996) Methods of Test, AMD 8379. Part 2: 1990 (1996) Specification for chimneys with stainless steel flue linings for use with solid fuel fired appliances. Part 3: 1990 (1996) Specification for chimneys with stainless steel flue lining for use with oil-fired appliances.	BS 4543
Chimneys. Clay/ceramic flue liners.	BS EN 1457: 1999
Chimneys. Metal chimneys.	BS EN 1859: 2000
Chimneys. Clay/ceramic flue blocks for single wall chimneys – requirements for test methods.	BS EN 1806: 2006
Chimneys. Execution standards of metal chimneys. Part 1. Chimneys for non-room sealed heating appliances.	BS EN 12391-1: 2004
Chimneys. Performance requirements for metal chimneys – metal liners and connecting flue pipe products.	BS EN 1856-2: 2005

(Continued)

Title	Standard
Chimneys. Thermal and fluid dynamic calculation methods. Chimneys serving one appliance.	BS EN 13384-1: 2002
Clay and calcium silicate modular bricks.	BS 6649: 1985
Clay bricks.	BS 3921: 1985
Clay flue linings and flue terminals.	BS 1181: 1999
Clay pavers – requirements and test methods.	BS EN 1344: 2002
Code of practice for protection of buildings against water from the ground.	CP 102: 1973
Code of practice for sheet roof and wall coverings.	CP 143: Part 5: 1964 CP 143: Part 10: 1973 CP 143: Part 12: 1970 CP 143: Part 15: 1973
Components for residential sprinkler systems. Specification and test methods for residential sprinklers.	DD 252: 2002
Components for smoke and heat control systems. Part 1: 1999 Specification for natural smoke and heat exhaust ventilators. Part 2: 1990 Specification for powered smoke and heat exhaust ventilators. Part 6: 2005 Specifications for cable systems.	BS 7346
Components of automatic fire alarm systems for residential premises: Part 1: 2000 Specification for self-contained smoke alarms and point-type smoke detectors.	BS 5446
Concrete. Part 1: 1990 Guide to specifying concrete. Part 2: 1990 Method for specifying concrete mixes. Part 3: 1990 Specification for the procedures to be used in producing and transporting concrete. Part 4: 1990 Specification for the procedures to be used in sampling, testing and assessing compliance of concrete.	BS 5328
Constituent materials and for mixtures – coated macadam for roads and other paved areas.	BS 4987-1: 2005
Construction and testing of drains and sewers.	BS EN 1610: 1998
Control of condensation in buildings.	BS 5250: 2002
Copper and copper alloys. Plumbing fittings: Part 1: Fittings with ends for capillary soldering or capillary brazing to copper tubes. Part 2: Fittings with compression ends for use with copper tubes. Part 3: Fittings with compression ends for use with plastics pipes. Part 4: Fittings combining other end connections with capillary or compression ends. Part 5: Fittings with short ends for capillary brazing to copper tubes.	BS EN 1254: 1998

(Continued)

Title	Standard
Copper and copper alloys. Seamless, round copper tubes for water and gas in sanitary and heating applications.	BS EN 1057: 1996
Copper and copper alloys. Tubes. Part 1: 1971 Copper tubes for water, gas and sanitation.	BS 2871
Copper direct cylinders for domestic purposes.	BS 699: 1984
Copper hot water storage combination units for domestic purposes.	BS 3198: 1981
Copper indirect cylinders for domestic purposes: Part 1: 2002 Open vented copper cylinders – requirements and test measures. Part 2: 1990 Specification for single feed indirect cylinders.	BS 1566
Damp-proof courses in masonry construction. Code of practice for design and installation.	BS 8215: 1991
Dedicated liquefied petroleum gas appliances. Domestic flueless space heaters (including diffusive catalytic combustion heaters).	BS EN 449: 1997
Design and construction of fully supported lead sheet roof and wall coverings.	BS 6915: 2001
Design of buildings and their approaches to meet the needs of disabled people – Code of practice.	BS 8300: 2001
Design of non-load bearing external vertical enclosures of buildings. Code of practice.	BS 8200: 1985
Design, installation, testing and maintenance of services supplying water for domestic use within buildings and their curtilages.	BS 6700: 1987
Direct surfaced wood chipboard based on thermosetting resins.	BS 7331: 1990
Discharge and ventilating pipes and fittings, sand-cast or spun in cast iron. Part 1: 1990 Specification for spigot and socket systems. Part 2: 1990 Specification for socket less systems.	BS 416
Drain and sewer systems outside buildings – hydraulic design and environmental considerations.	BS EN 752-4: 1998
Drain and sewer systems outside buildings: Part 1: 1996 Generalities and definitions. Part 2: 1997 Performance requirements. Part 3: 1997 Planning. Part 4: 1997 Hydraulic design and environmental aspects. Part 5: 1997 Rehabilitation. Part 6: 1998 Pumping installations. Part 7: 1998 Maintenance and operations.	BS EN 752
Ductile iron pipes, fittings, accessories and their joints for sewerage applications. Requirements and test methods.	BS EN 598: 1995
Earth retaining structures.	BS 8002: 1994
Electrical controls for household and similar general purposes.	BS 3955: 1986

(*Continued*)

Title	Standard
Electrical Installations (IEE Wiring Regulations 17th Edition).	BS 7671: 2008
Electrically powered hold-open devices for swing doors. Requirements and test methods.	BS EN 1155: 1997
Emergency lighting – lighting applications.	BS 5266-7: 1999
External renderings. Code of practice.	BS 5262: 1991
Fabrics for curtains and drapes – flammability requirements.	BS 5867-2: 1980 (1993) BS EN ISO 1182: 2002
Fabrics for curtains and drapes: Part 2: 1980 Flammability requirements.	BS 5867
Fibre boards.	BS 1142: 1989
Fibre cement flue pipes, fittings and terminals, Part 1: 1991 (1998) Specification for light quality fibre cement flue pipes, fittings and terminals. Specifications for heavy quality cement flue pipes, fittings and terminals: Part 2 1991.	BS 7435
Fire classification of construction products and building elements. Part 4: 1970 Classification using data from fire resistance tests, excluding ventilation services. Part 6: 1989 Classification using data from fire resistance tests on products and elements used in building service installations: fire resisting ducts and fire dampers. Part 8: 1972 Specification for smoke alarms.	BS 476
Fire classification of construction products and building elements. Part 1: 2002 Classification using test data from reaction to fire tests. Part 2: 2003 Classification using data from fire resistance tests, excluding ventilation services. Part 3: 2005 Classification using data from fire resistance tests on products and elements used in building service installations: fire resisting ducts and fire dampers. Part 5: 2005 Classification using data from external fire exposure to roof tests.	BS EN 13501
Fire detection and alarm systems for buildings: Part 1: 2002 Code of practice for system design, installation and servicing. Part 3: 1983 Specification for automatic release mechanisms for certain fire protection equipment. Part 8: 1998 Code of practice for the design, installation and servicing of voice alarm systems.	BS 5839
Fire detection and fire alarm devices for dwellings. Part 1: 2000. Specification for smoke alarms. Part 2: 2003 Specification for heat alarms.	BS 5446
Fire detection and fire alarm systems – manual call points.	BS EN 54-11: 2001
Fire detection and fire alarm systems for buildings. Code of practice for system design, installation, commissioning and maintenance.	BS 5266-1: 2005

(Continued)

Title	Standard
Fire detection and fire alarm systems for buildings. Code of practice for the design, installation and maintenance of fire detection and fire alarm systems in dwellings.	BS 5306-2: 1990
Fire detection and fire alarm systems for buildings. Code of practice for the design, installation, commissioning, and maintenance of voice alarm systems.	BS 5395-2: 1984
Fire door assemblies with non-metallic leaves (Sections 1 and 2). Code of practice.	BS 8214: 1990
Fire extinguishing installations and equipment on premises: Part 1: 1976 (1988) Hydrant systems, hose reels and foam inlets. Part 2: 1990 Specification for sprinkler systems. Part 4: 2004 Specification for carbon dioxide systems. Part 6.1: 1988 Foam system – specification for low foam expansion foam systems. Part 6.2: 1989 Foam system – specification for medium and high foam expansion foam systems. Part 7: 1988 Specification for powder systems.	BS 5306
Fire hydrant systems equipment: Part1:1975 Specification for boxes for landing valves for dry risers. Part 5: 1975 Specification for/boxes for foam inlets and dry riser inlets.	BS 5041
Fire performance of external cladding systems. Test methods for non-loadbearing external cladding systems applied to the face of a building.	BS 5499-1: 2002
Fire precautions in the design, construction and use of buildings: Part 0: 1996 Guide to fire safety codes of practice for particular premises. Part 1: 1990 Code of practice for residential buildings. Part 4: 1998 Code of practice for smoke control using pressure differentials. Part 5: 2005 Code of practice for fire-fighting stairs and lifts. Part 6: 1991 Code of practice for places of assembly. Part 7: 1997 Code of practice for the incorporation of atria in buildings. Part 8: 1999 Code of practice for means of escape for disabled people. Part 9: 1999 Code of practice for ventilation and air conditioning ductwork. Part 10: 1991 Code of practice for shopping complexes. Part 11: 1997 Code of practice for shops, offices, industrial, storage and other similar buildings. Part 12: 2004 Managing fire safety.	BS 5588
Fire protection measures: Part 1: 2006 Electrical actuation of gaseous total flooding extinguishing systems.	BS 7273

(*Continued*)

Title	Standard
Part 3: 2000 Electrical actuation of pre-action sprinkler systems.	
Fire resistance tests – alternative and additional procedures.	BS EN 1363-2: 1999
Fire resistance tests – general requirements.	BS EN 1363-1: 1999
Fire resistance tests – verification of furnace performance.	BS EN 1363-3: 1999
Fire resistance tests for door and shutter assemblies – fire doors and shutters.	BS EN 1634-1: 2000
Fire resistance tests for door and shutter assemblies – smoke control doors.	BS EN 1634-3: 2001
Fire resistance tests for door and shutter assemblies. Fire doors and shutters.	BS 5720: 1979
Fire resistance tests for door and shutter assemblies. Smoke control doors and shutters.	BS 5839-1: 2002
Fire resistance tests for loadbearing elements: Part 1: 1999 Walls. Part 2: 1999 Floors and roofs. Part 3: 2000 Beams. Part 4: 1999 Columns.	BS EN 1365
Fire resistance tests for loadbearing elements. Part 6: 2004 Beams. Part 8: 2003 Floors and roofs. Part 9: 2003 Columns.	BS 5839
Fire resistance tests for loadbearing elements – beams.	BS EN 1365-3: 2000
Fire resistance tests for loadbearing elements. Walls.	BS 7273-1: 2006
Fire resistance tests for non-loadbearing elements. Part 1: 1999 Walls. Part 3: 2006 Curtain walling. Full configuration (complete assembly).	BS EN 1364
Fire resistance tests for non-loadbearing elements – Ceilings.	BS EN 1364-2: 1999
Fire resistance tests for non-loadbearing elements. Walls.	BS 7346
Fire resistance tests for service installations – Ducts.	BS EN 1366-1: 1999
Fire resistance tests for service installations – Fire dampers.	BS EN 1366-2: 1999
Fire resistance tests for service installations. Part 3: 2004 Penetration seals. Part 5: 2003 Service ducts and shafts.	BS EN 1366
Fire resistance tests for service installations. Part 1: 2002 Penetration seals. Part 2: 2005 Raised access and hollow core floors.	BS 8414
Fire resistance tests for service installations. Ducts.	BS 8214: 1990
Fire resistance tests for service installations. Fire dampers.	BS 7974: 2001
Fire resistance tests for service installations. Linear joint seals.	BS EN 1366-4: 2006
Fire resistance tests for service installations. Raised access and hollow core floors.	BS EN 1366-6: 2004

(Continued)

Title	Standard
Fire resistance tests, Verification of furnace performance.	DD EN 1363-3: 1999
Fire tests for building products.	BS EN ISO 1716: 2000
Fire tests for building products – non-combustibility test.	BS EN ISO 1182: 2002
Fire tests for building products. Building products excluding footings exposed to thermal attack by a single burning item.	BS EN 13823: 2002
Fire tests for building products. Conditioning procedures and general rules for selection of substrates.	BS EN 13238: 2001
Fire tests for building products. Determination of the heat of combustion.	BS EN ISO 1716: 2002
Fire tests for building products. Non-combustibility test.	BS EN 150 1182: 2002
Fire tests for building products. Building products excluding floorings exposed to the thermal attack by a single burning item.	BS EN 13823: 2002
Fire tests for building products. Conditioning procedures and general rules for selection of substrates.	BS EN 13238: 2001 BS EN 1634-1: 2000
Fire tests for building products. Determination of the heat of combustion.	BS EN 1634-2: xxxx
Fire tests for building products. Ignitability when subjected to direct impingement of a flame.	BS EN ISO 11925-2: 2000
Fire tests for building products. Non-combustibility test.	BS EN 1634-2: xxxx BS EN 1634-3: 2001 BS EN ISO 1182: 2002
Fire tests on building materials and structures. Part 2: 2003 Methods for determination of the fire resistance of loadbearing elements of construction. Part 3: 2005 Methods for determination of the fire resistance of non-loadbearing elements of construction. Part 5: 2005 Methods for determination of the contribution of components to the fire resistance of a structure.	BS EN 13501
Fire tests on building materials and structures. Classification and method of test for external fire exposure to roofs.	BS 9990: 2006 BS EN 1125: 1997
Fire tests on building materials and structures. Method for assessing the heat emission from building materials.	BS EN 1155: 1997
Fire tests on building materials and structures. Method for determination of the fire resistance of elements of construction (general principles).	BS EN 1155: 1997 BS EN 12101-3: 2002
Fire tests on building materials and structures. Method for determination of the fire resistance of ventilation ducts.	BS' EN 12101-3:2002
Fire tests on building materials and structures. Method of test for fire propagation for products.	BS EN 12845: 2004 BS EN 13238: 2001
Fire tests on building materials and structures. Methods for determination of the fire resistance of non-loadbearing elements of construction.	BS EN 13501-4n000c

(Continued)

Title	Standard
Fire tests on building materials and structures. Non-combustibility test for materials.	BS EN 1364: 1999
Fire tests on building materials and structures.	
Part 3: 2004 Classification and method of test for external fire exposure to roofs.	
Part 4: 1970 (1984) Non-combustibility test for materials.	
Part 6: 1981 Method of test for fire propagation for products.	
Part 6: 1989 Method of test for fire propagation for products.	
Part 7: 1971 Surface spread of flame tests for materials.	
Part 8: 1972 Test methods and criteria for the fire resistance of elements of building construction (withdrawn).	
Part 11: 1982 (1988) Method for assessing the heat emission from building materials.	
Part 20: 1987 Method for determination of the fire resistance of elements of construction (general principles).	
Part 21: 1987 Methods for determination of the fire resistance of loadbearing elements of construction.	
Part 22: 1987 Methods for determination of the fire resistance of non-loadbearing elements of construction.	
Part 23: 1987 Methods for determination of the contribution of components to the fire resistance of a structure.	
Part 24: 1987 Method for determination of the fire resistance of ventilation ducts.	
Part 31: Methods for measuring smoke penetration through doorsets and shutter assemblies: Section 31.1: 1983 Measurement under ambient temperature conditions.	
Fire tests. Ignitability of building products subjected to direct impingement of flame. Single-flame source test.	BS EN ISO 11925-2: 2002 BS EN 50200: 2006
Fire, tests. On building materials and structures. Method of test to determine the classification of the 'surface spread of flame of products'.	BS EN 1365-1: 1999 BS EN 1365-2: 2000
Fire resistance tests. Fire dampers for air distribution systems. Classification, criteria and field of application of test results.	BS EN 1365-2: 2000
Fire resistance tests. Fire dampers for air distribution systems. Intumescent fire dampers.	BS EN 1365-3: 2000 BS ISO 10294-5: 2005
Fire resistance tests. Fire dampers for air distribution systems. Classification, criteria and field of application of test results.	BS ISO 10294-2: 1999
Fixed fire-fighting systems. Automatic sprinkler systems. Design, installation and maintenance.	BS EN 12845: 2004 BS EN 1365-3: 2000

(Continued)

Title	Standard
Flammability of textile fabrics when subjected to a small igniting flame applied to the face or bottom edge of vertically oriented specimens.	BS EN 1366-3: 2004
Flat roofs with continuously supported coverings – code of practice.	BS 6229: 2003
Flexible joints for grey or ductile cast iron drain pipes and fittings (BS 437) and for discharge and ventilating pipes and fittings (BS 416).	BS 6087: 1990
Flue blocks and masonry terminals for gas appliances: Part 1: 1986 Specification for precast concrete flue blocks and terminals. Part 2: 1989 Specification for clay flue blocks and terminals.	BS 1289-1: 1986
Flues and flue structures in buildings. Code of practice.	BS 5854: 1980 (1996)
Foundations. Code of practice.	BS 8004: 1986
Fuel oils for agricultural, domestic and industrial engines and boilers.	BS 2869: 2006
Galvanized corrugated steel: metric units.	(1988)
General criteria for certification bodies operating certification of personnel.	BS EN ISO/IEC 17024: 2003
General criteria for supplier's declaration of conformity.	BS EN ISO/IEC 17050-2: 2004
General criteria for the assessment of testing laboratories.	BS 7502: 1989
General criteria for the assessment of testing laboratories.	BS EN 45002: 1989
General criteria for the operation of various types of bodies performing inspection.	BS EN 450-1: 2005
General requirements for bodies operating assessment and certification/registration of quality systems.	BS EN 7512: 1989
General requirements for bodies operating product certification systems.	BS EN 45011: 1998
General requirements for the competence of testing and calibration laboratories.	BS EN ISO/IEC 17025: 2005
Glazing for buildings. Code of practice for safety related to human impact.	BS 6262-: 2005
Glossary of terms relating to solid fuel burning equipment. Domestic appliances.	BS 1846-1: 1994
Graphical symbols and signs:	BS 5499
Part 1: 2002 Safety signs, including fire safety signs. Specification for geometric shapes, colours and layout. Part 4: 2000 Safety signs, including fire safety sign. Code of practice for escape route lighting. Part 5: 2002 Graphical symbols and signs. Safety signs, including fire safety signs. Signs with specific safety meanings (specification for additional signs to those given in BS 5378: Part 1).	

(*Continued*)

Title	Standard
Graphical symbols and signs. Safety signs, including fire safety signs. Specification for geometric shapes, colours and layout.	BS EN 1365-4: 1999
Gravity drainage systems inside buildings: Part 1: Scope, definitions, general and performance requirements. Part 2: Wastewater systems, layout and calculation. Part 3: Roof drainage layout and calculation. Part 4: Effluent lifting plants, layout and calculation. Part 5: Installation, maintenance and user instructions.	BS EN 12056: 2000
Gravity drainage systems inside buildings: Part 2: 2000 Sanitary pipework, layout and calculation. Part 3: 2003 Roof drainage, layout and calculation. Part 4: 2000 Wastewater lifting plants. Layout and calculation.	BS EN 12056
Gravity drainage systems inside buildings. Gravity drainage systems inside buildings. General and performance requirements.	BS EN 12056-1: 2000
Grease separators. Selection of nominal size, installation, operation and maintenance.	BS EN 1825-2: 2002
Guide for design, construction and maintenance of single-skin air supported structures.	BS 6661: 1986
Guide to assessment of suitability of external cavity walls for filling with thermal insulants: Part 1: 2004 Existing traditional cavity construction.	BS 8208
Guide to development and presentation of fire tests and their use in hazard assessment.	BS 6336: 1998
Guide to durability of buildings and building elements, products and components.	BS 7543: 2003
Guide to the development of fire tests, the presentation of test data and the role of tests in hazard assessment.	BS EN 1366-1: 1999
Heat exchangers – methods of measuring the parameters necessary for establishing the performance.	BS EN ISO 306: 1997
Heating boilers for solid fuels, hand and automatically fired, nominal heat output of up to 300 kW – terminology, requirements, testing and marking.	BS EN 303-5: 1999
Heating boilers with forced draught burners. Terminology, general requirements, testing and marketing.	BS EN 303-1: 1999
Historic buildings. Guide to the principles of the conservation.	BS 7913: 1998
Hygrothermal performance of building components and building elements. Internal surface temperature to avoid critical surface humidity and interstitial condensation. Calculation methods.	BS EN ISO 13788: 2001
Ignitability of fabrics used in the construction of large tented structures.	BS EN 1366-1: 1999

(Continued)

Title	Standard
Impact performance requirements for flat safety glass and safety plastics for use in buildings.	BS 6206: 1981
Indicator plates for fire hydrants and emergency water supplies.	BS 3251: 1976 BS EN ISO 306: 2004
Installation and maintenance of flues and ventilation for gas appliances of rated input not exceeding 70 kW net: Part 1: 2000 Specification for installation and maintenance of flues. Part 2: 2000 Specification for installation of ventilation for gas appliances.	BS 5440
Installation in domestic premises of gas-fired ducted-air heaters of rated input not exceeding 60 kW.	BS 5864: 1989
Installation of chimneys and flues for domestic appliances burning solid fuel (including wood and peat). Part 1: 1984 (1998) Code of practice for masonry chimneys and flue pipes.	BS 6461
Installation of domestic gas cooking appliances (1st, 2nd and 3rd family gases).	BS 6172: 1990
Installation of domestic heating and cooking appliances burning solid mineral fuels: Part 1: 1994 Specification for the design of installations. Part 2: 1994 Specification for installing and commissioning on site. Part 3: 1994 Recommendations for design and on site installation.	BS 8303
Installation of factory-made chimneys to BS 4543 for domestic appliances: Part 1: 1992 (1998) method of specifying. Installation design information. Part 2: 1992 (1998) Specification for installation design. Part 3: 1992 Specification for site installation. Part 4: 1992 (1998) Recommendations for installation design and installation.	BS 7566
Installation of gas fires, convector heaters, fire/back boilers and decorative fuel effect gas appliances: Part 1: 2005 Gas fires, convector heaters and fire/back Boilers and heating stoves (1st, 2nd and 3rd family gases). Part 2: 2005 Inset live fuel effect gas fires of heat input not exceeding 15 kW (2nd and 3rd family gases). Part 3: 2005 Decorative fuel effect gas appliances of heat input not exceeding 20 kW (2nd and 3rd family gases), AMD 7033.	BS 5871
Installation of gas-fired hot water boilers of rated input not exceeding 60 kW.	BS 6798: 2000
Installation of hot water supplies for domestic purposes, using gas-fired appliances of rated input not exceeding 70 kW.	BS 5546: 2000

(*Continued*)

Title	Standard
Installations for separation of light liquids (e.g. petrol or oil): Part 1 Principles of design, performance and testing, marking and quality control.	BS EN 858: 2001
Internal and external wood doorsets, door leaves and frames: Part 1: 1980 (1985) Specification for dimensional requirements.	BS 4787
Investigation of potentially contaminated land. Code of practice.	BS 10175: 2001
Isolating transformers and safety isolating transformers. Requirements.	BS EN 60742: 1996
Ladders for permanent access to chimneys, other high structures, silos and bins.	BS 4211: 2005
Lifts and service lifts: Part 1: 1986 Safety rules for the construction and installation of electric lifts. (Part 1 to be replaced by BS EN 81-1, when published). Part 2: 1988 Safety rules for the construction and installation of hydraulic lifts. (Part 2 to be replaced by BS EN 81-2, when published). Part 5: 1989 Specifications for dimensions for standard lift arrangements. Part 7: 1983 Specification for manual control devices, indicators and additional fittings Amendment slip.	BS 5655
Lighting applications – Emergency lighting.	BS EN 1838: 1999
Lighting for buildings. Code of practice for daylighting.	BS 8206-2: 1992
Loading for buildings: Part 1: 1996 Code of practice for dead and imposed loads. Part 2: 1997 Code of practice for wind loads. Part 3:1988 Code of practice for imposed roof loads.	BS 6399
Low-voltage switchgear and controlgear assemblies. Particular requirements or low-voltage switchgear and control assemblies intended to be installed in places where unskilled persons have access to their use.	BS EN 60439-3: 1991
Maintained lighting in cinemas.	CP 1007: 1955
Masonry units: Part 1: 2003 Clay masonry units Part 2: 2001 Calcium silicate masonry units. Part 3: Aggregate concrete masonry units. Part 4: 2001 Autoclaved aerated concrete masonry units. Part 5: Manufactured stone masonry units. Part 6: 2001 Natural stone masonry units.	BS EN 771
Masonry units. Clay masonry units.	BS EN ISO 1182: 2002 BS EN ISO 11925-2: 2002

(Continued)

Title	Standard
Mastic asphalt roofing. Code of practice.	BS 8218: 1998
Measurement of sound insulation in buildings and of building elements: Part 1: 1980 Recommendations for laboratories. Part 3: 1980 Laboratory measurement of airborne sound insulation of building elements. Part 4: 1980 Field measurement of airborne sound insulation between rooms. Part 6: 1980 Laboratory measurement of impact sound insulation of floors. Part 7: 1980 Field measurements of impact sound insulation of floors.	BS 2750
Mechanical actuation of gaseous total flooding and local application extinguishing systems. Code of practice for the operation of fire protection measures.	BS 7273-2: 1992
Mechanical ventilation and air conditioning in buildings. Code of practice.	BS 5720: 1979
Metal flue pipes, fittings, terminals and accessories for gas-fired appliances with a rated input not exceeding 60 kW, AMD 8413.	BS 715: 1993
Metal ties for cavity wall construction.	BS 1243: 1978
Modular co-ordination in building.	BS 6750: 1986
Mortar for masonry: Part 2: 2002 Masonry mortar	BS EN 998
Natural stone cladding and lining. Code of practice for design and installation.	BS 8298: 1994
Non-automatic fire-fighting systems in buildings. Code of practice for.	BS 9990: 2006
Non-loadbearing precast concrete cladding. Code of practice for design and installation.	BS 8297: 2000
Oil burning equipment, Part 5: 1987 Specification for oil storage tanks.	BS 799
Oil firing: Part 1: 1977 Installations up to 44 kW output capacity for space heating and hot water supply purposes. Part 2: 1978 Installations of 44 kW or above output capacity for space heating, hot water and steam supply purposes. Part 3: 1975 Specification for inlet breechings for dry riser inlets. Part 3: 2004 Range hoods for residential use devices. Part 4: 2004 Fans used in residential ventilation systems. Part 6: 2004 Exhaust ventilation system packages used in a single dwelling. Part 7: 2003 Performance testing of a mechanical supply and exhaust ventilation units (including heat recovery) for mechanical ventilation systems intended for single family dwellings.	BS 5410

(Continued)

Title	Standard
Particleboards. Specifications. Requirements for load-bearing boards for use in humid conditions.	BS EN 312-5: 1997
Pavements constructed with clay, natural stone or concrete pavers – guide for the structural design of lightly trafficked pavements constructed of precast paving blocks.	BS 7533: Part 2: 2001
Plastic inspection chambers for drains.	BS 7158: 2001
Plastic piping systems for non-pressure underground drainage and sewerage. Unplasticized poly(vinylchloride) (PVC-U). Specifications for pipes, fittings and the system.	BS EN 1401-1: 1998
Plastic piping systems for soil and waste (low and high temperature) within the building structure. Acryionitrilebutadiene-styrene (ABS). Specifications for pipes, fittings and the system.	BS EN 1455-1: 2000 BS EN 1329-1: 2000
Plastic piping systems for soil and waste discharge (low and high temperature) within the building structure.	BS EN 1565-1: 2000
Plastic piping systems for soil and waste discharge (low and high temperature) within the building structure. Polyethylene (PE). Specifications for pipes, fittings and the system.	BS EN 1519-1: 2000
Plastic waste traps.	BS 3943: 1979 (1988)
Plastic. Thermal properties: Methods 120A to 12OE: 1990 Determination of the Vicat softening temperature (VST) of thermoplastics.	BS 2782
Plastic. Thermoplastic materials. Determination of Vicat softening temperature.	BS EN 150 306: 2004 BS EN 1366-5: 2003 BS EN 1366-6: 2004
Plastic. Introduction	BS EN 1366-4: 2006
Portland cements.	BS 12: 1989
Powered doors for pedestrian use. Code of practice for safety.	BS 7036: 1996
Powered lifting platforms for use by disabled persons. Code of practice.	BS 6440: 1999
Powered stairlifts.	BS 5776: 1996
Precast concrete masonry units: Part 1: 1981 Specification for precast concrete masonry units.	BS 6073
Precast concrete pipes fittings and ancillary products. Part 2: 1982 Specification for inspection chambers and street gullies. Part 100: 1988 Specification for unreinforced and reinforced pipes and fittings with flexible joints. Part 101: 1988 Specification for glass composite concrete (GCC) pipes and fittings with flexible joints. Part 120: 1989 Specification for reinforced jacking pipes with flexible joints. Part 200: 1989 Specification for unreinforced and reinforced manholes and soakaways of circular cross section.	BS 5911

(Continued)

Title	Standard
Precast, unreinforced concrete paving blocks. Requirements and test methods.	BS 6717: 2001
Prefabricated drainage stack units in galvanized steel.	BS 3868: 1995
Pressure sewerage systems outside buildings.	BS EN 1671: 1997
Profiled fibre cement. Code of practice.	BS 8219: 2001
Protection of buildings against water from the ground.	BS CP 102: 1973
Protection of structures against water from the ground. Code of practice.	BS 8102: 1990
Protective barriers in and about buildings. Code of practice.	BS 6180: 1999
Quality management and quality assurance standards.	BS EN ISO 9001: 2000
Rating the sound insulation in building elements: Part 1: 1984 Method for rating the airborne sound insulation in buildings and interior building elements. Part 2: 1984 Method for rating the impact sound insulation.	BS 5821
Recommendations for the storage and exhibition of archival documents.	BS 5454: 2000
Refrigerating systems and heat pumps – safety and environmental requirements: Installation site and personal protection.	BS EN 378 Part 3: 2000
Reinforced bitumen membranes for roofing. Code of practice.	BS 8217: 2005
Remote centres for alarm.	BS 5979: 2000
Resistance to fire of unprotected small cables for use in emergency circuits.	BS EN 50200: 2006
Roofs and paved areas. Code of practice for drainage.	BS 6367: 1983
Safety and control devices for use in hot water systems: Part 2: 1991 Specification for temperature relief valves for pressures from 1 bar to 10 bar. Part 3: 1991 Specification for combined temperature and pressure relief valves for pressures from 1 bar to 10 bar.	BS 6283
Safety aspects in the design, construction and installation of refrigerating appliances and systems.	BS 4434: 1989
Safety rules for the construction and installation of escalators and passenger conveyors	BS EN 1115: 1995
Safety rules for the construction and installation of lifts. Electric lifts.	BS EN 81-1: 1998
Safety rules for the construction and installation of lifts. Hydraulic lifts.	BS EN 771-1: 2003
Safety rules for the construction and installation of lifts. Particular applications for passenger and goods AMD 14751 passenger lifts. Accessibility to lifts for persons including persons with disability.	BS EN 81-70: 2003
Safety rules for the construction and installation of lifts. Particular applications for passenger and goods passenger lifts. Fire-fighters' lifts.	BS EN 81-72: 2003

(Continued)

Title	Standard
Safety rules for the construction and installation of lifts. Particular applications for passenger and goods AMD 14751 passenger lifts. Accessibility to lifts for persons including persons with disability.	BS EN 81-70: 2003
Sanitary installations: Part 1: 2006 Code of practice for scale of provision, selection and installation of sanitary appliances.	BS 6465
Sanitary pipework.	BS 5572: 1978
Sanitary tapware. Low pressure thermostatic mixing valves. General technical specification.	BS EN 1287: 1999
Sanitary tapware. Thermostatic mixing valves (PN 10). General technical specification.	BS EN 1111: 1999
Sanitary tapware. Waste fittings for basins, bidets and baths. General technical specifications.	BS EN 274: 1993
Sawn and processed softwood.	BS 4471: 1987
Separator systems for light liquids (e.g. oil and petrol). Principles of product design, performance and testing, marking and quality control.	BS EN 858-1: 2002
Separator systems for light liquids (e.g. oil and petrol). Selection of nominal size, installation, operation and maintenance.	BS EN 858-2: 2003
Sheet roof and wall coverings. Corrugated asbestos-cement.	BS 5247: 1975
Site investigations. Code of practice.	BS 5930: 1999
Slating and tiling – Code of practice for design.	BS 5534: 2003
Small sewage treatment works and cesspools. Code of practice for the design and installation of.	BS 6297: 1983
Small wastewater treatment plants less than 50PE.	BS EN 12566-1: 2000
Small wastewater treatment systems for up to 50PT. Prefabricated septic tanks.	BS EN 12566-1: 2000
Smoke and heat control systems. Specification for powered smoke and heat exhaust ventilators.	BS EN 12101-3: 2002 BS EN 81-2: 1998
Smoke and heat control systems. Specification for pressure differential systems. Kits.	BS EN 12101-6: 2005 BS EN 81-72: 2003
Softwood grades for structural use.	BS 4978: 1988
Sound insulation and noise reduction for buildings – Code of practice.	BS 8233: 1999
Specification for installation of gas-fired catering Appliances for use in all types of catering establishments (1st, 2nd and 3rd family gases).	BS 6173: 2001
Specification for open fireplace components.	BS 1251: 1987
Specification for performance requirements for cables required to maintain circuit integrity under fire conditions.	BS 6387: 1994
Specification for performance requirements for domestic flued oil burning appliances (including test procedures).	BS 4876: 1984

(Continued)

Title	Standard
Sprinkler systems for residential and domestic occupancies. Code of practice.	BS 9251: 2005 BS EN ISO 306: 2004
Stainless steels. List of stainless steels.	BS EN 10088-1: 1995
Stairs, ladders and walkways: Part 1: 1977 Code of practice for stairs. Part 2: 1984 Code of practice for the design of helical and spiral stairs. Part 3: 1985 Code of practice for the design of industrial type stairs, permanent ladders and walkways.	BS 5395
Stairs, ladders and walkways. Code of practice for the design of helical and spiral stairs.	BS ISO 10294-2: 1999
Stairs, ladders and walkways. Code of practice for the design, construction and maintenance of straight stairs and winders.	BS 5395-1: 2000
Steel plate, sheet and strip: Part 2: 1983 Specification for stainless and heat-resisting steel plate, sheet and strip.	BS 1449
Stone masonry.	BS 5390: 1976 (1984)
Storage and on-site treatment of solid waste from buildings.	BS 5906: 1980 (1987)
Structural design of buried pipelines under various conditions of loading. General requirements.	BS EN 1295-1: 1998
Structural design of low-rise buildings: Part 1: 1995 Code of practice for stability, site investigation, foundations and ground floor slabs for housing. Part 2: 2005 Code of practice for masonry walls for housing. Part 3: 1996 Code of practice for timber floors and roofs for housing. Part 4: 1995 Code of practice for suspended concrete floors for housing.	BS 8103
Structural fixings in concrete and masonry. Part 1: 1993 Method of test for tensile loading.	BS 5080: 1993
Structural steel in building: Part 2: 1969 – Metric units.	BS 449
Structural use of aluminium: Part 1: 1991 Code of practice for design. Part 2: 1991 Specification for materials, workmanship and protection	BS 8118
Structural use of concrete: Part 1: 1997 Code of practice for design and construction. Part 2: 1985 Code of practice for special circumstances. Part 3: 1995 Design charts for single reinforced beams, doubly reinforced beams and rectangular columns.	BS 8110
Structural use of steelwork in building: Part 1: 2000 Code of practice for design. Rolled and welded sections.	

(Continued)

Title	Standard
Part 2: 2001 Specification for materials, fabrication and erection. Rolled and welded sections. Part 3.1: 1990 Code of practice for design of simple and continuous beams. Part 4: 1994 Code of practice for design of composite slabs with profiled steel sheeting. Part 5: 1998 Code of practice for design of cold formed thin gauge sections. Part 6: 1995 Code of practice for design of light gauge profiled steel sheeting. Part 7: 1992 Structural use of steelwork in building – specification for materials and workmanship: cold-formed sections: Part 8: 2003 Structural use of steelwork in building – code of practice for fire resistant design.	BS 5950
Structural use of steelwork in building. Code of practice for fire resistant design.	BS 5950-8: 2003 BS ISO 10294-5: 2005
Structural use of timber: Part 2: 2002 Code of practice for permissible stress design, materials and workmanship. Part 3: 2006 Code of practice for trussed rafter roofs Part 4.1: 1978 Fire resistance of timber structures. Recommendations for calculating fire resistance of timber members. Part 4.2: 1990 Fire resistance of timber structures. Recommendations for calculating fire resistance of timber stud walls and joisted constructions. Part 6: 1996 Code of practice for timber framed walls. Part 6.1: 1996 Dwellings not exceeding three storeys.	BS 5268
Test methods for external fire exposure to roofs.	ENV 1187: 2002, test 4
Thermal bridges in building construction – calculation of heat flows and surface temperatures: Part 1: 1996 General methods. Part 2: 2001 Linear thermal bridges.	BS EN ISO 10211
Thermal insulating materials for pipes, tanks, vessels, ductwork and equipment operating within the temperature range $-40°C$ to $+70°C$.	BS 5422: 2001
Thermal insulation – determination of steady-state thermal transmission properties – calibrated and guarded hot box.	BS EN ISO 8990: 1996
Thermal insulation for use in pitched roof spaces in dwellings – Part 5: specification for installation of man-made mineral fibre and cellulose fibre insulation.	BS 5803-5: 1985 (as amended 1999)
Thermal insulation of cavity walls (with masonry or concrete inner and outer leaves) by filling with urea-formaldehyde (UF) foam systems.	BS 5618: 1985 (1992)
Thermal insulation of cavity walls by filling with blown man-made mineral fibre: Part 1: 1982 Specification for the performance of installation systems.	

(Continued)

Title	Standard
Part 2: 1982 Code of practice for installation of blown man-made mineral fibre in cavity walls with masonry and/or concrete leaves.	BS 6232
Thermal insulation of cavity walls using man-made mineral fibre batts (slabs) – specification for man-made mineral fibre batts (slabs).	BS 6676-1: 1986 (1994)
Thermal insulation. Determination of steady-state thermal transmission properties. Calibrated and guarded hot box.	BS EN ISO 8990: 1996
Thermal performance of building materials and products determination of thermal resistance by means of guarded hot plate and heat flow meter methods – dry and moist products of low and medium thermal resistance.	BS EN 12664: 2001
Thermal performance of building materials and products determination of thermal resistance by means of guarded hot plate and heat flow meter methods – products of high and medium thermal resistance.	BS EN 12667: 2000
Thermal performance of building materials and products. determination of thermal resistance by means of guarded hot plate and heat flow meter methods – thick products of high and medium thermal resistance.	BS EN 12939: 2001
Thermal performance of buildings – heat transfer via the ground – calculation methods.	BS EN ISO 13370: 1998
Thermal performance of buildings. Transmission heat loss co-efficient – calculation method.	BS EN ISO 13789: 1999
Thermal performance of windows and doors – determination of thermal transmittance by hot box method: Part 1: Complete windows and doors.	BS EN ISO 12567-1: 2000
Thermal performance of windows, doors and shutters Calculation of thermal transmittance.	BS EN ISO 10077-1: 2003 BS EN ISO 10077-2: 2003
Thermoplastics piping systems for non-pressure underground drainage and sewerage – structure walled piping systems of unplasticized polyvinyl chloride) (PVC-U), polypropylene (PP) and polyethylene (PE) – Part 1: specification for pipes, fittings and the system: Part 1: 1997 Guide to specifying concrete. Part 2: 1997 Methods for specifying concrete mixes. Part 3: 1990 Specification for the procedures to be used in producing and transporting concrete. Part 4: 1990 Specification for the procedures to be used in sampling, testing and assessing compliance of concrete.	BS EN 13476-1: 2001
Thermoplastics waste pipe and fittings.	BS 5255: 1989 BS EN ISO 1716: 2002
Thermostats for gas-burning appliances.	BS 4201: 1979 (1984)

(Continued)

Title	Standard
Tongued and grooved softwood flooring.	BS 1297: 1987
Topsoil.	BS 3882: 1994
Unplasticized PVC soil and ventilating pipes, fittings and accessories.	BS 4514: 1983 (1998)
Unplasticized polyvinyl chloride (PVC-U) pipes and plastics fittings of nominal sizes 110 and 160 for below ground drainage and sewerage.	BS 4660: 1989
Unplasticized PVC pipe and fittings for gravity sewers.	BS 5481: 1977 (1989)
Unplasticized PVC soil and ventilating pipes of 82.4 mm minimum mean outside diameter, and fittings and accessories of 82.4 mm and of other sizes. Specification.	BS 4514: 2001 ENV 1187: 2002, test 4
Unvented hot water storage units and packages.	BS 7206: 1990
UF foam systems suitable for thermal insulation of cavity walls with masonry or concrete inner and outer leaves.	BS 5617: 1985
Use of masonry: Part 1: 2005 Structural use of unreinforced masonry. Part 2: 2005 Structural use of reinforced and prestressed masonry. Part 3: 2005 Materials and components, design and workmanship.	BS 5628
Vacuum drainage systems inside buildings.	BS EN 12109: 1999
Vacuum sewerage systems outside buildings.	BS EN 1091: 1997
Ventilation for buildings – performance testing of components/products for residential ventilation: Part I: 2004 Externally and internally mounted air transfer devices.	BS EN 13141
Ventilation for buildings – performance testing of components/products for residential ventilation – Part 8: performance testing of unducted mechanical supply and exhaust ventilation units [including heat recovery] for mechanical ventilation systems intended for a single room.	prEN 13141-8: 2004
Ventilation for buildings – performance testing of components/products for residential ventilation – Part 9 humidity controlled external air inlet.	prEN 13141-9: 2004
Ventilation for buildings – performance testing of components/products for residential ventilation – Part 10 performance testing of unducted mechanical supply and exhaust ventilation units [including heat recovery] for mechanical ventilation systems intended for a single room.	prEN 13141-10: 2004
Ventilation principles and designing for natural ventilation.	BS 5925: 1991(1995)

(Continued)

Title	Standard
Vitreous china sanitary appliances.	BS 3402: 1969
Vitreous china washdown watercloset (WC) pans with horizontal outlet. Specification for WC pans with horizontal outlet for use with 7.5 L maximum flush capacity cisterns.	BS 5503-3: 1990
Vitreous-enamelled low-carbon-steel fluepipes, other components and accessories for solid-fuel-burning appliances with a maximum rated output of 45 kW.	BS 6999: 1989 (1996)
Vitrified clay pipes and fittings and pipe joints for drains and sewers: Part 1: 1991 Test requirements. Part 2: 1991 Quality control and sampling. Part 3: 1991 Test methods. Part 6: 1996 Requirements for vitrified clay manholes.	BS EN 295
Vitrified clay pipes, fittings and ducts, also flexible mechanical joints for use solely with surface water pipes and fittings.	BS 65: 1991
Wall hung WC pans. Specification for WC pans with horizontal outlet for use with 7.5 L maximum flush capacity cisterns.	BS 5504-4: 1990
Wastewater lifting plants for buildings and sites – principles of construction and testing: Part 1: Lifting plants for wastewater containing faecal matter. Part 2: Lifting plants for faecal-free wastewater. Part 3: Lifting plants for wastewater containing faecal matter for limited application.	BS EN 12050: 2001
Windows, doors and rooflights. Part 1: 2004 Design for safety in use and during cleaning of windows, including door-height windows and roof windows. Code of practice.	BS 8213
Wood preservatives. Guidance on choice, use and application.	BS 1282: 1999
Wood stairs Part 1: 1989 Specification for stairs with closed risers for domestic use, including straight and winder flights and quarter and half landings.	BS 585
Wood-based panels for use in construction – characteristics, evaluation of conformity and marking.	BS EN 13986: 2002
Workmanship on building sites.	BS 8000

Note: Copies of British Standards and British Standards Codes of Practice, European Standards, Drafts for Development and International Standards may be purchased from the British Standards Institution (BSI, PO Box 16206, Chiswick, London W4 4ZL. Website: www.bsonline.techindex.co.uk).

1.2 Publications specific to the Technical Handbooks

Section 1 – structure

Title	Reference	Publisher
Proposed revision of the simplified roof snow load map for Scotland (2003)	BRE Client Report no. 211–878	BRE
Small buildings guide – second edition (1994)	ISBN 0114952469	TSO
Wind loading for traditional dwellings – amendment of simplified design guidance for the Scottish Office small buildings guide (1999)	CV 4071	BRE

Section 2 – fire

Title	Reference	Publisher
Code of Practice – Hardware for Timber Fire Escape Doors (2000)		Building Hardware Industry Federation
Code of practice for fire resisting metal door sets (1999)		Door Shutters Manufacturers Association
Construction fire safety	CIS51	HSE
Council Directive 89/106/EEC Construction Products Directive (CPD)	94/611/EC 96/603/EC	EC
External Fire Spread: Building Separation and Boundary Distances (1991)	BR 187	BRE
Fire Performance of external thermal insulation for walls of multi-storey buildings (2002)	BR 135	BRE
Fire safe design: a new approach to multi-storey steel framed buildings (2000).	P288	Steel Construction Institute
Guidance on the mandatory licensing of Houses in Multiple Occupation (2000)	ISBN	TSO
Guidelines for the Construction of Fire Resisting Structural Elements	BR 128	BRE

(Continued)

Title	Reference	Publisher
International Fire Engineering Guidelines 2005		Australian Building Codes Board
Mandatory Licensing of Houses in Multiple Occupation: Guidance for Licensing Authorities		Scottish Executive
Smoke shafts protecting fire-fighting shafts: their performance and design (2002)		BRE

Section 3 – environment

Title	Reference	Publisher
Advice on Flues for Modern Open Flued Oil-Fired Boilers (2001)	Technical Information Sheet TI/129	Oil Firing Technical Association
Air Supply Requirements (2001)	Technical Information Sheet TI/132	Oil Firing Technical Association
Assessment of the risk of environmental damage being caused by spillage from domestic oil storage tanks (1999)	Technical Information Sheet TI/133	Oil Firing Technical Association
CIBSE Guide B: 1986: section B2 (1986)	–	Chartered Institution of Building Services Engineers
Code of practice for ground floor, multi-storey and underground car parks, section 4 (1994)	–	Association for Petroleum and Explosive Administration
Contaminants in soils, collation of toxicological data and intake values for humans	CLR9	Environment Agency
Contaminated land exposure assessment (CLEA) model, technical basis and algorithms	CRL10	Environment Agency
Continuous mechanical ventilation in dwellings: design, installation and operation (1994)	Digest 398	BRE
Control of legionella bacteria in water systems – approved code of practice	HSE L8	HSE
Dangerous Substances Directive	76/464/EEC	EC

(*Continued*)

Title	Reference	Publisher
Design Guidance on Flood Damage to Dwellings (1996)	–	Scottish Executive
Development and Flood Risk	C624	CIRIA
Development of Contaminated Land – Planning Advice Note	PAN 33	Scottish Executive
Drainage Assessment: a guide for Scotland	–	SEPA
Fire Protection of Oil Storage Tanks (2001)	Technical Information Sheet TI/136	Oil Firing Technical Association
Flows and Loads – 2, Code of practice	–	British Water
Garage installations (1999)	Technical Information Sheet TI/127	Oil Firing Technical Association
Gas installation in timber frame and light steel framed buildings (2006)	IGE/UP/7 (Edition 2)	Institution of Gas Engineers
Good Building Guide, Parts 1 and 2	GBG 42	BRE
Groundwater Directive 80/68/EEC		EC
Guidance for the safe development of housing on land affected by contamination (2000)		National House Building Council and Environment Agency
Harvesting rainwater for domestic use: – an information guide		Environment Agency
Housing For Varying Needs, 1999	–	Communities Scotland
Installing Oil Supply Pipes Underground (2001)	Technical Information Sheet TI/134	Oil Firing Technical Association
Land contamination risk assessment tools: an evaluation of some of the commonly used methods	Technical Report P260	Environment Agency
Lifetime Homes Standards		Joseph Rowntree Foundation
Manual for Scotland and Northern Ireland (2000)		
Mound filter systems for domestic wastewater	BR 478	BRE
National Waste Plan, 1999	–	SEPA
Non-liquid saturated treatment systems (1999)	NSF/ANSI 41–1999	National Sanitation Foundation (USA)

(Continued)

Title	Reference	Publisher
Oil-fired appliances and extract fans (1996)	Technical Information Sheet TI/112	Oil Firing Technical Association
Oil Firing Equipment Standard – Flues for use with Oil-Fired Boilers with Outputs not above 50 kW (2001)	Standard OFS E106	Oil Firing Technical Association
Oil Firing Equipment Standard – Steel Oil Storage Tanks and Tank Bunds for use with Distillate Fuels, Lubrication Oils and Waste Oils (2002)	Technical Standard OFS T200	Oil Firing Technical Association
Oil firing industry technical advice on fire valves	Technical Information Sheet TI/138.	Oil Firing Technical Association
Oil Firing Technical Association	Applied Standard OFS A101	Oil Firing Technical Association
Oil-fired appliance standard heating boilers with atomizing burners, output up to 70 kW and maximum operating pressures of 3 bar (1998)	Applied Standards A100	Oil Firing Technical Association
Passive stack ventilation systems (1994)	IP 13/94	BRE
Planning and Building Standards Advice on Flooding	PAN 69	Scottish Executive
Planning and Flooding, Scottish Planning Policy (2003)	SPP7	Scottish Executive
Planning and Sustainable Urban Drainage Systems	PAN61	Scottish Executive
PE oil tanks and bunds for distillate fuel	Technical	Oil Firing Technical
Positioning of flue terminals	Technical Information Sheet T1/135	Oil Firing Technical Association
Preparing for Floods (2003)	–	ODPM
Prevention of Environmental Pollution from Agricultural Activity, Code of Practice (2005)	–	Scottish Executive
Priority contaminants report	CLR 8	Environment Agency
Radon: guidance on protection measures for new dwellings in Scotland (1999)	BR376	BRE
Rainwater and greywater use in buildings: best practice guidance	C539	CIRIA

(*Continued*)

Title	Reference	Publisher
Reed beds, BRE Good Building Guide 42, Parts 1 and 2 (2000)	GBG 42	BRE
Roofs and roofing – performance, diagnosis, maintenance, repair and avoidance of defects		BRE
Room heaters with atomizing or vaporizing burners with or without boilers, heat output up to 25 kW	Applied Standard A102	Oil Firing Technical Association
Secondary model procedure for the development of appropriate soil sampling strategies for land contamination	R&D Technical Report P5	Environment Agency
Sewers for Scotland (2001)		Water Research Council
Soakaway design (1991)	BRE Digest 365	BRE Digest 365
Spillage of flue gases from solid fuel combustion appliances, Information Paper (1994)	IP 7/94	BRE
Standards for the repair of buildings following flooding	C623	CIRIA
Standards of Training in Safe Gas Installations, Approved Code of Practice		Health and Safety Commission
SUDS Advice Note – Brownfield Sites		SEPA
Sustainable Urban Drainage Systems: Design	ISBN	CIRIA
Technical aspects of site investigation	R&D Technical report P5	Environment Agency
The official guide to approved solid fuel products and services (2004–2005)		H ETAS
Thermal Insulation: Avoiding Risks, Report (2002)	BR 262	BRE
Underground storage tanks for liquid hydrocarbons		Scottish Executive
Wastewater recycling/reuse and water conservation devices (1996)	NSF 41	National Sanitation Foundation (USA)
Water Regulatory Advisory Scheme: Information and Guidance Note	9-02-04 9-02-05	WRAS

Section 4 – safety

Title	Reference	Publisher
Accessible Thresholds in New Housing		DETR
Building Sight (1995)		RNIB
Code for Lighting (2002)		CIBSE
Code of Practice 1: 'Bulk LPG Storage at Fixed Installations' – Part 4 – Buried/ Mounded LPG		Liquid Petroleum Gas Association
Storage Vessels, as amended		
Code of Practice 1: 'Bulk LPG Storage at Fixed Installations' – Part 1 – Design, Installation and Operation of Vessels Located Above Ground, as amended	–	Liquid Petroleum Gas Association
Code of Practice 1: 'Bulk LPG Storage at Fixed Installations' – Part 2 – Small bulk Propane Installations for Domestic and Similar Purposes', as amended		Liquid Petroleum Gas Association
Code of Practice 24: 'Use of LPG cylinders': Part 1 – The Use of Propane in Cylinders at Residential Premises		Liquid Petroleum Gas Association
Guidance on the use of Tactile Paving Surfaces (1998)		The Scottish Office/DETR
Guidance to the Water Supply (Water Fittings) Regulations 1999		DEFRA
Housing for Varying Needs, Parts 1 and 2		Communities Scotland
Inclusive Design – Planning Advice Note (2006) Inclusive Mobility (2002)	PAN 78	Scottish Executive Department for Transport
Preventing hot water scalding in bathrooms: using TMVs	IP 14/03	BRE
Safety in window cleaning using portable ladders	MISC 613	HSE

Section 5 – noise

Title	Reference	Publisher
Housing and sound insulation: Improving attached dwellings and designing for conversions (2006)		Arcamedia
Planning and Noise, Planning Advice Note (1999)	PAN56	Scottish Executive
Review of Sound Insulation Performance in Scottish Domestic Construction		Scottish Executive
Scottish House Condition Survey, Scottish Homes (1996)		Communities Scotland
Sound Advice on Noise: don't suffer in silence (2001)		Scottish Executive

Section 6 – energy

Title	Reference	Publisher
Accredited Construction Details (Scotland)		SBSA
Assessing Condensation Risk and Heat loss at Thermal Bridges around Openings (1994)	I P 12/94	BRE
Building Standards Circular on Energy, 2004	–	SBSA
CIBSE Guide A: Design Data – Section A3: Thermal Properties of Building Structures (1999)	–	Chartered Institution of Building Services Engineers
Conventions for U-value calculations (2002)	BR 443	BRE
Energy Efficiency Best Practice in Housing publication – Effective use of insulation in dwellings, September 2003	CE23	Energy Saving Trust (EST)
Good Practice Guide 302 published by Energy Efficiency Best Practice in Housing	GPG 302	EST
Guide for assessment of the thermal performance of aluminium curtain wall framing, September 1996	–	Council for Aluminium in Building

1.3 Other publications

Air Tightness Testing and Measurement Association (ATTMA)
www.attma.org

- Measuring Air Permeability of Building Envelopes, 2006.

BRE
www.bre.co.uk

- BR 262 *Thermal Insulation: Avoiding Risks*, Report (2002).
- BR 364 *Solar Shading of Buildings*, 1999.
- BRE Digest 498 *Selecting lighting controls*, 2006. ISBN 1 86081 905 2.
- BRE Digest 465 *U-values for light steel frame construction*.
- BR 443 *Conventions for U-value calculations*, 2006. (Available at: www. bre.co.uk/uvalues)
- Information Paper 1P1/06 *Assessing the effects of thermal bridging at junctions and around openings in the external elements of buildings*, 2006. ISBN 1 86081 904 4.
- *Delivered energy emission factors for 2003.* (Available at: www.bre.co.uk/filelibrary/2003emissionfactorupdate.pdf)
- CO_2 *emission figures for policy analysis*, July 2005. (Available at: www. bre.co.uk/filelibrary/co2emissionfigures2001.pdf)
- *Simplified Building Energy Model (SBEM) user manual and calculation tool.* (Available at: www.odpm.gov.uk)
- I P 10/02 *Metal Cladding: assessing the performance of built-up systems which use Z-spacers, Information Paper.*
- SAP 2005 *The Government's Standard Assessment Procedure for energy rating of dwellings.*

Centre for Window and Cladding Technology
www.cwct.co.uk

- Thermal assessment of window assemblies, curtain walling and non-traditional building envelopes, 2006. ISBN 1 87400 338 6.

CIBSE
www.cibse.org

- CIBSE Commissioning Code M *Commissioning Management*, 2003. ISBN 1 90328 733 2.
- CIBSE Guide A *Environmental Design*, 2006. ISBN 1 90328 766 8.
- AM 1 0 *Natural ventilation in non-domestic buildings*, 2005. ISBN 1 90328 756 1.
- TM 31 *Building Log Book Toolkit*, CIBSE, 2006. ISBN 1 90328 771 5.

- TM 36 *Climate change and the indoor environment. 28 750 2 impacts and adaptation,* 2005. ISBN 1 903.
- TM 37 *Design for improved solar shading control,* 2006. ISBN 1 90328 757 X.
- TM 39 *Building energy metering,* 2006. ISBN 1 90328 707 7.

Department for Education and Skills (DFES)
www.dfes.gov.uk

- Building Bulletin 1 01 *Ventilation of School Buildings, School Building and Design Unit,* 2005. (Download from: www.teachernet.gov.uk/iaq)

Department of the Environment, Food and Rural, Affairs (Defra)
www.defra.gov.uk

- The Government's Standard Assessment Procedure for energy rating of dwellings, SAP 2005. (Available at: www.bre.co.uk/sap2005)

Department of Transport, Local Government and the Regions (DTLR)
www.idea.gov.uk

- *Limiting thermal bridging and air leakage: Robust construction details for dwellings and similar buildings,* Amendment 1. Published by TSO, 2002. ISBN 0 1 1 753 631 8. (Available to download from EST website at: http://portal.est.org.uk/housingbuildings/calculators/robustdetails/)

Electrical Contractors' Association (ECA) and National Inspection Council for Electrical Installation Contracting (NICEIC)
www.eca.co.uk and www.niceic.org.uk

- *ECA comprehensive guide to harmonized cable colours, BS 7671:2001 Amendment No. 2,* ECA, March 2004.
- *Electrical Installers' Guide to the Building Regulations,* NICEIC and ECA, August 2004. Available at: www.niceic.org.uk and www.eca.co.uk
- *New fixed wiring colours – a practical guide,* NICEIC, Spring 2004.

Energy Saving Trust (EST)
www.est.org.uk

- CE66 *Windows for new and existing housing, 2006.*
- CE129 *Reducing overheating – a designer's guide, 2006.*

- GPG268 *Energy efficient ventilation in dwellings – a guide for specifiers*, 2006.
- GIL20 *Low energy domestic lighting*, 2006.

English Heritage
www.english-heritage.org.uk

- *Building Regulations and Historic Buildings*, 2002 (revised 2004).

Health and Safety Executive (HSE)
www.hse.gov.uk

- L24 *Workplace Health, Safety and Welfare: Workplace (Health, Safety and Welfare) Regulations 1992, Approved Code of Practice and Guidance, The Health and Safety Commission*, 1992. ISBN 0 71760 413 6.

Heating and Ventilating Contractors Association
www.hvca.org.uk

- DW/143 *A practical guide to ductwork leakage testing*, 2006. ISBN 0 90378 330 4.
- DW/144 *Specification for sheet metal ductwork*, 1998. ISBN 0 90378 327 4.

Institution of Engineering Technology
www.hvca.org.uk

- *Electrician's guide to the Building Regulations*, 2005. ISBN 0 86341 463 X. Available from: www.iee.org
- *IEE Guidance Note 1: Selection and erection of equipment*, 4th edition, 2002. ISBN 0 85296 989 9.
- *IEE Guidance Note 2: Isolation and switching*, 4th edition, 2002. ISBN 0 85296 990 2.
- *IEE Guidance Note 3: Inspection and testing*, 4th edition, 2002. ISBN 0 85296 991 0.
- *IEE Guidance Note 4: Protection against fire*, 4th edition, 2003. ISBN 0 85296 992 9.
- *IEE Guidance Note 5: Protection against electric shock*, 4th edition, 2003. ISBN 0 85296 993 7.
- *IEE Guidance Note 6. Protection against overcurrent*, 4th edition, 2003. ISBN 0 85296 994 5.
- *IEE Guidance Note 7: Special locations*, 2nd edition (incorporating the 1st and 2nd amendments), 2003. ISBN 0 85296 995 3.
- *IEE On-Site Guide* (BS 7671 IEE Wiring Regulations, 16th edition), 2002. ISBN 0 85296 987 2.
- *New wiring colours*, 2004. Leaflet available to download at: www.iee.org/cablecolours

Metal Cladding and Roofing Manufacturers Association
www.mcrma.co.uk

- *Guidance for design of metal cladding and roofing to comply with Approved Document L2.*

Modular and Portable Buildings Association (MPBA)
www.mpba.biz

- *Energy performance standards for modular and portable buildings, 2006.*

National Association of Rooflight Manufacturers
www.narm.org.uk

- *Use of rooflights to satisfy the 2002 Building Regulations for the Conservation of Fuel and Power.*

NBS (on behalf of ODPM)
www.thebuildingregs.com

- *Domestic Heating Compliance Guide, 2006.* ISBN 1 85946 225 1.
- *Low or Zero Carbon Energy Sources: Strategic Guide,* 2006. ISBN 1 85946 224 3.

Scottish Building Standards Agency
www.sbsa.gov.uk

- *SBSA Technical Handbook – 'Conservatories'.*
- *SBSA Technical Guide: 'U-values'.*

Steel Construction Institute
www.steel-sci.org

- *Metal Cladding: U-value calculation: Assessing thermal performance of built-up metal roof and wall cladding systems using rail and bracket spacers, 2002.*

Thermal Insulation Manufacturers and Suppliers Association (TIMSA)
www.timsa.org.uk

- *HVAC Guidance for Achieving Compliance with Part L of the Building Regulations, 2006.*

1.4 Legislation

- Boiler (Efficiency) Regulations, 1993.
- Building (Procedure) (Scotland) Regulations, 2004.
- Building (Scotland) Act, 2003.
- Civic Government (Scotland) Act, 1982 – Order 2000.
- Construction (Design and Management) Regulations, 1994.
- Construction (Health, Safety and Welfare) Regulations, 1996.
- Control of Pollution Act 1974.
- Electricity Act 1989.
- Electricity Safety, Quality and Continuity Regulations 2002.
- Energy Act 1983.
- Environment Act 1995.
- Environmental Protection Act, 1990.
- Factories Act 1961.
- Fire (Scotland) Act 2005 as amended.
- Fire Safety (Scotland) Regulations 2006.
- Gas Appliance (Safety) Regulations, 1995.
- Gas Safety (Installation and Use) Regulations, 1998.
- Groundwater Regulations 1998.
- Health & Safety at Work etc Act 1974.
- Manual Handling Operations Regulations, 1992.
- Mines and Quarries Act 1954.
- Regulation of Care (Scotland) Act 2001.
- Sewage (Scotland) Act, 1968.
- Technical Standards for compliance with the Building Standards (Scotland) Regulations, 1990, as amended.
- Water Byelaws 2004.
- Water Environment (Controlled Activities) (Scotland) Regulations 2005.
- Water Environment (Oil Storage) (Scotland) Regulations 2006.

1.5 Other relevant UK statutory instruments

- SI 1991/1620 Construction Products Regulations 1991.
- SI 1992/2372 Electromagnetic Compatibility Regulations 1992.
- SI 1994/3051 Construction Products (Amendment) Regulations 1994.
- SI 1994/3080 Electromagnetic Compatibility (Amendment) Regulations 1994.
- SI 1994/3260 Electrical Equipment (Safety) Regulations 1994.
- SI 2001/3335 Building (Amendment) Regulations 2001.
- SI 2005/1726 Energy Information (Household Air Conditioners) (No. 2) Regulations 2005.
- SI 2006/652 Building and Approved Inspectors (Amendment) Regulations 2006.

1.6 EU relevant legislation

- 94/61 1/EC implementing Article 20 of the Council Directive 89/1 06, IEEC on construction products.
- Commission Decision 2000/1 47/EC of 8th February 2000 implementing Council Directive 89/106/EEC.
- Commission Decision 2000/55&EC of 6th September 2000 implementing Council Directive 89/106/EEC.
- Commission Decision 20001367/EC of 3rd May 2000 implementing Council Directive 89/106/EEC.
- Commission Decision 2001/671/EC of 21 August 2001 implementing Council Directive 89/106/EC as regards the classification of the external fire performance of roofs and roof coverings.
- Commission Decision 2005/823iEC of 22 November 2005 amending Decision 2001/671/EC regarding the classification of the external fire performance of roofs and roof coverings.
- Commission Decision 961 603/EC of 4th October 1996.
- Construction Product (Amendment) Regulations 1 994 (SI 1 994 No. 3051).
- Construction Products Regulations 1 991 (SI 1 991 No. 1620).
- Disability Discrimination Act 1995.
- Education Act 1996.
- Electrical Equipment (Safety) Regulations 1994 (SI 1994 No. 3260).
- Electromagnetic Compatibility (Amendment) Regulations 1994 (SI 1994 No. 3080).
- Electromagnetic Compatibility Regulations 1992 (ISI 1992 No. 2372).
- European tests: Commission Decision 20001367,1 @ C of 3rd May 2000 implementing Council Directive 89/106/EEC.
- Health and Safety (safety signs and signals) Regulations 1996.
- Pipelines Safety Regulations 1996, SI 1996 No. 825 and the Gas Safety (Installation and Use) Regulations 1998 SI 1998 No. 2451.
- The Workplace (Health, Safety and Welfare) Regulations 1992.

1.7 Useful websites

Architectural Heritage Society of Scotland
http://www.ahss.org.uk

Architecture + Design Scotland
http://www.ads.org.uk

Association of Scottish Community Councils
http://www.ascc.org.uk

Bat Conservation Trust
http://www.bats.org.uk

Built Environment Forum Scotland
http://www.befs.org.uk

Faculty of Advocates
http://www.advocates.org.uk

For advice on planning decision appeals from the Scottish Executive
http://www.scotland.gov.uk/Topics/Planning/Appeals

For e-mail alerts about planning applications in your area
http://www.planningalerts.com

For further information on planning guidance
http://www.scotland.gov.uk/planning

For information on charities and Gift Aid
http://www.cafonline.org

For information on the Voluntary sector as a whole
http://www.scvo.org.uk

Forestry Commission Scotland
http://www.forestry.gov.uk/scotland

Historic Scotland
http://www.historic-scotland.gov.uk

Information on planning enforcement regulations
http://www.scotland.gov.uk/Publications/2007/10/31093316/0

Planning Aid for Scotland
http://www.planning-aid-scotland.org.uk

Royal Institute of British Architects
http://www.architecture.com

Royal Institute of Chartered Surveyors
http://www.rics.org.uk

Scottish Building Standards Agency
http://www.sbsa.gov.uk

Scottish Civic Trust
http://www.scottishcivictrust.org.uk

Scottish Environmental Protection Agency
http://www.sepa.org.uk

Scottish National Heritage
http://www.snh.org.uk

Shelter: Advice & Support
http://www.shelter.org.uk/adviceonline

Sustainable Scotland Network
http://www.sustainable-scotland.net/

The Law Society of Scotland
http://www.lawscot.org.uk

The National Playing Fields Association Scotland
http://www.npfa.co.uk/content/npfascotland/index.html

The Royal Town Planning Institute in Scotland
http://www.scotland.rtpi.org.uk

The Scottish Parliament
http://www.scottish.parliament.uk/home.htm

The Scottish Public Services Ombudsman
http://www.spso.org.uk

Scottish Stationery Office (TSO)
www.tsoscotland.com

1.8 Abbreviations

ABS	Acrylonitrilebutadiene-Styrene
AD	Approved Document
AFILS	Audio-Frequency Induction-Loop System
BMS	Building Management System
BRE	Building Research Establishment
CCA	Copper Chrome Arsenic
CHP	Combined Heat and Power
CHP(QI)	CHPQA Quality Index
CITB	Construction Industry Training Board
CLEA	Contaminated Land Exposure Assessment
COP	Coefficient of Performance
CPD	Construction Products Directive
DDA	Disability Discrimination Act
DHW	Domestic Hot Water (system)
DPC	Damp-Proof Course
DQRA	Detailed Quantitative Risk Assessment
EC	European Commission
EER	Energy Efficiency Ratio
EPC	Energy Performance Certificate
GQRA	Generic Quantitative Risk Assessment
GRP	Glass Reinforced Plastic

HETAS	Heating Equipment Testing Approval Scheme
HSE	Health and Safety Executive
HVAC	Heating Ventilation and Air Conditioning
HVCA	Heating and Ventilating Contractors' Association
LPG	Liquefied Petroleum Gases
LPGA	Liquefied Petroleum Gas Association
LZCT	Low and Zero Carbon Technologies
MEV	Mechanical Extract Ventilation
MVAC	Mechanical Ventilation and Air Conditioning
MVHR	Mechanical Supply and Extract With Heat Recovery
OSB	Oriented Strand Board
PE	Polyethylene
PIR	Passive InfraRed (Sensor)
PSV	Passive Stack Ventilation
PVC-U	Unplasticized Poly (Vinylchloride).
SABCO	Scottish Association of Chief Building Control Officers
SABSM	Scottish Association of Building Standards Managers
SBS	Scottish Building Regulation
SBSA	Scottish Building Standards Agency
SEDBUK	Seasonal Efficiency of Domestic Boilers
SEPA	Scottish Environment Protection Agency
SFP	Specific Fan Power
SHEVS	Smoke and Heat Exhaust Ventilation System
SNH	Scottish National Heritage

1.9 Books by the same authors

Title	Details	Publisher
Building Regulations in Brief (England, Wales & Northern Ireland) (5th Edition)	This handy and affordable guide is a time-saver for both professionals and enthusiasts. The information is sensibly organized by building element rather than by regulation, so that you can quickly lay your hands on whatever you need to know from whichever document. This new edition includes: • The new Regulatory Reform (Fire Safety) Order and what this means for Part B (Fire Safety) • Updates to Part L (Energy Efficiency) • An improved user-friendly index • Annexes covering: Access and facilities for disabled people; Conservation of fuel and power; Sound insulation and Electrical Safety provided online.	Butterworth-Heinemann ISBN: 978-0-7506-8444-6
Wiring Regulations in Brief	This essential reference will prove an invaluable guide for anyone based in the electrical industry working on electrical systems (design, installation, inspection and testing), who requires a comprehensive source of information on the specific requirements of the IEE Wiring Regulations (published by the IET), without having to trawl through the lengthy, complicated coverage of the Regulations themselves.	Butterworth-Heinemann ISBN: 978-0-7506-8973-1

ISO 9001:2000 for Small Businesses (3rd Edition)

The top-selling ISO quality management handbook.

A guide to cost-effective compliance with the requirements of ISO 9001:2000.

Small- and medium-sized companies face many challenges today including the demand by larger customers for ISO 9001 compliance. Four years into the current version of ISO 9001, the new edition of this life-saving book incorporates the hard-won field experience of actually working with the standard, along with a thoroughly updated and customizable generic Quality Manual with audit checklists for developing a complete Quality Management System.

Butterworth-Heinemann
ISBN-: 978-0-7506-6617-6

ISO 9001:2000 in Brief (2nd Edition)

A 'hands-on' book providing practical information on how to cost-effectively set up an ISO 9001:2000 Quality Management System.

With comprehensive coverage of the meaning, history and requirements of the current ISO 9001 standard, the book explains how businesses can easily and efficiently satisfy customer requirements for quality control and quality assurance. Four years into the current version of ISO 9001, the new edition of this valuable book incorporates the hard-won experiences of working with the standard, together with direct, accessible and straightforward guidance that is proven to work.

Butterworth-Heinemann
ISBN: 978-0-7506-6616-9

(Continued)

Title	Details	Publisher
ISO 9001:2000 Audit Procedures (2nd Edition)	Revised, updated and expanded, ISO 9001:2000 Audit Procedures describes the methods for completing management reviews and quality audits, and outlines the experiences of working with 9001:2000 since its launch in 2000. It also includes essential new material on process models, generic processes, the requirements for mandatory documented procedures, and detailed coverage of auditors, questionnaires.	Butterworth-Heinemann ISBN: 978-0-7506-6615-2
Quality Management System for ISO 9001:2000	'Quality Management System for ISO 9001: 2000' and accompanying CD is probably the most comprehensive set of ISO 9001:2000 compliant documents available worldwide. Fully customizable, it can be used as a basic template for any organization wishing to work in compliance with, or gain registration to, ISO 9001:2000.	ISBN: 978-0-9548-6474-3

Auditing Quality Management Systems

ISBN: 978-0-9548-6475-0

'Auditing Management Systems' and accompanying CD is the result of 5 years, field experience of the international standard for quality management (i.e. ISO 9001:2000) and is capable of being used to conduct an internal, external or third party audit of ANY Management System.

Optoelectronics and Fiber Optic Technology

Newnes
ISBN: 978-0-7506-5370-1

An introduction to the fascinating technology of fibre-optics.

Topical areas such as optoelectronics in LANs and WANs, cable TV systems, and the global fibre-optic highway make this book essential reading for anyone who needs to keep up with the technology of modern data communications.

(Continued)

Title	Details	Publisher
CE Conformity Marking 	Essential information for any manufacturer or distributor wishing to trade in the European Union. Practical and easy to understand.	Butterworth-Heinemann ISBN: 978-0-7506-4813-4
Environmental Requirements for Electromechanical and Electronic Equipment 	Definitive reference containing all the background guidance, ranges, test specifications, case studies and regulations worldwide.	Butterworth-Heinemann ISBN: 978-0-7506-3902-6

MDD Compliance using
Quality Management Techniques

The Medical Devices Directive (MDD) is an
all-encompassing document legislating for the
manufacture of any medical device or material
used either temporarily or permanently on or in
the human body. To achieve its main objectives
the MDD requires the manufacturer of all products
covered by the Directive to possess a fully
auditable Quality Management System consisting
of Quality Policies, Quality Procedures and Work
Instructions based on the ISO 9000 standard.
The book is based on the sound principles of
ISO 9000 and will guide to the reader, if required,
to eventually set up an ISO 9000 fully compliant
system.

Butterworth-Heinemann
ISBN: 978-0-7506-4441-9

Quality & Standards in Electronics

A manufacturer or supplier of electronic
equipment or components needs to know the
precise requirements for component certification
and quality conformance to meet the demands of
the customer. This book ensures that the professional
is aware of all the UK, European and International
necessities, and knows the current status of these
regulations and standards, and where to obtain
them.

Butterworth-Heinemann
ISBN: 978-0-7506-2531-9

CDs by the same author

Title	Details	ISBN
ISO 9001:2000 Quality Manual & Audit Checksheets 	A CD containing a soft copy of the generic Quality Management System featured in ISO 9001:2000 for Small Businesses (3rd edition) Plus – a soft copy of all the checksheets and example audit forms contained in ISO 9001:2000 Audit Procedures (2nd edition)	ISBN: 0 9548647 2 7
ISO 9001:2000 Audit Checklists 	A CD containing all of the major audit checksheets and forms required to conduct: • a simple internal review of a particular process; • an external (e.g. third party) assessment of an organization against the formal requirements of ISO 9001: 2000; • an assessment of an organization against the formal requirements of other associated management standards such as ISO 140001 and OHSAS.	ISBN: 0 9548647 1 9

… and if you are looking for something completely different

ISBN-13: 978-0954864767

A unique combination of a historical overview of cider making through the ages, the cider making process plus a collection of recipes using cider and cider apples (from Soups to Sweets from Pies to Preserves), including such tempting ideas as

- 'Onion and Cider Soup'
- 'Salmon and Oats in a Cider Sauce'
- 'Ham in Cider'
- 'Scrumpy Pheasant'
- 'Cider Scalloped Potatoes'
- 'Apple Strudel'
- 'Apple and Red Chilli Chutney'
- 'Tangy BBQ Sauce'
- 'Witches' Cauldron'

THE CYDER BOOK

An historical overview of Cider Making through the ages
Together with a unique collection of recipes using cider
and cider apples

**Written and compiled by
Ray & Claire Tricker**

1.10 Useful contact names and addresses

The following professional body is willing to provide general and informal advice about the Act. **However**, any advice given should **not** be seen as being endorsed by the Office of the Deputy Prime Minister!

The Royal Institution of Chartered Surveyors (RICS)
Technical Services Unit
12 Great George Street
London SW1P 3AD
Tel: 020 7222 7000 (extension 492)
Fax: 020 7222 9430

The following bodies hold lists of their members who may be willing to provide professional advice or act as a 'surveyor' under the Act – again with the provison that any advice given should **not** be seen as being endorsed by the Office of the Deputy Prime Minister!

Architecture and Surveying Institute
Register of Party Wall Surveyors
St. Mary House
15 St. Mary Street
Chippenham
Wiltshire SN15 3WD
Tel.: 01249 444505
Fax: 01249 443602

The Association of Building Engineers (ABE)
Private Practice Register
Lutyens House
Billing Brook Road
Weston Favell
Northampton NN3 8NW
Tel.: 01604 404121
Fax: 01604 784220

The Pyramus & Thisbe Club
Florence House
53 Acton Lane
London NW10 8UX
Tel.: 020 8961 3311
Fax: 020 8963 1689

The Royal Institute of British Architects (RIBA)
Clients Advisory Service
66 Portland Place
London W1N 4AD
Tel.: 020 7307 3700
Fax: 020 7436 9112

The Royal Institution of Chartered Surveyors (RICS)
Information Centre
12 Great George Street
London SW1P 3AD
Tel.: 020 7222 7000
Fax: 020 7222 9430

Professional contacts

Asbestos specialists

Asbestos Information Centre Ltd
PO Box 69
Widnes
Cheshire
WA8 9GW
0151 420 5866

Concrete specialist

British Ready Mixed Concrete Association
The Bury
Church Street
Chesham
Buckinghamshire HP5 1JE
01494 791050

Damp, rot, infestation

British Wood Preserving & Damp Proofing Association
Building No. 6
The Office Village
4 Romford Road
Stratford
London E15 4EA
020 8519 2588

English Nature
Northminster House
Northminster Road
Peterborough PE1 1VA
01733 340345

Countryside Council for Wales
Plaspenrhos
Penrhos Road
Bangor
Gwynedd LL57 2LG
01248 370444

Scottish Natural Heritage
12 Hope Terrace
Edinburgh EH9 2AS
0131 447 4784

Local Department of Environmental Health
Refer to your local directory

Decorators

British Decorators Association
32 Coton Road
Nuneaton

Warwickshire CV11 5TW
01203 353776

Scottish Decorators Federation
41A York Place
Edinburgh EH1 3HT
0131 557 9345

Electricians

National Inspection Council for Electrical Installation Contracting
37 Albert Embankment
London SE1 7UJ
020 7582 7746

Fencing erectors

Fencing Contractors Association
Warren Rd
Trellech
Monmouthshire NP25 4PQ
07000 560722

Glazing specialist

Glass and Glazing Federation
44–48 Borough High Street
London SE1 1XB
020 7403 7177

Heating installers

British Gas Regional Office
Refer to your local directory

Electricity Supply Company
Refer to your local directory

British Coal Corporation
Hobart House
Grosvenor Place
London SW1X 7AE
020 7235 2020

Heating and Ventilating Contractors Association

Esca House
34 Palace Court
London W2 4JG
020 7229 2488

National Association of Plumbing, Heating and Mechanical Services Contractors

Ensign House
Ensign Business Centre
Westwood Way
Coventry CV4 8JA
01203 470626

Home security

Local Crime Prevention Officer

Refer to your local directory

Local Fire Prevention Officer

Refer to your local directory

National Approval Council for Security Systems

Queensgate House
14 Cookham Road
Maidenhead S16 8AJ
01628 37512

Master Locksmiths Association

Units 4–5
Woodford Halse Business Park
Great Central Way
Woodford Halse
Daventry NN1 6PZ
01327 62255

British Security Industry Association

Security House
Barbourne Road
Worcester WR1 1RS
01905 21464

Insulation installers

Draught Proofing Advisory Association Ltd, External Wall Insulation Association, National Cavity Insulation Association, National Association of Loft Insulation Contractors

PO Box 12
Haslemere
Surrey GU27 3AH
01428 654011

Plasterers

Federation of Master Builders

14 Great James Street
London WC1N 3DP
020 7242 7583

Plumbers

National Association of Plumbing, Heating and Mechanical Services Contractors

Ensign House
Ensign Business Centre
Westwood Way
Coventry CV4 8JA
01203 470626

Roofers

Builders' Merchants Federation

15 Soho Square
London W1V 5FB
020 7439 1753

National Federation of Roofing Contractors

24 Weymouth Street
London W1G 7LX
020 7436 0387

Ventilation

**Heating and Ventilating
Contractors Association**
Esca House
34 Palace Court
London W2 4JG
020 7229 2488

Other useful contacts

British Board of Agrément (BBA)
PO Box 195
Bucknalls Lane
Garston
Watford WD2 7NG
Tel.: 01923 665300
Fax: 01923 665301
E-mail: bba@btinternet.com
Internet: http://www.bbacerts.co.uk

BSI
British Standards Institution
389 Chiswick High Road
London W4 4AL
Tel.: 020 8996 9001
Fax: 020 8996 7001
E-mail: info@bsi.org.uk
Internet: www.bsi.org.uk

Fensa Ltd
Fenestration Self-Assessment
Scheme
44–48 Borough High Street
London SE1 1XB
Tel.: 020 7207 5874
Fax: 020 7357 7458

HETAS Ltd
HETAS Ltd
12 Kestrel Walk
Letchworth
Hertfordshire SG6 2TB
Tel.: 01462 634721
Fax: 01462 674329

 Note: Registration scheme for companies and engineers involved in the installation and maintenance of domestic solid fuel fired equipment.

HMSO
The Stationery Office
The Publications Centre
PO Box 29
Norwich NR3 1GN
Telephone orders/General enquiries:
0870 600 5522
Fax orders: 0870 600 5533
Internet: www.thestationeryoffice.com

Institute of Plumbing
64 Station Lane
Hornchurch
Essex RM12 6NH
Tel.: 01708 472791
Fax: 01708 448987

Note: Approved Contractor Person Scheme (Building Regulations).

OFTEC
Oil Firing Registration Scheme
Century House
100 High Street
Banstead
Surrey SM7 2NN
Tel.: 01737 373311
Fax: 01737 373553

Robust Details Ltd
PO Box 7289
Milton Keynes MK14 6ZQ
Business Line: 0870 240 8210
Technical Support Line: 0870 240 8209
Fax: 0870 240 8203

UKAS
United Kingdom Accreditation
Service
21–47 High Street
Feltham, Middlesex TW3 4UN
Tel.: 0208 917 8400
Fax: 0208 917 8500

WIMLAS
WIMLAS Ltd
St. Peter's House
6–8 High Street
Iver, Buckinghamshire SL0 9NG
Tel.: 01753 737744
Fax: 01753 792321
E-mail: wimlas@compuserve.com

Useful websites

The Building Act and Building Regulations www.communities.gov.uk
(the new DCLG site)

Approved Documents
www.planningportal.gov.uk/england/professionals/en11531410382.html

Air Tightness Testing and Measurement Association (ATTMA)	www.attma.org
BRE	www.bre.co.uk
Centre for Window and Cladding Technology (CWCT)	www.cwct.co.uk
Chartered Institution of Building Services Engineers (CIBSE)	www.cibse.org
Department for Education and Skills (DFES)	www.dfes.gov.uk
Department of the Environment, Food and Rural Affairs (Defra)	www.defra.gov.uk
Department of Transport, Local Government and the Regions (DTLR)	www.dtlr.gov.uk
Energy Saving Trust (EST)	www.est.org.uk
English Heritage	www.english-heritage.org.uk
Health and Safety Executive (HSE)	www.hse.gov.uk
Heating and Ventilating Contractors Association (HVCA)	www.hvca.org.uk
Metal Cladding and Roofing Manufacturers Association (MCRMA)	www.mcrma.co.uk
Modular and Portable Buildings Association (MPBA)	www.mpba.biz
National Association of Rooflight Manufacturers (NARM)	www.narm.org.uk
NBS (on behalf of ODPM)	www.thebuildingregs.com
The Planning Portal	www.planningportal.gov.uk/england/genpub/en/
Thermal Insulation Manufacturers and Suppliers Association (TIMSA)	www.timsa.org.uk
TrustMark	www.trustmark.org.uk

Assessable thresholds	www.tso.co.uk
British Standards	www.bsonline.techindex.co.uk
Building near trees	www.nhbc.co.uk
Building Research	www.bre.co.uk
Carbon dioxide from natural sources and mining areas	www.bgs.ac.uk
	www.tso.co.uk
	www.defra.gov.uk
	www.ciria.org.uk
Cladding	www.mcrma.co.uk
Concrete in aggressive ground	www.bre.co.uk
Contaminated land	ww.defra.gov.uk
	www.ciria.org.uk
	www.hse.co.uk
Contamination in disused coal mines	www.tso.co.uk
Demolition	www.ciria.org.uk
Electrical safety	www.theiet.org
Environmental aspects	www.arup.com
	www.environment-agency.gov.uk
Excavation and disposal	www.defra.gov.uk
	www.ciria.org.uk
Flood protection	www.ciria.org.uk
	www.safety.odpm.gov.uk
Flooding from sewers	www.defra.gov.uk
	www.ciria.org.uk
Foundations	www.bre.co.uk
Gas contaminated land	www.defra.gov.uk
	www.bre.co.uk
	www.ciria.org.uk
Geoenvironmental and geotechnical investigations	www.ags.org.uk
Glass and glazing	www.ggf.org.uk
Hardcore	www.bre.co.uk
Health and safety	www.hse.co.uk
	www.defra.gov.uk
Land quality	www.environment-agency.gov.uk
Landfill gas	www.ciwm.co.uk
	www.defra.gov.uk
	www.gassim.co.uk
Laying water pipelines in contaminated ground	www.fwr.org
Low-rise buildings	www.bre.co.uk
Materials and workmanship	www.tso.co.uk
Methane	www.bgs.ac.uk
	www.tso.co.uk
	www.defra.gov.uk
	www.ciria.org.uk

Oil seeps from natural sources and mining areas	www.bgs.ac.uk
	www.tso.co.uk
	www.defra.gov.uk
	www.ciria.org.uk
Petroleum retail sites	www.petroleum.co.uk
Pollution control	www.odpm.gov.uk
Protection of ancient buildings	www.spab.org.uk
Radon	www.bre.co.uk
Robust construction details	www.tso.co.uk
Roofing design	www.mcrma.co.uk
Shrinkable clay soils	www.bre.co.uk
Soil sampling	www.defra.gov.uk
Soils, sludge and sediment	www.ciria.org.uk
Subsidence	www.bre.co.uk
Thermal bridging	www.bre.co.uk
	www.tso.co.uk
Thermal insulation	www.bre.co.uk
Timbers	www.bre.co.uk

Index

A

Absorbent blanket
  platform floor with 309
  ribbed floor with 310–311
Absorbent curtain, timber frames with
      428–430
Access
  to basements 625
  bathrooms 644–645
  to buildings 73
  within buildings 73–4
  desludging 236
  entrances 538–9, 606–630
  to fire services 68, 628–629
  to flues 493
  to kitchen 660–661
  to manual controls of windows
      520–521
  roofs 476–477, 626–627
  sanitary facilities 638–640
  shower rooms 336
  single stair 625, 626
  and usability 21, 23
Access decks
  escape routes across 394–395, 396
  flat roofs and 468–469
Accidents, danger from 75
Activity spaces, height of 440
Adjacent buildings and services,
      safeguarding of 12
Advertisements
  fee 124
  of planning application 97, 104
  planning permission for 142–3
Advice service, planning permission
      122, 126
Aggregate width 567
Agricultural buildings 81–2
  planning permission for extensions
      to 151
Airborne sound 591, 799
  minimum values 431

Air conditioning, and mechanical
      ventilation 78, 264–266
Air for combustion
  gas-fired appliances 687, 688
  supply 672–673
  to solid fuel appliance 695, 696
Air leakage limitation 192
Air-tightness testing 193
Alarm 562
Alarm systems
  in hospitals 814–815
  in residential care buildings 801–802,
      809
  in shopping centres 819
Alterations, to insulation envelope 194,
      435–436
Ancient Monuments and Archaeological
      Areas Act 1979 15, 52
Apartments
  kitchen height 564
  natural lighting 788
Appeals, against decisions 5, 11
Appliance efficiency, hot water storage
      systems 700
Approved Certifier of Construction 7,
      42–3
Approved Certifier of Design 7,
      42–3
Approved Certifiers 7
  appointment 43
  audit of work 43–4
  availability 43
  benefit of use 42
  Building (Procedure) (Scotland)
      Regulations 2004 42–4
  of construction 7, 42–3
  of design 7, 42–3
  information required 42–3
Approved Documents 21
Area Committee 106, 108
Area measurements 90
Artificial lighting 78, 758–759

Auditoriums
escape routes 396
exit 535
Automatic fire detection and alarm
systems 740
enclosed shopping centres 744
hospitals 743
installation of 569
mixed-use buildings 742
residential buildings 741
residential care buildings 742
in shopping centres 819
Automatic life safety fire suppression
systems 69, 822
residential care buildings 809–810
Automatic smoke detection 562
Axial fans 268

B
Balconies 630–631
fire precautions 633
means of escape 632–633
requirements 631–632
safety precautions 634
Barn owls
and planning permission 118–19
Basements 290–299
compartmentation 291–293
in domestic buildings 297–298
escape routes in 398
escape stair 565, 573
escape stairs in 396
in non-domestic buildings 293–297
requirements 290–291
meeting 291–299
venting of heat and smoke from 295–
296
windows in 515
Bathrooms
access 644–645
electrical safety 657
fire resistance duration 395
floors 336
gas-fired appliances 657
heating circuits 658
oil-firing appliances 657
Bats and their roosts, and planning
permission 117–18
Battens along top of joists 312
in conversions 313

Biomass fuel, woody, bulk storage of
415
Block plan, planning permission 103
Boilers 724–726
controls 758
plant control 725–726
solid fuel 724
Boundary wall, planning permission to
alter 138
Brick-built house, walls in 344
Brickwork/blockwork
coursed
cavity walls in 351
solid walls in 350
internal load-bearing walls in 352
Building Act 1984 15–16, 21
Part A: Structure 16
Part B: Fire safety 16
Part C: Site preparation 16
Part D: Toxic substances 16
Part E: Resistance to the passage of
sound 16
Part F: Ventilation 16
Part G: Hygiene 16
Part H: Drainage and waste
disposal 16
Part J: Combustion appliances 16
Part K: Protection from falling,
collision and impact 16
Part L: Conservation of fuel and
power 16
Part M: Access and facilities for
disabled people 16
Part P: Electrical safety 16
Building design and construction
container, woody biomass fuel 526
energy performance 525
fire prevention 524, 526
insulation envelope 526
U-values for building elements
of 544
safety 525
Building design and construction, under
Building (Scotland) Act 2003
additional requirements 184
mandatory requirements 168
meeting requirements 168–197
requirements 167–168
Building documentation and information
energy performance certificates 827

carbon dioxide emissions,
    calulation 827–828
dwellings, information 828–829
location of 830–831
non-domestic buildings,
    information 829–830
metering-non-domestic buildings 832
    existing buildings 832
    new buildings 832
requirements 824–826
written information 826–827
    domestic building 826
    non-domestic buildings 826
    work on existing building 827
Building (Procedure) (Scotland)
    Regulations 2004
appeals provision 49–50
Approved Certifiers 42–4
authorities needs to be consulted 55–7
Building Warrants 27–42
Care commission 57
completion certificate 46–8
compliance notices 50–1
conversion 20
coverage 19–20
Crown buildings 51
dangerous buildings 52–5
employment of professional
    builder 45
energy 21, 24–5
Energy Performance Certificates
    44–5
environment 21, 23, 25
exemptions 25–6
fire 21–2
fire authorities 55–6
Highways department 56
historic buildings 52
Historic Scotland 57
local authority records 57
need 18–19
noise 21, 23–4
non-compliance with Technical
    Handbook 26
planning officers 45
police 57
purpose 18
relaxations 45–6, 48–9
requirements 20
safety 21, 23

Scottish Environment Protection
    Agency 56
Scottish Ministers involvement 48–9
Scottish Water 56–7
Section 1: Structure 21
Section 2: Fire 21–2
Section 3: Environment 21, 23, 25
Section 4: Safety 21, 23
Section 5: Noise 21, 23–4
Section 6: Energy 21, 24–5
sharing of information 55
structure 21
Technical Handbooks 20–5
Verifiers 26–7
    impose continuing requirements 51
work not requiring a warrant and
    exempt 26
Building regulation act
    regulation 4, schedule 2, 545
Building Regulation Completion
    Certificate 36
Building Regulations
    apply to Crown buildings 51
    approval 129–30
        for planning permission 105
    building notice 6
    in Building (Scotland) Act 2003 2–3,
        12–13
    buildings exempted from 130
    compliance of 4, 6
    defective and dangerous buildings 2,
        4, 10
    exemptions from 6, 130
    guidance documents 3, 7
    interoperability of 16
    non-compliance 6
    power to make 2–3
    relaxation in 3, 6
    Section 1: Structures 16
    Section 2: Fire 16
    Section 3: Environment 16
    Section 4: Safety 16
    Section 5: Noise 16
    Section 6: Energy 16
Building Regulations 1965 19
Building Regulations 2000 16
Building Regulations compliance
    notice 6–7, 50–1
Building Regulations (Northern Ireland)
    2000 17

Building Regulations (Northern Ireland)
    Order 1979 15–16
  Technical Booklet B: Materials and
    workmanship 16
  Technical Booklet C: Preparation of
    site and resistance to moisture 16
  Technical Booklet D: Structure 16
  Technical Booklet E: Fire Safety 16
  Technical Booklet F: Conservation of
    fuel and power 16
  Technical Booklet G: Sound
    installation of dwellings 16
  Technical Booklet H: Stairs, ramps
    and protection from impact 16
  Technical Booklet J: Solid waste in
    buildings 16
  Technical Booklet K: Ventilation 16
  Technical Booklet L: Heat-producing
    appliances 16
  Technical Booklet N: Drainage 16
  Technical Booklet P: Sanitary
    appliances 16
  Technical Booklet R: Access for
    facilities and disabled people 16
Buildings
  ancillary to
    flats or maisonettes 84
    houses 84
  controlled by other legislation 81
  description of 86–9
  energy performance 191
  evacuation of 4–5, 10, 12, 14
  interpretation 5
  not frequented by people 81
  not requiring warrant 86–9
  performance testing 190
  planning permission for
    extension 149–51
  plans, planning permission 103
  small 83
  specialist nature 82–3
  structural safety 21
  temporary 83
  without planning permission 119
Building safety 186–189
  from fire 180
  from noise 189
Building (Scotland) Act 1959 15
Building (Scotland) Act 1970 15
Building (Scotland) Act 2003 19
  aim 2

alternative solutions 164
approval of construction work 2–3, 18
building materials
  short-lived 165
  unsuitable 165
Building Regulations in 2–3, 12–13
  see also Building Regulations
Building warrants 3, 7–8
Certifications of Design 8
civil liability 11
Completion Certificates 3, 8–9
compliance and enforcement 2, 4, 6
compulsorily purchase building 10
contents 2–5
defective and dangerous buildings 4,
  9–10
ENV 166–167
evacuation of buildings 5, 12, 14
exemptions 6
independent certification schemes 166
local authority 6, 10
mixture of domestic and non-domestic
  building 164
modification of enactments 5, 12, 15
non-compliance 6–7
overview 162–163
Part 1: Building Regulations 2–3
Part 2: Approval of construction work,
  etc. 2–3
Part 3: Compliance and
  enforcement 2, 4, 6
Part 4: Defective and dangerous
  buildings 2, 4, 9–10
Part 5: General 2, 4–5
Part 6: Supplementary 2, 5
policing of 6
powers of entry, inspection and
  testing 5, 12, 14
procedure regulations 4–5, 12–14
provisions 2
relaxation 6
Royal Assent to 2
Schedule 1 5, 12–13
Schedule 2 5, 12–13
Schedule 3 5, 12–13
Schedule 4 5, 12, 14
Schedule 5 5, 12, 14
Schedule 6 5, 12, 15
scheduled monuments and listed
  buildings 4, 10
sections and subsections 3–5

sell off materials from demolished
    buildings 11
supplementary regulations 5, 12–15
Technical Handbooks 162–163
technical specifications 165–166
Verifiers and Certifiers 3–5, 7–8,
    12–13
Building (Scotland) Regulations
    2004 1–2, 28
building, works, services and
    equipment not requiring
    warrant 61
Building Standards 61–2
changes in occupation or use of
    building 60
clearing of footpaths 65
continuing requirements 66
durability, workmanship and fitness of
    materials 61
exempted buildings and services,
    fittings and equipment 60
limited-life buildings 61
measurements 61
provision of protective work 63–5
relaxations 65
requirements 59–91
Section 0: General Provisions 60–6
Section 1: Structure 67
    disproportionate collapse 67
    functional standard 67
Section 2: Fire 67–9
    access to fire service 68
    automatic life safety fire
        suppression systems 69
    cavities 68
    communication 68
    compartmentation 67
    escape 68
    escape lighting 68
    fire service facilities 68–9
    internal linings 68
    separation 67
    spread from neighbouring
        buildings 68
    spread on external walls 68
    spread to neighbouring buildings 68
    structural protection 67
    water supply to fire service 68
Section 3: Environment 69–73
    combustion appliances 71–2
    condensation 71

existing drains 69
facilities in dwellings 71
flooding and ground water 69
fuel storage 72–3
heating 71
infiltration systems 70
moisture from ground 69
natural lighting 71
precipitation 70
private wastewater treatment
    systems 70
sanitary facilities 71
site preparation 69
solid waste storage 73
surface water drainage 70
treatment plants 70
ventilation 71
wastewater drainage 70
Section 4: Safety 73–5
    access to buildings 73
    access within buildings 73–4
    aids to communication 74–5
    danger from accidents 75
    danger from heat 75
    electrical fixtures 74
    electrical safety 74
    fixed seating 75
    liquefied petroleum gas storage 75
    pedestrian protective barriers 74
    stairs and ramps 74
    vehicle protective barriers 75
Section 5: Noise 76
    resisting sound transmission 76
Section 6: Energy 76–80
    artificial and display lighting 78
    building insulation envelope 77
    carbon dioxide emissions 76
    commissioning building services
        78
    Energy Performance
        Certificates 79–80
    heating systems 77
    insulation of pipes, ducts and
        vessels 77
    mechanical ventilation and air
        conditioning 78
    metering 80
    writing information 78–9
securing of unoccupied and partially
    completed buildings 65
Building services commissioning 78

Building Standards
  applicable to
    construction 61
    conversions 62
    demolition 62
    provision of services, fittings and
      equipments 62
  registers 4
Building Standards Advisory
      Committee 4, 11, 49
Building Standards Assessment 11–12
Building Warrants 3, 7–8, 27–42
  action for non-compliance 8
  for advertisement 142–3
  application
    for amendment 37
    to Verifier 32–3
  architect requirement 29
  assistance in understanding
      requirements 36
  for building
    conservatory 146–7
    extension 149–51
    garden wall or fence 137–8
    hardstanding for vehicle or
      boat 140–1
    new house 156–7
    porch 143–4
  Building (Procedure) (Scotland)
      Regulations 2004 27–42
  building work does not require 33–4
  for central heating 136
  Certifications of Design 8
  conditional 37
  conservatories 36
  contravention of regulations 41
  for conversions and change of
      use 151–6
  for decoration and repairs 131–2
  for defective building 160–1
  definition 27–8
  for demolition 36, 157–60
  for dormer windows 148–9
  for electrical work 134–5
  enforcement notices 4, 8, 50–1
  to erect satellite dishes, television and
      radio aerials, wind turbines and
      flagpoles 141–2
  extensions 34–5
  fee discounts 32

  fee for 29–31
  for felling or lopping trees 138
  for garages 145–6
  grant of 37
  for infilling 157
  information
    availability on 9
    to neighbours 28
    required in application 28
  inspection while work progress 40
  for installing swimming pool 141
  issuance of 7, 27
  late application for 38
  for laying path or driveway 138–9
  limited-life warrants 39
  for loft conversions 148–9
  obtaining 28
  for oil storage tank 136–7
  for outbuildings 144–5
  for planting hedge or tree 137
  for plumbing 135–6
  refusal to grant 40
  for repair work 35–6
  for replacement windows and
      doors 133–4
  response time 33
  for roof extensions 148–9
  staged warrants 38–9
  for structural alterations inside 132–3
  submission of completion
      certificate. 41–2
  validity 37–8
  Verifiers and Certifiers for 7–8, 27, 37
  withdrawal application 40
  works on site differ from approved
      plans 40–1
Bulk storage, of woody biomass fuel
      415
Business Improvement Districts 95
Buttressing walls 359–360, 483

C
Carbon dioxide emissions 76
  calulation 827–828
Care commission 57
Car ports 12
Catwalk 571
Cavities 68
  high-rise domestic buildings 399
  wall 387–389

Cavity barriers 387–388, 471–473, 530
  ceilings 439
  floors 330
  hospitals 812
  residential care buildings 803, 805
  roofs 471–473
  windows 512
Cavity masonry
  wall constructions 407–408
  wall type 2, 422–425
Cavity walls
  in coursed brickwork/blockwork 351
  insulation 133
Ceilings 436–450
  automatic smoke detection fixed to 562
  energy saving 438
  entrance hall or stair 561
  environment/health requirements 437
  extensions 447–448
  fire precautions 440–443
  fire-resistant 441–442
  fire safety 436–437
  imposed loads on 438
  internal linings 443
  maximum U-values for 448
  reaction to fire 389
  rooflights 446–450
  safety requirements 437–438
  smoke alarms 442
  smoke outlets 443
  structure 438–440
  suspended 471
  thermoplastic materials in 443–445
  ventilation 446
  wall junction 439
  wall type 1, junctions at 421
  wall type 2, junctions at 424
  wall type 3, junctions at 427
  wall type 4, junctions at 430
Cellars 290–299
Central heating, planning permission
      for 136
Centrifugal fans 268
Certificates
  of Agricultural Holding 102–3
  of lawful use, fee for 125
  of Site Ownership 102
Certification
  of construction 9
  of Design 8

Certifiers 3–5, 7–8
  functions 4, 7–8
Change of use see also Conversions and
      change of use
  fee 124–5
Changes, in occupation or use of
      building 60
Chases, walls 360
Chimneys 89, 412–413, 414, 481–509
  air for combustion 496–497
  ceilings 440
  combustible materials 497–507
  combustion appliances 491–492,
      497–509
  construction 481, 487–490
  environment/health requirements 482
  escape routes 482
  fire safety 481
  flue-pipes 470, 493–494
    serving combustion appliances
      508
  flues 491–496
  hearths 505–507
  labelling 486–487
  masonry 484–485, 489
  metal 476, 489–490
  noise control for 418
  precipitation 486
  roofs 475–476
  structures and 483–485
  supporting 484
CHPQA Quality Index (CHP(QI)) 728
CHP(QI) see Combined Heat and Power
      Quality Index (CHP(QI))
Chronically Sick and Disabled Persons
      Act 1970 15
Civic Government (Scotland) Act
      1982 15
Civil engineering construction works 82
Civil liability, Building (Scotland) Act
      2003 11
Cladding, planning permission for 133
Clean Air Act 1993 15
Cleaning windows 519–520
Clear opening width of doorway 531
Collared roof 466–467
Collision, with projections 519
Combined escape routes 567
Combined Heat and Power Quality Index
      (CHP(QI)) 195

Combustibility
  in compartment walls 379
  external wall 392
  in separating floors 327
Combustible materials 413, 497–507
  flue-pipes and 501
  gas-fired appliances and 504–505
  hearths and 505–507
  masonry chimneys and 498–499
  metal chimneys and 500
  oil-firing appliances and 503–504
  relationship of hearths to 414
  relationship of metal chimneys to 414
  solid fuel appliances and 502–503
Combustion appliances 71–2, 496–509
  air for 672–673
  air for combustion 496–497
  combustible materials 662–667
  component parts 497–498
  fire safety 661–662
  flues 669
    design 669–670
    extract fans 671
    liners 670–671
    as passive stack ventilation 671
  fuel storage 673–675
  gas-fired appliances 683–691
  heating and cooking 672
  installation 86–7
  labelling of 486–487
  metering 675–676
  oil-firing appliances 676–683
  protection from 667–668
  relationship with combustible
    materials 497–507
  removal of products of
    combustion 507–509
  requirements 658–661
  safety of 413
  solid fuel appliances 692–698
Commercial satellite dishes 142
Commissioning building services 78
Common entrance see Doors
Communication 68
  aids to 74–5
Compartmentation 67
  basement 291–293
  for fire control in hospitals 403–404
  for fire control in residential care
    buildings 402

hospitals 810–811
  in residential care buildings 802, 808
  roofs 470
Compartment floors 324–326, 565
  openings through 529
Compartment walls 375–382, 565
  combustibility 379
  fire shutters 380
  junctions 381–382
  lift wells 379
  openings and service
    penetrations 379–380
  openings through 529
  service opening 380–381
  smoke venting shafts 379
  special fire risk 377, 379
  supporting structure 379
  ventilating ducts 381
Completion Certificates 3, 8–9
  acceptance and rejection 48
  Building (Procedure) (Scotland)
    Regulations 2004 46–8
  information availability on 9
  late submission 47
  need 46
  submission 46–7
  Verifier inspection 48
Concrete base
  with floating layer 304–307
  soft covered 301–304
Concrete floors
  suspended 335
Condensation 71
  control in roofs 474–475
  of floor surface 336–337
  inner surface, walls 412
  interstitial 474
Conservation areas
  consent and planning permission 104
  and planning permission 114–15
  property and planning
    permission 109, 132
  trees in 114–15
Conservation of fuel and power 21, 24–5
Conservatories
  construction over windows 786
  dividing elements 545
  escape windows 788
  external walls 786–787
  floors 341

heating 788
locations 785
natural lighting 788
planning permission for building 146–7
requirements 785
roofs and ventilation of 475
stair landings 542–3
U-values 436
ventilation 268–270, 787
wall 406
windows 517, 522–523
Conservatory Guidance 36
Construction and development buildings 83
Construction materials, for timber-framed buildings 179
Construction Products Directive (CPD) 162
Constructions
floor type 1 301–302
floor type 2 305–306
floor type 3 308–313
wall type 1 418–420
wall type 2 422–423
wall type 3 425–426
wall type 4 429
Continuing requirements
Building (Scotland) Regulations 2004 66
Control of Pollution Act 1974 15
Controls, door entry system 543
Convector 87
Conversions 85, 91
floor
battens along top of joists in 313
platform floor with absorbent blanket 309
garages 800
of heated buildings 433, 771
of historic building 193
Conversions and change of use
buildings suitable for 154
difference between 152
historic building 155–6
isolated building 154
material changes of use 154
old building 155–6
planning permission for 151–6
residential conversions 155

standards apply to 152–3
structural condition 155
workshop conversions 155
Corridors 534, 591
automatic opening ventilators 562
fire and smoke control in 533, 593–595
fire control in 394
requirements 592
safety measures 595–597
smoke control 562
smoke control in 394
Council's website
applying online for planning permission 102
Covenants, and planning permission 113
CPD see Construction Products Directive (CPD)
Crèches, occupancy capacity 818
Crime prevention, and planning permission 112
Criminal Procedure (Scotland) Act 1995 15
Crown application 5
Crown buildings, regulations apply 51
Curtilage of listing 113

D
Dangerous buildings 9–10, 52–5
building regulations for 4
notices 54–5
Day lighting 13
Dead loads
on floors 321, 322
roofs 464
Decoration, planning permission for 131–2
Defective Building Notice 160–1
Defective buildings 9–10
building regulations for 4
planning permission for 160–1
Demolition
dangerous buildings 158–60
planning permission for 157–60
Depth measurements 90
Design, and planning permission 112
Detached single-storey building 86–7
Development application, fee for 125
Development buildings 83
Development Management 95

Development Plans 94–6, 98, 107–8
Direct transmission 416
Discharge
  flues terminal 496
    windows 516
  plume from condensing boilers 509
Disproportionate collapse 67
Documentation and information
    see Building documentation and
    information
Domestic buildings *see* Dwellings
Domestic hot water (DHW)
    systems 700–702
  electric system 701
  thermal efficiencies 700
Domestic Technical Handbook 1
Doors 88–9
  accessible thresholds 539–40
  cavity barriers 530
  corridors
    fire and smoke control in 533
    glazed vision panel 541
  domestic building 524
  entry systems 543, 755
  fire doors 529–30
  fire-fighting shaft 566
  fire prevention 524, 526
  hospitals 537
  internal 540
  locking device 543
  locks on 531
    electrically powered 532
  means of access 525
  non-domestic buildings 524
  openings 531, 541, 542
  planning permission for replacement
    133–4
  powered 539–40
  protected zone 534
  residential care buildings 537
  sanitary facilities 537
  unframed glazed 544
  unisex sanitary accommodation 538
Dormer windows
  planning permission for 148–9
Downlighters, floor type 3 constructions
  308, 310, 311
Drainage 12, 223–225
  drains 227–230
  dungsteads and farm effluent tanks
    248–249

fuel storage 246–247
greywater recycling 243–244
gutters 249
requirements 225–227
safety 227
sanitary appliances 243
sanitary facilities 245
sewers 233–234
solid waste storage 247–248
surface water 230–233
ventilation 243
wastewater 233
  discharge 244–245
  treatment plants 234–242
Drains 227–230
Driveway, planning permission for
  laying 138–9
Dry central heating systems 717–718
Dry fire 574
Drying indoors 767
Dry space heating 342
Ducted warm air heating 297–298, 332,
  396, 445, 561
Durability 12
Dwangs, roof bracing with 375
Dwellings 561
  access between storeys 575
  accessibility 542
  accessible entrance of 575, 576
  in basements 297–298
  drying washing indoors 767
  electrical safety 763–765
  energy performance certificates
    828–829
    location of 830
  environment/health requirements
    for 347–348
  escape routes in high-rise 399–400
  extensions to
    dimensions of 776
    enhanced apartment 781
    over existing windows 782
    over glazed opening to a room
      782–783
    sanitary facilities 781–782
    wastewater drainage 781
  facilities in 765–768
  fire door 529
  fire safety 21–2, 346, 524, 546
  fire service facilities 574
  form and fabric of 512

handrails 555
information 828–829
with limited entrance storey
    accommodation 575–6
location of Energy Performance
    Certificate 343
means of access 525, 546
means of escape 546
noise control 337–338, 416–417
opening areas in 545
resisting sound transmission to
    477–478
roofs 450, 451
separating floors in 326
    combustibility 327
    junctions 329
    supporting structure 328
separating walls in 382–383
    junctions between 386–387
    openings and service
        provisions 385–386
    support for 384
smoke clearance 333
surface water drainage from 473
total opening areas 545
transmission of noise from 547
unassisted access 575, 576
ventilation 274–276, 762–763
vertical circulation in common areas
    574–5
written information 826

E
Electrical fixtures 74, 747–754
    in dwellings 754–759
    outlets and controls of 747–748
Electrical installation 89
Electrically operated hold-open
        devices 566
    deactivation 529
Electrically powered locks 569
Electrical planning permission for work
        134–5
Electrical safety 74, 730–731,
        763–765
    automatic fire detection and alarm
        systems *see* Automatic fire
        detection and alarm systems
    electrical fixtures *see* Electrical
        fixtures
    emergency exits 737

electrically powered locks for
        737–738
    obstacles 738
emergency lighting 734–735
requirements 731–733
smoke alarms 738–740
Electric underfloor heating 342
Emergency lighting 298, 580
    in shopping centres 823
Enclosed shopping centres, fire control
    in 404
End restraint walls 373
Energy 76–80
    artificial and display lighting 78
    building insulation envelope 77
    carbon dioxide emissions 76
    commissioning building services 78
    conservation provisions 21, 24–5
    Energy Performance Certificates
        79–80
    heating systems 77
    insulation of pipes, ducts and
        vessels 77
    mechanical ventilation and air
        conditioning 78
    metering 80
    writing information 78–9
Energy efficiency, hot water storage
    systems 709–710
Energy performance 191
Energy Performance Certificate
    (EPC) 79–80, 343
    Building (Procedure) (Scotland)
        Regulations 2004 44–5
    carbon dioxide emissions,
        calulation 827–828
    dwellings, information 828–829
    location of 830–831
    non-domestic buildings,
        information 829–830
Energy savings 301, 348
    ceilings 438
    floors 340–343
    roofs 453
    walls 431–436
    windows 511, 521–523
England and Wales, building regulations
    in 15–16
Enhanced apartment floors 336
Entrances
    access 606–630

Entrances (*continued*)
  doors 614
    powered 616–617
    lobbies 618–619
  requirements 607–608
  routes to 611–612
    length 613
    surfaces 612–613
    width 613
  setting-down point 610–611
  signs, hearing enhancement
    systems 617–625
  warm air heating to 619
ENV *see* European Pre-standards (ENV),
    under Building (Scotland) Act
    2003
Environment 69–73
  combustion appliances 71–2
  condensation 71
  and dwellings 21, 23, 25
  existing drains 69
  facilities in dwellings 71
  flooding and ground water 69
  fuel storage 72–3
  heating 71
  infiltration systems 70
  moisture from ground 69
  natural lighting 71
  precipitation 70
  private wastewater treatment
    systems 70
  sanitary facilities 71
  site preparation 69
  solid waste storage 73
  surface water drainage 70
  treatment plants 70
  ventilation 71
  wastewater drainage 70
Environmental aspects, of
    buildings 182–184
Environmental discharges, garages
    794–795
Environmental Impact Assessment
    (Scotland) Regulations 1999
    127
Environmental safety, for buildings
    404–406
Environment/health requirements 300,
    347–348
  basement and cellers 291

Errors, correction of 95
Escape lighting 68
  hospitals 814
  residential care buildings 809
Escape routes 294, 561, 566, 567
  across access decks 394–395, 396
  across flat roofs 394–395, 396
  across protected lobby 395
  auditoriums 396
  basements 398
  in central core 569, 570
  chimneys 482
  combined 295
  domestic buildings 530
    obstacles 535
  doorway widths 531
  escape stairs, combination of 567
  exits 530
  flat roofs 468–469
  floor 331–332
  locks on 531–3
  protected area 396–398
  reduce smoke spreading 569
  in residential building 394
  revolving doors and automatic
    doors 535
  walls 394–397
Escape stairs 294–295, 560
  in basements 396
  calculated method, appropriate
    capacity of 568
  effective width of 566
  electrically powered locks 569
  external 395, 398
  facilities in hospitals with hospital
    streets 295, 296–297
  fire-fighting facilities 295, 296–297,
    577
  fire-fighting shafts 578
  fire protection 565
  hospital street 580, 581
  independence of 569
  maisonette 561
  ventilation of 578–9
  windows and external 514–515
Escape windows 514
  conservatories 788
  roofs 469
EU Directive (2002/91/EC) 827
European Commission (EC) 162

European Directive on the Control
        of Major Accident Hazards
        Involving Dangerous
        Substances 128
European Pre-standards (ENV),
        under Building (Scotland) Act
        2003 166–167
Exempted buildings and services 60, 81–4
    agricultural and related buildings 81–2
    ancillary to flats or maisonettes 84
    ancillary to houses 84
    buildings or work not frequented by
        people 81
    construction and development
        buildings 83
    controlled by other legislation 81
    paved areas 84
    protective works 81
    small buildings 83
    specialist nature buildings 82–3
    temporary buildings 83
    works of civil engineering
        construction 82
Exhaust ceiling level 562
Existing drains 69
Exit doors
    auditoriums 535
    locks on 531
        electrically powered 532–3
        secured against entry 532
Extensions
    ceilings 447–448
    conversion, of heated buildings 771
    to dwellings 781–783
        dimensions 776
    insulation envelope 771–773
        alterations to 773–774
    to insulation envelope 193, 434
    insulation envelope and ceiling
        447–448
    to non-domestic buildings 783–784
    over existing windows 518
    requirements 769–771
    roofs 458
    site preparation 780
    structure 774–780
    wall 406
External escape stairs 395, 398, 535,
        562, 564
    fire safety 564, 572

External walls 564
    cladding 392, 393–394
    fire resistance of 390–394
    fire spread in high-rise domestic
        buildings 400
    wall type 1, junctions at 421
    wall type 2, junctions at 424–425
    wall type 3, junctions at 427–428
    wall type 4, junctions at 430
External water storage tanks, planning
        permission for 145
Extract fans 267–268
Extractor fan 88
Extra-low-voltage installations 748

F
Fans, for ventilation
    axial 268
    centrifugal 268
    extract 267–268
    in-line 268
Felling trees, planning permission for 138
Fence, planning permission for
        building 137–8
Fire 67–9
    alarm and detection systems 22
    authorities 55–6
    automatic life safety fire suppression
        systems 69
    cavities 68
    communication 68
    compartmentation 67
    conversions and 91
    doors 529–30
    escape 68
    internal linings 68
    precautions 12
    safety in domestic buildings 21–2
    separation 67
    service
        access 68
        facilities 68–9
        water supply 68
    spread
        on external walls 68
        from neighbouring buildings 68
        to neighbouring buildings 68
    structural protection 67
    suppression systems 69
    water supply to fire service 68

Fire
  reaction to
    in external walls more than 1 m
      from boundary 389
    in wall and ceiling surfaces 389
Fire alarm system, roofs 470
Fire detection
  roofs 470
  in shopping centres 819
Fire evacuation 573
Fire-fighting facilities, in enclosed
    shopping centres 586–587
Fire-fighting lift 576
Fire-fighting lobby 565
Fire-fighting shaft 536, 565
Fire hazard rooms, in residential care
    buildings 402–403
Fireplaces see also Flues
  labelling of combustion appliances
    and 486–487
  opening 491
  recesses 507
Fire precautions
  ceilings 440–443
  in hospitals 814–816
  in residential care buildings
    automatic life safety fire
      suppression systems 809–810
    compartment walls 802
    detection and alarm systems 801–
      802, 809
    fire service access 807
  roofs 467–471
  shopping centres 818
    automatic fire detection and alarm
      systems 819
    automatic life safety fire
      suppression systems 822
    smoke and heat exhaust ventilation
      system 820–822
Fire Precautions Act 1971 15
Fire protection 564
  fire-fighting shaft 565
Fire resistance
  ceilings 440–441
  duration 333, 334, 536–7, 564, 565
  walls 400–402
  of external walls 390–392
  garages 793–794
  roofs 470

Fire-resistant ceilings 441–442
Fire safety 564
  ceilings 436–437
  chimneys 481
  roofs 451
  windows 510
Fire Safety and Safety of Places of Sport
    Act 1987 15
Fire safety engineering
  detached non-domestic buildings 181
  spread to neighbouring buildings 181
  structural protection 180
Fire safety requirements 346–347
  for basements and cellars 290
  in corridors 394
  in enclosed shopping centres 404
  floors 299–300
  in hospitals 403–404
  residential care buildings 402–403
  timber-framed buliding floors 324
Fire (Scotland) Act 2005 55
Fire service facilities 536, 573, 574,
    578, 585
  domestic buildings 574
  in high-rise domestic buildings 400,
    585
  in hospitals 586
  hospitals with hospital streets 296–297
  shopping centres 823–824
Fire shutters 398, 566
  compartment floors 325
  compartment walls 380
Fire spread, roofs 467–468
Fire-stopping, separating floors 329
Fit-out buildings, U-value 432
Fixed glazing 525, 527
Fixed ladders 557
Fixed seating 75
Flagpoles, planning permission to
    erect 141–2
Flanking transmission 416, 417
Flat
  height of 563
  single stair access 563
  storey height 561
Flat roofs
  constructions 454–455
    condensation 474
  escape routes 394–395, 396,
    468–469

Floating layer
  concrete base with 304–307
  timber base with 308–315
Flooding and ground water 69
  in buildings 185
Floor heating 757–758
Floor joists, timber-framed
      buildings 321–323
Floor penetrations
  floor type 1 304
  floor type 2 307
  floor type 3 314–315
  floor type 4 317–318
Floors 299–343
  accessible bathrooms and shower
      rooms 336
  adjacent to ground 333, 335
  cavity barriers 330
  compartment 324–326
  energy requirements 340–343
  enhanced apartment 336
  galleries 331
  loudspeakers 343
  means of escape 331–332
  noise control 337–338
  openings in 331
  performance testing 339–340
  requirements 299–343
  separating 326–329
  smoke clearance 332–333
  structural protection 330
  structural stability 318–320
  surface condensation 336–337
  timber-framed buildings 320–324
  type 1 301–304
  type 2 304–307
  type 3 308–315
  type 4 315–318
  wall type 1, junctions at 421
  wall type 2, junctions at 424
  wall type 3, junctions at 427
  wall type 4, junctions at 430
Floor surfaces, in common areas of
      buildings 337
Flue liners 87, 413, 494–495
Flue-pipes 413, 470, 493–494
  combustible materials and 501
  serving combustion appliances 508
Flues 491–496 see also Fireplaces
  access to 493

design of 492–493
flue liners 494–495
flue-pipes 493–494
openings in 493
outlets 495–496
size of 491–492
terminal discharging 496
Footpaths, clearing of 65
Foundations
  floors 318–319
  walls 349
Framed solid masonry
  wall constructions 409–410
Free-standing advertisements
  planning permission for 143
Fuel storage 72–3, 722–724
  containment 415
  planning permission for tanks 145
Full applications fee 123–4
Full planning permission 100
Functional standard 67
Future installation 590

G
Gable wall strapping 361, 459–460
Galleries
  floor 331
  occupancy capacity 571
Garages
  conversions 800
  environmental discharges 794–795
  fire resistance 793–794
  noise 799–800
  planning permission for 145–6
  requirements 789–791
  stairs 798
  structure 791–793
    dimensions 792
  ventilation 271–273, 411, 538,
      795–798
Garden sheds 12
Garden wall
  planning permission for building
      137–8
  planning permission to alter 138
Gas and oil 715
Gas-fired appliances 683–691
  air
    for combustion 687, 688
    for cooling 687

Gas-fired appliances (*continued*)
  in bathrooms and bedrooms 509, 684
  chimneys 508, 685
  combustible material 684
    separation 685
  combustible materials and 504–505
  flue outlets 508, 685–687
  on hearth 684
  LPG
    storage 688
    storage cylinders 691
    storage tanks 689–691
  size of flues 492, 683
Gas Safety (Installation and Use)
    Regulations 1984 136
Glass and Glazing Federation (GGF) 134
Glazing
  vulnerability to human impact 543–4
Government's restrictions, on planning
    permission 126–8
Greywater 656
Ground storey floor level 565
Ground supported floors 333, 335
Guarding windows 520
Guidance documents 3, 7
Gutters, drainage 249

H
Hardstanding 88
  planning permission for 140–1
Hazard identification 185
Health and Safety at Work, etc. Act
    1974 15
Hearths
  combustible materials and 505–507
  functions of 505
  gas-fired appliances 504–505
  location of 506
  oil-fired appliances 503–504
  relationship to combustible
    materials 414
  solid fuel appliances 502–503
Heat, danger from 75
Heated buildings
  conversion of 433
Heated buildings, conversion 771
Heaters 87
Heating 71, 185
  in buildings 185
  conservatories 788
Heating appliance 136

Heating Equipment Testing Approval
    Scheme's (HETAS) Guide 495
Heating installation 86–7
Heating systems 77, 449
  appliances 713
  boilers 724–726
  commissioning 728–730
  conservatories 714
  dry central heating 717–718
  efficiency 726–728
  fuel storage 722–724
  metering 730
  requirements 711–712
  underfloor 721–722
  warm air systems 718–721
  wet central heating
    controls for 716–717
    electric 716
    gas and oil 715
    solid fuel 715
Heat loss resistance, through thermal
    bridging 192
Heat pumps 719–721
  efficiency 341
Heat systems
  artificial lighting 758–759
  boiler controls 758
  under floor heating 757–758
  warm air systems 757
Heat ventilation systems, in shopping
    centres 820–822
Heavy masonry leaf, junctions at 314
Hedge, planning permission for
    planting 137
Height measurements 90
Heritage Scotland 107
High-rise buildings 570
  escape routes 399–400
  fire-fighting lift 576
Highways department 56
Historic buildings 46, 52
  conversion 193
Historic Scotland 57, 108, 114
Hoarding, planning permission for 143
Hold-open devices 529
Home Energy Efficiency Scheme
    (Scotland) Regulations 2006 136
Hospitals 579 *see also* Residential care
    buildings
  automatic fire detection and alarm
    systems 743

cavity barriers 812
compartmentation 810–811
doors 537
escape lighting 814
fire
  detection and alarm systems
    814–815
  service facilities 815–816
fire control in 403–404
hospital street 813–814
locks 814
means of escape 812
protected lobbies 814
Hot water 342
  discharge 708
  storage installation 136
Hot water storage systems
  appliance efficiency 700
  DHW systems 700–702
  energy efficiency 709–710
  hot water, discharge 708
  insulation 703
  requirements 698–699
  solar water heating 708–709
  solid fuel boilers 699
  steam, discharge 708
  unvented hot water systems 703–706
    discharge from 706–708
  written information 710
Hot water underfloor heating 342
Housing (Scotland) Act 1986 15
Housing (Scotland) Act 1987 15
Humidity, control of 767

I

IET Wiring Regulations 134
Illuminated signs 143
Independent certification schemes 166
Indoor air pollutants 252
Industrial stairs 557
Infestation, resistance to 12
Infilling, planning permission for 157
Infiltration systems 70
Infringes right of way, and planning
  permission 109–10
In-line fans 268
Inner rooms 565
Insulation envelope 348, 526
  alterations 194
  alterations to 435–436, 545
  ceiling extension 447–448

extensions 193, 771–773
  alterations to 773–774
extensions to 434
floors 340–341
roofs and alterations to 479–480
U -values for building elements
  of 479, 521, 544
walls 432, 433
windows and alterations to 522
Intercom 753
Internal doors 540
Internal linings 68
  ceilings 443
  in residential care buildings 802–805
  walls 389
Internal load-bearing walls, in brickwork/
  blockwork 352
Interstitial condensation
  walls 412

J

Joists, floor
  timber-framed buildings 321–323
Junctions
  cavity barrier 388–389
  in ceilings 439
  compartment floors 326
  compartment walls 381–382
  floor type 1 303
  floor type 2 307
  floor type 3 313–314
  floor type 4 317
  with roofs, fire spread 467
  separating floors 329
  between separating walls 386–387
  wall type 1 421
  wall type 2 424–425
  wall type 3 427–428
  wall type 4 430

K

Kitchens
  accessibility 760–761
  electrical safety 763–765
  requirements 759–760
  space provision in 765–767
  ventilation 761–763

L

Labelling, combustion appliances
  486–487

Land Compensation (Scotland) Act
1973 15
Lateral restraints
chimneys 483
at roof level 460
Lateral supports
floors 338, 484
by floors 319
walls 374
roofs 338, 458–459, 462, 484
by roofs, walls 374–375
of walls 354–355, 458
Licensing (Scotland) Act 1976 15
Life safety fire suppression systems 69
Lifting devices 752–753
Lifts and platforms 581
meeting the requirement
escape routes 583
fire safety 584
separating walls 584
structure 583
requirements 582–583
fire 582
noise 583
safety 582–583
Lift wells
compartment walls 379
Lighting 763–764
light fittings 445
and planning permission 112
windows and controls for artificial 523
Lighting diffusers, ceilings 445
Light masonry leaf
junctions at 314
Limited lifespan buildings 12, 61
Limited-life warrants 39
Limiting, air infiltration 192
Linings
ceilings and internal 443
internal walls 389
Lintels 344
Liquefied Petroleum Gas (LPG) 75, 87,
723
fixed tanks and cylinders 298–299
safety requirements 291
storage 688
storage cylinders 691
storage tanks 689–691
separation distance 689
Listed buildings 4, 10

consent and planning permission 104–5
owner's responsibilities for 114
planning permission for 108, 113–14,
132
Load-bearing walls 345–346
brickwork/blockwork in internal 352
Loadings, foundations 318–319
Loads
ceilings 438
on roof 458, 464–465
Lobbies
dimensions 599
energy certificates 606
escape stairs 603
fire-fighting shafts 603–606
protection 599–603
requirements 598
smoke clearance 604
Local authority
Building (Scotland) Act 2003 6, 10
compulsorily purchase of building 10
functions 4, 10
records 57
rights of entry, inspection and
testing 5, 10, 12, 14
sell off materials from demolished
buildings 11
Local Development Plan, non
compliance and planning
permission 109
Local Government, etc. (Scotland) Act
1994 15
Local Government Act 1988 15
Local Government in Scotland Act
2003 15
Local Government (Scotland) Act
1973 15
Local Plans 96
Location plan 103
Locks, hospitals 814
Loft conversions, planning permission
for 148–9
Lopping trees, planning permission
for 138
Loudspeakers
floors 343
walls 418
Low and Zero Carbon Technologies
(LZCT) 24, 446, 521
LPG see Liquefied Petroleum Gas (LPG)

M

Maisonettes 561
Mandatory requirements, for design and
        construction, under Building
        (Scotland) Act 2003 168
Masonry, cavity
    wall constructions 407–408
    wall type 2, 422–425
Masonry, solid
    wall constructions 407–408
        framed 409–410
    wall type 1 418
    wall type 3 425–428
Masonry chimneys 412, 484–485, 489
    combustible materials and 498–499
Masonry cladding, walls 369–371
Masonry leaf, junctions at 314
Masonry walls 350–354
    compressive strength of 353–354
Materials, durability 61
Maximum floor area, structural
        stability 319, 320
Maximum span of floors
    structural stability 319, 320
    timber-framed buildings 321
Measurements 61, 90
Mechanical ventilation 261–264
    and air conditioning 78
    ceilings 446
    MVAC 264–266
    windows 516
Mechanical ventilation and air
        conditioning (MVAC) 264–265
    control of 265
    efficiency of 266
Meeting requirements, for building
        design and construction
    conversions 170
    disproportionate collapse 170
    foundations 173
    stability 173
    structural 168–170
Metal chimneys 413, 476, 489–490
    combustible materials and 500
    protection 490
    relationship to combustible
        material 414
Metering 80
Metering-non-domestic buildings
    existing buildings 832

new buildings 832
Mines and Quarries (Tips) Act 1969 15
Mixed use buildings 22, 561
    automatic fire detection and alarm
        systems 742
Model Form Q 43
Moisture and decay, resistance to 12
Moisture from ground 69
Multi-storey buildings
    compartmentation 292–293
    compartmentation and 376, 377–378

N

Nailing and fixing, walls panels
        372–373
National Care Standards 57
National Planning Framework 94
Natural lighting 71
    provision 412
Nature conservation issues, and planning
        permission 117–19
Nature Conservation (Scotland) Act
        2004 117
Neighbouring properties plan
    and planning permission 103
Neighbour notification certificate 97
New house, planning permission for
        building 156–7
Noise
    garages 799–800
    resisting sound transmission 76
    transmission 21, 23–4
Noise control 301, 348
    floors 337–338
    roofs 453, 477–480
    walls 415–418
Non-combustible material 573
Non-domestic buildings 22, 561
    basements in 293–297
    cavity barrier 388
    external wall cladding 393–394
    fire door 529–30
    fire resistance of external walls
        390–391
    fire safety 300, 524
    fire safety requirements in 346
    fire service facilities 577
    glazed vision panel 541
    means of access 527, 546
        to roof 547

Non-domestic buildings (*continued*)
  planning permission for extensions
      to 150–1
  protected zone 534
  separating floors in 327
    junctions 329
    supporting structure 328
  separating walls in 383–384
    junctions between 387
    openings and service
        provisions 386
    supporting structure for 384
  vertical circulation 575
Non-domestic Technical Handbook 1
Non-load-bearing partitions support,
    timber-framed buliding floors 323
Non-residential building 86
Northern Ireland, building regulations
    in 15–17
Notches and holes 463–464
Notice to Neighbour Form 97

O

Occupancy capacity 569
  buildings 569
  shopping centres 816
    crèches 818
    entire shopping centre 817
    shops 817–818
Offenses
  by bodies corporate 5
  penalties for 5
Oil-firing appliances 676–677
  air for cooling 680
  chimneys 678
  chimneys and flue-pipes serving 508
  and combustible material 677–678
  combustible materials 503–504
  flue outlets 678–680
  location of 677
  oil storage tanks 680–683
    construction 681
    installation 682
  size of flues 492
Oil storage 723
  tanks 136–7
Open access balconies 632–633
Openings *see also* Windows
  in building shell 544
  in buttressed walls 439

  ceilings 439
  chimneys 485
  compartment walls 379–380
  fireplaces 491
  in flues 493
  separating floors 328–329
  size and proportions 178
  in small single-storey, single-leaf
      buildings 362
  structures
    size and proportions of 528
    through compartment walls and
        compartment floors 529
  through separating walls 385–386
  walls 358–359, 373
    framing 374
  windows 512–513
Open raised external decking 88
Open risers 551
Open space, provision of 13
Openwork floor
  unenclosed escape stair 571
Orders and regulations 5
Ordnance Survey map, and planning
    permission 104
Outbuildings *see* Garages
Outdoor sign 88
Outline applications fee 123
Outline planning permission 99–100
Overhangs, walls 361

P

PAN 56, 24
Panels, walls 367–368, 371–373
Parapet walls 352
Parking garage 563
Partially completed buildings 65
Partitions, wall type 3, junctions at 428
Passenger lifts 589, 756
Passive stack ventilation (PSV) 259–261
  ceilings 446
  roofs 475
Path, planning permission for
    laying 138–9
Paved areas 84, 88
Pedestrian protective barriers 74, 518,
    559
Performance testing 190
  floors 339–340
  walls 431

Permitted development, and planning
    permission 149–50
Pipework 87
Pitched roofs
    construction 456–457
    condensation 474
Planning Aid for Scotland 122
Planning and Building Regulations
    (Amendment) (NI) Order
    1990) 16
Planning application
    advertisement of 97
    advice service 122, 126
    cost 122
    fees 123–5
    finding 98
    government's restrictions on 126–8
    outline applications 123
Planning authorities, assessment of
    performance of 95
Planning (Consequential Provisions)
    (Scotland) Act 1997 15, 126
Planning control
    breaches of 119–20
    Breach of Condition Notice 120–1
    enforcement 119–22
    Enforcement Notice 120–2
    Listed Building Enforcement
        Notice 121–2
    needs 120
    public involvement 120
    Stop Notice 122
Planning (Control of Major Accident
        Hazards) (Scotland) Regulations
        2000 128
Planning decisions 93
Planning etc. (Scotland) Act 2006 94–5
Planning (Hazardous Substances)
        (Scotland) Act 1997 126
Planning (Hazardous Substances)
        (Scotland) Regulations 1993 127
Planning (Listed Buildings and
        Conservation Areas) (Scotland)
        Act 1997 52, 126
Planning matters, public involvement
        in 98
Planning Officers 45, 96–7, 105, 107–8
Planning permission
    action for building without 119
    for advertisement 142–3

advertisement applications 97, 104
advice service 122, 126
to alter garden wall or boundary
    wall 138
to alter position of heating
    appliance 136
application 100–1
applying online 102
approval 129–30
authority to take final decision on
    appeal 111
block plan 103
for building
    conservatory 146–7
    extension 149–51
    garden wall or fence 137–8
    hardstanding for vehicle or
        boat 140–1
    new house 156–7
    plans 103
    porch 143–4
Building Regulations approval 105
for cavity wall insulation 133
for central heating 136
Certificates of Agricultural
        Holding 102–3
Certificates of Site Ownership 102
for cladding 133
Conservation Area consent 104
controls 93–4
for conversions and change of
        use 151–6
council refusal to 111
for decoration and repairs 131–2
for defective building 160–1
for demolishing dangerous
        buildings 158–60
for demolition 157–60
for dormer windows 148–9
for electrical work 134–5
for extensions to
    agricultural buildings 151
    non-domestic buildings 150–1
fees and exemptions 100
for felling or lopping trees 138
finding planning application 98
for flagpoles 141–2
gaining 99
for garages 145–6
government's restrictions on 126–8

Planning permission (*continued*)
  for hot water storage installation 136
  important areas taken into
      consideration 104–5
  for infilling 157
  information in application for 101–2
  for installing swimming pool 141
  for laying path or driveway 138–9
  listed building 132
  listed building consent 104–5
  location plan 103
  for loft conversions 148–9
  matters cannot be taken into
      account 110
  necessity 98
  need 96–7
  neighbouring properties plan 103
  notifying neighbours regarding 97
  objections against granting 108–10
    existing permission 109
    infringes right of way 109–10
    listed building 108
    non compliance with local
      Development Plan 109
    property in conservation area 109
    property subject to covenant 109
  for oil storage tank 136–7
  Ordnance Survey map 104
  for outbuildings 144–5
  permitted development and 149–50
  planning application cost 122
  planning control enforcement 119–22
  plans to be submitted 103–4
  for planting hedge 137
  for plumbing 135–6
  process 105–8
  property in conservation area 132
  public involvement in planning
      matters 98
  for radio aerials 141–2
  reaching decision 108
  for replacement
    demolished building 160
    windows and doors 133–4
  requirement for 96
  for roof extensions 148–9
  for satellite dishes 141–2
  Scottish Ministers involvement 94
  site layout plan 103
  for structural alterations inside 132–3

  stumbling blocks 110–11
  for television aerials 141–2
  think about before start work 112–19
    barn owls 118–19
    bats and their roosts 117–18
    conservation areas 114–15
    covenants 113
    crime prevention 112
    design 112
    lighting 112
    listed buildings 113–14
    nature conservation issues 117–19
    Tree Preservation Order 115–17
  time taken by council to decide 107–8
  for tree plantation 137
  trees protection 105
  types available 99–100
  validity 111
  for wind turbines 141–2
Planning questionnaires 100
Planning system
  assessment of performance 95
  Business Improvement Districts 95
  correction of errors 95
  development management 95
  development plans 94
  financial provisions 95
  national planning framework 94
  need for 93
  Planning etc. (Scotland) Act 2006
    94–5
  trees 95
Platform floor with absorbent
    blanket 309
Plumbing
  planning permission for 135–6
Police 57
Pollution
  external 250–251
  indoor air pollutants 252
Porch, planning permission for
    building 143–4
Powered doors 751–752
  manual control 540
  uses of 539
Powered lifting platforms 589–590
Precipitation 70, 780
  chimneys 486
  roofs 473
  in walls 405–406

Prior approval application fee 125
Private stair
  effective width of 550
  geometry of 548
  stair landing 552
Private wastewater treatment systems 70
Procedure regulations 4–5, 12–14
Professional builder, employment of 45
Property
  right to enter 10
  subject to covenant and planning
    permission 109
Protected enclosures 535
  escape routes 398
Protected lobbies 534
  automatic opening ventilators 563
  escape routes 395, 396–397
Protected zone 534
  escape routes 397
  escape stair 564, 570, 571
    in basement 573
  external wall 564
  hospital street 579
  protected lobby 562, 570
  rooms, toilets and washrooms in 571
Protective barriers 558
  balconies 634
Protective works 81
  provision of 63–5
PSV see Passive Stack Ventilation (PSV)
Public Health Act of 1875 19
Public involvement, in planning
    matters 98

R
Racking resistance, walls 365–366
Radiators 87
Radio aerials, planning permission to
    erect 141–2
Rainwater harvesting 656
Raised tie roof 465–466
Ramps 89
  fire precautions 547
  handrails to 555–6
  headroom 556
Reasonably practicable, definition of 35
Recesses, walls 358–359
Refuse facilities 768
Relaxations, under Building (Scotland)
    Regulations 2004 65

Repairs, planning permission for 131–2
Replacement work 89
Requirements
  for building design and construction,
      under Building (Scotland) Act
      2003 167–168
  for domestic buildings 184
  for non-domestic buildings 184
Rescue operations 565
Reserved matters applications fee 124
Reserved planning permission 100
Residential buildings
  automatic fire detection and alarm
      systems 741
  door 534
  escape routes in 394
  means of access, wheelchair user 525
Residential care buildings see also
      Hospitals
  automatic fire detection and alarm
      systems 742
  compartmentation in 808
  doors 537
  escape lighting 809
  fire control 402–403
  fire precautions 806
    automatic life safety fire
        suppression systems 809–810
    compartment walls 802
    detection and alarm systems
        801–802, 809
    fire service access 807
    internal linings 802–805
    means of escape 806, 808–809
    requirements 801
Restraint
  on concrete floor 355
  joist hanger 355
  vertical lateral 356
  walls, end 373
Ribbed floor with absorbent
      blanket 310–311
RICS Building Cost Information Surveys
      of Tender Prices 31
Roads (Scotland) Act 1984 15
Roof bracing, with dwangs 375
Roof coverings 468
  ceilings 448–449
Roof extensions, planning permission
      for 148–9

Rooflights 88–9
  ceilings 446–450
  cleaning of 519–520
  maximum *U*-values 446–447
  thermoplastic materials in 444–445,
    449
Roofs 450–480
  access decks 468–469
  bracing with dwangs 462
  cavity barriers 471–473
  chimneys 475–476
  collared 466–467
  compartmentation 470
  constructions 453
    ceilings and 439–440
    flat 454–455
    pitch 456–457
  control of condensation in 474–475
  conversion of unheated building 480
  coverings 448–449, 468
  in domestic buildings 451
  energy savings 453
  environment/health requirements
    451–452
  escape routes 468–469
    windows 469
  extensions 458
  fire precautions 467–471
  fire resistance 470
  fire safety 451
  fire spread 467–468
  flat 468–469
  gable wall strapping 459–460
  lateral restraints 460
  lateral supports 458–459, 462
  loadings 453
    dead 464
    imposed 458, 464–465
  noise control/insulation 453,
    477–480
  notches and holes 463–464
  precipitation 473
  raised tie 465–466
    connection details 467
  safety requirements 452
  secure access to 476–477
  SHEVS 470–471
  structures 450
  timber-framed walls 461–462
  vehicle protective barriers 480
  ventilation 475

  wall type 1, junctions at 421
  wall type 2, junctions at 424
  wall type 3, junctions at 427
  wall type 4, junctions at 430
Room, escape stairs 569
Room-sealed appliance 87

S
Safe access 300
Safety 73–5, 186–189
  access
    to buildings 73
    within buildings 73–4
  aids to communication 74–5
  basement and cellers 291
  ceilings 437–438
  in common areas of domestic
    buildings 586–587
  danger
    from accidents 75
    from heat 75
  electrical fixtures 74
  electrical safety 74
  from fire 180
  fixed seating 75
  liquefied petroleum gas storage 75
  from noise 189
  pedestrian protective barriers 74
  roofs 452
  stairs and ramps 74
  vehicle protective barriers 75
  windows 510–511, 518–521
Safety of Sports Grounds Act 1975 15
Sandwich panels 367–368
Sanitary discharge stack systems 333,
    335, 336
Sanitary facilities 71, 88–9, 768,
    781–782
  accessibility 760–761
  accessible toilets 643
  bathrooms 644–645
  changing facilities 646–647
  doors 537
  drainage 245
  dwellings 637–640
  internal doors 643
  non-domestic buildings 641
  number of 647–648
  for public in entertainment and
    assembly 650, 652
  in residential buildings 648–649

in shops and shopping malls
        650–651
shower rooms 644–645
ventilation 271, 653–655, 762
walls 411
Sanitary pipework 653–655
Satellite dishes, planning permission to
        erect 141–2
Scheduled monuments 4, 10
Scotland Act 1998 15
Scottish Association of Building
        Standards Managers
        (SABSM) 49
Scottish Association of Chief Building
        Control Officers (SACBCO) 49
Scottish Building Standards Agency
        (SBSA) 21, 28, 36
Scottish Commission for the Regulation
        of Care 57
Scottish Environment Protection Agency
        (SEPA) 55, 56, 137
Scottish Ministers
    involvement under Building
        (Procedure) (Scotland)
        Regulations 2004 48–9
    rights of entry, inspection and
        testing 5, 10, 12, 14
Scottish Natural Heritage (SNH) 117
Scottish Public Services
        Ombudsman 119
Scottish Statutory Instrument 2004
        No. 406 1–2
Scottish Water 56–7
Secure by Design initiative 57
Self-closing fire doors 564, 566
    domestic buildings 529
    non-domestic buildings 529–30
Self-closing fire doors 566
Separating floors 326–329
    loudspeakers 343
    performance testing 340
Separating walls 382–389
    high-rise domestic buildings 399
    performance testing 340
Separation 67
Service openings
    compartment floors 325
    compartment walls 380–381
    separating floors 328–329
Service penetrations, compartment
        walls 379–380

Service provisions
    separating floors 328–329
    through separating walls 385–386
Sewerage (Scotland) Act 1968 15
Shell buildings
    U-value 432
Sheriff Court 49
SHEVS see Smoke and Heat Exhaust
        Ventilation System (SHEVS)
Shopping centres 581
    automatic fire detection and alarm
        systems 744
    fire control in 404
    fire precautions 818
        automatic fire detection and alarm
            systems 819
        automatic life safety fire
            suppression systems 822
        smoke and heat exhaust ventilation
            system 820–822
    fire service facilities 823–824
    means of escape 823
    occupancy capacity 816
        crèches 818
        shops 817–818
    requirements 816
    separating walls 822
Shops, occupancy capacity 817–818
Shower rooms
    access 644–645
    electrical safety 657
    floors 336
Single-leaf buildings, single-storey
    openings in small 362
Single-leaf external walls 352
Single-storey buildings
    compartmentation 292, 376
    single-leaf, openings in small 362
Site layout plan, and planning
        permission 103
Site preparation 12, 404
    extensions 780
    harmful and dangerous substances
        69
    protection
        from harmful and dangerous
            substances 185
        from radon gas 69, 185
Size of buildings
    apartments 175–177
    domestic 173–174

Size of buildings (*continued*)
   single-storey, single-leaf 175
   timber-framed 173–174
Size of extensions, to domestic
   buildings 174
Small buildings 83
Smoke alarms 763
  ceilings 442
Smoke and Heat Exhaust Ventilation
   System (SHEVS) 282–284,
   470–471, 744–746
Smoke clearance
  capability 577
  floors 332–333
Smoke control 562
  ceilings 442
  in corridors 394
Smoke detection 566
Smoke outlets 443
  basement 295–296
Smoke spreading, reduction 569
Smoke ventilation systems, in shopping
   centres 820–822
Smoke venting shafts, compartment
   walls 379
Soakaways 780
Socket outlets 756–757, 764
Soft covered concrete base 301–304
Solar water heating 708–709
Solid fuel 715
  boilers 724
    hot water storage systems 699
Solid fuel appliances
  air for combustion 695, 696
  air supply for combustion 497
  chimneys and flue-pipes serving 508
  and combustible material 692–693
  combustible materials and 502–503
  flue outlets 495–496, 693–695
   minimum areas of 693–694
  location of 692
  size of flues 491
  woody biomass fuel, storage 695–698
Solid masonry
  wall constructions 407–408
   framed 409–410
  wall type 1 418
  wall type 3 425–428
Solid walls
  in coursed brickwork/blockwork 350
  in uncoursed stone and flints 350

Solid waste storage 73
Solum 404
Sound
  insulation 12, 24, 76
  resistance 417
Specialist nature buildings 82–3
Staged warrants 38–39
Stair and landing configuration 580
Stair flights 554–5
Stair landings 591
  clear space 542, 552
  corduroy tactile paving 553
  effective width of 550
  external door 552
  minimum length of 552
  outward-opening fully glazed
   doors 542–3, 554, 788, 789
Stairlift 550, 575
  dwelling 88
Stairs
  in agricultural buildings 557
  airborne sound 559
  construction and measurement of
   rise, going, tread and pitch of
   548–9
   stair flights and landings 550–1
  corduroy tactile paving 553
  escape *see* Escape stairs
  external escape 395, 398
  fire precautions 546, 547
  garages 798
  handrails to 555–6
  headroom on 556
  interruption of lateral support 557
  lighting 559
  opening in floors for 323
  openings for 558
  protective barriers 558
  and ramps 74
  risers and treads 551
  stair landing 552–3
  tapering treads 555
  unassisted 575
Stand-alone buildings 545
  ceilings 449
  roofs 478
  windows 523
Statutory Instrument 2004 No. 428 19
Steam, discharge 708
Stone/flints, uncoursed
  solid walls in 350

Storage buildings 565
Storage system, electric 342
Strapping, gable wall 459–460
Strength and stability 12
Structural alterations inside, planning
        permission for 132–3
Structural protection 67
    from fire 180
    floors 330
Structural stability
    floors 318–320
    walls 348–349
Structure openings
    size and proportions of 528
    through compartment walls and
        compartment floors 529
Structures 67
    ceilings 338–340
    chimneys 483–485
    disproportionate collapse 67
    functional standard 67
    plans 94, 96
    roofs 450
    windows 511–513
Stumbling blocks, and planning
        permission 110–11
Supply water temperatures 341
Supporting structure
    compartment walls 379
    fire resistance of external wall 392
    separating floors 328
    for separating walls 384
Surface condensation, floors 336–337
Surface water drainage 70
    from dwellings 473
Suspended concrete floors 335
Suspended timber floors 335
Sustainable development 93
Swimming pool, planning permission
        for installing 141

T
Technical Documents 2007 21
Technical Handbooks
    Building (Procedure) (Scotland)
        Regulations 2004 20–25
    mixture of domestic and non-domestic
        building 164
    overview 162–163
    scope
        domestic buildings 163

non-domestic buildings 163
    Section 1: Structure 21
    Section 2: Fire 21–2
    Section 3: Environment 21, 23, 25
    Section 4: Safety 21, 23
    Section 5: Noise 21, 23–4
    Section 6: Energy 21, 24–5
Telecommunication equipment satellite
        dishes 142
Television aerials, planning permission
        to erect 141–2
Temporary buildings 83
Temporary premises 12
Temporary waiting spaces 572
Thermal bridging 336, 412
    for heat loss resistance 192
Thermal efficiencies, for DHW
        systems 700
Thermal insulating material 88
Thermoplastic materials, in
        ceilings 443–445
Thermostatic controls 87
Timber base
    with floating layer 308–315
    with independent ceiling 315–318
Timber floors
    floor type 4, incorporating
        deafening 316
    suspended 335
Timber-framed buildings 320–324,
        461–462
    absorbent curtain 428–430
    roofs 461–462
    walls in 345
    windows 512
Timber-frame walls 363–365
    junctions at 313
    masonry cladding 369
    panel 367
Topmost storey 561
Town and Country Planning (Control
        of Advertisements) (Scotland)
        Regulations 1984 88
Town and Country Planning
        (Development by Planning
        Authorities) (Scotland)
        Regulations 1981 127
Town and Country Planning
        (Development Contrary to
        Development Plans) (Scotland)
        (No. 2) Direction 1994 128

Town and Country Planning
    (Enforcement of Control) (No. 2)
    (Scotland) Regulations 1992 128
Town and Country Planning (General
    Development Procedure)
    (Scotland) Order 1992 127
Town and Country Planning (General
    Permitted Development)
    (Scotland) Order 1992 127
Town and Country Planning (Notification
    of Applications) (Scotland)
    Amendment Direction 1997 –
    Notification of Planning
    Applications 128
Town and Country Planning (Notification
    of Applications) (Scotland)
    Amendment Direction 1998 –
    Notification of Planning
    Applications 128
Town and Country Planning (Notification
    of Applications) (Scotland)
    Amendment (No. 2) Direction
    1998 128
Town and Country Planning (Notification
    of Applications) (Scotland)
    Direction 1997 128
Town and Country Planning (Scotland)
    Act 1997 126
Town and Country Planning (Simplified
    Planning Zones) (Scotland)
    Regulations 1987 127
Town and Country Planning (Structure
    and Local Plans) (Scotland)
    Regulations 1983 126
Town and Country Planning (Use
    Classes) (Scotland) Order
    1997 127
Transfer grilles 561
Treatment plants 70, 234–242
Tree Preservation Orders (TPOs) 95,
    105, 115–17, 138
Trees
  in conservation areas 114–15
  planning permission for planting
    137
  protection 95, 105, 115–17
    and planning permission 105
Tree Work Notice application 105
Trickle ventilators 516
  ventilation 257–258

Trustees
  civil liability 5
  criminal liability of 5

U
Unauthorized operational
    development 121
Underfloor heating 721–722
Undivided corridor 569
Unframed glazed doors 544
Unheated buildings
  roof conversion in 480
Unoccupied completed buildings 65
Unprotected area
  external walls in 390–391
Unvented hot water storage system
    750–751
U-values, maximum 432–436
  for floors 340, 341
  for rooflights 446–447
  roofs 478, 479
  for walls 432, 433, 434, 435
  windows 522

V
Vehicle protective barriers 75
  roofs and 480
Ventilated protected lobby 563
Ventilating ducts
  compartment floors 325
  compartment walls 381
  separating floors 329
Ventilation 13, 71, 186, 653–654
  background ventilators 258–259
  ceilings 446
  combustion appliances 276–277
    air for combustion 278–280
    air for cooling 280
    safe operation 281
  commissioning 288–290
  conservatories 268–270
  of conservatories 787
  defined 250
  ducts 572
  of dwelling 257
  dwellings 274–276, 654–655
  environment 285–286
  external pollution 250–251
  fans 266–268
  garages 271–273, 795–798

indoor air pollutants 252
of large garages 411
mechanical 261–264
MVAC 264–266
natural ventilation 256–257
non-domestic buildings 655
PSV 259–261
requirements 252–255
roofs 475
safety 286–288
sanitary facilities 271
SHEVS 282–284
smoke and heat ventilation 281–284
trickle ventilators 257–258
wall cavities 273
of wall cavities 406–411
windows 516
Ventilators, automatic opening 562
Verifiers 3–5, 7–8
    Building (Procedure) (Scotland)
        Regulations 2004 26–27
    functions 4, 7–8
    imposition of continuing requirements
        by 3
    list of 27
    performance criteria for 27
    responsibilities 26–27
Vertical circulation
    in common areas of domestic
        buildings 587–588
    between storeys of a non-domestic
        building 588
Vertical fire shutters 566
Vertical lateral restraint to walls 356
Vertical support
    walls 351
Voice alarms 747
Voice alarm system, installation of 569

W
Walls 344–436
    in brick-built house 344
    buttressing 359–360, 483
    cavities 387–389
        ventilation 273, 406–411
    chases 360
    compartmentation 375–382
    condensation in inner surface 412
    conservatories 406, 786–787
    end restraint 373

escape routes 394–397
extensions 406, 778–779
    single leaf 776
fire resistance 390–392, 400–402
heights 349
lateral support 354–355
    by floors 356, 374
    at roof level 356, 357, 374–375
masonry 350–354
masonry cladding 369–371
openings 358–359, 373, 528
    framing 374
overhangs 361
panels 367–368, 371–373
protected lobby, division of 570
racking resistance 365–366
reaction to fire 389, 400, 402
recesses 358–359
requirements 345–436
sanitary facilities 411
separating 382–389
sheathing 365
shopping centres 822
structural stability 348–349
stud sizing 365–366
thickness 350, 351
ties 356–357, 367–368
timber-frame 363–365
timber-framed 179, 461–462
in timber-framed houses 345
type 1, constructions 418–420
type 1, junctions 421
type 1, solid masonry 418
type 2, cavity masonry 422–425
type 4, timber frames with absorbent
    curtain 428–430
vertical lateral restraint to 356
with/out major openings 363
Warm air heating systems, ducted 445
Warm air systems 757
    ducted warm air heating 718
    electrical 718
    gas-fired warm air systems 719
    heat pumps 719–721
Warrant of ejection 14
Warrant service, fee for 29–31
Washing down provisions
    floors 337, 415
    walls 415
Washing indoors 767

Washrooms
  fire resistance duration 395
Wastewater 233
  discharge 244–245
  drainage 70, 781
  treatment plants 234–242
Wastewater treatment systems 70
Water (earth) closets
  greywater 656
  hot water discarge 657
  rainwater harvesting 656
  requirements 635–636
  sanitary facilities 637–652
  sanitary pipework 653
  ventilation 653–655
Water Environment (Controlled
    Activities) (Scotland)
    Regulations' 2005 56
Water Environment (Oil Storage)
    (Scotland) Regulations 2006 56,
    137
Water Supply (Water Quality) (Scotland)
    Regulations 2001 13
Wet central heating systems
  controls for 716–717
  electric 716
  gas and oil 715
  solid fuel 715
Wildlife and Countryside Act 1981 117
Windows 88–9, 509–523 *see also*
    Openings
  access to manual controls of 520–521
  in basements 515
  cavity barriers 512
  cleaning of 519–520
  conservatories 517

  energy saving 511, 521–523
  environment/health requirements 510,
    515–517
  escape routes 514–515
  extensions over existing 518
  fire safety 510
  and flues discharge 516
  guarding for cleaning 520
  openable 514
  openings 512–513
  planning permission for
      replacement 133–4
  roofs and escape 469
  safety requirements 510–511,
    518–521
  stand-alone buildings 523
  structures 511–513
  ventilation 516
Wind turbines
  planning permission to erect 141–2
Woody biomass fuel
  bulk storage of 415, 538, 697
    heating 697, 698
    secondary heating 698
  storage container 526
Works
  of civil engineering construction 82
  description of 86–9
  execution of 4
  not frequented by people 81
  not requiring warrant 86–9
Written information 78–9
  domestic building 826
  hot water storage systems 710
  non-domestic buildings 826
  work on existing building 827